AF711804

DUXBURY

A First Course in Monte Carlo

George S. Fishman

University of North Carolina at Chapel Hill

THOMSON

BROOKS/COLE

Australia • Brazil • Canada • Mexico • Singapore
Spain • United Kingdom • United States

THOMSON
BROOKS/COLE

A First Course in Monte Carlo
George S. Fishman

Publisher: Curt Hinrichs
Assistant Editor: Ann Day
Editorial Assistant: Fiona Chong
Technology Project Manager: Earl Perry
Marketing Manager: Stacy Best
Marketing Assistant: Brian Smith
Marketing Communications Manager: Nathaniel Bergson-Michelson
Project Manager, Editorial Production: Andy Marinkovich
Creative Director: Rob Hugel
Art Director: Lee Friedman
Print Buyer: Doreen Suruki
Permissions Editor: Stephanie Lee
Production Service, Copyediting, Composition: Interactive Composition Corporation
Text Designer: Carolyn Deacy
Cover Artwork: Gia Giasullo, Studio eg
Cover Printer: Coral Graphics
Printer: R.R. Donnelley/Crawfordsville

Printed in the United States of America

1 2 3 4 5 6 7 09 08 07 06 05

For more information about our products, contact us at:
Thomson Learning Academic Resource Center
1-800-423-0563
For permission to use material from this text or product, submit a request online at
http://www.thomsonrights.com.
Any additional questions about permissions can be submitted by e-mail to thomsonrights@thomson.com.

Library of Congress Control Number: 2005932540

Student Edition: ISBN 0-534-42046-X

THIS BOOK PRINTED ON ACID-FREE RECYCLED PAPER

Thomson Higher Education
10 Davis Drive
Belmont, CA 94002-3098
USA

Thomson Learning
5 Shenton Way
#01-01 UIC Building
Singapore 068808

Australia/New Zealand
Thomson Learning Australia
102 Dodds Street
Southbank, Victoria 3006
Australia

Canada
Thomson Nelson
1120 Birchmount Road
Toronto, Ontario M1K 5G4
Canada

UK / Europe / Middle East / Africa
Thomson Learning
High Holborn House
50/51 Bedford Row
London WC1R 4LR
United Kingdom

Latin America
Thomson Learning
Seneca, 53
Colonia Polanco
11560 Mexico
D.F. Mexico

Spain (includes Portugal)
Thomson Paraninfo
Calle Magallanes, 25
28015 Madrid, Spain

Contents

Preface

The Monte Carlo method is both broad and deep in content. It consists of a collection of techniques for generating random samples on a computer and then using them to solve problems, most notably estimating solutions to integrals, summations, and minima of functions. My motivation for writing this book was to provide an intermediate-level account that would both familiarize readers with the principles and techniques of the Monte Carlo method and offer them a taste of how the techniques perform in practice. Moreover, I wanted to do this in a way that would focus on concept, illustration, and practice. With these objectives, the book presents relatively concise descriptions of each major Monte Carlo technique, usually followed by an example that demonstrates why that technique has become a valued computational tool of quantitative analysis in so many fields. The examples cover diverse applications, some of which introduce readers to problems that might not otherwise come to mind being solvable by Monte Carlo techniques.

Since the Monte Carlo method by definition is an applied tool, the book also provides hands-on exercises that enable students to try each technique and, in particular, to wrestle with the decisions that need to be made to employ it effectively. As new concepts are introduced, subsequent exercises invite students to apply them to problems tackled in earlier exercises. Carrying problems forward in this way provides opportunities for students to compare solutions obtained with different methods, especially with regard to computational and statistical efficiency.

Being a computer-based methodology, the Monte Carlo method relies heavily on software to implement its techniques. So much applications software exists today for this purpose that it is often considered a utility. For example, computing an average or a critical value of the normal distribution requires merely a few clicks or keystrokes. Although much of this convenience is commendable, easy availability often leads the potential user to conclude that little, if any, benefit would result from putting in the time to acquire more than a superficial understanding of how the Monte Carlo method works. Although this may be true for some relatively small problems, there are at least two motivations for acquiring an understanding of these techniques.

First, general-purpose Monte Carlo applications software rarely, if ever, exploits the full range of methodologies that can facilitate solution. This is especially true when a problem contains many constraints that have to be satisfied and when computing time required to obtain an estimated solution with acceptable error of approximation is substantial. Second, it is not true that what you do not know will not hurt you, as the following event shows.

In spring 2004, a major applications software vendor made available a new implementation of a particular pseudorandom-number generator. It was intended to replace a long-used generator whose output had properties that were known to depart from the theoretical ideal for a sequence of independent random numbers. Although theory underlying the new generator was sound, its implementation in this setting was not. It occasionally produced negative integers although theory allows only nonnegative integers. Although a patch was eventually made available to repair the code, users of the new generator prior to the announcement of the patch would ultimately bear responsibility, at least partially, for incorrect solutions that the faulty implementation generated. That is, blame cannot be assigned entirely to the generator or the vendor. Being aware of the existence of alternative pseudorandom-number generators with well-known proven track records of performance (as described in Chapter 4) might have reduced the severity of this unfortunate experience.

The book is written at an intermediate level and is intended for upper-level undergraduates and graduate students. Chapter 1 describes alternative plans for reading it, based on students' backgrounds. A basic plan focuses on techniques that apply to generating independent samples and using these data to estimate solutions. It assumes a knowledge of multivariable calculus, intermediate probability theory, and intermediate statistics. A more comprehensive plan includes Markov chain Monte Carlo (MCMC), including its application to optimization by simulated annealing and related techniques. MCMC provides a way to generate dependent samples from a multivariable distribution when no convenient method is available to generate independent samples. Simulated annealing provides a way of using MCMC to estimate minima. In addition to the previously mentioned prerequisites, this plan also assumes familiarity with Markov chains and the analysis of sequences of dependent observations.

Writing a book of this kind inevitably forces an author to compromise mathematical rigor. The book proves some, but not all, theorems. The terminology maximum (max) and minimum (min) replace supremum (sup) and infimum (inf), respectively throughout the book. In the present setting, these indiscretions carry no major penalty for the reader. Occasionally, the book describes techniques with a substantial literature and several levels of sophistication. Some parts of this are described in place. Other parts are relegated to appendices. More challenging extensions are deferred until Chapter 9, which is intended for readers whose curiosity has been aroused. They principally focus on concepts. The examples in the chapter illustrate particular aspects of these concepts.

The book is intended for an audience that relies on sequential computers, personal or mainframe, to do computing. Needless to say, the structure of many problems makes them good candidates for performing Monte Carlo in a parallel or distributed computing environment to improve their computational efficiency. Swapping, sometimes called parallel tempering in Chapter 8, offers one setting in which this is most easily seen.

Many instructive conversations with colleagues have enriched the content of this book. I am grateful to Russell Cheng of the University of Southampton, Paul Glasserman of Columbia University, Pierre L'Ecuyer of the University of Montreal, Faming Liang of Texas A & M. University, Makoto Matsumoto of Hiroshima University, and David Rubin and Jon Tolle, both of the University of North Carolina at Chapel Hill, for their helpful comments and guidance. I am especially gratefully to Michael Cory, formerly of GlaxoSmithKline, for his guidance and patience as I prepared the example showing how Monte Carlo can contribute to the solution of the protein folding problem.

Writing this book was the project I chose to see me through the transition in July, 2001, from the faculty of the Department of Operations Research at the University of North Carolina at Chapel Hill to retirement. From observing how this change in status affected other retired faculty, I knew that I would benefit from a professional challenge that would constructively occupy some of my time during this transition. In retrospect, I believe writing the book was a good choice. I have enjoyed every minute of its preparation. However, this satisfaction would not have been possible without the help of Vidyadhar Kulkarni, the chair of the department, who provided office space and other amenities to me. For his many kindnesses, I am most grateful to him.

Chapel Hill, North Carolina
January 11, 2005

CHAPTER 1

Introduction

Mention of Monte Carlo usually brings to mind the famous gambling casinos in Monaco. The sheer excitement that the name Monte Carlo evoked motivated scientists working on the Manhattan Project near the end of World War II to appropriate this terminology for a newly developed technique of approximation. The technique relied on statistical sampling on a computer to provide solutions to problems that arose during the research effort to develop the first atomic bomb. Conceptually, casino gambling and computer-based sampling are both randomizing procedures. Since Monte Carlo, Monaco, had long ago become associated with the "luck of the draw," assigning the name Monte Carlo to the collection of newly developed techniques for sampling on a computer was an apt way of describing the collection's most essential feature. Besides, its connotation of excitement and glamor added pizzazz to an otherwise drab accumulation of formulae and computer instructions.

Given the Monte Carlo (MC) method's origin at the beginning of the digital computer age, it is no exaggeration to say that the method provided one, if not the first, example of the computer's potential for solving large-scale complex problems. Today that potential has been realized in many areas. As testament, the January/February 2000 issue of *Computing in Science and Engineering,* a joint publication of the American Institute of Physics and the IEEE Computer Society, lists the Monte Carlo method's *Metropolis algorithm* among the "10 algorithms with the greatest influence on the development and practice of science and engineering on the 20th century."[1] Summarizing its significance, an accompanying article states that

> The Metropolis algorithm has been the most successful and influential of all members of the computational species that used to be called the "Monte Carlo

[1]Dongarra and Sullivan (2000).

Method." Today, topics related to this algorithm constitute an entire field of computational science supported by deep theory and having applications ranging from physical simulations to the foundation of computational complexity.[2]

Frequently employed to solve problems in business and industry, the physical and social sciences, and many fields of engineering, MC applications include:

Area	Example	Exercise
Association between eye and hair colors	6.11.2	
Microchip layout	8.3.1	
Network reliability	2.7.3	
Pricing options	2.7.4	
Pricing mortgage-backed securities		2.H.11–12
Repair warranties		2.H.13–15
Power generation		2.H.16–17
Restoring environments		8.H.1–2
Selecting an inventory system	2.8.1	2.H.20
Sharing resources	6.16.1	6.H.13
Queueing systems	6.14.1	6.H.13
Sizing animal populations	6.4.1	
Protein structures	8.3.2	

This short list hardly does justice to the breadth of application of the Monte Carlo method. Access to the method is now literally "fingertips away." Software for implementing its techniques is available on many personal computers.

Random sampling is an essential component of the Monte Carlo method. That fact often leads to the conclusion that the method applies exclusively to problems with inherent probabilistic structure. To the contrary, it applies equally well to problems that are completely deterministic in content. Examples 2.5.1 and 2.5.2 illustrate this for the estimation of π. This ability to provide solutions to both probabilistic and deterministic problems gives the Monte Carlo method a breadth of application greater than that of most other techniques of numerical approximation.

1.1 ABOUT THIS BOOK

This book provides an introduction to the Monte Carlo (MC) method. It describes the collection of techniques that MC employs to approximate the values of mathematical integrals and summations, and the minima of functions and the points at which the minima occur. Regardless of application, the principal steps in every

[2]Beichl and Sullivan (2000).

MC technique are the same, namely:

- Generate a random sample
- Convert the sample data to an estimate of the solution
- Assess the error of approximation

The availability of so much related applications software, conceptually simple to apply, makes it possible for a newcomer to use the most elementary forms of the MC method after a relatively small investment in preparatory time.

1.2 INTEGRATION AND SUMMATION

If an analytical expression exists for a multivariable integral as a function of its parameters, its evaluation for given parametric values provides a numerical solution to the integral subject only to an inevitable error of approximation that a finite word-size computer induces. However, the availability of an analytical expression is more the exception than the rule, especially when the dimension or the number of coordinates, m, over which integration is to occur is moderate to large. This limitation arises from two sources, the form of the *integrand* (the function to be integrated) and the shape of the *domain of integration* (the region in the m-dimensional space over which integration is to occur). This shape is a consequence of the constraints a problem imposes on the relationships among the m variables.

For a bounded and continuous integrand in $m = 1$ dimension, well-established numerical techniques exist that dictate n locations in the domain of integration at which the integrand is to be evaluated. Each technique also specifies some form of averaging of the n evaluations to come up with an approximate solution. At worst, the error of approximation is n^{-1}. Deterministic numerical techniques also exist for $m > 1$. However, their worst-case error tends to decrease as n^{-m}, implying a decreasing convergence rate as m increases for a fixed number of points n.

Although this rate can be improved if the integrand is known to meet specified regularity conditions, detecting whether or not these properties are present becomes increasingly difficult as m increases. Nevertheless, the need in some fields to evaluate the same multivariable integral repeatedly has made it worthwhile to perform a one-time pre-analysis to check for the presence of these properties and then repeatedly benefit from an approximating procedure that exploits them. Similar issues arise when approximating summations in the discrete case.

MC offers an alternative environment for solution. Well-established MC techniques exist, each of which dictates a scheme for randomly choosing n points in the domain of integration at which evaluation is to occur and specifies the form of averaging to be employed. Most importantly,

- For a considerably wider range of integrands and summations, MC induces a standard error of approximation that decreases as $n^{-1/2}$, regardless of dimension size.
- In contrast to the worst-case error bound for most deterministic methods, the standard error is an average measure of error.
- MC has the added convenience of facilitating the evaluation of a numerical estimate of this error from the same sample data.

The $n^{-1/2}$ convergence rate makes clear that although deterministic methods are preferable for one dimension, MC is favored for higher-dimensional integrands and summands, especially when little is known about their properties.

1.2.1 Improving Efficiency

The most elementary form for approximating integrals and summations is called *standard Monte Carlo,* or *standard MC* (see Section 2.4). Although simplicity of concept and ease of execution are two of standard MC's greatest strengths, many problems call for more sophisticated techniques to give answers with acceptable accuracy and with the timeliness that so much of computing demands today. For example, suppose that standard MC is applied to estimating the value of an integral or summation and takes computing time T to produce an estimate with sufficiently small error of approximation to instill considerable confidence in using the estimate for subsequent decision making.

If the error is unacceptably large, additional sampling can reduce it at the cost of increasing T. However, when the T required to reduce the error to within an acceptable bound is itself unacceptably large, standard MC is not adequate. More in-depth MC techniques can modify the recipe for sample data generation (see Chapter 3) or the form of the estimator (see Chapter 5) in ways that produce a smaller error than standard MC for the same amount of computing time. Both individually and collectively, these sophisticated MC techniques provide a remarkably powerful set of tools for solving problems.

1.3 MINIMIZING A FUNCTION

Deterministic optimization techniques exist for finding the global minimum of a function of m variables, subject to constraints on how the variables relate to each other. Many procedures iteratively move through a succession of locations or sites in the *feasible space,* the region in which the constraints hold, and apply a deterministic decision rule that selects the next location to visit on the way to solution. The rule makes its choice based on properties of the *objective function,* the function to be minimized. For example, steepest descent methods, frequently employed for continuous functions, require the values of the derivatives at each location as well as the value of the function.

The MC approach to optimization also exploits local properties to gain its computational and statistical efficiency. It does this in two ways:

- **Random Deterministic (RD).** On each iteration, it
  - RD1. Randomly generates a location in the feasible space.
  - RD2. Applies a *deterministic rule* to decide on the next location on the path to solution.
- **Random Random (RR).** On each iteration, it
  - RR1. Randomly generates a location in the feasible space.
  - RR2. Applies a *randomized rule* to decide whether to move to the new location on the path to solution.

In an RD procedure, MC merely serves as a utility, providing input to the deterministic rule directing the path to solution. In an RR procedure, the decision rule also incorporates elements of randomness. Most notable among the latter is the technique of *simulated annealing* (see Chapter 8). Its appeal arises from how little knowledge it requires about the objective function. In particular, no derivatives are needed. Although it sometimes calls for an exceptionally large number of iterations to produce an estimate, ease of implementation has made it a staple in the tool chest of *stochastic optimization techniques*. Often, it is employed as a first step in the search for a minimum.

1.3.1 Improving Efficiency

As in the case of integration and summation, simplicity of concept and ease of execution have given simulated annealing its appeal. However, a problem whose objective function exhibits large variations from maxima to local minima over relatively small regions can consume an inordinate amount of computer time in the search for a global minimum. Again, more sophisticated MC techniques can speed up the search. These include *simulated tempering* (Section 8.5), *swapping* (Section 8.6), *stochastic tunneling* (Section 8.7), and *genetic algorithms* (Section 8.8). These exploit special structure. When carefully applied, they can significantly reduce the number of iterations required to obtain an estimate of the minimum with acceptable accuracy.

1.4 READING PLANS

The MC method is the totality of all random sampling techniques on a computer and all simple and in-depth techniques for converting these data into (random) approximate solutions to problems. They exploit special structure for the purpose of generating estimates with acceptable errors in computationally efficient ways. This book describes both the sampling and estimating techniques. They include:

Chapter	Topic
2	Generating independent trials (independent Monte Carlo)
3	Sampling from diverse distributions
4	Generating numbers in the interval [0, 1) that closely resemble a sequence of independent samples drawn from the uniform distribution on [0, 1) (pseudorandom number generation)
5	Reducing variance
6	Generating sample paths of dependent observations (Markov chain Monte Carlo)
7	Assessing how accurately a Monte Carlo estimate approximates the unknown value of the quantity of interest
8	Estimating the global minimum of a function and the points at which the minimum occurs
9	Advanced concepts in Monte Carlo

A course based on Chapters 2, 3, and selected sections of Chapters 4 and 5 introduces students to MC sampling experiments based on independent trials. It assumes a familiarity with probability theory and statistics, including continuous and discrete random variables, expectation, conditional probability, and elementary statistical inference for independent identically distributed observations. A remarkably large number of problems are amenable to approximate solution by these methods.

A course based on Chapters 2 through 8 deepens and broadens this introduction. Most notably, when independent MC cannot provide a recipe for generating *independent* samples from a specified distribution, Markov chain Monte Carlo (MCMC) usually can meet the need by generating *dependent* samples from the distribution. The presence of this dependence adds to the challenge of evaluating the accuracy of MCMC-generated estimates. In addition to the background assumed above, this more comprehensive course calls for a familiarity with the concept of Markov chains and with the statistical analysis of sample paths of dependent observations. Chapters 6 and 7 provide this background, focusing on the most basic of issues that affect accuracy, convergence to equilibrium, and standard error.

Chapter 9 extends the depth and breadth of the material covered in these two chapters. Intended for readers whose curiosity has been aroused, it describes advanced concepts in MCMC that reveal many additional facets of the method. Its technical level presents a greater challenge to the reader than earlier chapters.

Every reader should become familiar with some parts of Chapter 4. Principally intended to serve as a reference on the many alternative techniques for generating pseudorandom numbers on a computer, it describes the strengths and weaknesses of each technique. Since linear congruential generators continue to be the most commonly encountered generators in applications software, at a minimum, a student needs to be aware of how they work in practice. The chapter's remaining sections can be assigned in accordance with the emphasis that the instructor chooses to place on other individual techniques.

1.5 EXERCISES

The MC method is an applications-oriented concept. Accordingly, most exercises in this book invite students to perform sampling experiments on a computer. These hands-on exercises sensitize them to the many practical considerations that influence the design and implementation of a sampling plan. Some exercises call for applying different MC methods to produce approximate solutions to the same problem. Then subsequent exercises may invite students to compare the quality of these solutions. Carrying the same problem forward in different exercises allows students to develop an appreciation for how well each individual technique works on a problem and how different techniques compare with regard to computational and statistical efficiency.

Some hands-on exercises are elementary. They describe in detail what is to be done. Others are more demanding, focusing on describing the problem's setting and asking the student to formulate the plans for successful execution and solution. The more direct exercises will develop a student's facility with the mechanics of the method. The more descriptive exercises will challenge the student to develop a deeper perspective on what it takes to solve a problem. As guidance, the answer section at the back of the book gives partial solutions to selected hands-on exercises in Chapters 2, 5, 6, 7, and 8.

Several chapters also contain exercises that require derivations and proofs of assertions made in the text. They provide opportunities to trace the steps that give many a MC formulation its rationale.

1.6 SOFTWARE

All MC methodologies depend on the availability of software for generating samples from distributions (Chapter 3), which depend on software for generating pseudorandom numbers (Chapter 4). Much of this is available in commonly encountered PC applications software. In addition, Internet links to selected sources of applications software for executing MC are cited throughout the book. Using them will save considerable time when solving the hands-on exercises.

An Internet link is also cited to software for analyzing sample paths of dependent observations. Using this software facilitates solution to many hands-on exercises in Chapter 7.

1.7 NOTATION

The glossary contains definitions of much of the mathematical notation used in this book. In particular, boldface type denotes a row of numbers or variables, called a row vector. Occasionally, a matrix is used. This is an array of numbers or variables arranged in rows of equal numbers of entries and columns of equal numbers of entries. The vector and matrix notations are used principally for concise description. The few matrix operations employed in the text are explained in context.

1.8 REFERENCES

Beichl, I., and F. Sullivan (2000). The Metropolis algorithm, *Computing in Science and Engineering,* **2,** 65–69.

Dongarra, J., and F. Sullivan (2000). Guest editors' introduction, *Computing in Science and Engineering,* **2,** 22–23.

CHAPTER 2

Independent Monte Carlo

Problems encountered in many areas of science, technology, and business frequently rely for their solutions on the evaluation of multivariable integrals, summations, and global minima of multivariable functions and the points at which these minima occur. When evaluation by deterministic numerical methods is computationally intractable, the Monte Carlo (MC) method, a collection of techniques based on random sampling on a computer, can provide approximate solutions. Formally, these problems include the following elements.

Integral Evaluation. For each dimension $i = 1, \ldots, m$, let x_i denote a point in the interval $\Re := (-\infty, \infty)$ so that $\mathbf{x} := (x_1, \ldots, x_m)$ is a point in the m-dimensional region $\Re^m := (-\infty, \infty)^m$. Let $g := \{g(\mathbf{x}),\ \mathbf{x} \in \Re^m\}$ denote a function and $\mathcal{X}$ a region in $\Re^m$ for which the integral

$$\lambda := \int_{\mathcal{X}} g(\mathbf{x})\mathrm{d}\mathbf{x} \tag{2.1}$$

exists. Sometimes $\mathcal{X}$ is called a *subset* of $\Re^m$ and is denoted by $\mathcal{X} \subseteq \Re^m$. The problem is to evaluate (2.1).

When $\mathcal{X}$ is a collection of discrete points in $\Re^m$, the problem becomes the evaluation of the summation

$$\lambda := \sum_{\mathbf{x} \in \mathcal{X}} g(\mathbf{x}), \tag{2.2}$$

where the notation $\mathbf{x} \in \mathcal{X}$ denotes that $\mathbf{x}$ is a point in $\mathcal{X}$. We refer to a discrete set $\mathcal{X}$ as *countably finite* if it contains a finite number of points and as *countably infinite* if it contains an infinite number of points.

Function Minimization. For a function $H := \{H(\mathbf{x}),\ \mathbf{x} \in \mathcal{X}\}$, it is of interest to evaluate

$$\begin{aligned}\xi(H) &:= \min_{\mathbf{x}\in\mathcal{X}} H(\mathbf{x}) \\ &= \text{minimum of the function } H,\end{aligned} \tag{2.3}$$

when it exists. Occasionally, computing elements of

$$\mathcal{X}_{\text{min}} := \{\mathbf{x} \in \mathcal{X} : H(\mathbf{x}) = \zeta(H)\}, \tag{2.4}$$

which is the set of all points at which H has its global minimum, is also desirable, especially when H has a unique minimum. When an exact solution eludes computational methods based on deterministic algorithms, MC procedures, in particular, *simulated annealing* (Chapter 8), can provide approximate ones.

Some problems require for solution both integral evaluation and function minimization. For example, it is occasionally of interest to evaluate

$$\zeta(H) := \min_{\theta\in\Theta} H(\theta),$$

where $\Theta \subseteq \Re$ and, for some function $\{g(\mathbf{x}, \theta),\ \mathbf{x} \in \mathcal{X},\ \theta \in \Theta\}$,

$$H(\theta) := \int_{\mathcal{X}} g(\mathbf{x}, \theta)\mathrm{d}\mathbf{x}. \tag{2.5}$$

In one approach, a sequence of MC experiments gives estimates of $H(\theta)$ for selected θ in Θ and uses them to estimate $\zeta(H)$. Section 2.9 describes a methodology that provides an approximate solution to this problem.

2.1 INDEPENDENT MONTE CARLO

At its heart, MC randomly selects a finite number of points from the region $\mathcal{X}$ and uses them to compute an approximation of λ in (2.1). Assume that

$$\int_{\mathcal{X}} g^2(\mathbf{x})\mathrm{d}\mathbf{x} < \infty$$

and let $\omega := \{\omega(\mathbf{x}) \geq 0 : \mathbf{x} \in \mathcal{X};\ \int_{\mathcal{X}} \omega(\mathbf{z})\mathrm{d}\mathbf{z} = 1\}$ denote a probability density function (p.d.f.) with the property

$$\int_{\mathcal{X}} \frac{g^2(\mathbf{x})}{\omega(\mathbf{x})}\mathrm{d}\mathbf{x} < \infty. \tag{2.6}$$

As our account evolves, the basis for choosing ω will become clear. For conciseness, we hereafter call a probability distribution a distribution.

For a random m-dimensional point $\mathbf{X} := (X_1, \ldots, X_m)$ with p.d.f. ω, the ratio $\frac{g(\mathbf{X})}{\omega(\mathbf{X})}$ has moments

$$\mathrm{E}_\omega \left[\frac{g(\mathbf{X})}{\omega(\mathbf{X})}\right]^l := \int_{\mathcal{X}} \left[\frac{g(\mathbf{x})}{\omega(\mathbf{x})}\right]^l \omega(\mathbf{x})\mathrm{d}\mathbf{x} \qquad l = 1, 2, \ldots.$$

In particular, $\frac{g(\mathbf{X})}{\omega(\mathbf{X})}$ has mean

$$\mathrm{E}_{\omega}\left[\frac{g(\mathbf{X})}{\omega(\mathbf{X})}\right] = \lambda \tag{2.7}$$

and variance

$$\mathrm{var}_{\omega}\left[\frac{g(\mathbf{X})}{\omega(\mathbf{X})}\right] = \sigma^2 := \int_{\mathcal{X}} \left[\frac{g(\mathbf{x})}{\omega(\mathbf{x})}\right]^2 \omega(\mathbf{x})d\mathbf{x} - \lambda^2. \tag{2.8}$$

Whenever (2.7) holds, we say that $\frac{g(\mathbf{X})}{\omega(\mathbf{X})}$ is an *unbiased estimator* of λ.

Algorithm IMC describes Monte Carlo in its conceptually simplest form. It generates n independent samples called *replications* and then computes an average $\bar{\lambda}_n$ as an approximation to λ. The superscript $^{(i)}$ denotes replication i.

ALGORITHM

IMC

Purpose: To estimate λ.

Given: $n :=$ number of independent replications.

Method:

$i = 1$

While $i \le n$:

randomly generate $\mathbf{X}^{(i)} := \left(X_1^{(i)}, \ldots, X_m^{(i)}\right)$ independently from ω.

evaluate $\frac{g(\mathbf{X}^{(i)})}{\omega(\mathbf{X}^{(i)})}$

$i = i + 1$.

Compute:

$$\bar{\lambda}_n := \frac{1}{n}\sum_{i=1}^{n} \frac{g(\mathbf{X}^{(i)})}{\omega(\mathbf{X}^{(i)})}. \tag{2.9}$$

As an approximation to σ^2, compute:

$$\begin{aligned} s_n^2 &:= \frac{1}{n-1}\left\{\sum_{i=1}^{n}\left[\frac{g(\mathbf{X}^{(i)})}{\omega(\mathbf{X}^{(i)})} - \bar{\lambda}_n\right]^2\right\} \\ &= \frac{1}{n-1}\left\{\sum_{i=1}^{n}\left[\frac{g(\mathbf{X}^{(i)})}{\omega(\mathbf{X}^{(i)})}\right]^2 - n\bar{\lambda}_n^2\right\}. \end{aligned} \tag{2.10}$$

■

The estimate (2.9) has important statistical properties. They include:

P1. $\mathrm{pr}\left(\bar{\lambda}_n \to \lambda\right) \to 1$ as $n \to \infty$ (Law of Large Numbers)

P2. $\mathrm{var}_{\omega}\left(\bar{\lambda}_n\right) = \frac{\sigma^2}{n}$ (independent and identically distributed data)

P3. $\frac{\bar{\lambda}_n - \lambda}{\sqrt{\sigma^2/n}} \xrightarrow{\mathrm{d}} \mathcal{N}(0, 1)$ as $n \to \infty$ (Central Limit Theorem), where $\mathcal{N}(0, 1)$ denotes the standard normal distribution with mean 0, variance 1, and p.d.f.

$$\phi(z) = (2\pi)^{-1/2} e^{-z^2/2} \qquad -\infty < z < \infty. \tag{2.11}$$

The right-pointing arrows in properties P1, P2, and P3 denote *convergence*. In particular, property P1 asserts that $\bar{\lambda}_n$ approaches λ in value with probability 1 as the sample size n increases. An equivalent notation is

$$\bar{\lambda}_n \to \lambda \qquad \text{w.p.1} \qquad \text{as } n \to \infty,$$

where w.p.1 denotes "with probability 1."

As n increases, property P2 reveals that the variance of the estimator decreases at rate n^{-1}. Property P3 asserts that as n increases, the probability distribution of the standardized random variable $\frac{\bar{\lambda}_n - \lambda}{\sqrt{\sigma^2/n}}$ approaches the normal distribution with mean 0 and variance 1. Note that the Law of Large Numbers does not require (2.6) to hold, but the other two properties do.

Since σ^2 is unknown, the normal limit characterizes behavior but does not allow implementation. If the condition

$$\int_{\mathcal{X}} \frac{g^4(\mathbf{x})}{\omega^3(\mathbf{x})} d\mathbf{x} < \infty \tag{2.12}$$

is also satisfied, then

P4. $s_n^2 \to \sigma^2$ w.p.1 as $n \to \infty$

and

P5. $\dfrac{\bar{\lambda}_n - \lambda}{\sqrt{s_n^2/n}} \xrightarrow{d} \mathcal{N}(0, 1)$ as $n \to \infty$.

The estimator $\bar{\lambda}_n$ has *standard error* $\sigma/n^{1/2}$, approximated by

$$\text{s.e.} := \frac{s_n}{n^{1/2}}, \tag{2.13}$$

relative error $\sigma/\lambda n^{1/2}$, approximated by

$$\text{r.e.} := \frac{s_n}{\bar{\lambda}_n n^{1/2}}, \tag{2.14}$$

and for fixed $\delta \in (0, 1)$ the *asymptotically valid* $100 \times (1 - \delta)$ percent confidence interval for λ

$$\mathcal{I}_n(\delta) := \left[\bar{\lambda}_n - \Phi^{-1}(1 - \delta/2)s_n/n^{1/2},\ \bar{\lambda}_n + \Phi^{-1}(1 - \delta/2)s_n/n^{1/2}\right], \tag{2.15}$$

where

$$\begin{aligned} \Phi^{-1}(u) &:= \min[x \in \Re : \Phi(x) = u,\ 0 \le u \le 1] \\ &= \text{smallest value } x \text{ for which } \Phi(x) = u \end{aligned} \tag{2.16}$$

denotes the *inverse* of the standard normal distribution function (d.f.)

$$\Phi(y) := \int_{-\infty}^{y} \phi(z)\mathrm{d}z \qquad -\infty < y < \infty. \tag{2.17}$$

By asymptotically valid we mean that the interval $\mathcal{I}_n(\delta)$ includes the unknown constant λ with probability approaching $1-\delta$ as n becomes large. Formally,

$$\mathrm{pr}\,[\mathcal{I}_n(\delta) \supseteq \{\lambda\}] \to 1-\delta \qquad n \to \infty,$$

where the notation $\mathcal{A} \supseteq \mathcal{B}$ denotes that the set $\mathcal{A}$ includes the set $\mathcal{B}$. As n increases, the error measures, s.e., r.e., and the length of $\mathcal{I}_n(\delta)$, all tend to decrease as $n^{-1/2}$. This convergence rate characterizes a substantial number of MC techniques that rely on independent replications. Hereafter, we refer to independent MC sampling as IMC.

EXAMPLE 2.1.1

ONE-DIMENSIONAL INTEGRATION

A relatively straightforward problem in one dimension illustrates how IMC works and, in particular, how accuracy improves with sample size. Consider the integral

$$\lambda = \int_0^1 g(x)\mathrm{d}x,$$

where

$$g(x) := \begin{cases} \mathrm{e}^{-x\cos\pi x} & 0 \le x \le 1 \\ 0 & \text{elsewhere.} \end{cases} \tag{2.18}$$

Although the exact value of λ is not available analytically, deterministic numerical methods could easily be employed to approximate λ with relatively high accuracy. Alternatively, we describe how IMC provides an estimate.

Suppose we choose the uniform distribution

$$\omega(x) = \begin{cases} 1 & 0 \le x \le 1 \\ 0 & \text{elsewhere} \end{cases} \tag{2.19}$$

for sampling, denoted by $\mathcal{U}(0, 1)$. Then Table 2.1 shows the results of applying Algorithm IMC for successively increasing sample sizes $n_1 = 20$, $n_i = 2n_{i-1}$ for $i = 2, \ldots, 15$. Theory in Section 2.1 suggests that s.e., r.e., and the 0.99 confidence-interval length should decrease as $2^{-1/2} \approx 0.7071$ for successive sample sizes. Except for sampling variation for small n_i, the corresponding sample s.e. values in the table are in agreement with theory. The most accurate point estimate of λ is $1.3287 \pm .0026024$ for $n_{15} = 327,680$. ■

Since its selection required no extensive analysis on the part of the IMC user before experimentation, the uniform distribution was a convenient choice for this example. Moreover, highly efficient methods exist for generating samples from $\mathcal{U}(0, 1)$ (see Chapter 3). Since total computing time for all 15 experiments took a fraction of a second to produce an acceptable final level of accuracy, there

TABLE 2.1 | *Estimating* $\lambda = \int_0^1 \mathbf{e}^{-x\cos\pi x}\mathbf{d}x$ *using independent Monte Carlo*

Experiment i	n_i	$\bar{\lambda}_n \pm \Phi^{-1}(.995)$ s.e.	s.e.	r.e.
1	20	1.4844 ± .36473	1.4160×10^{-1}	9.5390×10^{-2}
2	40	1.2953 ± .23733	9.2137×10^{-2}	7.1133×10^{-2}
3	80	1.3887 ± .19299	7.4923×10^{-2}	5.3952×10^{-2}
4	160	1.3467 ± .12064	4.6835×10^{-2}	3.4778×10^{-2}
5	320	1.2292 ± .076526	2.9709×10^{-2}	2.4169×10^{-2}
6	640	1.3511 ± .058267	2.2621×10^{-2}	1.6742×10^{-2}
7	1280	1.3652 ± .042944	1.6672×10^{-2}	1.2212×10^{-2}
8	2560	1.3181 ± .029445	1.1431×10^{-2}	8.6728×10^{-3}
9	5120	1.3208 ± .020741	8.0522×10^{-3}	6.0963×10^{-3}
10	10240	1.3386 ± .014959	5.8073×10^{-3}	4.3384×10^{-3}
11	20480	1.3193 ± .010341	4.0148×10^{-3}	3.0431×10^{-3}
12	40960	1.3235 ± .0073264	2.8443×10^{-3}	2.1490×10^{-3}
13	81920	1.3264 ± .0052021	2.0196×10^{-3}	1.5226×10^{-3}
14	163840	1.3293 ± .0036850	1.4306×10^{-3}	1.0762×10^{-3}
15	327680	1.3287 ± .0026024	1.0103×10^{-3}	7.6035×10^{-4}

is no need to delve into a more elaborate analysis beforehand to identify an ω that would exploit the structure of g in order to reduce variance. However, these favorable circumstances are not always present. Therefore, the reader should regard Example 2.1.1 as merely the simplest and most elementary use of MC for evaluating an integral.

The choice of ω is at the discretion of the IMC user. Although (2.7) makes clear that the choice strongly influences $\text{var}_\omega g(\mathbf{X})$ and therefore $\text{var}_\omega \bar{\lambda}_n$, the cost or computing time to generate a random sample from ω is an equally important consideration. Section 2.3 begins to formalize the tradeoff between variance and cost, and Section 2.4 introduces *importance sampling*, a method of selecting ω so as to smooth out the variation in $\{g(\mathbf{x})/\omega(\mathbf{x}),\ \mathbf{x} \in \mathcal{X}\}$, which reduces variance in the sampling experiment.

2.1.1 Why Monte Carlo?

MC sampling is one among many numerical techniques for evaluating an integral. One may well ask: Why resort to IMC, which adds sampling error to the overall error of approximation? Both conceptually and practically, the appeal of IMC stems from how little knowledge of the properties of g it requires to produce a numerical estimate of λ with error of approximation decreasing as $n^{-1/2}$. IMC merely requires (2.12) to hold. This rate contrasts with the $n^{-1/m}$ convergence rate that deterministic (nonrandom) numerical methods induce for the more limited class of bounded continuous functions. With qualification, this rate also holds if g is bounded with a finite set of discontinuities. To improve on the $n^{-1/m}$ rate, these deterministic methods require additional information on the smoothness properties of g. More often than not, obtaining this extra knowledge takes time and effort, beyond what MC demands, to produce a usable result. This is especially true as the dimension m increases. Consequently MC, with its relatively meager requirements, has considerable appeal.

2.2 GENERATING SAMPLES

Generating random samples is an essential part of every MC experiment. Although Chapter 3 concisely describes the principal techniques for doing so, it is instructive at this early point in our account to provide the reader with at least one conceptualization of how this generation is performed. Accordingly, we describe a special case of the *inverse-transform method.*

Suppose Z is a random variable with absolutely continuous p.d.f. $\{f(z),\ a \le z \le b\}$ and d.f.

$$F(z) = \int_a^z f(y)\mathrm{d}y \qquad a \le z \le b$$

Then the random variable $F(Z)$ is from $\mathcal{U}(0, 1)$ (see Theorem 3.3.1). To generate a random sample, we execute the steps:

1. Randomly sample U from $\mathcal{U}(0, 1)$.

2. Solve $F(Z) = U$.

Because $\{F(z)\}$ is strictly increasing, there is a unique solution. Section 3.3 generalizes the method to apply to piecewise continuous p.d.f.s and probability mass functions (p.m.f.s) for discrete random variables. Hereafter, we frequently refer to a p.d.f. or a p.m.f. by the generic term *distribution.*

EXAMPLE 2.2.1 SAMPLING FROM THE LOGISTIC DISTRIBUTION

Suppose Z has the *logistic distribution*

$$f(z) = \begin{cases} \frac{\mathrm{e}^{-(z-\alpha)/\beta}}{\beta[1+\mathrm{e}^{-(z-\alpha)/\beta}]^2} & -\infty < \alpha < \infty,\ \beta > 0, -\infty < z < \infty \\ 0 & \text{elsewhere} \end{cases}$$

with d.f.

$$F(z) = \frac{1}{1+\mathrm{e}^{-(z-\alpha)/\beta}},$$

denoted by $\mathcal{L}(\alpha, \beta)$. If U is from $\mathcal{U}(0, 1)$, then

$$Z = \alpha + \beta \ln\left(\frac{U}{1-U}\right)$$

is from $\mathcal{L}(\alpha, \beta)$. ■

In principle, random sampling from any univariate distribution can be performed in an analogous way. However, computational efficiency sometimes dictates that an alternative technique be employed; hence, the diversity of methods described in Chapter 3.

EXAMPLE 2.2.2

SAMPLING IN A DETERMINISTIC PROBLEM

Suppose g has the representation

$$g(\mathbf{x}) = g(x_2, \ldots, x_m \mid x_1)g(x_1) \qquad a \le x_1 \le b, \ (x_2, \ldots, x_m) \in \mathcal{X}_{-1},$$

where $\{g(x_1)\}$ is positive and absolutely continuous on $[a, b]$, and $\mathcal{X}_{-1}$ denotes $\mathcal{X}$ omitting dimension 1. Then (2.1) has the equivalent form

$$\lambda = \int_a^b g(x_1) \left[\int_{\mathcal{X}_{-1}} g(x_2, \ldots, x_m \mid x_1) \mathrm{d}x_2 \cdots \mathrm{d}x_m \right] \mathrm{d}x_1,$$

and $\{Cg(x_1),\ a \le x_1 \le b\}$, where $C = \left[\int_a^b g(x_1)\mathrm{d}x_1\right]^{-1}$, is a p.d.f. In practice as well as principle, one can use this distribution to generate a random sample for dimension 1 via the inverse-transform method.

As illustration, let

$$g(x_1) := \begin{cases} \frac{1}{x_1^7} & 0 < a \le x_1 \le b < \infty \\ 0 & \text{elsewhere.} \end{cases}$$

If U is from $\mathcal{U}(0, 1)$, then

$$X_1 = \frac{1}{[a^{-6} - U(a^{-6} - b^{-6})]^{1/6}}$$

has the p.d.f. $\{Cg(x_1),\ a \le x_1 \le b\}$ with $C = 6a^6b^6/(b^6 - a^6)$. ■

2.3 CHOOSING A MONTE CARLO SAMPLING PLAN

Many alternative MC sampling plans can provide answers to a particular problem. Naturally, we prefer the one that gives the greatest statistical accuracy for the least cost. Two characterizations of the relative desirability of two alternative sampling plans are the *cost ratio,*

$$\text{CR} := \frac{\text{computer generating time per replication for MC plan 1}}{\text{computer generating time per replication for MC plan 2}}, \tag{2.20}$$

and the *variance ratio,*

$$\text{VR} := \frac{\text{variance per replication for MC plan 1}}{\text{variance per replication for MC plan 2}}. \tag{2.21}$$

For $i = 1$ and 2, let c_i, σ_i^2, and n_i denote the unit generating time, the variance per replication, and the number of independent replications, respectively, for IMC plan i. Then $\bar{\lambda}_n$ in (2.9) has variance σ_i^2/n_i and consumes $c_i n_i$ computing time. Two alternative scenarios reveal the significance of the product CR $\times$ VR.

Equal Variances. To achieve a specified variance b takes

$$n_i = \frac{\sigma_i^2}{b}$$

observations on plan i with unit generating time c_i. Total computing time is

$$c_i n_i = c_i \frac{\sigma_i^2}{b}.$$

Therefore, achieving the same variance for both $\bar{\lambda}_{n_1}$ and $\bar{\lambda}_{n_2}$ takes

$$\frac{c_1 n_1}{c_2 n_2} = \frac{c_1 \sigma_1^2}{b} \frac{b}{c_2 \sigma_2^2} = \text{CR} \times \text{VR}$$

as much time on plan 1 as on plan 2. If CR $\times$ VR > 1 (< 1), then plan 1 (2) requires more (less) computing time than plan 2 (1).

Equal Computing Times. Conversely, equal computing times a expended on each plan induce sample sizes

$$n_1 = \frac{a}{c_1} \quad \text{and} \quad n_2 = \frac{a}{c_2}$$

so that the ratio of variances of $\bar{\lambda}_{n_i}$ for n_1 and n_2 independent replications is

$$\frac{\sigma_1^2}{n_1} \times \frac{n_2}{\sigma_2^2} = \frac{c_1 \sigma_1^2}{a} \times \frac{a}{c_2 \sigma_1^2} = \text{CR} \times \text{VR}.$$

If CR $\times$ VR > 1 (< 1), then the same computing time induces a greater variance for $\bar{\lambda}_{n_1}$ under plan 1 than for $\bar{\lambda}_{n_2}$ under plan 2. Hereafter, we occasionally refer to the benefits of plan 2 in the following ways. If VR > 1, we say that plan 2 improves *statistical efficiency* or is *variance reducing* relative to plan 1. If CR > 1, we say that plan 2 improves *computational efficiency*. If CR $\times$ VR > 1, we say that plan 2 improves *efficiency*.

Since each plan i also incurs a setup time, s_i, the plan's total computing time is actually $s_i + c_i n_i$. The foregoing analysis assumes that n_i is considerably greater than s_i/c_i so that the error in omitting this setup time is negligible.

Hereafter, we refer to computer generating time per replication as *unit generating time*. It has two components:

1. The computer time to sample $\mathbf{X} = (X_1, \ldots, X_m)$ from its distribution
2. The computer time to evaluate $g(\mathbf{X})$, given $\mathbf{X}$ as input

We call the time to generate X_i its *sample generating time*. Component 1 is the sum of the sample generating times for $X_1, \ldots, X_m$. Often, but not always, this sum accounts for a substantial part of the unit generating time.

Most techniques that reduce variance inevitably require more time and effort to perform a presampling analysis. These are also costs. With the current availability of so much low-cost computing power, there is a natural tendency merely to resort to the most immediately available MC technique, thus avoiding the expenditure of user time and effort that a presampling analysis demands. This strategy makes considerable sense when the solution requires relatively moderate accuracy and the longer computing time does not impede overall progress. Nevertheless, many problems justify a pre-analysis. In particular:

- Sometimes, performing a presampling analysis suggests that $\mathrm{CR} \approx 1$ and also provides a nontrivial lower bound (> 1) for VR. Example 2.7.3 on network reliability illustrates this case. Other times, pre-analysis merely shows that $\mathrm{VR} > 1$. After sampling, it is not unusual to estimate CR and VR from the data and timing information to demonstrate the benefit of the technique and to encourage its application to problems with similar structures. Used in this way, the technique reduces the presampling analysis time and effort that a potential MC user needs to devote to a collection of different but related problems whose solutions are to be estimated from MC data.
- A high-accuracy requirement, a premium on fast reporting of results, or both make the presampling investment in time and effort, to effect a variance reduction, well spent. This is especially true in Examples 2.7.4 and 5.4.4, where sampling experiments need to be executed repeatedly to compute option prices for many different stocks at many different times each day.

MC techniques that reduce variance ($\mathrm{VR} > 1$) often increase unit generating times ($\mathrm{CR} < 1$). Since two different techniques frequently rely on different distributions from which to generate samples, the costs of generating these samples may and often do differ. Because sample generating times for all distributions in a problem are frequently a major component of unit generating time, it is always desirable to choose sample generating algorithms that are computationally efficient for the particular distributions on which an MC technique relies. When comparing two IMC methods, CR reflects the difference in these sample generating times. Chapter 3 describes many techniques for reducing sample generating cost.

2.4 IMPORTANCE SAMPLING

Chapter 5 describes many alternative methods for reducing variance. Here we describe one particular method, importance sampling, which has two especially important characteristics:

- It offers considerable discretion to an MC user who chooses the distribution from which to generate samples that lead to increased statistical efficiency.
- It can frequently be combined with other methods to enhance statistical efficiency further.

Importance sampling's all pervasive contribution to designing an MC experiment becomes increasingly apparent in the examples and hands-on exercises in this and subsequent chapters.

Consider two candidate distributions for sampling, $\{\omega_1(\mathbf{x}),\ \mathbf{x} \in \mathcal{X}\}$ and $\{\omega_2(\mathbf{x}),\ \mathbf{x} \in \mathcal{X}\}$. Since the sign of

$$\operatorname{var}_{\omega_1}\left[\frac{g(\mathbf{X})}{\omega_1(\mathbf{X})}\right] - \operatorname{var}_{\omega_2}\left[\frac{g(\mathbf{X})}{\omega_2(\mathbf{X})}\right] = \int_{\mathcal{X}} g^2(\mathbf{x})\left[\frac{1}{\omega_1(\mathbf{x})} - \frac{1}{\omega_2(\mathbf{x})}\right] d\mathbf{x} \tag{2.22}$$

determines which distribution induces the greater statistical accuracy for the same sample size, we have a way of anticipating which distribution is preferable. In particular, if $\frac{\omega_2(\mathbf{x})}{\omega_1(\mathbf{x})} < 1$ where $g^2(\mathbf{x})$ is large and $\frac{\omega_2(\mathbf{x})}{\omega_1(\mathbf{x})} > 1$ where $g^2(\mathbf{x})$ is small, then ω_2 is likely to induce smaller variance than ω_1. Generating samples from a distribution chosen by this form of comparative analysis is called *importance sampling*.

Every integrand g admits the representation

$$g(\mathbf{x}) = h(\mathbf{x})\pi(\mathbf{x}) \qquad \mathbf{x} \in \mathcal{X} \subseteq \Re^m, \tag{2.23}$$

where

$$\pi(\mathbf{x}) \geq 0 \qquad \mathbf{x} \in \mathcal{X}$$

and

$$\int_{\mathcal{X}} \pi(\mathbf{x}) d\mathbf{x} = 1.$$

That is, $\pi := \{\pi(\mathbf{x}),\ \mathbf{x} \in \mathcal{X}\}$ is a probability distribution. There is always the trivial factorization $\{\pi(\mathbf{x}) = |\mathcal{X}|^{-1},\ \mathbf{x} \in \mathcal{X}\}$ and $\{h(\mathbf{x}) = |\mathcal{X}|g(\mathbf{x}),\ \mathbf{x} \in \mathcal{X}\}$. Also, stochastic problems often have a natural factorization, where $\mathbf{X} := (X_1, \ldots, X_m)$ is a random vector with distribution π and the objective is to evaluate the expectation

$$\mathrm{E}_\pi h(\mathbf{X}) = \int_{\mathcal{X}} h(\mathbf{x})\pi(\mathbf{x}) dx = \lambda.$$

If $\mathcal{X}$ is a countable set of points, we write

$$\mathrm{E}_\pi h(\mathbf{X}) = \sum_{\mathbf{x} \in \mathcal{X}} h(\mathbf{x})\pi(\mathbf{x}) = \lambda.$$

The factorization (2.23) frequently prompts the choice $\omega = \pi$ for sample generation, and its use in practice is often called *standard MC*. For n independent replications $\mathbf{X}^{(1)}, \ldots, \mathbf{X}^{(n)}$,

$$\bar{\lambda}_n = \frac{1}{n}\sum_{i=1}^{n} h(\mathbf{X}^{(i)})$$

gives an unbiased estimate of λ with standard error $\sqrt{n^{-1}\operatorname{var}_\pi h(\mathbf{X})}$, where

$$\operatorname{var}_\pi h(\mathbf{X}) = \int_{\mathcal{X}} h^2(\mathbf{x})\pi(\mathbf{x}) d\mathbf{x} - \lambda^2.$$

However, $\omega = \pi$ is not necessarily the best choice. Provided that ω satisfies restrictions (2.6) and (2.12) and $g(\mathbf{x}) \geq 0 \; \forall \mathbf{x}$, the distribution

$$\omega(\mathbf{x}) := \begin{cases} \frac{g(\mathbf{x})}{\lambda} & \mathbf{x} \in \mathcal{X} \\ 0 & \text{elsewhere,} \end{cases} \tag{2.24}$$

implying $\text{var}_\omega h(\mathbf{X}) = 0$, would be ideal. Since λ is unknown, this assignment has only academic interest. Nevertheless, sampling from a carefully selected ω other than the optimal (2.24) can also reduce variance. In particular, importance sampling exploits the factorization (2.23) to identify an ω for which

$$\mathrm{E}_\omega \left[\frac{g(\mathbf{X})}{\omega(\mathbf{X})} \right] = \mathrm{E}_\pi h(\mathbf{X})$$

and

$$\text{var}_\omega \left[\frac{g(\mathbf{X})}{\omega(\mathbf{X})} \right] \leq \text{var}_\pi h(\mathbf{X}).$$

Rewriting (2.22) with $\omega_1(\mathbf{x}) = \pi(\mathbf{x})$ and $\omega_2(\mathbf{x}) = \omega(\mathbf{x})$ gives

$$\begin{aligned} \text{var}_\pi h(\mathbf{X}) - \text{var}_\omega \left[\frac{g(\mathbf{X})}{\omega(\mathbf{X})} \right] &= \mathrm{E}_\pi h^2(\mathbf{X}) - \mathrm{E}_\omega \left[\frac{g^2(\mathbf{X})}{\omega^2(\mathbf{X})} \right] \\ &= \mathrm{E}_\pi \left\{ h^2(\mathbf{X}) \left[1 - \frac{\pi(\mathbf{X})}{\omega(\mathbf{X})} \right] \right\}. \end{aligned} \tag{2.25}$$

This reveals that a choice of ω that makes $\pi(\mathbf{x})/\omega(\mathbf{x})$ small (< 1) when $h^2(\mathbf{x})$ is large tends to reduce variance.

In general, realizing the benefits of any technique that reduces variance inevitably depends on how well one exploits a problem's properties and structure. Examples 2.5.1, 2.5.2, 2.7.3, and 2.7.4 demonstrate this for importance sampling. The next section formalizes a setting that frequently arises in MC and creates the basis for Examples 2.5.1 and 2.5.2.

2.5 ESTIMATING VOLUME

Let $\mathcal{R}$ denote a continuous region in $\Re^m$ and let

$$g(\mathbf{x}) := \begin{cases} 1 & \text{if } \mathbf{x} \in \mathcal{R} \\ 0 & \text{elsewhere.} \end{cases} \tag{2.26}$$

Then

$$\begin{aligned} \lambda &= \int_{\Re^m} g(\mathbf{x}) d\mathbf{x} \\ &= \int_{\mathcal{R}} d\mathbf{x} \end{aligned}$$

is called the *volume* of $\mathcal{R}$. Alternatively, if $\mathcal{R}$ is countably finite, then

$$\lambda = \sum_{\mathbf{x}\in\mathcal{R}} g(\mathbf{x})$$
$$= \text{number of points in } \mathcal{R}$$

is called the *size* of $\mathcal{R}$. Hereafter, we write $|\mathcal{A}|$ to denote the volume or size of a set $\mathcal{A}$.

Many problems reduce to the evaluation of volume or size. IMC provides a simple method of estimation. Let $\mathcal{X}$ denote a region in $\Re^m$ that includes $\mathcal{R}$. That is, $\Re^m \supseteq \mathcal{X} \supseteq \mathcal{R}$. Suppose the shape of $\mathcal{X}$ is chosen so that sampling from the uniform distribution

$$\omega(\mathbf{x}) := \begin{cases} \frac{1}{|\mathcal{X}|} & \text{if } \mathbf{x} \in \mathcal{X} \\ 0 & \text{elsewhere} \end{cases}$$

is relatively easy and inexpensive. As illustration, this would be so if $\mathcal{X}$ were an m-dimensional hypercube or hyperrectangle. If $\mathbf{X}$ is an m-dimensional random point with distribution ω, and $\frac{g(\mathbf{X})}{\omega(\mathbf{X})}$ in Algorithm IMC is evaluated as

$$\frac{g(\mathbf{X})}{\omega(\mathbf{X})} = \begin{cases} |\mathcal{X}| & \text{if } \mathbf{X} \in \mathcal{R} \\ 0 & \text{elsewhere,} \end{cases}$$

then it is easily seen that

$$\mathrm{E}_\omega \left[\frac{g(\mathbf{X})}{\omega(\mathbf{X})}\right] = \lambda$$

is an unbiased estimator of λ with

$$\mathrm{var}_\omega \left[\frac{g(\mathbf{X})}{\omega(\mathbf{X})}\right] = \lambda(|\mathcal{X}| - \lambda)$$

that depends on how close $|\mathcal{X}|$ is to the volume $\lambda = |\mathcal{R}|$. Moreover,

$$\begin{aligned} \text{probability that } \mathbf{X} \text{ falls in } \mathcal{R} &= \mathrm{E}_\omega g(\mathbf{X}) \\ &= \int_{\mathcal{R}} \frac{d\mathbf{x}}{|\mathcal{X}|} \\ &= \frac{|\mathcal{R}|}{|\mathcal{X}|}, \end{aligned} \tag{2.27}$$

hereafter called the *hitting probability*. Thus, Algorithm IMC gives

$$\begin{aligned} \bar{\lambda}_n &:= \frac{1}{n}\sum_{i=1}^{n} \frac{g(\mathbf{X})}{\omega(\mathbf{X})} \\ &= |\mathcal{X}| \times \frac{\text{number of points that fall in } \mathcal{R}}{\text{number of points } n \text{ sampled from } \mathcal{X}} \end{aligned}$$

with

$$\mathrm{var}_\omega \bar{\lambda}_n = \frac{\lambda}{n}\left(|\mathcal{X}| - \lambda\right). \tag{2.28}$$

For some $c > 0$, suppose the objective is to obtain $\bar{\lambda}_n$ satisfying $\text{var}_\omega \bar{\lambda}_n \le c$. This implies a sample size n no smaller than

$$n^* := \frac{\lambda}{c}(|\mathcal{X}| - \lambda). \tag{2.29}$$

We now see that in addition to sample generating time per replication, the choice of $\mathcal{X}$ affects the required number of replications to meet a specified accuracy. Often, choosing an $\mathcal{X}$ that reduces generating time per replication (variance) increases variance (generating time per replication). The need to balance these considerations is a recurrent theme in MC.

EXAMPLE 2.5.1

ESTIMATING π

Recall our earlier assertion that the choice of sampling distribution ω is at the discretion of the IMC user. Examples 2.5.1 and 2.5.2 on estimating volume illustrate how this choice affects both variance and computing time. Let

$$\mathcal{R}(a, b, r) := \{(x_1, x_2) \in \Re^2 : (x_1 - a)^2 + (x_2 - b)^2 \le r^2\}$$

denote the set of all points that lie inside the circle centered at $x_1 = a,\ x_2 = b$ with radius r. Its area,

$$|\mathcal{R}(a, b, r)| = \int_{\mathcal{R}(a,b,r)} \mathrm{d}x_1 \mathrm{d}x_2 = \pi r^2, \tag{2.30}$$

provides the basis for an IMC sampling experiment to estimate π.

In the notation of (2.1), let

$$g(x_1, x_2) := \begin{cases} \frac{1}{r^2} & \text{if } (x_1, x_2) \in \mathcal{R}(a, b, r) \\ 0 & \text{elsewhere,} \end{cases} \tag{2.31}$$

so that

$$\lambda = \int_{\Re^m} g(x_1, x_2) \mathrm{d}x_1 \mathrm{d}x_2 = \pi.$$

Let $\mathcal{X} := \{(x_1, x_2) \in \Re^2 : a - r \le x_1 \le a + r,\ b - r \le x_2 \le b + r\}$ denote the square of the smallest side length enclosing the circle, as in Figure 2.1. The square shape makes it convenient to sample from the uniform distribution $\{\omega(x_1, x_2) := \omega_1(x_1)\omega_2(x_2),\ (x_1, x_2) \in \mathcal{X}\}$, where

$$\omega_1(x_1) := \begin{cases} \frac{1}{2r} & \text{if } a - r \le x_1 \le a + r \\ 0 & \text{elsewhere} \end{cases}$$

and

$$\omega_2(x_2) := \begin{cases} \frac{1}{2r} & \text{if } b - r \le x_2 \le b + r \\ 0 & \text{elsewhere.} \end{cases}$$

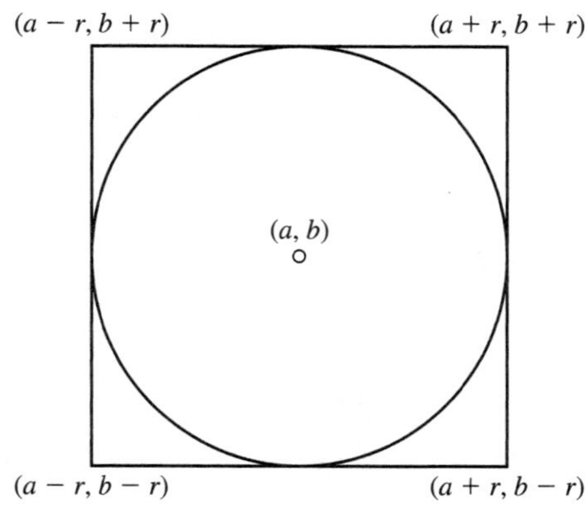

FIGURE 2.1 | *Circle in a square*

If for $i = 1$ and 2, X_i is randomly sampled from ω_i, then

$$\mathrm{E}_\omega\left[\frac{g(X_1, X_2)}{\omega(X_1, X_2)}\right] = \int_{\mathcal{X}}\left[\frac{g(x_1, x_2)}{\omega(x_1, x_2)}\right]\omega(x_1, x_2)\mathrm{d}x_1\mathrm{d}x_2$$
$$= \pi$$

and

$$\mathrm{E}_\omega\left[\frac{g(X_1, X_2)}{\omega(X_1, X_2)}\right]^2 = \int_{\mathcal{X}}\left[\frac{g(x_1, x_2)}{\omega(x_1, x_2)}\right]^2\omega(x_1, x_2)\mathrm{d}x_1\mathrm{d}x_2$$
$$= 4\pi,$$

so that

$$\mathrm{var}_\omega\left[\frac{g(X_1, X_2)}{\omega(X_1, X_2)}\right] = \pi(4 - \pi)$$

is independent of $a, b,$ and r. Therefore, n independent replications using IMC give

$$\bar{\lambda}_n := \frac{1}{n}\sum_{i=1}^{n}\frac{g(X_1, X_2)}{\omega(X_1, X_2)}$$

as an unbiased estimator of π with standard error s.e. $= \sqrt{\pi(4 - \pi)/n} = 1.642/\sqrt{n}$ and relative error r.e. $= \sqrt{(4/\pi - 1)/n} = 0.5227/\sqrt{n}$.

To sample (X_1, X_2), we note that if U is from $\mathcal{U}(0, 1)$ (see [2.19]), then for constants $c < d$, $Z = c + (d - c)U$ has the uniform distribution

$$f_Z(z) := \begin{cases} \frac{1}{d-c} & \text{if } c \le z \le d \\ 0 & \text{elsewhere,} \end{cases}$$

denoted by $\mathcal{U}(c, d)$. As Section 3.3 shows, this is a straightforward application of the inverse-transform method of sample generation. Algorithm CIRSQ summarizes the steps for sample generation. ■

TABLE 2.2 | ***IMC estimation of π using Algorithm CIRSQ***
$\left(\pi = 3.14159265\ldots,\ a = b = r = \frac{1}{2}\right)$

n	$\bar{\lambda}_n$	$\pm\Phi^{-1}(.995)\times$s.e.	s.e.	r.e.
10^2	3.12	$\pm$ .4268	1.657×10^{-1}	5.311×10^{-2}
10^3	3.096	$\pm$ .1363	5.290×10^{-2}	1.709×10^{-2}
10^4	3.1416	$\pm$ .04230	1.642×10^{-2}	5.227×10^{-3}
10^5	3.14588	$\pm$ .01335	5.184×10^{-3}	1.648×10^{-3}
10^6	3.142556	$\pm$ .004228	1.642×10^{-3}	5.223×10^{-4}
10^7	3.1417744	$\pm$ .001338	5.193×10^{-4}	1.653×10^{-4}

ALGORITHM | **CIRSQ**

Purpose: To generate (X_1, X_2) randomly from the uniform distribution on $\mathcal{X}$.

Given: $a,\ b,\ r.$

Method:

Randomly generate U_1 and U_2 independently from $\mathcal{U}(0, 1)$.
$X_1 = a - r + 2rU_1.$
$X_2 = b - r + 2rU_2.$
Return (X_1, X_2). ■

Table 2.2 displays results generated by Algorithm PI using Algorithm CIRSQ on six independent IMC experiments with successively increasing n. They include estimates of $\lambda = \pi$, and their error assessments, where

$$\left[\bar{\lambda}_n - \Phi^{-1}(.995) \times \text{s.e.}(\bar{\lambda}_n),\ \bar{\lambda}_n + \Phi^{-1}(.995) \times \text{s.e.}(\bar{\lambda}_n)\right]$$

gives the 99% confidence interval for λ. Note that although the standard error decreases as $n^{-1/2}$, the confidence interval, $\{3.1404364, 3.1431124\}$, for $n = 10^7$ gives only three-digit accuracy.

ALGORITHM | **PI**

Purpose: To estimate π.

Given: $a,\ b,\ r,\ n$, and $R = 4$.

Method:

$i = 1.$
$V_1 = 0$ and $V_2 = 0$.
While $i \le n$:
  randomly sample X_1 and X_2 using Algorithm CIRSQ
  if $(X_1 - a)^2 + (X_2 - b)^2 \le r$:
    $V_1 = V_1 + 1$ and $V_2 = V_2 + 1$
  $i = i + 1.$

$$\bar{\lambda}_n = \frac{RV_1}{n} \text{ and } s_n^2 = \frac{1}{n-1}\left(R^2 V_2 - n\bar{\lambda}_n^2\right).$$

$$\text{s.e.} = \sqrt{\frac{s_n^2}{n}} \text{ and r.e.} = \frac{\text{s.e.}(\bar{\lambda}_n)}{\bar{\lambda}_n}.$$

■

EXAMPLE 2.5.2

ESTIMATING π EFFICIENTLY

Sampling (X_1, X_2) in Example 2.5.1 from the uniform distribution on the smallest square $\mathcal{X}$ enclosing $\mathcal{R}(a, b, r)$ offers conceptual simplicity and computational efficiency per replication in the sampling step. However, the probability that (X_1, X_2) falls in the circle is $\pi/4 = 0.7854$, which determines the variance as $\pi(4 - \pi)$ per replication. An alternative choice of $\mathcal{X}$ increases this *hitting probability*, thereby reducing variance. It does so by exploiting the many symmetries inherent in the problem.

A *polygon* is a bounded two-dimensional region, all of whose sides are straight lines. A *regular polygon* is a polygon whose sides are of equal length. A *regular octagon* is an eight-sided regular polygon. In what follows, a regular octagon will be referred to without the modifier "regular," for conciseness of exposition.

Instead of the square, consider the smallest octagon $\mathcal{X}$ enclosing the circle, as depicted in Figure 2.2. Since it has smaller area than the square, with side lengths $2r$, (2.28) implies that sampling from the octagon instead of the square induces a smaller variance. Although sampling from the octagon is more difficult, symmetry allows us to reduce the severity of the difficulty. Observe that triangle

$$\mathcal{T} := \left\{ (x_1, x_2) \in \mathcal{X} : a + \frac{d}{r}(x_2 - b) \le x_1 \le a - \frac{d}{r}(x_2 - b) \right\}, \qquad (2.32)$$

where $d := r \tan 22.5°$, encloses one-eighth of the circle with area $\pi r^2/8$. Since Figure 2.2 shows that $\mathcal{T}$ is composed of two right triangles of equal area, $\mathcal{T}$ has area $2\left(\frac{dr}{2}\right) = r^2 \tan 22.5°$.

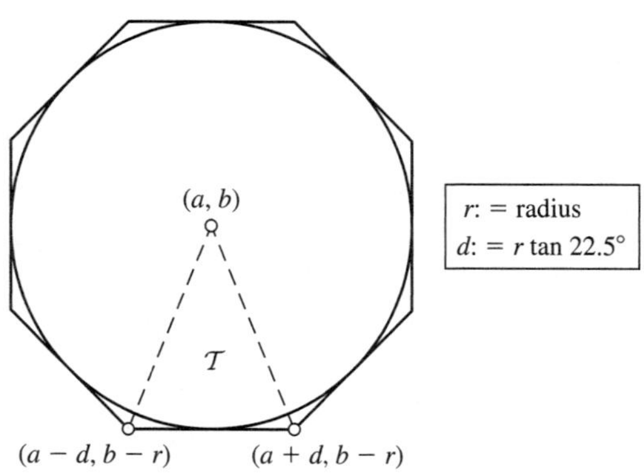

FIGURE 2.2 | *Circle in an octagon*

Because of this symmetry, it suffices to restrict sampling to $\mathcal{T}$, from which it is much easier to sample than from the octagon $\mathcal{X}$. Let

$$g(x_1, x_2) := \begin{cases} \frac{8}{r^2} & \text{if } (x_1, x_2) \in \mathcal{R}(a, b, r) \\ 0 & \text{elsewhere,} \end{cases} \tag{2.33}$$

and suppose we sample (X_1, X_2) from the uniform distribution

$$\omega(x_1, x_2) := \begin{cases} \frac{1}{r^2 \tan 22.5^\circ} & \text{if } (x_1, x_2) \in \mathcal{T} \\ 0 & \text{elsewhere.} \end{cases}$$

The hitting probability (2.28) is now $\pi/8 \tan 22.5^\circ = 0.9481$. Also,

$$\mathrm{E}_\omega \left[\frac{g(X_1, X_2)}{\omega(X_1, X_2)} \right] = \pi,$$

and

$$\mathrm{E}_\omega \left[\frac{g(X_1, X_2)}{\omega(X_1, X_2)} \right]^2 = 8\pi \tan 22.5^\circ,$$

so that

$$\begin{aligned} \mathrm{var}_\omega \left[\frac{g(X_1, X_2)}{\omega(X_1, X_2)} \right] &= \pi(8 \tan 22.5^\circ - \pi) \\ &\le \pi(4 - \pi). \end{aligned}$$

Recall that Example 2.5.1 had variance $\pi(4-\pi)$ per replication when sampling from the square. Therefore, the variance ratio for Algorithm CIRSQ (plan 1) in Example 2.5.1 and Algorithm CIROCT in the current example (plan 2) is

$$\mathrm{VR} = \frac{4 - \pi}{8 \tan 22.5^\circ - \pi} = 4.987, \tag{2.34}$$

implying that IMC would require about five independent replications to achieve the same variance as one replication when sampling from $\mathcal{T}$. Although this observation clearly favors sampling from an octagon rather than a square, there is more to the story.

Whereas X_1 and X_2 are independent random variables in Example 2.5.1, when sampled from the uniform distribution on the enclosing square, they are not independent when sampled from the uniform distribution on $\mathcal{T}$. Perusal of (2.32) makes this dependence apparent. More generally, dependence among the coordinates $X_1, \ldots, X_m$ in an m-dimensional problem frequently makes sampling difficult from their joint distribution ω. Then the conditional sampling techniques of the Markov chain Monte Carlo (MCMC) method in Chapter 6 allow one to generate samples that have the correct joint distribution, but at the cost of inducing dependence among successive samples. Fortunately, we can exploit additional

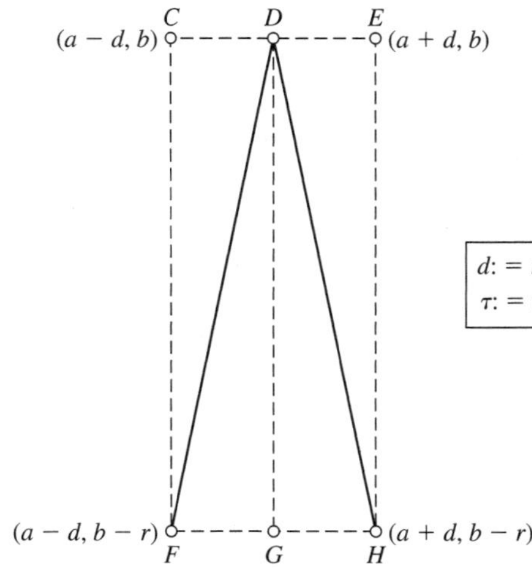

FIGURE 2.3 | *Triangle in a rectangle*

properties in the present problem to generate (X_1, X_2) efficiently without resorting to MCMC.

A relatively straightforward way to accommodate the dependence between X_1 and X_2 is to sample (X_1, X_2) repeatedly from rectangle $CEHF$ in Figure 2.3 until the sample falls into the triangle $\mathcal{T}$. Algorithm CIROCT describes the formal steps. ■

ALGORITHM

CIROCT

Purpose: To generate (X_1, X_2) randomly from the uniform distribution on $\mathcal{T}$ in (2.32).

Given: $a, b, r, d := r \tan 22.5^\circ$.

Method:

Until $X_1 - a > \frac{d}{r}(X_2 - b)$ and $X_1 - a < -\frac{d}{r}(X_2 - b)$:
- randomly generate U_1 and U_2 independently from $\mathcal{U}(0, 1)$
- $X_1 = a - d + 2dU_1$
- $X_2 = b - r + rU_2$.

Return (X_1, X_2). ■

Algorithm CIROCT induces a random number of trials until success, I, thus introducing us to the *acceptance-rejection method* of sampling (see Section 3.6). Let p denote the probability of success on a trial. Since the trials are independent, I has the geometric distribution

$$\text{pr}(I = i) = \begin{cases} p(1-p)^{i-1} & i = 1, 2, \ldots \\ 0 & \text{elsewhere,} \end{cases} \tag{2.35}$$

TABLE 2.3 | *Estimating π using Algorithm CIROCT*
$\left(\pi = 3.14159265\ldots,\ a = b = r = \frac{1}{2}\right)$

n	$\bar{\lambda}_n$	$\Phi^{-1}(.995)\times$s.e.	s.e.	r.e.	VR
10^2	3.15	.1860	7.222×10^{-2}	2.294×10^{-2}	5.26
10^3	3.158	.05713	2.218×10^{-2}	7.023×10^{-3}	5.67
10^4	3.1500	.01850	7.181×10^{-3}	2.280×10^{-3}	5.23
10^5	3.14007	.006015	2.335×10^{-3}	7.436×10^{-4}	4.93
10^6	3.141412	.001895	7.357×10^{-4}	2.342×10^{-4}	4.98
10^7	3.1415438	.0005990	2.326×10^{-4}	7.403×10^{-5}	4.98

denoted by $\mathcal{G}e(p)$, and

$$\begin{matrix}\text{mean number of} \\ \text{trials until success}\end{matrix} = \sum_{i=1}^{\infty} ip(1-p)^{i-1} = \frac{1}{p}.$$

Recall that triangles CDF, DGF, DEH, and DHG all have equal areas. Since rectangle $CEHF$ has twice the area of $\mathcal{T}$, it follows that $p = \frac{1}{2}$ and, therefore, the mean number of trials until success is $\frac{1}{p} = 2$.

Table 2.3 shows results for Algorithm PI with $R = 8\tan 22.5^\circ$ and Algorithm CIROCT (plan 2) replacing Algorithm CIRSQ (plan 1) for sample generation. The empirical variance ratios are as expected theoretically (see [2.34]). The cost ratio (2.20) is CR $= 1/2.25$, revealing that the approximate five-fold variance reduction comes at the cost of more than a doubling in computing time. In particular, to attain the same variance as Algorithm CIROCT does for n independent replications, Algorithm CIRSQ would require CR $\times$ VR $= 4.987/2.25 = 2.216$ times as much computing time.

Symmetry also offers an alternative single-iteration method of sampling uniformly from $\mathcal{T}$. Recall that triangles CDF, DGF, DEH, and DHG in Figure 2.3 all have equal area. If (X_1, X_2) is uniformly distributed on triangle CDF, then $(2a - d - X_1, 2b - r - X_2)$ is uniformly distributed on triangle DGF. Similarly, if (X_1, X_2) is uniformly distributed on triangle DEH, then $(2a+d-X_1, 2b-r-X_2)$ is uniformly distributed on triangle DHG. These properties suggest sampling from the uniform distribution on the rectangle $CEHF$ and then, if necessary, transforming the location of the points. Algorithm TRIRECT describes the formal steps.

ALGORITHM **TRIRECT**

Purpose: To generate (X_1, X_2) randomly from the uniform distribution on $\mathcal{T}$ in (2.32).

Given: $a, b, r, d := r\tan 22.5^\circ$.

Method:

Randomly generate U_1 and U_2 independently from $\mathcal{U}(0, 1)$.
$X_1 = a - d + 2dU_1$.

$X_2 = b - r + rU_2$.
If $X_1 < a + \frac{d}{r}(X_2 - b)$:
  $X_1 = 2a - d - X_1$
  $X_2 = 2b - r - X_2$
  return (X_1, X_2). (point is in DGF)
If $X_1 > a + \frac{d}{r}(b - X_2)$:
  $X_1 = 2a + d - X_1$
  $X_2 = 2b - r - X_2$
  return (X_1, X_2). (point is in DHG)
Return (X_1, X_2). (point is in $DHF = DGF \cup DHG$) ■

Algorithm PI with $R = 8 \tan 22.5°$ and Algorithm TRIRECT replacing Algorithm PI give results similar to those in Table 2.3, but the cost ratio (2.20) now is CR $= 1/1.19$, a 47% reduction compared to the ratio for Algorithm CIROCT. To achieve the same variance as Algorithm TRIRECT does for n independent replications, Algorithm CIRSQ would require CR $\times$ VR $= 4.987/1.19 = 4.1908$ times as much computing time. Moreover, the advantage of TRIRECT over CIROCT is apparent.

Comparing the performance of Algorithms CIROCT and TRIRECT reveals the benefit that accrues from careful attention to algorithmic design before sampling begins. This is especially true when many experiments are to be executed and estimates are required with high relative accuracy. Chapter 5 describes additional techniques that allow this benefit to be realized in practice.

2.6 INTERPRETING RELATIVE ERROR

Except for setup time, total computing time for an IMC experiment is directly proportional to the number of replications n. Moreover, n^*, the sample size required to satisfy a specified accuracy as, for example, in (2.29), increases as c decreases. Since one rarely has a good estimate of n^* prior to sampling, IMC experimenters usually choose an n, execute the experiment for that number of replications, and then rely on confidence intervals, standard errors, and relative errors to infer whether or not $\bar{\lambda}_n$ has attained an accuracy adequate for the purposes of the problem under study.

Table 2.4 provides a gauge for interpreting the numerical values that arise for the relative error (r.e.) when the objective is to estimate λ so as to satisfy

$$\text{pr}\left(|\bar{\lambda}_n - \lambda| \le \lambda\theta\right) \ge 1 - \delta \tag{2.36}$$

for specified $\delta \in (0, 1)$ and $\theta > 0$.

From property P3, it is clear that for large n,

$$\text{pr}\left[|\bar{\lambda}_n - \lambda| \le \Phi^{-1}(1 - \delta/2)\frac{\sigma}{n^{1/2}}\right] \approx 1 - \delta. \tag{2.37}$$

TABLE 2.4 | *Required relative error*

Relative error (r.e.) $= \frac{\theta}{\Phi^{-1}(1-\delta/2)}$

δ	$\Phi^{-1}(1-\delta/2)$	$\theta = .01$	$\theta = .02$	$\theta = .05$	$\theta = .10$
.05	1.95996	5.102×10^{-3}	.01020	.02551	.05102
.01	2.57583	3.882×10^{-3}	7.764×10^{-3}	.01941	.03882
.001	3.29053	3.035×10^{-3}	6.069×10^{-3}	.01517	.03034

Collectively, (2.36) and (2.37) imply

$$\Phi^{-1}(1-\delta/2)\frac{\sigma}{n^{1/2}} \le \lambda\theta.$$

Since r.e. is an estimate of $\frac{\sigma}{\lambda n^{1/2}}$, we can estimate accuracy by comparing its numerical value computed from sample data with $\frac{\theta}{\Phi^{-1}(1-\delta/2)}$.

Table 2.4 shows this ratio for selected δ and θ. If the objective had been to estimate π to within $100 \times \theta = 1$ percent with probability $1-\delta = 0.99$, then Table 2.2 reveals that the sample size $n = 10{,}000$ would have sufficed when sampling from a square. More interestingly, Table 2.3 reveals that $n = 1{,}000$ would have sufficed when sampling from an octagon.

Of course, absolute accuracy is also important. Table 2.2 shows four-digit accuracy for $n \ge 10^7$, whereas Table 2.3 shows it for $n \ge 10^6$. More generally, MC techniques frequently show two-digit accuracy at modest sample sizes but require substantially larger ones to achieve high levels of absolute accuracy.

2.7 | PRODUCT AND NONPRODUCT SPACES

The shape of the region $\mathcal{X}$ and the functional form of the integrand g determine whether (2.1) can be evaluated exactly. If it cannot, these same considerations determine the MC methods most suitable for estimating (2.1). This section describes several characteristics of region and integrand that influence the choice of sampling plan. We begin with the difference between product spaces and nonproduct spaces.

The region $\mathcal{X}$ over which integration occurs is a compilation of all constraints imposed on the m variables. For each variable $i = 1, \ldots, m$, a univariate constraint specifies all values that x_i can assume. A multivariate constraint specifies restrictions on the values that two or more variables can assume simultaneously. A collection of univariate constraints forms a *product space,* whereas a collection of multivariate and, possibly, univariate constraints forms a *nonproduct space.*

As an illustration, suppose there are $m = 2$ one-dimensional regions

$$\mathcal{X}_1 := \{x_1 \in \Re : a_1 \le x_1 \le b_1\} \qquad b_1 > a_1$$

and

$$\mathcal{X}_2 := \{x_2 \in \Re : a_2 \le x_2 \le b_2\} \qquad b_2 > a_2.$$

These univariate constraints tell us that the 2-tuples (x_1, x_2) take values in the two-dimensional product space

$$\mathcal{X}' := \{(x_1, x_2) \in \Re^2 : \ x_1 \in \mathcal{X}_1 \text{ and } x_2 \in \mathcal{X}_2\},$$

sometimes denoted by $\mathcal{X}' = \mathcal{X}_1 \times \mathcal{X}_2$. However, adding the multivariate constraint $\min(a_1, a_2) \leq x_1 + x_2 \leq \min(b_1, b_2)$ restricts x_1 and x_2 to the region

$$\mathcal{X} := \{(x_1, x_2) \in \mathcal{X}_1 \times \mathcal{X}_2 : \ \min(a_1, a_2) \leq x_1 + x_2 \leq \min(b_1, b_2)\},$$

which is a nonproduct space. Note that $\mathcal{X} \subseteq \mathcal{X}'$. As additional illustrations, sampling takes place over a product space in Example 2.5.1, but over nonproduct spaces in Algorithms CIRCOT and TRIRECT in Example 2.5.2. This distinction contributes to the increased sampling cost for the latter over the former. In practice, sampling from a nonproduct space is usually more costly than sampling from a product space.

These space characterizations immediately generalize to $m > 2$ dimensions in continuous space. Also, analogous characterizations hold for discrete spaces where each variate takes values in some subset of $\mathbb{Z}$:= set of all integers. Note that $\mathbb{Z}$ is a subset of $\Re$, denoted by $\mathbb{Z} \subset \Re$.

Suppose that $\mathcal{X}$ is a product space and that g admits the factorization

$$g(\mathbf{x}) := \prod_{i=1}^{m} g_i(x_i) \qquad \mathbf{x} \in \mathcal{X} = \mathcal{X}_1 \times \cdots \times \mathcal{X}_m,$$

where $\mathcal{X}_i$ denotes the set of all values that x_i can assume, for $i = 1, \ldots, m$. Then (2.1) easily converts into the product of m one-dimensional integrals, usually solvable analytically or by highly efficient numerical techniques. If g cannot be factored in this way, the simplest IMC procedure samples X_i from a distribution $\{\omega_i(x_i),\ x_i \in \mathcal{X}_i\}$ for each i, yielding $\frac{g(\mathbf{X})}{\omega(\mathbf{X})}$ as an estimate of λ with variance (2.8), where $\omega(\mathbf{X}) := \prod_{i=1}^{m} \omega_i(X_i)$. Although the next section emphasizes that an alternative choice for ω may induce a smaller variance, the salient point here is that a product space simplifies sample generation.

By contrast, a nonproduct space with its multivariate constraints restricts the choice of sampling procedures as Example 2.5.2 demonstrates. Sometimes, the restrictions are so severe as to make IMC computationally intractable within the context of the nonproduct space. Then Markov chain Monte Carlo (see Chapter 6) offers a more indirect sampling method that lowers unit generating cost but usually induces dependence among the sample outcomes on successive trials. This dependence frequently inflates the variance of the point estimator of λ. Occasionally, a second alternative facilitates IMC by transforming the nonproduct space into a product space. The two following examples illustrate the concept.

EXAMPLE 2.7.1 INTEGRATION OVER A SIMPLEX

For $i = 1, \ldots, m$, let $\mathcal{X}_i = [0, \infty)$ and suppose that (2.2) is to be evaluated over the *simplex*

$$\mathcal{X} := \{\mathbf{x} := (x_1, \ldots, x_m) \in \mathcal{X}_1 \times \cdots \times \mathcal{X}_m : \ 0 \leq x_1 + \cdots + x_m \leq 1\},$$

which is a nonproduct space. For g everywhere continuous and differentiable on $\mathcal{X}$, it can be shown that the transformations

$$\begin{aligned} x_1 &= y_1 \\ x_2 &= (1-y_1)y_2 \\ x_3 &= (1-y_1)(1-y_2)y_3 \\ \vdots &= \vdots \\ x_m &= (1-y_1)\times\cdots\times(1-y_{m-1})y_m \end{aligned}$$

make (2.2) equivalent to

$$\lambda = \int_{\mathcal{X}_1}\cdots\int_{\mathcal{X}_m} g_*(y_1,\ldots,y_m)\left[\prod_{i=1}^{m-1}(1-y_i)^{m-i}\right]\mathrm{d}y_1\cdots\mathrm{d}y_m,$$

where $g_*(y_1,\ldots,y_m) = g(x_1,\ldots,x_m)$. Integration is now over the m-dimensional product space $[0,1]^m$. This and the next example rely on the standard theory of variable transformation in multivariable calculus. ■

EXAMPLE 2.7.2 INTEGRATION OVER THE UNIT HYPERSPHERE

For $i = 1,\ldots,m$, let $\mathcal{X}_i = [0,\infty)$ and suppose that (2.2) is to be evaluated over the m-dimensional unit hypersphere

$$\mathcal{X} := \left\{\mathbf{x} := (x_1,\ldots,x_m) \in \mathcal{X}_1\times\cdots\times\mathcal{X}_m :\ x_1^2+\cdots+x_m^2 \le 1\right\},$$

which is a nonproduct space. For g everywhere continuous and differentiable on $\mathcal{X}$, it can be shown that the transformation to *polar coordinates,*

$$\begin{aligned} x_1 &= r\prod_{l=1}^{m-1}\cos\theta_l \\ x_2 &= r\sin\theta_1\prod_{l=2}^{m-1}\cos\theta_l \\ \vdots &= \vdots \\ x_{m-1} &= r\sin\theta_{m-2}\cos\theta_{m-1} \\ x_m &= r\sin\theta_{m-1}, \end{aligned}$$

where $0\le r\le 1$, $0\le\theta_1\le 2\pi$, and $-\frac{\pi}{2}\le\theta_i\le\frac{\pi}{2}$ for $i = 2,\ldots,m-1$, leads to the equivalent integral

$$\lambda = \int_{\mathcal{Z}} g_{**}(r,\theta_1,\ldots,\theta_{m-1})r^{m-1}\left[\prod_{l=2}^{m-1}(\cos\theta_l)^{l-1}\right]\mathrm{d}\theta_1\mathrm{d}\theta_2\cdots\mathrm{d}\theta_{m-1}\mathrm{d}r,$$

where

$$\mathcal{Z} := [0,1]\times[0,2\pi]\times\left[-\frac{\pi}{2},\frac{\pi}{2}\right]^{m-2}$$

and $g_{**}(r,\theta_1,\ldots,\theta_{m-1}) = g(x_1,\ldots,x_m)$. Clearly, $\mathcal{Z}$ is a product space. ■

EXAMPLE 2.7.3

NETWORK RELIABILITY

A network provides a convenient representation for designing a system and evaluating its performance, especially in areas such as transportation, telecommunications, and command and control. *Connectedness,* the ability to communicate between a system's component subsystems, is a frequently employed criterion of performance.

Graphs are often employed to characterize a network conceptually. A *graph* $\mathcal{G}$ is a collection or set of *vertices* or *nodes* $\mathcal{V}$ and a collection of *edges* or *arcs* $\mathcal{E} = \{1, \ldots, m\}$ that connect pairs of nodes. Figure 2.4 shows an example. In an *undirected graph,* as in Figure 2.4, edges have no directional orientation. In a *directed graph,* each edge points from one of its nodes to the other.

Nodes and edges can be thought of as representing the components of a system. For present purposes, assume that node components are perfect but edge components fail randomly and independently. Let $1 - p_i$ denote the probability that edge i fails for each $i \in \mathcal{E}$. Let $x_i = 1$ if edge i operates and $x_i = 0$ if it fails. Then $\mathbf{x} := (x_1, \ldots, x_m)$ denotes a state of the m edges and $\mathcal{X} = \{0, 1\}^m$, the set of all m-dimensional state vectors.

Consider two nodes s and t in $\mathcal{V}$ and define the subset

$$\begin{aligned}\mathcal{X}(s, t) &:= \{\mathbf{x} \in \mathcal{X} : \text{nodes } s \text{ and } t \text{ are } \textit{not} \text{ connected}\} \\ &= \text{subset of all state vectors for which nodes } s \text{ and } t \text{ are connected.}\end{aligned}$$

Nodes s and t are said to be connected if at least one path of operating edges exists between them. For example, if all arcs on any of the paths $\{1, 5\}$, $\{1, 3, 6\}$, $\{1, 3, 4, 7\}$, $\{2, 7\}$, $\{2, 4, 6\}$, or $\{2, 4, 3, 5\}$ operate in Figure 2.4, s and t are connected. If no such path exists, we say that nodes s and t are not connected or, equivalently, the system has *failed.* Let

$$\pi_i(x_i, p_i) := \begin{cases} p_i & \text{if } x_i = 1 \\ 1 - p_i & \text{if } x_i = 0, \end{cases} \tag{2.38}$$

which is called a Bernoulli distribution and denoted by $\mathcal{B}er(p_i)$. Let

$$g(\mathbf{x}, \mathbf{p}) := h(\mathbf{x})\pi(\mathbf{x}, \mathbf{p}),$$

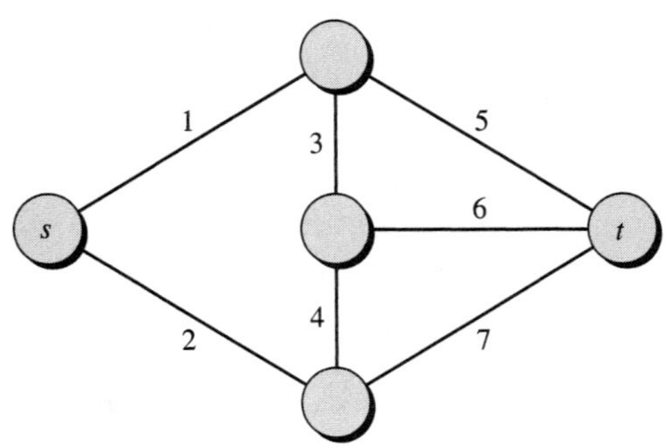

FIGURE 2.4 | *Network* ($m = 7, r = 2$)

where

$$h(\mathbf{x}) := \begin{cases} 1 & \text{if } \mathbf{x} \in \mathcal{X}(s,t) \\ 0 & \text{otherwise,} \end{cases}$$

$\mathbf{p} = (p_1, \ldots, p_m)$, and

$$\begin{aligned} \pi(\mathbf{x}, \mathbf{p}) &:= \prod_{i=1}^{m} \pi_i(x_i, p_i) \\ &= \prod_{i=1}^{m} p_i^{x_i}(1-p_i)^{1-x_i} \qquad\qquad \mathbf{x} = (x_1, \ldots, x_m) \in \{0,1\}^m. \\ &= \text{probability of observing state } \mathbf{x} \end{aligned}$$

Here $\{\pi(\mathbf{x}, \mathbf{p}),\ \mathbf{x} \in \mathcal{X}\}$ is the joint distribution of m independent Bernoulli distributed random variables and

$$\begin{aligned} \lambda(\mathbf{p}) &:= \sum_{\mathbf{x} \in \mathcal{X}} g(\mathbf{x}, \mathbf{p}) \\ &= \sum_{\mathbf{x} \in \mathcal{X}} h(\mathbf{x})\, \pi(\mathbf{x}, \mathbf{p}) \\ &= \sum_{\mathbf{x} \in \mathcal{X}(s,t)} \pi(\mathbf{x}, \mathbf{p}) \\ &= \text{probability that the system fails} \end{aligned} \tag{2.39}$$

is called the *s-t* reliability. The objective is to evaluate $\lambda(\mathbf{p})$.

For the example in Figure 2.4,

$$h(\mathbf{x}) = (1-x_1x_5)(1-x_1x_3x_6)(1-x_1x_3x_4x_7)(1-x_2x_7)(1-x_2x_4x_6)(1-x_2x_3x_4x_5). \tag{2.40}$$

More generally,

$$h(\mathbf{x}) = \prod_{\mathcal{P} \in \mathcal{H}(\mathcal{G})} \left(1 - \prod_{i \in \mathcal{P}} x_i\right),$$

where $\mathcal{H}(\mathcal{G})$ denotes the collection of all paths in $\mathcal{G}$ connecting s and t.

Two computational issues need to be addressed. One concerns the evaluation of $h(\mathbf{x})$ and the other, the evaluation of $\lambda(\mathbf{p})$. Since $|\mathcal{H}(\mathcal{G})|$, the size of $\mathcal{H}(g)$, can grow exponentially in $|\mathcal{E}| = m$, direct product evaluation of $h(\mathbf{x})$ can be costly for large networks. However, a *depth-first search* reduces the cost of evaluating $h(\mathbf{x})$ to being proportional to $\max(|\mathcal{E}|, |\mathcal{V}|)$ in the worst case. (See Aho, et al. 1974, p. 176.)

If $\mathbf{p}$ is known, then one would surmise that $\lambda(\mathbf{p})$ in (2.39) is computable with no more than numerical round-off error. However, for moderate-sized networks $\mathcal{X}$ can be so large that exact evaluation as in (2.39) becomes intractable. For example, if $\mathcal{G}$ is a *complete graph,* each node of which shares an edge with every other node, then $|\mathcal{X}| = 2^m$. For $m = 100$, this implies $|\mathcal{X}| \doteq 10^{30.1}$. IMC offers a relatively simple procedure for estimating $\lambda(\mathbf{p})$.

Since $\{\pi(\mathbf{x}, \mathbf{p}),\ \mathbf{x} \in \mathcal{X}\}$ is a distribution, standard MC uses the sampling distribution $\{\omega(\mathbf{x}, \mathbf{p}) := \pi(\mathbf{x}, \mathbf{p}),\ \mathbf{x} \in \mathcal{X}\}$. Algorithm ZERONE describes the corresponding sample generation. ■

ALGORITHM **ZERONE**

Purpose: To generate X_i randomly from $\mathcal{B}er(p_i)$ for $i = 1, \ldots, m$.

Given: $p_1, \ldots, p_m$.

Method:

```
i = 1.
While i ≤ n:
    randomly generate U_i from U(0, 1)
    if U_i ≤ p_i, X_i = 1
        otherwise, X_i = 0
    i = i + 1.
```

Then

$$\mathrm{E}_{\omega(\cdot,\mathbf{p})}\left[\frac{g(\mathbf{X},\mathbf{p})}{\omega(\mathbf{X},\mathbf{p})}\right] = \lambda(\mathbf{p})$$

and

$$\mathrm{E}_{\omega(\cdot,\mathbf{p})}\left[\frac{g^2(\mathbf{X},\mathbf{p})}{\omega^2(\mathbf{X},\mathbf{p})}\right] = \lambda(\mathbf{p})$$

so that

$$\mathrm{var}_{\omega(\cdot,\mathbf{p})}\left[\frac{g(\mathbf{X},\mathbf{p})}{\omega(\mathbf{X},\mathbf{p})}\right] = \lambda(\mathbf{p})[1-\lambda(\mathbf{p})].$$

■

Standard MC with $\omega = \pi$ and n independent replications leads to

$$\bar{\lambda}_n = \frac{1}{n}\sum_{i=1}^{n}\frac{g(\mathbf{X}^{(i)},\mathbf{p})}{\omega(\mathbf{X}^{(i)},\mathbf{p})} = \frac{1}{n}\sum_{i=1}^{n} h\left(\mathbf{X}^{(i)}\right)$$

as an estimate of the s-t reliability $\lambda(\mathbf{p})$ with standard error $\sqrt{\lambda(\mathbf{p})[1-\lambda(\mathbf{p})]/n}$ and relative error $\sqrt{[1-\lambda(\mathbf{p})]/\lambda(\mathbf{p})n}$. Algorithm STREL describes the sampling experiment.

If components are highly reliable so that the p_i are close to unity, then $\lambda(\mathbf{p})$ is close to zero and the relative error can be very large for large n. As an example, for $\lambda(\mathbf{p}) = 10^{-4}$, Table 2.4 implies that

$$n = \frac{1}{(.03882)^2} \times \frac{1-\lambda(\mathbf{p})}{\lambda(\mathbf{p})} \approx 6.6 \times 10^6 \text{ independent replications}$$

are required to obtain an estimate $\bar{\lambda}_n$ within 10% of $\lambda(\mathbf{p})$ with probability $1 - \delta = 0.99$. This observation motivates the search for a sampling distribution $\omega \neq \pi$ that substantially reduces variance and, thereby, required sample size. In what follows, we show how importance sampling exploits known properties of networks to effect these reductions.

ALGORITHM **STREL**

Purpose: To estimate λ in (2.39).

Given: Network $\mathcal{G}(\mathcal{V}, \mathcal{E})$ with $m = |\mathcal{E}|$, nodes s and t, $p_1, \ldots, p_m$, n.

Method:

$i = 1$.
$V_1 = 0$ and $V_2 = 0$.
While $i \leq n$:
  randomly sample $(X_1, \ldots, X_m)$ using Algorithm ZERONE
  if nodes s and t are not connected
    $V_1 = V_1 + 1$ and $V_2 = V_2 + 1$
  $i = i + 1$.

$\bar{\lambda}_n(\mathbf{p}, \mathbf{p}) = \dfrac{V_1}{n}$ and $s_n^2 = \dfrac{1}{n-1}\left[V_2 - n\bar{\lambda}_n^2(\mathbf{p}, \mathbf{p})\right]$.

$\text{s.e.}\left(\bar{\lambda}_n(\mathbf{p}, \mathbf{p})\right) = \sqrt{\dfrac{s_n^2}{n}}$ and $\text{r.e.}\left(\bar{\lambda}_n(\mathbf{p}, \mathbf{p})\right) = \dfrac{\text{s.e.}\left(\bar{\lambda}_n(\mathbf{p}, \mathbf{p})\right)}{\bar{\lambda}_n(\mathbf{p}, \mathbf{p})}$. ■

Suppose $\mathbf{X}$ is randomly drawn from $\{\omega(\mathbf{x}, \mathbf{q})\}$ for some vector $\mathbf{q} := (q_1, \ldots, q_m)$ of component-operating probabilities not necessarily equal to $\mathbf{p}$. Then

$$\mathrm{E}_{\omega(\cdot,\mathbf{q})}\left[\frac{g(\mathbf{X}, \mathbf{p})}{\omega(\mathbf{X}, \mathbf{q})}\right] = \lambda(\mathbf{p}),$$

as before, but

$$\begin{aligned}\mathrm{E}_{\omega(\cdot,\mathbf{q})}\left[\frac{g^2(\mathbf{X}, \mathbf{p})}{\omega^2(\mathbf{X}, \mathbf{q})}\right] &= \mathrm{E}_{\pi(\cdot,\mathbf{p})}\left\{h(\mathbf{X}, \mathbf{p})\left[\frac{\pi(\mathbf{X}, \mathbf{p})}{\pi(\mathbf{X}, \mathbf{q})}\right]\right\} \\ &= \sum_{\mathbf{x} \in \mathcal{X}(s,t)} \left[\frac{\pi(\mathbf{x}, \mathbf{p})}{\pi(\mathbf{x}, \mathbf{q})}\right] \pi(\mathbf{x}, \mathbf{p})\end{aligned}$$

so that

$$\mathrm{var}_{\omega(\cdot,\mathbf{q})}\left[\frac{g(\mathbf{X}, \mathbf{p})}{\omega(\mathbf{X}, \mathbf{q})}\right] = \sum_{\mathbf{x} \in \mathcal{X}(s,t)} \left[\frac{\pi(\mathbf{x}, \mathbf{p})}{\pi(\mathbf{x}, \mathbf{q})}\right] \pi(\mathbf{x}, \mathbf{p}) - \lambda^2(\mathbf{p}).$$

This last expression reveals that randomly sampling $\mathbf{X}$ from $\{\omega(\mathbf{x}, \mathbf{q})\}$ for a $\mathbf{q}$ such that $\pi(\mathbf{x}, \mathbf{p}) < \pi(\mathbf{x}, \mathbf{q})$ for all $\mathbf{x} \in \mathcal{X}(s, t)$ leads to a variance reduction. As illustration, let $p_i = p > 1/2$ and $q_i = q < p$ for all $i = 1, \ldots, m$. Then choosing q

such that

$$\frac{\pi(\mathbf{x},p)}{\pi(\mathbf{x},q)} = \prod_{i=1}^{m} \left(\frac{p}{q}\right)^{x_i} \left(\frac{1-p}{1-q}\right)^{1-x_i} < 1 \tag{2.41}$$

reduces variance. A more ambitious objective is to find the q that minimizes

$$P_{\max} := \max_{\mathbf{x}\in\mathcal{X}(s,t)} \frac{\pi(\mathbf{x},p)}{\pi(\mathbf{x},q)}, \tag{2.42}$$

subject to $1/2 < q < p < 1$. Let

$$\begin{aligned} x &:= x_1 + \cdots + x_m \\ &= \text{number of operating arcs.} \end{aligned}$$

Because

$$\frac{\pi(\mathbf{x},p)}{\pi(\mathbf{x},q)} = \left(\frac{p}{q}\right)^{x} \left(\frac{1-p}{1-q}\right)^{m-x}$$

is an increasing function of x, it is of interest to determine the smallest x for which $h(\mathbf{x}) = 0$ for the corresponding $\mathbf{x} \in \{0,1\}^m \backslash \mathcal{X}(s,t)$, the set of all states in which the system fails.

For each reliability network $\mathcal{G}$, there exists an integer $r \in \{1,\dots,m\}$ such that for $x > m-r$, nodes s and t are connected. That is, at least one path of operating edges exists between s and t. For example, if at least six arcs operate in the seven-arc network in Figure 2.4, then s and t are connected. Here $r = 2$.

More formally, r is the size of the *minimal s-t cutset* of $\mathcal{G}$. A cutset denotes a set of edges that disconnect s and t when all the edges fail, and a minimal s-t cutset is one containing no subset of edges that disconnects s and t. Because $\{1,2\}$ is the only minimal s-t cutset in the network in Figure 2.4, $r = 2$.

Because $x > m-r$ implies that the corresponding $\mathbf{x}$ is not in $\mathcal{X}(s,t)$, $h(\mathbf{x}) = 0$. However, for $x \le m-r$, the corresponding $\mathbf{x}$ may be in $\mathcal{X}(s,t)$, so that $h(\mathbf{x}) = 1$. Therefore, it is worthwhile to employ a q that keeps the ratio in (2.41) less than 1 for all $x \le m-r$.

Because for all $\mathbf{x} \in \mathcal{X}(s,t)$, $h(\mathbf{x}) = 1$ and $x \le m-r$, it follows that

$$P_{\max} = \left(\frac{p}{q}\right)^{m-r} \left(\frac{1-p}{1-q}\right)^{r}. \tag{2.43}$$

To minimize (2.43), we have

$$\begin{aligned} \frac{\partial \ln P_{\max}}{\partial q} &= -\frac{m-r}{q} + \frac{r}{1-q} \\ &= 0 \end{aligned}$$

and

$$\frac{\partial^2 \ln P_{\max}}{\partial q^2} > 0,$$

so that $q = 1 - r/m$ is the unique solution that minimizes $P_{\max}$.

Because $\frac{\pi(\mathbf{x},p)}{\pi(\mathbf{x},q)} \leq 1$, the variance ratio

$$\text{VR}(p,q) := \frac{\text{var}_{\mathbf{p}}\left[\frac{g(\mathbf{X},\mathbf{p})}{\omega(\mathbf{X},\mathbf{p})}\right]}{\text{var}_{\mathbf{q}}\left[\frac{g(\mathbf{X},\mathbf{q})}{\omega(\mathbf{X},\mathbf{q})}\right]} \tag{2.44}$$

has, for $p \geq 1 - r/m$, the lower bound

$$\text{VR}(p, 1 - r/m) \geq 1. \tag{2.45}$$

For p_i not all equal, a variance reduction also accrues provided that $\min_{1\leq i\leq m} p_i > q = 1 - r/m > 1/2$.

Importance sampling with $q = 1 - r/m$ offers at least one additional benefit. Because $g(\mathbf{X},\mathbf{p})/\omega(\mathbf{X},\mathbf{q})$ has *coefficient of variation*

$$\gamma(\mathbf{p},\mathbf{q}) := \frac{\sqrt{\text{var}_{\mathbf{q}}\left[g(\mathbf{X},\mathbf{p})/\omega(\mathbf{X},\mathbf{q})\right]}}{\lambda(\mathbf{p})}, \tag{2.46}$$

it is easily seen that for $\mathbf{q} = \mathbf{p}$

$$\gamma(\mathbf{p},\mathbf{p}) = \frac{1}{\sqrt{\lambda(\mathbf{p})}} - 1 \to \infty \qquad \text{as } \min_{1\leq i\leq m} p_i \to 1,$$

and is thus unbounded as the components become increasingly reliable. By contrast, for $p_i = p$ and $q_i = q$ for all i with $p > q$,

$$\gamma(\mathbf{p},\mathbf{q}) \to \sqrt{\binom{2r}{r}\frac{1}{a_{m-r}q^{m-r}(1-q)^r} - 1} < \infty \qquad \text{as } p \to 1, \tag{2.47}$$

where a_{m-r} denotes the number of minimal s-t cutsets of $\mathcal{G}$ and is the coefficient of $p^{m-r}(1-p)^r$ in the system reliability (2.39). The limit follows from applying L'Hospital's rule to (2.46). Thus, importance sampling induces a bounded coefficient of variation as p increases. Moreover, $q = 1 - r/m$ minimizes the bound. As illustration, the network in Figure 2.4 has a single minimal s-t cutset ($a_5 = 1$), and the limit (2.47) is 19.86 as $p \to 1$.

Table 2.5 compares standard MC (plan 1) and importance sampling (plan 2) for estimating $\lambda(\mathbf{p})$ for the network in Figure 2.4 for $p_1 = \cdots = p_7 = p$ and $n = 10^6$ independent replications. As implied by theory, the variance ratio (VR) increases with p. Since $n^{1/2}$r.e. is an estimate of $\gamma(p, 1 - r/m)$, (2.47) gives the limiting value of $19.86/n^{1/2} = 1.986 \times 10^{-2}$, which bounds the relative errors in the table.

For MC in general, the appeal of a bounded coefficient of variation is that a finite sample size n satisfies a specified bound on relative error, regardless of the inherent parametric values in the problem. Frequently, although not always, an importance sampling procedure can be designed to induce a finite bound.

Standard MC uses Algorithm STREL with $p_1 = \cdots = p_m = p$, and importance sampling uses the same algorithm but replaces p by $q = 1 - r/m$ and the steps

TABLE 2.5 | *Variance ratio (VR) for estimating s-t network reliability by importance sampling with* $\{\omega(\mathbf{x}, \mathbf{q})\}$

	Network reliability				
p	True $\lambda(\mathbf{p})$	Estimated $\bar{\lambda}_n(\mathbf{p}, \mathbf{p})$	r.e.$(\bar{\lambda}_n(\mathbf{p}, \mathbf{p}))$	VR	Theoretical VR lower bound in (2.45)
.80	.06242	.05993	2.806×10^{-3}	1.992	1.158
.90	.01299	.01278	4.103×10^{-3}	4.587	2.571
.95	2.878×10^{-3}	2.865×10^{-3}	5.549×10^{-3}	11.30	7.846
.99	1.030×10^{-4}	1.032×10^{-4}	7.431×10^{-3}	175.6	159.6

$m = 7,\ r = 2,\ q = 1 - 2/7 = 5/7,\ n = 10^6$

$V_1 = V_1 + 1$ and $V_2 = V_2 + 1$ by

$$R = \prod_{j=1}^{m} \left(\frac{p}{q}\right)^{X_i} \left(\frac{1-p}{1-q}\right)^{m-X_i}$$
$$V_1 = V_1 + R$$
$$V_2 = V_2 + R^2.$$

This importance-sampling application demonstrates at least one other attractive feature. Although all sampling was performed for a single fixed q, the data from the n independent replications can be employed to estimate the s-t reliability function for any value of $p > q$. Table 2.5 shows results for merely four such values. Also, the function could have been estimated for distinct $p_i > q$. Therefore, importance sampling, as applied here, eliminates the need for multiple experiments for different $\mathbf{p}$. However, it is important to note that it is the special structure of this problem that guarantees the variance reduction. Network reliability $\lambda(\mathbf{p})$ is also amenable to estimation by the Rao-Blackwellization variance-reducing technique in Section 5.2. In particular, it fits within the framework of exceedance probability analysis in Section 5.2.1.

Although the benefits of importance sampling are evident here, this sort of sampling cannot be applied unequivocally to all problems. As a safeguard, any inclination to use importance sampling to estimate a λ in this way should, if possible, be preceded by an inquiry as to what is the least variance ratio that can be expected. If that lower bound does not exceed unity, then there is no guarantee that variance inflation will be avoided.

EXAMPLE 2.7.4

PRICING AN ASIAN OPTION I

Whereas Example 2.7.3 demonstrated importance sampling in the context of (discrete) Bernoulli $0 - 1$ sampling, the current example describes an application of this variance-reducing technique in the context of sampling from the (continuous) normal distribution.

Pricing options is of considerable interest to securities dealers and investors. An option gives an investor the right to purchase one unit of a security at a specified *strike price* K at a future time T. Given K and T, it is of interest to determine the price to offer for the option at the present time 0. Let $S(t)$ denote the price of one share of the security at time $t \geq 0$. A common characterization of price fluctuations assumes that $S(0)$ is fixed and $\{S(t),\ t \geq 0\}$ is a stochastic process whose random variation is characterized by the *stochastic differential equation*

$$\frac{\mathrm{d}S(\tau)}{S(\tau)} = \xi_1 \mathrm{d}\tau + \xi_2 \mathrm{d}W(\tau), \tag{2.48}$$

where:

- The parameter ξ_1 denotes the *risk-free, continuously compounded interest rate.*
- The parameter ξ_2 denotes the asset's *volatility*. Low (high) volatility implies small (large) fluctuations.
- For $\tau_0 = 0 < \tau_1 < \tau_2 < \cdots$,

  $$W(\tau_j) - W(\tau_{j-1}) := \sqrt{\tau_j - \tau_{j-1}} X_j,$$

  where $X_1, X_2, \ldots$ are independent and identically distributed (i.i.d.) random variables from $\mathcal{N}(0, 1)$, the standard normal distribution with mean 0 and variance 1.

Parameter ξ_1 characterizes the market for funds, whereas ξ_2 characterizes the particular asset.

The stochastic process $\{W(\tau),\ \tau \geq 0\}$ is called a *Wiener process* or *Brownian motion,* and the stochastic process $\{S(\tau),\ \tau \geq 0\}$ is called *geometric Brownian motion.* It is known that (2.48) has the solution

$$S(\tau) = S(0)\exp\left[\left(\xi_1 - \frac{1}{2}\xi_2^2\right)\tau + \xi_2 W(\tau)\right] \qquad \tau \geq 0. \tag{2.49}$$

The *Asian option* is one among several types of options. Over the time interval $[0, T]$, the Black-Scholes options-pricing formulation gives the *payoff* on an arithmetic Asian option (e.g., Hull 1993) as

$$\left[\frac{1}{T}\int_0^T S(\tau)\mathrm{d}\tau - K\right]^+,$$

where

$$(z)^+ := \begin{cases} z & \text{if } z \geq 0 \\ 0 & \text{elsewhere.} \end{cases}$$

The *discounted payoff*

$$\mathrm{e}^{-\xi_1 T}\left[\frac{1}{T}\int_0^T S(\tau)\mathrm{d}\tau - K\right]^+ \tag{2.50}$$

takes into account the fact that the option is to be purchased at time 0 but exercised at fixed future time $T > 0$.

Because price history over the interval $[0, T]$ is not available at time 0, it is not possible to evaluate the discounted payoff at that time. However, the Black-Scholes formulation gives the expected discounted payoff

$$\mathrm{E}\left\{ \mathrm{e}^{-\xi_1 T} \left[\frac{1}{T} \int_0^T S(\tau)\mathrm{d}\tau - K \right]^+ \right\} \tag{2.51}$$

as the price that an investor at time 0 should be willing to pay for an Asian option on one unit of the underlying asset.

Since the expectation (2.51) is taken with respect to the Brownian motion $\{W(\tau)\}$, it does not admit a closed form for direct evaluation. However, deterministic numerical methods are available (e.g., Linetsky 2004). MC sampling can also provide an approximation, and this has been the commonly employed approach. For fixed $T > 0$ and integer $m > 0$, define $\Delta := T/m,\ \tau_0 = 0$, and $\tau_j = \tau_{j-1} + \Delta$ for each $j = 1, \ldots, m$. Then (2.49) gives the price of the asset at the end of period j at time $j\Delta$ as

$$\begin{aligned} S(\tau_j) &= S(0)\exp\left[\left(\xi_1 - \frac{\xi_2^2}{2} \right) j\Delta + \xi_2\sqrt{\Delta} \sum_{l=1}^{j} X_l \right] \\ &= S(\tau_{j-1})\exp\left[\left(\xi_1 - \frac{\xi_2^2}{2} \right) \Delta + \xi_2\sqrt{\Delta} X_j \right], \end{aligned} \tag{2.52}$$

so that the average

$$\overline{S}_m := \frac{1}{m} \sum_{j=1}^{m} S(\tau_j) \tag{2.53}$$

approximates $\frac{1}{T}\int_0^T S(\tau)\mathrm{d}\tau$ with error that becomes negligibly small as Δ decreases or, equivalently, m increases. Then for given $X_1, \ldots, X_m$,

$$h(X_1, \ldots, X_m) := \mathrm{e}^{-\xi_1 T}(\overline{S}_m - K)^+. \tag{2.54}$$

is an approximation to the discounted payoff (2.50), and

$$\lambda = \mathrm{E}_\pi h(X_1, \ldots, X_m) \tag{2.55}$$

approximates the expected discounted payoff (2.51), where π denotes the joint distribution of $X_1, \ldots, X_m$

$$\begin{aligned} \pi(\mathbf{x}) &:= \prod_{j=1}^{m} \phi(x_j) \qquad \infty < x_j < \infty,\ j = 1, \ldots, m, \\ &= (2\pi)^{-m/2} \mathrm{e}^{-\left(x_1^2 + \cdots + x_m^2\right)/2} \end{aligned} \tag{2.56}$$

and (2.11) defines the normal p.d.f. $\{\phi(x)\}$.

Algorithm ASOPT describes the estimation of (2.51), employing standard MC to generate n independent replications, each of $X_1, \ldots, X_m$. Table 2.6 presents results for $n = 10^6$ performed on independent experiments for each set of values for m, ξ_2, and K. Comparing the relative errors with the entries in Table 2.4 provides a basis for assessing the relative accuracy of the estimates. ■

TABLE 2.6 | *Asian options pricing based on standard Monte Carlo*

m	ξ_2	K	$\bar{\lambda}_n$	s.e.	r.e.
16	0.1	45	6.058	2.940×10^{-3}	4.853×10^{-4}
16	0.1	50	1.922	2.220×10^{-3}	1.155×10^{-3}
16	0.1	55	0.2036	7.450×10^{-4}	3.660×10^{-3}
16	0.3	45	7.162	7.704×10^{-3}	1.076×10^{-3}
16	0.3	50	4.174	6.321×10^{-3}	1.514×10^{-3}
16	0.3	55	2.21	4.778×10^{-3}	2.162×10^{-3}
64	0.1	45	5.997	2.846×10^{-3}	4.747×10^{-4}
64	0.1	50	1.846	2.140×10^{-3}	1.160×10^{-3}
64	0.1	55	0.1744	6.741×10^{-4}	3.865×10^{-3}
64	0.3	45	7.028	7.475×10^{-3}	1.064×10^{-3}
64	0.3	50	4.02	6.086×10^{-3}	1.514×10^{-3}
64	0.3	55	2.077	4.552×10^{-3}	2.191×10^{-3}

$S(0) = 50,\ \xi_1 = .05,\ T = 1.0,\ n = 10^6$

ALGORITHM ASOPT

Purpose: To estimate λ in (2.55) by standard MC.

Given: $\xi_1,\ \xi_2,\ K,\ m,\ S_0 := S(0),\ T$, and n.

Method:

$\Delta = \frac{T}{m}$.
$i = 1,\ V_1 = 0$, and $V_2 = 0$.
While $i \le n$:
    $j = 1$ and $Z = 0$.
    While $j \le m$:
        randomly sample X_j from $\mathcal{N}(0, 1)$
        $S_j = S_{j-1} \exp\left[\left(\xi_1 - \frac{1}{2}\xi_2^2\right)\Delta + \xi_2\sqrt{\Delta}X_j\right]$
        $Z = Z + S_j$
        $j = j + 1$.
    $Y = e^{-\xi_1 T}\left(\frac{Z}{m} - K\right)^+$
    $V_1 = V_1 + Y$ and $V_2 = V_2 + Y^2$
    $i = i + 1$.
$\bar{\lambda}_n = \frac{V_1}{n}$ and $s_n^2 = \frac{1}{n-1}\left(V_2 - n\bar{\lambda}_n^2\right)$.
s.e. $= \sqrt{\frac{s_n^2}{n}}$ and r.e. $= \frac{\text{s.e.}(\bar{\lambda}_n)}{\bar{\lambda}_n}$. ■

Importance Sampling. In practice, a securities dealer needs to estimate options prices for many different assets with sufficient speed to remain current price-wise in a continuously fluctuating securities market. Moreover, these evaluations are to be performed many times during a trading day. Although standard MC can provide the price for one option in negligible computing time, substantial time is required to evaluate a multiplicity of options prices on a frequent basis, and this total time expenditure inevitably erodes a dealer's ability to maintain an advantage. This reality prompts the search for techniques that can estimate options prices with specified accuracy in considerably less computing time than standard MC. The remainder of this example describes how importance sampling (Section 2.4) can serve this purpose. This objective can also be accomplished by stratified sampling (see Section 5.1) or by quasirandom sampling (see Section 5.5).

The p.d.f.

$$\phi(x,\theta) := (2\pi)^{-1/2}e^{-(x-\theta)^2/2} \qquad -\infty < x < \infty$$

defines the normal distribution with mean θ and variance 1, denoted by $\mathcal{N}(\theta, 1)$. Suppose that $X_1, \ldots, X_m$ are independently sampled from $\mathcal{N}(\nu_1, 1), \ldots, \mathcal{N}(\nu_m, 1)$, respectively, so that $\mathbf{X} := (X_1, \ldots, X_m)$ has the joint distribution

$$\begin{aligned}\omega(\mathbf{x},\nu) &:= \prod_{j=1}^{m}\phi(x_j,\nu_j) \qquad -\infty < x_j < \infty,\ j=1,\ldots,m, \qquad (2.57)\\ &= (2\pi)^{-m/2}e^{-[(x_1-\nu_1)^2+\cdots+(x_m-\nu_m)^2]/2}\end{aligned}$$

where $\mathbf{x} := (x_1, \ldots, x_m)$ and $\nu := (\nu_1, \ldots, \nu_m)$. For h in (2.54), π in (2.56), and ω in (2.57), the importance-sampling formulation in Section 2.4 gives

$$\mathrm{E}_{\omega(\cdot,\nu)}\left[\frac{h(\mathbf{X})\pi(\mathbf{X})}{\omega(\mathbf{X},\nu)}\right] = \lambda$$

with

$$\mathrm{var}_{\omega(\cdot,\nu)}\left[\frac{h(\mathbf{X})\pi(\mathbf{X})}{\omega(\mathbf{X},\nu)}\right] = \int_{\Re^m} Q(\mathbf{x},\nu)d\mathbf{x} - \lambda^2, \qquad (2.58)$$

where

$$\begin{aligned}Q(\mathbf{x},\nu) &:= \frac{h^2(\mathbf{X})\pi^2(\mathbf{X})}{\omega(\mathbf{X},\nu)}\\ &= h^2(\mathbf{x})\exp\left[-\sum_{j=1}^{m}\left(\frac{1}{2}x_j^2+\nu_j x_j-\frac{1}{2}\nu_j^2\right)\right].\end{aligned}$$

Because for each $(x_1, \ldots, x_m) \in \Re^m$, either

$$-\sum_{j=1}^{m}\nu_j x_j + \frac{1}{2}\sum_{j=1}^{m}\nu_j^2 \le 0$$

or

$$-\sum_{j=1}^{m}\nu_j x_j + \frac{1}{2}\sum_{j=1}^{m}\nu_j^2 \ge 0,$$

the choice of ν determines an m-dimensional *hyperplane*

$$\left\{\sum_{i=1}^{m} \nu_i x_i = \frac{1}{2}\sum_{i=1}^{m} \nu_i^2,\ \mathbf{x} \in \Re^m\right\}$$

that partitions $\Re^m$ into two subsets:

$$\mathcal{R}(\nu) := \left\{\mathbf{x} \in \Re^m : 0 \leq \frac{\pi(\mathbf{x})}{\omega(\mathbf{x}, \nu)} \leq 1\right\}$$

and

$$\Re^m \setminus \mathcal{R}(\nu) = \left\{\mathbf{x} \in \Re^m : \frac{\pi(\mathbf{x})}{\omega(\mathbf{x}, \nu)} > 1\right\}.$$

For $m = 2$ and fixed (ν_1, ν_2), $\left\{\nu_1 x_1 + \nu_2 x_2 = \frac{1}{2}\left(\nu_1^2 + \nu_2^2\right),\ (x_1, x_2) \in \Re^2\right\}$ defines a line in two-dimensional space. A hyperplane is merely a generalization of this concept in higher-dimensional ($m > 2$) space.

One would like to choose $\nu_1, \ldots, \nu_m$ so that $Q(\mathbf{x}, \nu)$ is large when $\mathbf{x} \in \mathcal{R}(\nu)$ and small when $\mathbf{x} \in \Re^m \setminus \mathcal{R}(\nu)$. Choosing ν to minimize (2.58) would do this, but this is beyond our capability. An alternative strategy (Glasserman et al. 1999) chooses $\nu_1, \ldots, \nu_m$ to minimize $\max_{\mathbf{x}\in\mathcal{X}} Q(\mathbf{x}, \nu)$ or, equivalently, to minimize $\max_{\mathbf{x}\in\mathcal{X}} \ln Q(\mathbf{x}, \nu)$. This is closely related to the strategy used in Example 2.7.3 for network reliability. Since

$$\ln Q(\mathbf{x}, \nu) = 2 \ln h(\mathbf{x}) - \sum_{j=1}^{m}\left(\frac{1}{2}x_j^2 + \nu_j x_j - \frac{1}{2}\nu_j^2\right), \tag{2.59}$$

the required first-order conditions are

$$\frac{\partial \ln Q(\mathbf{x}, \nu)}{\partial x_j} = 2\frac{\partial \ln h(\mathbf{x})}{\partial x_j} - x_j - \nu_j = 0 \tag{2.60}$$

and

$$\frac{\partial \ln Q(\mathbf{x}, \nu)}{\partial \nu_j} = -x_j + \nu_j = 0. \tag{2.61}$$

Combining (2.60) and (2.61) reduces the problem to solving the m, possibly nonlinear, equations

$$\left.\frac{\partial \ln h(\mathbf{x})}{\partial x_j}\right|_{x_j=\nu_j} = \nu_j, \tag{2.62}$$

which in the present case is equivalent to finding the $(\nu_1, \ldots, \nu_m) = (x_1, \ldots, x_m)$ in $\Re^m$ that solves

$$\max_{\mathbf{x}\in\mathcal{X}} \left[\ln h(\mathbf{x}) - \frac{1}{2}\sum_{j=1}^{m} x_j^2\right]. \tag{2.63}$$

The resulting ν is unique and optimal if and only if $\nu \in \Re^m$ and $Q(\nu, \nu) \geq Q(\mathbf{x}, \nu)$ for all $\mathbf{x} \in \Re^m$.

An iterative procedure determines the $\nu_1, \ldots, \nu_m$ that satisfy (2.60) and (2.61) (Glasserman et al. 1999). Noting that we need be concerned only with $\{\mathbf{y} := (y_1, \ldots, y_m) \in \Re^m : h(\mathbf{y}) > 0\}$, the first-order conditions reduce to

$$\frac{\partial \ln h(\mathbf{y})}{\partial y_j} = y_j \qquad j = 1, \ldots, m.$$

Therefore, it is easily seen that for time increment Δ,

$$y_j = \frac{\xi_2 \Delta^{1/2}}{m h(\mathbf{y})} \sum_{i=j}^{m} S_j \qquad j = 1, \ldots, m$$

or, recursively,

$$\begin{aligned} y_1 &= \frac{\xi_2 \Delta^{1/2} \left[h(\mathbf{y}) + K \right]}{h(\mathbf{y})} \\ y_j &= y_{j-1} - \frac{\xi_2 \Delta^{1/2} S_{j-1}}{m h(\mathbf{y})} \qquad j = 2, \ldots, m. \end{aligned} \tag{2.64}$$

Algorithm OPTNU applies a bisection search to find $(\nu_1, \ldots, \nu_m) = (y_1, \ldots, y_m)$ using these recursions. It begins with an initial value b to prime (2.64). Because h in (2.54) is monotone increasing in $x_1, \ldots, x_m$ and because $X_1, \ldots, X_m$ are i.i.d. from $\mathcal{N}(0, 1)$, $\text{pr}(X_j > 3) = 1 - \Phi(3)$ is small so that $\text{pr}(X_1 > 3, \ldots, X_m > 3) = [1 - \Phi(3)]^m$ is very small. Therefore, initializing the bisection search with $y_1, \ldots, y_m = 3$ and upper bound $b = h(3, \ldots, 3)$ is reasonable. Since no theorem establishes uniqueness for a ν bounded above by the initial value of b, a user is encouraged to repeat the computation for several initial values, say $h(3.5, \ldots, 3.5)$, $h(4, \ldots, 4), \ldots,$ and to compare the optimal solutions. Negligible differences support the contention of uniqueness. Glasserman et al. (1999) report this to have been their experience.

ALGORITHM OPTNU

Purpose: To solve for $\nu = (\nu_1, \ldots, \nu_m)$.

Given: $y_j = 3;\ j = 1, \ldots, m$, lower bound $a := 0$,
upper bound $b := h(3, \ldots, 3)$, and tolerance $\alpha > 0$.

Method:

Repeat:
- $h_{\text{new}} = (a + b)/2$
- use (2.64) to compute $y_1, \ldots, y_m$ with $h(\mathbf{y}) := h_{\text{new}}$ as input
- use (2.49) to compute $S_1, \ldots, S_m$
- $h_{\text{old}} = h_{\text{new}}$
- $h_{\text{new}} = \left(\bar{S}_m - K\right)^+$
- if $h_{\text{old}} < h_{\text{new}}$, $\quad a = h_{\text{old}}$ and $b = h_{\text{new}}$
- if $h_{\text{old}} \geq h_{\text{new}}$, $\quad a = h_{\text{new}}$ and $b = h_{\text{old}}$

until $|h_{\text{new}} - h_{\text{old}}| \leq \alpha$.

Return $\nu = (y_1, \ldots, y_m)$. ■

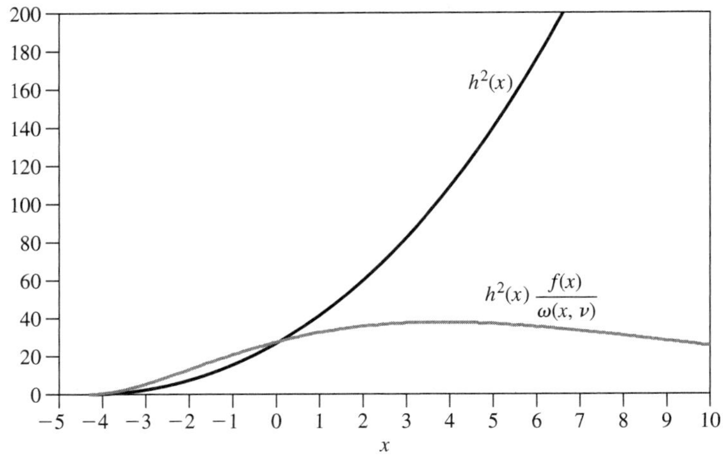

FIGURE 2.5 | *Importance sampling for Asian options pricing*
$m=1$, $\Delta=1/16$, $\xi_2=0.10$, $\rho=0.05$, $S(0)=50$, $K=45$, $T=1/16$, $\nu=0.2731$

Because standard MC gives $\text{var}_\pi h(\mathbf{X}) = \text{E}_\pi h^2(\mathbf{X}) - \lambda^2$ whereas importance sampling gives $\text{var}_{\omega(\cdot,\nu)} h(\mathbf{X}) = \text{E}_\pi\left[h^2(\mathbf{X})\frac{\pi(\mathbf{X})}{\omega(\mathbf{X},\nu)}\right] - \lambda^2$, it is of interest to compare $\{h^2(\mathbf{x})\}$ with $\left\{h^2(\mathbf{x})\frac{\pi(\mathbf{X})}{\omega(\mathbf{X},\nu)}\right\}$. Figure 2.5 does this for the one-dimensional case ($m = 1$) for parameter values that induce $\nu = 0.2731$ as the optimal shift factor. The benefit of importance sampling is apparent.

To perform importance sampling, one can use Algorithm ASOPT with sampling from $\mathcal{N}(\nu_1, 1), \ldots, \mathcal{N}(\nu_m, 1)$ in place of $\mathcal{N}(0, 1)$ and replace $V_1 = V_1 + Y$ and $V_2 = V_2 + Y^2$ with

$$R = \exp\left[-\sum_{j=1}^{m}\left(\nu_j X_j - \frac{1}{2}\nu_j^2\right)\right]$$

$$V_1 = V_1 + YR$$

$$V_2 = V_2 + (YR)^2.$$

Table 2.7 displays the variance ratios (VR) for a collection of IMC sampling experiments, each run for $n = 10^6$ independent replications. Because sample generating times from $\{\omega(\mathbf{x}, \nu)\}$ and $\{\omega(\mathbf{x}, \mathbf{0})\}$ are virtually the same and because computing $g(\mathbf{X})/\omega(\mathbf{X}, \nu)$ instead of $g(\mathbf{X})$ adds negligible time, the unit cost ratios (CR) are close to unity.

2.8 | BOOTSTRAP METHOD

Methods of statistical inference extract information from sample data to estimate unknown values of model parameters and to test hypotheses about these models. Among these, *resampling schemes* can improve the quality of these estimates and

TABLE 2.7 | *Importance sampling for Asian options pricing Monte Carlo***

m	ξ_2	K	Price	Variance ratio
16	0.10	45	6.05	11.0
		50	1.92	7.0
		55	0.20	21.0
16	0.30	45	7.15	8.3
		50	4.17	9.2
		55	2.20	12.0
64	0.10	45	6.00	11.0
		50	1.85	7.3
		55	0.17	23.0
64	0.30	45	7.02	8.3
		50	4.02	9.2
		55	2.08	12.0

$S(0)=50$, $\xi_1=0.05$, $T=1.0$, $r=100$ strata, $n=10^6$
*Source: Glasserman et al. (1999).

tests by extracting additional information not accessible by more traditional inferential methods. In particular, the *bootstrap method* does this by IMC sampling from the *empirical distribution functions* (e.d.f.s) constructed from the original sample data (e.g., Efron and Tibshirani 1993, Good 2001, and Hjorth 1994). The technique often provides point estimates with higher accuracy and, in particular, an assessment of this accuracy, unavailable with traditional methods.

Relatively simple to apply, the bootstrap method is best introduced by an example that illustrates its versatility. Example 2.8.1 describes a novel application of the method to a problem of selecting the best among several alternative inventory systems. After that, we describe a more general algorithm for applying the technique. Later, Example 6.11.2 describes how MCMC resampling reverses the conclusion regarding the hypothesis that eye color is independent of hair color.

EXAMPLE 2.8.1

CHOOSING THE BEST INVENTORY SYSTEM

Five alternative inventory systems have been proposed to replace a company's antiquated one.[1] The company wants to choose the system with the smallest long-run average operating cost for the inventory system. Since the averages are

[1]I am grateful to R. Cheng, who described to me his version of this resampling application.

TABLE 2.8 | *Sample average monthly operating costs for inventory systems**

Observation	System 1	2	3	4	5
1	126.97	126.21	120.77	131.64	141.09
2	124.31	128.22	120.32	137.07	143.86
3	126.68	130.45	120.61	129.91	144.30
4	122.66	130.68	123.65	129.97	141.72
5	127.23	127.40	127.34	131.08	142.61
Sample mean	125.57	128.59	124.34	131.93	142.72
Sample s.d.	2.00	1.94	3.90	2.96	1.37

*Source: Modified data set in Law and Kelton (1991).

unknown, MC sampling experiments are executed under each inventory system, yielding the data in Table 2.8. The observations are mutually independent and approximately normally distributed. Using the grand sample means as estimates of the unknown long-run average monthly costs, one might conclude that

$$\min(125.57, 128.59, 124.34, 131.93, 142.72) = 124.34 \tag{2.65}$$

points to system 3 as being the best choice. Although system 3 may indeed be the best, this method of inference provides no basis for assessing how credible the choice is. Since system 3 also has the largest sample standard deviation, the need to assess accuracy becomes especially compelling. In what follows, we describe a relatively general setting for identifying the best and study the inventory problem as a particular case.

Let $Z_1, \ldots, Z_M$ denote mutually independent continuous random variables, where Z_i has unknown p.d.f. $\{f_i(z), -\infty < z < \infty\}$ and unknown mean

$$\mu_i = \int_{-\infty}^{\infty} z f_i(z) dz.$$

It is of interest to identify

$$i_{\text{best}} := \text{the variable with the smallest mean.}$$

For each variable $i = 1, \ldots, M$, suppose N independent observations $Z_{i1}, \ldots, Z_{iN}$ are available. For example, variable i in Table 2.8 may denote inventory system i and Z_{ij} its jth sample average monthly operating cost. To relate the exposition to the example and without loss of generality, we hereafter refer to system i rather than variable i.

Let

$$\bar{Z}_i := \frac{1}{N}\sum_{j=1}^{N} Z_{ij} \qquad i = 1, \ldots, M$$

denote the sample long-run average monthly cost for system i. Then

$$\hat{i}_{\text{best}} = \min(\overline{Z}_1, \ldots, \overline{Z}_M) \tag{2.66}$$

estimates which of the M systems is best, as in (2.65). The accuracy of rule (2.66) can be measured by

$$\begin{aligned} P_N(i) &:= \text{pr}\left(\hat{i}_{\text{best}} = i\right) \\ &= \text{probability of selecting system } i \text{ given a sample of} \\ &\quad\ N \text{ observations on each system } 1, \ldots, M. \end{aligned}$$

In particular, it can be shown that

$$P_N(i) \to \begin{cases} 1 & \text{if } i = i_{\text{best}} \\ 0 & \text{otherwise} \end{cases} \qquad \text{as } N \to \infty, \tag{2.67}$$

implying

$$\hat{i}_{\text{best}} \to i_{\text{best}} \quad \text{w.p.1} \quad \text{as } N \to \infty.$$

In practice, we have only a finite number of empirical samples, N, and in this case $0 < P_N(i) < 1$. Since the limits (2.67) indicate that only one of these probabilities, $P_N(i_{\text{best}})$, converges to unity while the others converge to zero, estimates of these probabilities for fixed N would provide an assessment of the confidence to be placed in $\hat{i}_{\text{best}}$ as the correct choice.

Although estimates of the probabilities are not readily available by traditional methods, the bootstrap method allows us to estimate closely related *conditional probabilities* by exploiting information available in the e.d.f.s corresponding to each data sample $\mathcal{Z}_i := \{Z_{i1}, \ldots, Z_{iN}\}$. For system i, the e.d.f. is defined as

$$\begin{aligned} F_{iN}(z) &:= \frac{\text{number of points in } \mathcal{Z}_i \text{ less than or equal to } z}{\text{sample size } N} & -\infty < z < \infty \\ &= \frac{l}{N} & Z_{i,(l-1)} < z < Z_{i,(l)}, \end{aligned}$$

where the *order statistics* $Z_{i,(1)} < Z_{i,(2)} \cdots < Z_{i,(N)}$ denote $Z_{i1}, \ldots, Z_{iN}$ ordered from smallest to largest, and $Z_{i,(0)} := 0$. Also,

$$f_{iN}(z) := \begin{cases} \frac{1}{N} & \text{if } z \in \mathcal{Z}_i \\ 0 & \text{elsewhere} \end{cases} \tag{2.68}$$

is the *empirical probability mass function* (e.p.m.f.) for system i. If the Z_{ij} were discrete, then neighboring order statistics could be equal, thus requiring a redefinition of F_{iN} and f_{iN}. However, that need not affect the sampling procedure to be described.

A *matrix* is a collection of numbers or variables organized into a two-dimensional array of rows and columns. It provides a convenient way to describe data concisely. For each system $i = 1, \ldots, M$, let $X_{i1}, \ldots, X_{iN}$ denote N independent samples randomly drawn for the e.p.m.f. $\{f_{iN}(z),\ z \in \mathcal{Z}_i\}$. This is equivalent to drawing the $M \times N$ random matrix

$$\mathbf{X} := \begin{pmatrix} X_{11} & X_{12} & \cdots & X_{1N} \\ X_{21} & X_{22} & \cdots & X_{2N} \\ \vdots & \vdots & \ddots & \vdots \\ X_{M1} & X_{M2} & \cdots & X_{MN} \end{pmatrix}$$

from the product space

$$\begin{aligned} \mathcal{X} &:= \underbrace{\mathcal{Z}_1 \times \cdots \times \mathcal{Z}_1}_{N \text{ times}} \times \underbrace{\mathcal{Z}_2 \times \cdots \times \mathcal{Z}_2}_{N \text{ times}} \times \cdots \times \underbrace{\mathcal{Z}_M \times \cdots \times \mathcal{Z}_M}_{N \text{ times}} \\ &= \mathcal{Z}_1^N \times \cdots \times \mathcal{Z}_M^N \end{aligned}$$

with probability

$$\begin{aligned} \pi(\mathbf{x}) &:= \prod_{i=1}^{M} \prod_{l=1}^{N} f_{iN}(x_{il}) \\ &= \begin{cases} \frac{1}{N^{MN}} & \text{if } \mathbf{x} \in \mathcal{X} \\ 0 & \text{otherwise,} \end{cases} \end{aligned} \tag{2.69}$$

where $\mathbf{x}$ denotes the $M \times N$ matrix

$$\mathbf{x} := \begin{pmatrix} x_{11} & x_{12} & \cdots & x_{1N} \\ x_{21} & x_{22} & \cdots & x_{2N} \\ \vdots & \vdots & \ddots & \vdots \\ x_{M1} & x_{M2} & \cdots & x_{MN} \end{pmatrix}$$

in $\mathcal{X}$. Let

$$h_i(\mathbf{X}) := \begin{cases} 1 & \text{if } \overline{X}_i < \overline{X}_l \quad \text{for all } l \neq i, \\ 0 & \text{otherwise} \end{cases} \tag{2.70}$$

where

$$\overline{X}_i := \frac{1}{N} \sum_{j=1}^{N} X_{ij}.$$

That is, $h_i(\mathbf{X}) = 1$ if system i is selected and 0 otherwise. Then

$$\begin{aligned} \lambda_i &:= \sum_{\mathbf{x} \in \mathcal{X}} h_i(\mathbf{x}) \pi(\mathbf{x}) \\ &= \text{pr}[h_i(\mathbf{X}) = 1 | \mathcal{Z}_1, \ldots, \mathcal{Z}_M] \\ &= \text{pr(selecting system } i \text{ given } \mathcal{Z}_1, \ldots, \mathcal{Z}_M) \end{aligned} \tag{2.71}$$

is the conditional probability of selecting system i for a sample drawn randomly from the e.p.m.f.s based on the given data $\mathcal{Z}_1, \ldots, \mathcal{Z}_M$. For the inventory problem, the summation in (2.71) has $|\mathcal{X}| = N^{MN} \approx 3 \times 10^{17}$ matrices, demonstrating the intractability of direct evaluation.

Although $P_N(i) \neq \lambda_i$, we do have the desirable limiting property that

$$\lambda_i \to \begin{cases} 1 & \text{if } i = i_{\text{best}} \\ 0 & \text{elsewhere} \end{cases} \qquad \text{as } N \to \infty. \tag{2.72}$$

As before, $0 < \lambda_i < 1$ for finite N, their magnitudes indicating how often each system is selected when given a data set as in Table 2.8. The bootstrap method uses IMC sampling to obtain these estimates.

Suppose we repeat the aforementioned steps n times and let the superscript $^{(l)}$ denote replication l. Then for each system $i = 1, \ldots, M$,

$$\overline{\lambda}_{in} := \frac{1}{n} \sum_{l=1}^{n} h_i\left(\mathbf{X}^{(l)}\right)$$

has the property

$$\overline{\lambda}_{in} \to \lambda_i \qquad \text{w.p.1} \qquad \text{as } n \to \infty.$$

Moreover,

$$\text{var}\overline{\lambda}_{in} = \frac{\lambda_i}{n}(1 - \lambda_i),$$

so that we can control the accuracy of our estimate of λ_i merely by choice of sample size n. Algorithms FINDBEST and DISUNI describe how the bootstrap does all this. ■

ALGORITHM

FINDBEST

Purpose: To estimate $\lambda_1, \ldots, \lambda_M$ in (2.71).

Given: $M :=$ number of systems, $N :=$ number of samples, and data sets $\mathcal{Z}_1, \ldots, \mathcal{Z}_M$.

Method:

```
l = 1.
For i = 1, ..., M: C_i = 0 .
While l ≤ n:
    for i = 1, ..., M:
        S_i = 0
        for j = 1, ..., N:
            randomly generate J using Algorithm DISUNI.
            X_i = Z_iJ
            S_i = S_i + X_i
        X̄_i = S_i/N              (compute sample average for system i)
    I= system with smallest sample average X̄_1, ..., X̄_M
    C_I = C_I + 1
    l = l + 1.
For i = 1, ..., M:
    λ̄_in = C_i/n
```

$$s_{in}^2 = \frac{1}{n-1}(C_i - n\bar{\lambda}_{in}^2)$$
$$\text{s.e.}_i = \sqrt{\frac{s_{in}^2}{n}}$$
$$\text{r.e.}_i = \frac{\text{s.e.}_i}{\bar{\lambda}_{in}}.$$

■

ALGORITHM DISUNI

Purpose: To generate J randomly from the uniform distribution on integers $1, \ldots, N$.

Given: N.

Method:

Randomly generate U from $\mathcal{U}(0, 1)$.
$J = 1 + \lfloor NU \rfloor$.

■

Because for J and X_i in Algorithm FINDBEST,

$$\begin{aligned}\text{pr}(J = j) &= \text{pr}\left(\frac{j-1}{N} \leq U < \frac{j}{N}\right)\\ &= \frac{1}{N}\end{aligned}$$

and

$$\text{pr}(X_i = Z_{ij} | J = j) = 1,$$

it follows that

$$\begin{aligned}\text{pr}(X_i = Z_{ij}, J = j) &= \text{pr}(X_i = Z_{ij} | J = j)\text{pr}(J = j)\\ &= \frac{1}{N}\\ &= f_{iN}(Z_{ij}),\end{aligned}$$

as desired. The first three rows under $n = 10^3$ and $n = 10^6$ in Table 2.9 display estimates for systems 1, 2, and 3 based on the inventory data in Table 2.8. For $n = 10^6$ independent replications, they show that system 3 is selected most frequently with $\bar{\lambda}_{in} = 0.9622$, which has small relative error.

This approach easily extends to broader questions of ranking systems. For example, let

$$\lambda_{i_1,i_2} := \text{probability of selecting system } i_1 \text{ as having the smallest average, and system } i_2 \text{ as having the next smallest average, given } \mathcal{Z}_1, \ldots, \mathcal{Z}_M. \tag{2.73}$$

Then replacing $h_i(\mathbf{X})$ in (2.70) with

$$h_{i_1,i_2}(\mathbf{X}) := \begin{cases} 1 & \text{if } \overline{X}_{i_1} < \overline{X}_{i_2} < \overline{X}_{i_l} \qquad \forall l \neq i_1, i_2 \\ 0 & \text{otherwise} \end{cases}$$

TABLE 2.9 | *Selecting the best inventory system*

Order for selected $i_1, \dots, i_l$ $n = 10^3$	$\bar{\lambda}_{i_1,\dots,i_l;n}$	s.e. $\times 10^{-2}$	r.e.
1	.022	.4639	.2108
2	.020	.4427	.2214
3	.958	.6343	.1510
1,2,3,4,5	.001	.09995	.9995
1,3,2,4,5	.021	.4534	.2159
2,1,3,4,5	.005	.2231	.4461
2,3,1,4,5	.015	.3844	.2563
3,1,2,4,5	.795	1.277	.06227
3,1,4,2,5	.003	.1730	.5765
3,2,1,4,5	.160	1.159	.07246
$n = 10^6$		**$\times 10^{-4}$**	
1	.02196	1.466	.006673
2	.01529	1.227	.005048
3	.9622	1.906	.005048
1,2,3,4,5	.001171	.3420	.02921
1,3,2,4,5	.02076	1.426	.006869
1,3,4,2,5	.000035	.05916	.1690
2,1,3,4,5	.001679	.4094	.02438
2,3,1,4,5	.01361	1.159	.008511
3,1,2,4,5	.8047	3.964	.002030
3,1,4,2,5	.001767	.4200	.02377
3,2,1,4,5	.1558	3.626	.002328

allows us to estimate (2.73). Analogous definitions apply for $i_1, \dots, i_l$ for $l = 2, \dots, M$.

Table 2.9 also shows results for selected rankings. For $n = 10^6$ replications, the order 3,1,2,4,5 occurs with estimated probability $\lambda_{3,1,2,4,5;n} = 0.8047$ and small relative error, thereby suggesting this to be the most likely ordering among the five systems.

Other formulations of *ranking and selection problems* also exist. For example, it may be of interest to identify the system

$$i_{\text{best}} := \left(i : \mu_i \leq \min_{1 \leq l \leq M} \mu_l \text{ and } \mu_l - \mu_i > c, \ \forall l \neq i \right),$$

if it exists, for some specified $c > 0$. Exercise H.20 addresses this problem.

The bootstrap method applies to a considerable range of problems, each with its own definition of $\{h(\mathbf{x}),\ \mathbf{x} \in \mathcal{X}\}$. Algorithm BS describes the steps for IMC sampling. However, in each case its efficacy depends on the existence of limits analogous to those in (2.67) and (2.72) as N, the size of the original data set, increases. The reader is referred to Efron and Tibshirani (1993) for a discussion of this issue.

ALGORITHM BS

Purpose: To estimate $\lambda = \sum_{\mathbf{x}\in\mathcal{X}} h(\mathbf{x})\pi(\mathbf{x})$ where (2.69) defines π.

Given: e.p.m.f.s $\{f_{iN}(z_i)\}$ for $i = 1, \ldots, M, n$.

Method:

$l = 1,\ V_1 = 0$, and $V_2 = 0$.
While $l \le n$:
  $i = 1$
  While $i \le M$:
    $j = 1$
    While $j \le N$:
      randomly sample X_{ij} from $\{f_{iN}(x)\}$
      $j = j + 1$
    $i = i + 1$.
  $Y = h(\mathbf{X}),\ V_1 = V_1 + Y$, and $V_2 = V_2 + Y^2$
  $l = l + 1$.
$\bar{\lambda}_n = \frac{V_1}{n}$ and $s_n^2 = \frac{1}{n-1}\left(V_2 - n\bar{\lambda}_n^2\right)$.
s.e. $= \sqrt{\frac{s_n^2}{n}}$ and r.e. $= \frac{\text{s.e.}}{\bar{\lambda}_n}$. ■

2.9 REGRESSION ANALYSIS

MC estimates are rarely ends in themselves. Usually, they serve as input to a decision-making process or another mathematical analysis. For example, a need sometimes exists for both integral evaluation and function minimization in the same problem. Let

$$\lambda(\theta) := \int_{\mathcal{X}} h(\mathbf{x})\pi(\mathbf{x}, \theta)d\mathbf{x} \qquad \theta \ge 0,$$

where λ is an unknown continuous monotonically increasing function of θ. Suppose it is of interest to find θ such that $\lambda(\theta) = a$ for some fixed $a \ge 0$. Let

$$H(\theta) := |\lambda(\theta) - a|.$$

Then this problem is equivalent to finding the θ that minimizes $\{H(\theta)\}$ subject to $\theta \ge 0$. Let $\theta_{\min}$ denote the optimal value. This is a variation of the function minimization problem in the introduction to this chapter where we know the solution $H(\theta_{\min}) = 0$ in (2.3) and seek the solution $\mathcal{X}_{\min} = \{\theta_{\min}\}$ in (2.4).

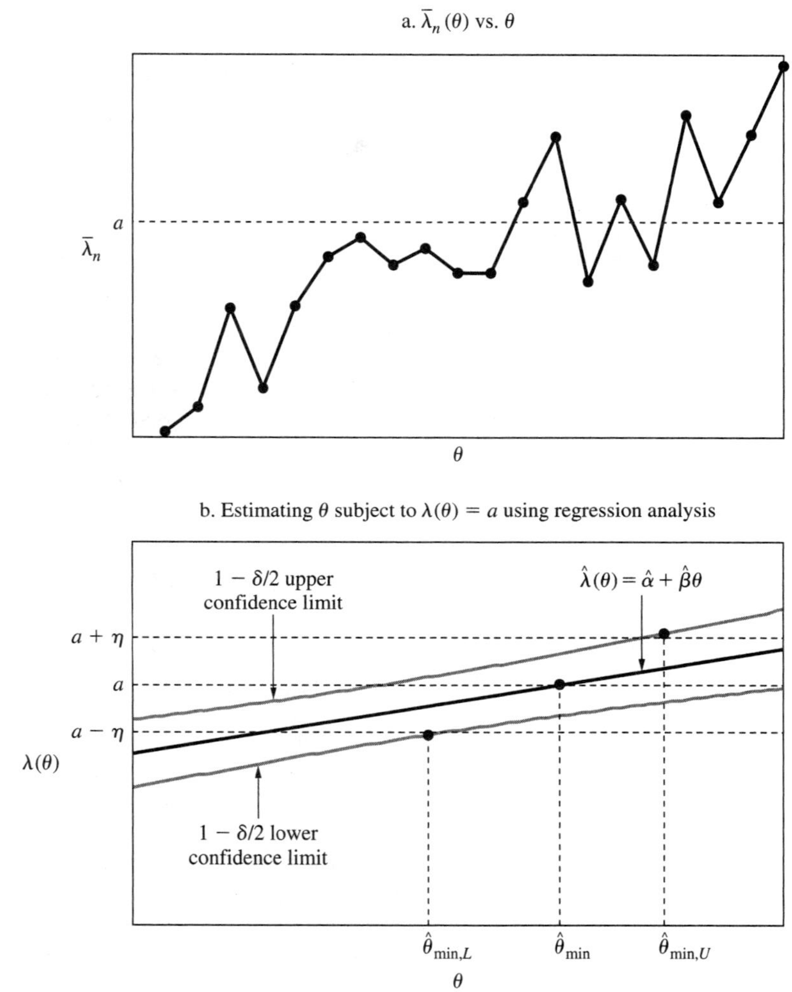

FIGURE 2.6 | *Estimating θ subject to $\lambda(\theta) = a$ using regression analysis*

However, we do not know how the function H or, equivalently, λ behaves as a function of θ.

One approach to solution selects a grid of points $\Theta := \{0 < \theta_1 < \cdots < \theta_r < \infty\}$, which presumably contains values of θ such that $\lambda(\theta) \le a$ and $\lambda(\theta) > a$. For each $\theta \in \Theta$, it then executes n independent replications and computes an unbiased estimate $\bar{\lambda}_n(\theta)$ of $\lambda(\theta)$. Because of sampling variation, $\bar{\lambda}_n(\theta_1), \ldots, \bar{\lambda}_n(\theta_r)$ are not necessarily monotone, as Figure 2.6a illustrates, especially in the neighborhood of $\theta = a$. This makes it difficult visually to infer a suitable estimate of $\theta_{\min}$. Regression analysis provides a way to overcome this limitation. It does this by fitting a line to the points $\{(\theta_1, \bar{\lambda}_n(\theta_1)), \ldots, (\theta_r, \bar{\lambda}_n(\theta_r))\}$ or to a subset of them that is likely to contain the point $(a, \lambda(a))$. It then uses the line and its error measures to estimate $\theta_{\min}$ and its accuracy.

For some $1 < N \le r$ and $0 \le l \le r - N$, suppose that $\theta_{l+1} \le a \le \theta_{l+N}$ and that a plot of the points $\{(\theta_{l+1}, \bar{\lambda}_n(\theta_{l+1})), \ldots, (\theta_{l+N}, \bar{\lambda}_n(\theta_{l+N}))\}$ suggests a linear relationship. This is roughly the case in Figure 2.6a.

This linearity encourages us to consider the two-variable regression model

$$Y_i = \alpha + \beta Z_i + \epsilon_i \qquad i = 1, \ldots, N, \tag{2.74}$$

where the intercept α and the slope β are unknown constants and $\epsilon_1, \ldots, \epsilon_N$ denote i.i.d. normally distributed random errors with mean zero and variance $\sigma^2_{y.z}$. In the present context, $Y_i := \bar{\lambda}_n(\theta_{l+i})$, $Z_i := \theta_{l+i}$, and the independence and asymptotic normality of the $\bar{\lambda}_n(\theta_i)$ justifies the independence and normality of the ϵ_i. The assumption of common variance for $\epsilon_1, \ldots, \epsilon_N$ implies that $\sigma^2_{y.z} = \text{var}\,\bar{\lambda}_n(\theta_{l+1}) = \cdots = \text{var}\,\bar{\lambda}_n(\theta_{l+N})$. The assumption is tenable if the s.e.(θ_{l+i}), the standard errors of the $\bar{\lambda}_n(\theta_{l+i})$ in the original IMC experiment for $\theta_1, \ldots, \theta_r$, are relatively close in value.

Because $\epsilon_1, \ldots, \epsilon_N$ have zero mean, it follows that the conditional mean of Y given Z is

$$\mathrm{E}(Y \mid Z) = \alpha + \beta Z.$$

In terms of the present problem,

$$\lambda(\theta) = \mathrm{E}\bar{\lambda}_n(\theta) \approx \alpha + \beta\theta \qquad \theta_{l+1} \le \theta \le \theta_{l+N}, \tag{2.75}$$

so that

$$\theta_{\min} = \frac{a - \alpha}{\beta}. \tag{2.76}$$

The approximation symbol is used here merely to stress that this linearity is strictly a local phenomenon, except when $\lambda(\theta)$ is indeed linear in θ for all $\theta \ge 0$.

Expression (2.74) provides a basis for getting a better estimate of $\theta_{\min}$. According to standard regression theory, minimizing the sum of squared errors

$$\sum_{i=1}^{N} \epsilon_i^2 = \sum_{i=1}^{N} (Y_i - \alpha - \beta Z_i)^2 \tag{2.77}$$

yields the *least-squares estimators* of α, β, and $\sigma^2_{y \cdot x}$ as

$$\begin{aligned} \hat{\alpha} &= \bar{Y}_N \\ &:= \frac{1}{N}\sum_{i=1}^{N} Y_i \\ \hat{\beta} &= \frac{\sum_{i=1}^{N} Y_i Z_i - N\bar{Y}_N \bar{Z}_N}{\sum_{i=1}^{N} Z_i^2 - N\bar{Z}_N^2} \\ \hat{\sigma}^2_{y.z} &= \frac{1}{N-2}\sum_{i=1}^{N} (Y_i - \hat{\alpha} - \hat{\beta} Z_i)^2, \end{aligned}$$

where

$$\overline{Z}_N := \frac{1}{N}\sum_{i=1}^{N} Z_i.$$

Then (2.76) suggests

$$\widehat{\theta}_{\min} = \frac{a - \widehat{\alpha}}{\widehat{\beta}} \tag{2.78}$$

as a point estimate of $\theta_{\min}$, in contrast to the previously mentioned approach that chooses the θ from the points $\theta_{l+1}, \ldots, \theta_{l+N}$ for which $|a - \lambda_n(\theta)|$ is minimal.

The present approach also allows us to assess the accuracy of $\widehat{\theta}_{\min}$ by constructing an interval estimate for $\theta_{\min}$. From standard regression theory,

$$\mathrm{pr}\left[\widehat{\alpha} + \widehat{\beta}Z - \Delta \le \mathrm{E}(Y|Z) \le \widehat{\alpha} + \widehat{\beta}Z + \Delta\right] = 1 - \delta \tag{2.79}$$

yields a $100 \times (1 - \delta)$ percent *prediction interval* for $\mathrm{E}(Y|Z)$, where

$$\Delta := t_{N-1}(1 - \delta/2)\sqrt{\frac{1}{N} + \frac{(Z - \overline{Z}_N)^2}{(N-1)s_Z^2}}, \tag{2.80}$$

$t_{n-1}(1 - \delta/2)$ denotes the $1 - \delta/2$ critical value of the Student t distribution with $n - 1$ degrees of freedom, and

$$s_Z^2 := \frac{1}{N-1}\sum_{i=1}^{N}(Z_i - \overline{Z}_N)^2.$$

For $\widehat{\beta} > 0$ the smallest Z that satisfies (2.79) for $\mathrm{E}(Y|Z) = a$ is

$$Z_L := \frac{a - \widehat{\alpha} - \Delta}{\widehat{\beta}},$$

and the largest is

$$Z_U := \frac{a - \widehat{\alpha} + \Delta}{\widehat{\beta}}.$$

In terms of the present problem, $[Z_L, Z_U]$ is a $1 - \delta$ interval estimate for $\theta_{\min}$. If $\widehat{\beta} < 0$, the assignments are reversed.

Suppose the objective is to estimate θ such that for some $\eta > 0$,

$$\mathrm{pr}\left[|\overline{\lambda}_n(\theta) - \lambda(\theta)| \le \eta\right] = 1 - \delta.$$

Then, based on (2.79), $\left[\widehat{\theta}_{\min,L}, \widehat{\theta}_{\min,U}\right]$ provides an interval estimate for $\theta_{\min}$, where

$$\widehat{\theta}_{\min,L} := \min(\theta \ge 0 : a - \widehat{\alpha} - \widehat{\beta}\theta - \Delta(\theta) \ge \eta)$$

and (2.81)

$$\widehat{\theta}_{\min,U} := \min(\theta \ge 0 : a - \widehat{\alpha} - \widehat{\beta}\theta + \Delta(\theta) \le \eta).$$

If calculations reveal that $\widehat{\theta}_{\min,L} \ge \widehat{\theta}_{\min,U}$, then the amount of data available for the regression is insufficient to meet the $\pm\eta$ error bound.

As illustration, suppose that

$$\begin{aligned}
N &= 20\\
\overline{Z}_N &= 54.10\\
\widehat{\sigma}_Z^2 &= 1.4\\
\widehat{\alpha} &= -173.56\\
\widehat{\beta} &= 5.006767\\
\widehat{\sigma}_{y\cdot z}^2 &= (3.7540)^2,
\end{aligned}$$

and the objective is to estimate θ such that $\lambda(\theta) = a = 100$ to within $\pm\eta$ for $\eta = 4$. Then (2.78) yields $\widehat{\theta}_{\min} = 54.59$ and (2.81), with the Student t critical value $t_{18}(0.995) = 2.878442$, yields $\widehat{\theta}_{\min,L} = 54.20$ and $\widehat{\theta}_{\min,U} = 54.84$. Figure 2.6b shows the graphs.

2.10 LESSONS LEARNED

- The Monte Carlo (MC) method applies to deterministic as well as stochastic problems.
- Independent Monte Carlo (IMC) with n independent replications results in an estimate of λ whose error decreases as $n^{1/2}$.
- When a sampling distribution ω is selected, exploiting the properties of the integrand can considerably increase statistical efficiency. Importance sampling provides one technique for doing this.
- Both standard error and sampling cost are principal considerations in selecting among several candidates for sampling distribution.
- The product CR $\times$ VR (see Section 2.3) allows an IMC user to identify which of two alternatives distributions gives the greatest benefit.
- Using IMC, the bootstrap method (see Section 2.8) provides a way to obtain improved point estimates. In particular, it often can estimate the accuracy when no such measure is available from more traditional methods.
- Applying simple linear regression analysis to a collection of independent unbiased IMC estimates of points on a function provides one way of estimating the minimum of the function and the accuracy of the estimated minimum.

2.11 HANDS-ON EXERCISES

Each hands-on exercise provides an opportunity to design and execute an IMC sampling experiment, estimate the values of unknown parameters of interest, and assess their statistical accuracy. The deterministic problems in Exercises H.1 through H.10 call for relatively straightforward estimation of mathematical expressions. Exercises in Chapter 5 again take up these problems, allowing the student to evaluate the benefits of applying several alternative variance-reducing techniques.

Exercises H.11 through H.17 are embedded in institutional settings and challenge the student to formulate more detailed probabilistic models before applying IMC. Some encourage the use of asymptotic distributional approximations that, when they apply, can significantly reduce sampling cost per replication. Others require sampling from the binomial, exponential, normal, Poisson, and uniform distributions. Exercises H.18 through H.20 provide hands-on experience with importance sampling, exploiting probabilistic structure, and bootstrapping.

Algorithms in Chapter 3 describe how to generate samples for these exercises. However, a more expeditious approach makes uses of applications software specifically developed for efficient sampling. For example, Microsoft Excel and MATLAB have the capability for generating the required samples. For C-language programming, the Internet website

http://www.or.unc.edu/~gfish/fcmc.html

also contains easily implementable software for efficiently generating these samples. In particular, it contains rng_afm.c, which provides a convenient front end for the *Mersenne twister pseudorandom number generator* (see Section 4.7.1), whose performance is known to be considerably superior to that of generators frequently encountered in more general applications software. Chapter 3 lists additional sources.

Significant Figures

All solutions should contain clearly labeled numerical results to at least four *significant figures*. The 1990 edition of *A Concise Dictionary of Mathematics*, C. Clapham editor, has the entry:

> **significant figures** To count the number of **significant figures** in a given number, start with the first non-zero digit form the left and, moving to the right, count all digits thereafter, counting final zeros, if they are to the right of the decimal point. For example, 1.2048, 1.2040, 0.012048, 0.0012040 and 1204.0 all have 5 significant figures. In *rounding* or *truncation* of a number to n significant figures, the original is replaced by a number with n significant figures.
>
> Note that final zeros to the left of the decimal point may or may not be significant: the number 1204000 has at least 4 significant figures, but without more information there is no way of knowing whether or not any more figures are significant. When 1203960 is rounded to 5 significant figures to give 1204000, an explanation that this has 5 significant figures is required. This could make it clear by writing it in *scientific notation*: 1.2040×10^6.
>
> To say that $a = 1.2048$ to 5 significant figures means that the exact value of a becomes 1.2048 after rounding to 5 significant figures: that is to say, $1.20475 \le a \le 1.20485$.

H.1. Describe and implement an IMC sampling plan with $n = 10^6$ independent replications to estimate λ in (2.1) for

$$g(x, y) := \begin{cases} x^2 y^2 \sin(x+y)\ln(x+y) & 0 \le x \le \frac{\pi}{4},\ \frac{\pi}{2} \le y \le \frac{3\pi}{4} \\ 0 & \text{elsewhere} \end{cases} \tag{2.82}$$

and sampling from

$$\omega(x, y) := \begin{cases} \left(\frac{4}{\pi}\right)^2 & 0 \le x \le \frac{\pi}{4},\ \frac{\pi}{2} \le y \le \frac{3\pi}{4} \\ 0 & \text{elsewhere.} \end{cases} \tag{2.83}$$

Compute $\bar{\lambda}_n$, its standard error, and its relative error. This particular sample size facilitates stratified sampling in Section 5.1.

H.2. Repeat the experiment in Exercise H.1 with an alternative sampling plan that increases computational efficiency (see Section 2.3).

H.3. Describe and implement an IMC sampling plan with $n = 10^6$ independent replications to estimate λ in (2.1) for

$$g(x, y) := \begin{cases} x e^{xy} \sin^2\left(x + \frac{\pi}{2} y\right) & 0 \le x \le \frac{\pi}{2},\ 0 \le y \le 1 \\ 0 & \text{elsewhere} \end{cases} \tag{2.84}$$

and sampling from

$$\omega(x, y) := \begin{cases} \frac{2}{\pi} & 0 \le x \le \frac{\pi}{2},\ 0 \le y \le 1 \\ 0 & \text{elsewhere.} \end{cases} \tag{2.85}$$

Compute $\bar{\lambda}_n$, its standard error, and its relative error.

H.4. Repeat the experiment in Exercise H.3 with an alternative sampling plan that increases computational efficiency (see Section 2.3).

H.5. Describe and implement an IMC sampling plan with $n = 10^6$ independent replications to estimate

$$\lambda := \int_0^\infty \int_0^\infty \left[\sum_{z=1}^{9} g(x, y, z)\right] \mathrm{d}x\mathrm{d}y, \tag{2.86}$$

where

$$g(x, y, z) := \begin{cases} \frac{x z^{3/2} e^{-(xz+x)}}{45(x^2+zy^2)} & x \ge 0,\ y \ge 0,\ z = 1, \ldots, 9 \\ 0 & \text{elsewhere} \end{cases} \tag{2.87}$$

and using the sampling distributions

$$\omega_X(x) := \begin{cases} e^{-x} & x \geq 0 \\ 0 & \text{elsewhere,} \end{cases} \tag{2.88}$$

$$\omega_Y(y) := \begin{cases} \frac{2}{\pi(1+y^2)} & y \geq 0 \\ 0 & \text{elsewhere,} \end{cases} \quad \text{(see Table 3.1)} \tag{2.89}$$

$$\omega_Z(z) := \begin{cases} \frac{1}{9} & z = 1, \ldots, 9 \\ 0 & \text{elsewhere.} \end{cases} \tag{2.90}$$

H.6. Repeat the experiment in Exercise H.5 with an alternative sampling plan that induces CR $\times$ VR > 1.

H.7. Describe and implement an IMC sampling plan with $n = 10^6$ independent replications to estimate λ with

$$g(x, y, z) := \begin{cases} e^{-x^{1/2}(y^{3/2}+2y)} & 0 \leq x \leq y < \infty \\ 0 & \text{elsewhere.} \end{cases} \tag{2.91}$$

H.8. Describe and implement an IMC sampling plan with $n = 10^6$ independent replications to estimate λ with

$$g(x, y, z) := \begin{cases} (3y - 2xz)^{1/3} & 0 \leq x \leq \frac{3y}{2z},\ 0 \leq y \leq 1,\ \frac{1}{4} \leq z \leq 1 \\ 0 & \text{elsewhere.} \end{cases} \tag{2.92}$$

Suppose the lower bound on z were 0 instead of $\frac{1}{4}$. How would that affect your sampling plan?

H.9. Describe and implement an IMC sampling plan with $n = 10^6$ independent replications to estimate λ with

$$g(x, y, z) := \begin{cases} xe^{-(y+z)} & 0 \leq x \leq \min(y, z),\ y \geq 0,\ z \geq 0 \\ 0 & \text{elsewhere.} \end{cases} \tag{2.93}$$

H.10. Use IMC with $n = 10^6$ independent replications to estimate the volume of the polytope

$$\begin{aligned} x_1 + 2x_2 + 2x_3 + 2x_4 &\leq b \\ a_{i1}x_1 + a_{i2}x_2 + a_{i3}x_3 + a_{i4}x_4 &\leq b \quad i = 1, \ldots, 15, \end{aligned} \tag{2.94}$$

where $b = 1$ and

i	a_{i1}	a_{i2}	a_{i3}	a_{i4}
1	−	−	−	−
2	−	−	−	+
3	−	−	+	−
4	−	+	−	−
5	+	−	−	−
6	−	−	+	+
7	−	+	−	+
8	+	−	−	+
9	−	+	+	−
10	+	−	+	−
11	+	+	−	−
12	−	+	+	+
13	+	+	+	−
14	+	+	−	+
15	+	−	+	+

Word Problems

H.11. Pricing a Mortgage-Backed Security. Consider a *security* that pays C dollars monthly for m months. For example, the underlying income-generating asset may be a collection of mortgages with guaranteed $m = 30 \times 12 = 360$ monthly payments of C dollars and with zero cash value thereafter. At the beginning of the payment period, the *present value* is

$$PV := C\left[\frac{1}{1+r/12} + \frac{1}{(1+r/12)^2} + \cdots + \frac{1}{(1+r/12)^{360}}\right], \tag{2.95}$$

under the assumption that the prevailing annual interest rate, r, remains constant over the m month period.

By way of interpretation, $\frac{C}{(1+r/12)^j}$ is the amount that one would have to invest today in an alternative riskless asset in order receive C dollars in month j in the future. By adding together $\frac{C}{1+r/12}, \frac{C}{(1+r/12)^2}, \ldots, \frac{C}{(1+r/12)^{360}}$, we derive the amount we would have to invest today to receive 360 monthly payments each of C dollars. In principle, an investor might contemplate buying the security if its price were no greater than PV and to decline to purchase it if its price were to exceed PV.

In reality, mortgage-backed securities are far more complex in structure. For example, they frequently allow prepayment at specific or arbitrarily selected times during the m payback months. Also, some securities allow holders to sell at selected or arbitrarily selected times. These and other options make the analysis of financial instruments a challenge. For present purposes, we assume that none of these options apply.

However, one property keeps the analysis of an option-free security from being as simple in form as in (2.95). Rather than being constant, the interest rate fluctuates randomly with time. As a consequence, present value is a random variable

$$PV(\theta) := C\sum_{i=1}^{m}\prod_{j=1}^{i}\left(\frac{1}{1+Q_j/12}\right), \tag{2.96}$$

where Q_j is the random interest rate realized in month j for $j = 1, \ldots, m$, and θ is a parameter that determines the volatility of interest rates. Since $PV(\theta)$ is a random variable, an informed purchase/no purchase decision relies on a knowledge of its probabilistic properties.

In addition to randomness, the rate in a particular month depends partially on the realized rates for previous months. For example, suppose that

$$Q_j = KQ_{j-1}e^{\theta X_j} \qquad j = 1, \ldots, m, \tag{2.97}$$

where

$$\begin{aligned} Q_0 &= \text{initial annual interest rate,} \\ K &:= e^{-\theta^2/2}, \end{aligned}$$

and X_j denotes a random variable from $\mathcal{N}(0, 1)$. By recursively applying (2.97), one has

$$\begin{aligned} Q_1 &= Q_0 K e^{\theta X_1} \\ Q_2 &= Q_0 K^2 e^{\theta(X_1+X_2)} \\ \vdots &= \vdots \\ Q_j &= Q_0 K^j e^{\theta(X_1+\cdots+X_j)}, \end{aligned}$$

revealing that Q_j is a lognormally distributed random variable (see Example 5.4.4) with mean $Q_0K^je^{j\theta^2/2} = Q_0$ and variance $Q_0^2K^{2j}e^{j\theta^2}\left(e^{j\theta^2}-1\right) = Q_0^2\left(e^{j\theta^2}-1\right)$. Because for $l > j$, Q_j and Q_l both depend on $X_1, \ldots, X_j$, Q_j and Q_l are clearly dependent random variables with

$$\operatorname{corr}(Q_j, Q_l) = \frac{e^{(j+l)\theta^2/2}-1}{\sqrt{(e^{j\theta^2}-1)(e^{l\theta^2}-1)}} \qquad j, l = 1, \ldots, m.$$

Because the probabilistic structure of $PV(\theta)$ does not admit analytical characterization in closed form, IMC becomes a methodology worth considering. In particular, IMC estimates of the mean $\lambda(\theta) := \mathrm{E}PV(\theta)$ and d.f. of $PV(\theta)$ are helpful in deciding on the price one is willing to pay for the asset.

a. For $PV(\theta)$, describe an IMC sampling plan based on n independent replications to estimate $\lambda(\theta)$ and $\text{pr}[PV(\theta) \le z]$ for selected z.

b. Let $m = 360$, $C = 1{,}000$, $\theta = 0.02$, and $Q_0 = 0.07$. Perform $n = 10{,}000$ independent replications and display estimates of $\lambda(\theta)$ and $\text{pr}[PV(\theta) \le z]$ for $z = 110{,}000 + 5{,}000(i-1)$ for $i = 1, \ldots, 15$, along with their respective standard errors and relative errors.

c. How confident are you that your estimate of $\text{pr}[PV(\theta) \le 120{,}000]$ is within ± 0.005 of its true value? Give a quantitative justification.

H.12. At best, we have a rough estimate of the *volatility* θ in Exercise H.11. As a consequence, a more judicious evaluation considers a range of values for θ in which its true value, presumably, lies. In particular, this enhanced experiment allows us to infer how $\lambda(\theta)$ and $\text{pr}[PV(\theta) \le z]$ vary as θ varies.

a. With these objectives in mind, perform additional IMC sampling experiments for $\theta = 0.02 + 0.002i$ for $i = 0, 1, \ldots, 15$ and compute estimates of $\lambda(\theta)$ and $\text{pr}[PV(\theta) \le z]$ for the z in Exercise H.11b. Use the same values for K, Q_0, and n as in Exercise H.11.

b. Use the point estimates and their standard errors to infer how $\lambda(\theta)$ and $\text{pr}[PV(\theta) \le z]$ change as θ varies, distinguishing between *apparent* and *real* changes at the 0.99 confidence level. Graphs as well as tables are helpful.

c. Suppose an investor is willing to consider purchase only if $\text{pr}[PV(\theta) \le 120{,}000] \le 0.05$. Given $Q_0 = 0.07$, what is the value of the volatility θ at which this bound is met? Assess the accuracy of the estimate.

Note: Solving this problem requires IMC sampling experiments with successively finer grids for θ than in part a and, most likely, larger numbers of replications. Also, regression analysis may be helpful (see Section 2.9).

Asymptotic Distributional Approximations

Although theoretically exact procedures exist for sampling from every univariate distribution, the values of distributional parameters occasionally allow a variate to be sampled from the normal distribution with negligible error of approximation, but with considerable reduction in computing time. Exercises H.13 and H.14 illustrate the concept.

H.13. Repair Warranties. Cold Zone, Incorporated plans to manufacture a new 20 cubic foot refrigerator and to offer a five-year warranty on all parts and labor. The company has contracted with Fast Fix, Incorporated to repair refrigerators under warranty and to pay Fast Fix D dollars for each serviced refrigerator. To underwrite these future expenditures, Cool Zone plans to set aside C dollars at the beginning of year one in a reserve account that earns interest at a fixed rate r compounded annually. The initial reserve is reduced annually by the amount paid to Fast Fix and the reserve balance continues to earn interest at rate r. If the reserve fund is exhausted, Cool Zone plans to pay Fast Fix directly from the current account annually thereafter. The objective of this exercise is to estimate how large Cool Zone's initial reserve C should be such that the mean net present value of all warranty costs is zero over a 20-year horizon.

In year i, let

$$\begin{aligned}
S_i &:= \text{number of refrigerators sold} \\
&= \text{random variable from the Poisson distribution} \\
p_z &:= e^{-\beta}\frac{\beta^z}{z!} \qquad \beta > 0, \qquad z = 0, 1, \ldots, \\
&\quad \text{denoted by } \mathcal{P}(\beta) \text{ (e.g., Example 3.7.4)} \\
T_i &:= \text{number of refrigerators under warranty} \\
&= \begin{cases} S_1 & i = 1 \\ T_{i-1} + S_i & i = 2, \ldots, 5 \\ T_{i-1} + S_i - S_{i-5} & i \geq 6 \end{cases} \\
V_i &:= \text{number of refrigerators repaired in year } i \\
&= \text{random variable from the binomial distribution} \\
q_z &:= \binom{T_i}{z} q^z (1-q)^{T_i - z} \quad 0 < q < 1, \; z = 0, 1, \ldots, T_i, \\
&\quad \text{denoted by } \mathcal{B}(T_i, q).
\end{aligned}$$

Cool Zone's repair cost in year i will be DV_i. If it exceeds the amount remaining in the reserve at the end of year $i-1$, Cool Zone plans to replenish the account with C dollars. Therefore, the amount remaining in the reserve fund at the end of year i is

$$\begin{aligned}
C_0 &= C \\
C_i &= \begin{cases} (C_{i-1} - DV_i)(1 + r) & \text{if } C_{i-1} > DV_i \\ 0 & \text{if } C_{i-1} \leq DV_i \end{cases} \qquad i \geq 1.
\end{aligned}$$

Then $\frac{(DV_i - C_{i-1})^+}{\prod_{j=1}^{i}(1+Q_j)}$ is the discounted warranty cost incurred in year i, and $\frac{C_{20}}{\prod_{j=1}^{20}(1+Q_j)}$ is the discounted value of the remaining initial reserve at the end of year 20, where Q_i denotes the prevailing interest rate in year i. It is of interest to estimate the C that makes the mean net present value of the sequence of warranty costs,

$$\lambda(C) := C + \mathrm{E}\left[\sum_{i=1}^{20} \frac{(DV_i - C_{i-1})^+}{\prod_{j=1}^{i}(1+Q_j)} - \frac{C_{20}}{\prod_{j=1}^{20}(1+Q_j)}\right], \tag{2.98}$$

equal to zero. Expression (2.97) defines Q_i with $Q_0 = 0.07$, and $\theta = 0.02$.

a. Describe an IMC sampling plan and statistical procedure based on n independent replications to estimate the C that makes $\lambda(C)$ zero within a statistical accuracy of ± 50 with probability 0.99.

b. For $\beta = 5{,}000$, $D = 50$, $p = 0.01$, and varying C, execute the sequence of IMC experiments, starting with sample sizes $n = 10{,}000$ and increasing them as needed to meet the specified accuracy. The estimate of C that satisfies $\lambda(C) = 0$ need not be a single value.

To reduce sampling cost in the present exercise, we can take advantage of *asymptotic properties* of the Poisson and binomial distributions.

Poisson Approximation. If Z is from $\mathcal{P}(\beta)$, then the standardized random variable $\frac{Z-\beta}{\sqrt{\beta}}$ converges in distribution to $\mathcal{N}(0, 1)$ as $\beta \to \infty$. Therefore, Z can be approximated by $\lfloor \max(0, \beta + 0.50 + \sqrt{\beta}X) \rfloor$, where X is from $\mathcal{N}(0, 1)$ for large β. In practice, the error of approximation is negligible for $\beta \geq 30$. The *continuity correction* 0.50 merely improves the approximation.

Binomial Approximation. If Z is from $\mathcal{B}(N, p)$, then the standardized variable $\frac{Z-Np}{\sqrt{Np(1-p)}}$ converges in distribution to $\mathcal{N}(0, 1)$ as $N \to \infty$. Then for X from $\mathcal{N}(0, 1)$, $\lfloor \max(0, Np + 0.50 + \sqrt{Np(1-p)}X) \rfloor$ approximates Z for large N. As to error of approximation, common practice is to use the approximation for $N \min(p, 1-p) \geq 5$.

Because these approximations, when they apply, make sampling from the normal distribution considerably less costly than sampling from the Poisson and binomial distributions, they increase computational efficiency while inducing negligible error.

H.14. In Exercise H.13, Fast Fix has offered Cool Zone a flat price of \$50 per unit repaired (option A). It now decides to offer an alternative option B. Past experience indicates that repair times (in hours) have the Weibull distribution

$$f(z) = \frac{\alpha z^{\alpha-1} e^{-(z/\theta)^\alpha}}{\theta^\alpha} \qquad \alpha > 0,\ \theta > 0,\ z \geq 0,$$

denoted by $\mathcal{W}(\alpha, \theta)$, with $\alpha = 0.80$, $\theta = 0.58841$, mean $\mu = \theta\Gamma(1 + 1/\alpha) = 2/3$, and variance $\nu^2 = \theta^2 \left[\Gamma(1 + 2/\alpha) - \Gamma^2(1 + 1/\alpha)\right] = 0.70618$. A graph of this distribution reveals that although many repair times are small, some are extremely large. If X is from the exponential distribution with mean 1, then $Z := (X/\theta)^{1/\alpha}$ is from $\mathcal{W}(\alpha, \theta)$.

Fast Fix estimates its direct hourly labor cost to be \$37.50, which it usually doubles to cover other costs and to build in a suitable profit margin. These cost considerations motivated option A offered to Cool Zone based on $\frac{40}{60} \times \$37.50 \times 2 = \$50$. Fast Fix now offers Cool Zone option B, which is a charge of \$75 per hour for time expended repairing its refrigerators. Repeat the experiment of Exercise H.13 to determine the C that makes the corresponding mean net present value of the cost stream vanish. Use the same accuracy as in Exercise H.13.

Repeat the IMC experiment using option B and determine whether the C for which $\lambda(C) = 0$ within an accuracy of ± 50 with probability 0.99 differs from the C estimated in Exercise H.13. Justify your conclusion based on the warranty model and the generated data.

Once again, an opportunity exists to reduce computing time by exploiting an approximation with small error.

> **Weibull Approximation.** If $Y_1, \ldots, Y_v$ are i.i.d. from $\mathcal{W}(\alpha, \theta)$, then as $v \to \infty$, the standardized variate $Z := \frac{Y - v\mu}{\sqrt{v\nu^2}}$, where $Y := Y_1 + \cdots + Y_v$, converges to a random variable from $\mathcal{N}(0, 1)$ with probability 1. Therefore, $\max(0, v\mu + \sqrt{v\nu^2} Z)$ approximates Y with an error that decreases as v increases. In the present problem, the annual number of failures v is sufficiently large to make this error negligible.

H.15 For options A and B in Exercises H.13 and H.14, Cool Zone's management regards the values of C that make $\lambda(C) = 0$ unacceptably large. As an alternative policy, management considers setting $C = 0$ and paying annual warranty cost $\max(0, Q_i - C_{i-1})$ from the current account, where Q_i denotes repair costs in year i.

a. Under each option, estimate $\lambda(0)$ and

$$\begin{aligned} s_l &:= \text{pr}\left[q_{l-1} \le \max(0, Q_i - C_{i-1}) < q_l\right] \\ &= \text{probability that cost is in } [q_{l-1}, q_l) \text{ in an arbitrarily chosen year,} \end{aligned}$$

where $q_l := 1000l$ for $l = 0, 1, \ldots, 30$.

b. If management is averse to large annual payments, which option would you recommend? Justify your answer. Histograms and graphs of exceedance probabilities may be helpful.

Benefits of Aggregation

Analysis prior to execution occasionally allows a reduction in the number of distributions that need to be sampled, which can decrease computing time devoted to sampling. Exercises H.16 and H.17 collectively illustrate the concept.

H.16. Power Generation. (Adapted from Iyengar and Mazumdar 1993.) Amass Energy Group, a power grid consortium, consists of r energy-generating stations. Station i has m_i generating units, each of capacity c_i megawatts. Every day the grid supplies energy to s cities, of which city j demands d_j megawatts. Contractual arrangements require station i to supply $100 \times a_{ij}$ percent of its daily available output to city j, where $0 \le a_{ij} \le 1$ for all i and j. For $l = 1, \ldots, m_i$ and $i = 1, \ldots, r$, define

$$Z_{il} := \begin{cases} 1 & \text{if unit } l \text{ at station } i \text{ is operating} \\ 0 & \text{otherwise,} \end{cases}$$

so that station i has $Z_i := Z_{i1} + \cdots + Z_{im_i}$ operating units putting out $c_i Z_i$ megawatts. The grid satisfies demand if and only if

$$\sum_{i=1}^{s} a_{ij} c_i Z_i \geq d_j \qquad \forall j. \tag{2.99}$$

That is, every city has its demand satisfied. Past history reveals that for all stations $i \neq l$, $Z_{i1}, \ldots, Z_{im_i}$ and $Z_{l1}, \ldots, Z_{lm_l}$ are independent. Amass Energy Group is interested in estimating

$$\lambda := \text{pr(demand exceeds supply)}. \tag{2.100}$$

a. Show that (2.100) can be represented as a multidimensional integral over a subset of $[0, 1]^m$, where $m := m_1 + \cdots + m_r$. In particular, define the subset.

b. Describe an IMC sampling plan based on Bernoulli sampling (2.38) and n independent replications for computing point estimates of (2.100) and of

$$\alpha_j := \text{pr(city } j\text{'s demand exceeds its allocated supply)} \quad j = 1, \ldots, s.$$

c. Perform the experiment in part b for $n = 10^6$ independent replications using the data in Tables 2.10 and 2.11. In addition to all point estimates, compute their standard and relative standard errors and record the execution time for the IMC experiment.

TABLE 2.10 | ***Power station characteristics***

Station i	m_i	c_i	$1 - p_i$
1	1	13.600	.16
2	1	11.990	.16
3	1	5.600	.14
4	9	.760	.02
5	6	.650	.01
6	6	1.115	.02
7	1	5.530	.17
8	3	5.530	.17
9	2	7.350	.12
10	8	.780	.02
11	11	1.090	.02
12	8	.780	.02
13	18	1.453	.02
14	6	7.737	.02
15	4	.945	.01

TABLE 2.11 | *Percentages of station energy supply allocated to each city* $100 \times a_{ij}$

City j	1	2	3	4	5	6	7	8	9	10	11	12	13	14	15	16	17	18	19	20
$100 \times d_j$	604	715	904	1,012	580	540	620	606	797	652	905	537	399	669	585	588	486	586	700	830
Station i																				
1	30	20	15			21	14													
2			8		20			41		15		7				9				
3		10					10		25	5	10		10		10			10	10	
4				80												20				
5	20						10			15		10				10	10	10		15
6		10	10			15			7	8					9	6	20		15	
7	3	6	9		15		11	7		10		12			14					13
8	7	5	6	8	4	5	5	5	5	8	4	10	5	5	4	4	2	2	3	3
9	3	3	2	2	4	4	5	5	10	4	8	5	5	5	5	4	8	6	5	7
10	10	10	10	10					10		15			15						20
11														14	21			15	20	30
12			20		8			15			41		9					7		
13	6	6	6	6	6	6	6	6	8	6		6		6		6	4	6	4	6
14	5	7	8	6	5	4	5	5	8	5										
15					10	10	10	10		15		10			15		20			

d. Compute a point estimate and a 0.99 confidence interval for the system's

$$\text{MTBF} := \text{mean time between failures} \\ = \frac{1-\lambda}{\lambda}.$$

e. At which station would you upgrade the generators to produce the largest increase in (2.100)? Use previous results and information about reliability, together with any additional independent IMC runs you execute, to justify your choice.

Upgrading. In the present problem, upgrading is defined as doubling a generator's mean number of operating days between failures (MTBF). If a generator has failure probability $1 - p_i$, then MTBF $= \frac{p_i}{1-p_i}$.

f. Management will seriously consider your recommendation only if the system MTBF increases by at least four days with, at least, 0.98 probability. Determine whether your proposal meets this criterion. Explain your answer in detail.

H.17. Generating the sample data in Exercise H.16b necessarily takes computing time that increases as m. Therefore, each replication has computing time that grows as ms. Because all units at station i have reliability p_i, we can exploit a relationship between Bernoulli random variables and a binomially distributed random variable to reduce the cost of sampling $Z_1, \ldots, Z_r$.

Bernoulli-Binomial Relationship. If $Y_1, \ldots, Y_N$ are i.i.d. from $\mathcal{B}er(p)$, then $Y := Y_1 + \cdots + Y_N$ is from $\mathcal{B}(N, p)$.

The inverse-transform method provides one way to generate Y.

a. Show that for $q \in (0, 1)$ and N a positive integer, the algorithm:

$Q_0 = (1-q)^N$.
$S = Q_0$.
Sample U from $\mathcal{U}(0, 1)$.
$X = 0$.
While $U > S$:
  $X = X + 1$
  $Q_X = Q_{X-1} \times (N - X + 1) \times q/X(1-q)$
  $S = S + Q_X$.
Return X.

generates X from $\mathcal{B}(N, q)$ with $1 + Nq$ comparisons on average.

b. Repeat the IMC experiment in Exercise H.16c using the algorithm in part a to generate $Z_1, \ldots, Z_r$. Compare results for point estimates, standard and relative standard errors, and execution times.

Because mean computing time grows linearly in Nq, the inverse-transform method is computationally efficient only when Nq is small. To increase computational efficiency, make use of the property that if Z is from $\mathcal{B}(n, q)$, then $n - Z$ is from $\mathcal{B}(n, 1-q)$.

Importance Sampling

H.18. Exercise H.16 is intrinsically a network-reliability problem.

a. Describe an adaptation of the importance sampling technique of Example 2.7.3 to the problem in Exercise H.16, focusing on stations 7 and 8 with failure probability $1 - q = 0.25$.

b. Repeat the IMC experiment in Exercise H.16b using the adaptation. In addition to displaying the estimate of λ in (2.100) and its standard and relative standard errors, show the variance reduction.

c. Explain the motivation for selecting $1 - q = 0.25$.

Exploiting Probabilistic Structure

H.19. Although binomial sampling offers a computational advantage in Exercise H.17, determining whether demand exceeds supply on all n replications takes computing time that grows as nrs. However, the low failure rates in Table 2.10 offer an opportunity to reduce this cost. Let

$$T_{ijl} := \begin{array}{l}\text{number of operating days that elapse between the} \\ l\text{th and } (l-1)\text{st failures of unit } j \text{ at station } i.\end{array}$$

Then it is known that, for each i and j, $\{T_{ijl},\ l \geq 1\}$ is a sequence of i.i.d. random variables from the geometric distribution $\mathcal{G}e(p_i)$ and, therefore, its MTFB is $\frac{p_i}{1-p_i}$. For example, $p_i = 0.02$ implies a 49-day mean. Consequently, a unit at station i contributes to a change in each of the city supplies $\sum_{i=1}^{r} a_{ij} c_i Z_i$, $1 \leq j \leq s$, once every $1 + \frac{p_i}{1-p_i} = 50$ replications on average. The i.i.d. property follows from the

Memoryless Property of the Geometric Distribution. If X is from $\mathcal{G}e(p)$, as in (2.35), with $p \in (0, 1)$, then

$$\text{pr}(X = x \mid X \geq x_0) = \begin{cases} (1-p)p^{x-x_0} & x \geq x_0 \geq 0 \\ 0 & \text{elsewhere.} \end{cases}$$

Algorithms A and B exploit this property. They keep a calendar $\mathcal{L}$ for the next T future days that allows the experiment to keep track of the next time each unit fails. If a unit fails more than T days in the future, then inserting this unit in an ancillary list $\mathcal{L}^*$ enables the experiment to resample for the next failure once T days have elapsed.

ALGORITHM A

For $k = 1, \ldots, s$:
  $V_k = \sum_{i=1}^{r} a_{ik} c_i m_i$.
  $W_k = V_k$.
$l = T$.
For $i = 1, \ldots, r$:
  For $j = 1, \ldots, m_i$: execute Algorithm B.

```
ϕ = 0, S = 0, l = 1.
While l ≤ n:
    M = L((l − 1) (mod T) + 1)
    L((l − 1) (mod T) + 1) = ∅                    [make list L(l − 1) empty]
    for each (i, j) ∈ M:
        for k = 1, ..., s: W_k = W_k − a_ik c_i
    if ∪_{k=1}^r (W_k < d_k) not empty, ϕ = 1; otherwise, ϕ = 0
    S = S + ϕ
    M = M ∩ [L*((l − 1) (mod T))]
    L*((l − 1) (mod T)) = ∅.                      [make list L*(l − 1) empty]
    for (i, j) ∈ M:
    remove (i, j) from M
        execute Algorithm B
    for k = 1, ..., s :  W_k = V_k
    l = l + 1.
Return S (number of replications in which demand exceeds supply). ■
```

ALGORITHM B

Input: $i =$ station, $j =$ unit, $l =$ day.

Sample T_{ij} from the geometric distribution $\mathcal{G}e(p_i)$ as in (2.35). If $T_{ij} \le T$, insert (i, j) in list $\mathcal{L}((l + T_{ij}) \pmod{T} + 1)$. Otherwise, insert (i, j) in list $\mathcal{L}^*q((l-1) \pmod{T} + 1)$. ■

a. Show that the condition $\min_{1 \le i \le r} p_i \ge 1 - r/\sum_{i=1}^{r} m_i$ is sufficient for Algorithms A and B to induce computing time per replication that increases more slowly than rs.

b. Repeat the IMC experiment in Exercise H.16c, using Algorithms A and B for $n = 10^6$ and $T = 100$. Compare the estimate of λ, its standard error, relative standard error, and execution time with those in Exercises H.16c and H.17b.

Bootstrap

H.20. For the inventory system in Example 2.8.1,

a. Formulate a resampling experiment to estimate the probability that inventory system 3 has a mean cost at least c units less than the mean cost for any other system.

b. Generate $n = 10^6$ independent replications to estimate the probability for each $c \in \{0.50i,\ i = 1, \ldots, 10\}$.

c. Describe the extent to which inventory system 3 dominates the other inventory systems with regard to cost.

2.12 REFERENCES

Aho, A. V., J. E. Hopcroft, and J. D. Ullman (1974). *The Design and Analysis of Computer Algorithms,* Addison-Wesley, Reading, Mass.

Efron, B., and R. J. Tibshirani (1993). *An Introduction to the Bootstrap Method,* Chapman and Hall, New York and London.

Glasserman, P., P. Heidelberger, and P. Shahabuddin (1999). Asymptotically optimal importance sampling and stratification for pricing path-dependent options, *Mathematical Finance,* **9,** 117–152.

Good, P. I. (2001). *Resampling Methods*, second edition, Birkhäuser, Boston.

Hjorth, J. S. U. (1994). *Compute Intensive Statistical Methods,* Chapman and Hall, New York and London.

Hull, J. C. (1993). *Futures, Options, and Other Derivative Securities*, second edition, Prentice-Hall, Englewood Cliffs, New Jersey.

Iyengar, S., and M. Mazumdar (1993). Computation of a reliability index for power systems, University of Pittsburgh, Pa.

Law, A. M., and W. D. Kelton (1991). *Simulation Modeling and Analysis*, second edition, McGraw-Hill, New York.

Linetsky, V. (2004). Spectral expansions for Asian (average price) options, *Operations Research,* to appear.

CHAPTER 3

Sample Generation

Generating samples from diverse probability distributions on a computer is an intrinsic part of every MC experiment. This chapter presents a relatively concise description of the theory and computational practices underlying this sampling. Readily available commercial software often includes the wherewithal to generate samples from many different distributions, making it convenient for MC users to rely on it. Alternatively, Internet sites provide links to sample-generating software when MC experiments are encoded in FORTRAN and C. For example, the following sites both link to CRAND, a library that contains computationally efficient C++ code by E. Stadlober for many commonly encountered continuous and discrete distributions.

- http://cgm.cs.mcgill.ca/%7Eluc/rng.html (maintained by L. Devroye)
- http://random.mat.sbg.ac.at/links/rando.html (maintained by P. Hellekalek)

Given the ease of accessing *off-the-shelf* software for generating samples, why devote a chapter to its theory and algorithmic design? There are several reasons:

- Problem context may call for sampling from a nonstandard distribution. For example, suppose it calls for sampling from the p.d.f. $\{ce^{-x\cos \pi x}, 0 \leq x \leq 1\}$, where the constant c is unknown. This chapter describes several ways to devise an algorithm to effect sampling from this and other nonstandard distributions.
- In some applications of the MC method, sampling from distributions accounts for the vast majority of computing effort. This is especially true in chemistry and physics, where total computing time can be substantial and sample generation accounts for more than 80% of this time. Given several alternative procedures for generating samples from a particular distribution, choosing the one that minimizes generating time from that distribution may significantly reduce total computing time. This chapter describes how to assess algorithms' computational efficiency.

- If an MC experiment has as input a sequence of i.i.d. random fractions from the uniform distribution $\mathcal{U}(0, 1)$, all sample-generating procedures in this book are theoretically exact. Nevertheless, when implemented on a computer, the procedures would generate samples whose distributions differ from the desired ones. One source of error, which is usually negligible, is the finiteness of the word size in computers.

  A second source of error can be a problem. All computer-based sampling algorithms rely for input on a sequence of fractions formed by mapping a sequence of *pseudorandom numbers* to the unit interval [0, 1). We call this fractional sequence *pseudorandom-number generation (PNG) input*. In principle, the PNG input is intended to resemble a sequence of i.i.d. random deviates drawn from the uniform distribution $\mathcal{U}(0, 1)$. This substitution of PNG input for an i.i.d. $\mathcal{U}(0, 1)$ sequence accounts for the second source of error. It reflects differences between the actual and the theoretical ideal. Moreover, since different sampling algorithms transform their PNG input in different ways, some algorithms for sampling from a particular distribution induce greater departures from their theoretical ideal than others do. That is, some are more sensitive to PNG input errors than others. Chapter 4 delves into this topic in detail and, in particular, Example 4.4.1 illustrates how these errors in input affect sample generation.

The principal methodologies for sample generation are:

- Inverse-transform method (Sections 3.3 and 3.4).
- Composition method (Section 3.5).
- Acceptance-rejection method (Section 3.6).
- Ratio-of-uniforms method (Section 3.7).

The chapter describes each methodology, relying heavily on examples to illustrate how the methods work in practice. It emphasizes concepts whose implementation leads to computationally efficient generation. More complete discussions and other algorithms are contained in Devroye (1986) and Fishman (1996).

3.1 SELECTING A SAMPLING ALGORITHM

At the outset, we emphasize that the particular details of each MC experiment dictate which of several alternative algorithms for sampling from a particular distribution offers the greatest computational efficiency. Nevertheless, because two different cost-related settings pervade all sample generation, it is instructive to begin by describing them. They are

A. Repeatedly sampling from a distribution whose parametric values are fixed.

B. Repeatedly sampling from a distribution whose parametric values change on successive draws.

In setting A, computing and storing all constants used in an algorithm prior to sampling eliminates the repeated setup cost of computing them on each draw. The savings can be substantial. Moreover, sampling from a discrete distribution is most efficiently done by using precomputed tables as described in Section 3.4, in computing time independent of the size of the *support* of the distribution. The support of a p.d.f. or p.m.f. $\{f(x),\ x \in \mathcal{X}\}$ denotes the set of all values $x \in \mathcal{X}$ such that $f(x) > 0$.

By contrast, setting B requires constants to be computed in line on each draw. Moreover, the required computing time for some algorithms is a function of their parametric values. However, there is a class of algorithms for each of which there exists a constant upper bound on mean sample generating time that is independent of these parametric values. These are in common use for many distributions. However, their use needs to be balanced against other considerations.

Although a bounded mean-sample-generating-time algorithm for a particular distribution keeps computing time from becoming excessive, it is not unusual for another algorithm, whose time depends on the parametric values, to yield smaller mean sample generating time for some values of the parameters. Therefore, a computationally efficient algorithm frequently incorporates more than one generating method and automatically chooses the one to use on a particular instance based on these parametric values. The present account repeatedly stresses this issue, most notably in Example 3.7.3 for Gamma generation and in Example 3.7.4 for Poisson generation.

Setting A also provides opportunities for improving computational efficiency when generating samples from a continuous distribution. One approach merely computes the setup before sampling begins and then proceeds with the same algorithm as would be used under setting B. A second approach realizes additional computational efficiency by designing the generating algorithm with due consideration to the parametric values that are to remain fixed on every draw. For example, in response to a user-specified distribution and parametric values, the following Internet site automatically designs and produces C, FORTRAN, or JAVA code that is computationally efficient in this respect as well as theoretically exact (Ledold et al. 2001):

http://statistik.wu-wien.ac.at/anuran

In addition to a sizable selection of standard and nonstandard continuous distributions, code can also be generated for these distributions with restricted or truncated support, thereby incorporating additional computational efficiency. This alternative approach generates code expeditiously with relatively little user effort.

3.2 INDEPENDENT AND DEPENDENT VARIATES

Suppose that $X_1, \ldots, X_m$ are independent random variables, where for each dimension $i = 1, \ldots, m$, X_i has distribution $\omega_i := \{\omega_i(x),\ x \in \mathcal{X}_i \subseteq \Re\}$ satisfying

$$\omega_i(x) \geq 0 \qquad \forall x \in \mathcal{X}_i$$

and

$$\int_{\mathcal{X}_i} \omega_i(x)\mathrm{d}x = 1.$$

To generate a random sample $\mathbf{X} := (X_1, \ldots, X_m)$, it suffices to sample each X_i randomly from its univariate distribution ω_i. Then the methodologies described in this chapter easily apply. However, suppose that $X_1, \ldots, X_m$ are dependent. Then for each $i = 1, \ldots, m$, X_i depends on the values that $X_1, \ldots, X_{i-1}, X_{i+1}, \ldots, X_m$ assume and $\omega_1, \ldots, \omega_m$ are not sufficient for sampling.

Some special cases allow one to overcome this limitation. If $\mathbf{X}$ has the multinormal or multinomial distribution, the previously mentioned Internet sites can provide C software for sample generation. Also, see Devroye (1986) and Fishman (1996). In other cases, the special structure of the dependence can be exploited. Example 2.5.2 illustrates the concept. However, many MC problems require sampling $\mathbf{X}$ from a joint distribution ω not amenable to the direct application of the techniques in this chapter. Then *Markov chain Monte Carlo sampling*, as in Chapter 6, overcomes the limitation but at the cost of inducing dependence among successive samples of $X_1, \ldots, X_m$. That chapter describes in detail how this sample generation is conducted.

3.3 INVERSE-TRANSFORM METHOD

We now begin our description of the principal methodologies for univariate generation. The inverse-transform (IT) method provides the most direct route for generating a random sample.

Theorem 3.3.1

Let $\{F(z),\ a \leq z \leq b\}$ denote a d.f. with inverse d.f.

$$F^{-1}(u) := \min\{z \in [a, b] : F(z) \geq u,\ 0 \leq u \leq 1\}.$$

Let U denote a random variable from $\mathcal{U}(0, 1)$. Then $Z = F^{-1}(U)$ has the d.f. F. ■

PROOF The proof follows from the observation that $\mathrm{pr}(Z \leq z) = \mathrm{pr}[F^{-1}(U) \leq z] = \mathrm{pr}[U \leq F(z)] = F(z)$. ■

Table 3.1 lists several continuous distributions whose inverse d.f.s are known analytically, thus allowing straightforward determination of Z. The preferred form in the table merely improves computational efficiency by taking advantage of the property that if U is from $\mathcal{U}(0, 1)$, then so is $1 - U$. Although other

TABLE 3.1 | *Selected continuous distributions*

	P.d.f.	Inverse d.f.	Preferred form
Uniform Distribution $\mathcal{U}(\alpha, \beta)$	$1/(\beta-\alpha)$ $\alpha \le z \le \beta$	$\alpha + (\beta-\alpha)u$	—
Beta Distribution $\mathcal{B}e(\alpha, 1)$	$\alpha z^{\alpha-1}$ $\alpha > 0,\ 0 \le z \le 1$	$u^{1/\alpha}$	—
Beta Distribution $\mathcal{B}e(1, \beta)$	$\beta(1-z)^{\beta-1}$ $\beta > 0,\ 0 \le z \le 1$	$1-(1-u)^{1/\beta}$	$1-u^{1/\beta}$
Exponential Distribution $\mathcal{E}(\beta)$	$\frac{1}{\beta}e^{-z/\beta}$ $\beta > 0, z \ge 0$	$-\beta\ln(1-u)$	$-\beta \ln u$
Logistic Distribution $\mathcal{L}(\alpha, \beta)$	$\frac{e^{-(z-\alpha)/\beta}}{\beta[1+e^{-(z-\alpha)/\beta}]^2}$ $\beta > 0,\ -\infty < z < \infty$	$\alpha + \beta\ln[u/(1-u)]$	—
Noncentral Cauchy Distribution $\mathcal{C}(\alpha, \beta)$	$\frac{\beta}{\pi[\beta^2+(z-\alpha)^2]}$ $\beta > 0,\ -\infty < z < \infty$	$\alpha + \beta \tan \pi(u-1/2)$	$\alpha + \beta/\tan \pi u$
Pareto Distribution $\mathcal{P}a(\alpha, \beta)$	$\alpha\beta^\alpha/z^{\alpha+1}$ $z \ge \beta > 0, \alpha > 0$	$\beta/(1-u)^{1/\alpha}$	$\beta/u^{1/\alpha}$
Weibull Distribution $\mathcal{W}(\alpha, \beta)$	$(\alpha/\beta^\alpha)z^{\alpha-1}e^{-(z/\beta)^\alpha}$ $\alpha, \beta > 0\ z \ge 0$	$\beta[-\ln(1-u)]^{1/\alpha}$	$\beta(-\ln u)^{1/\alpha}$

methods may generate samples more rapidly from a particular d.f. in Table 3.1, the availability of these closed-form inverse d.f.s gives the IT method considerable appeal.

EXAMPLE 3.3.1

SAMPLING FROM AN EXPONENTIAL DISTRIBUTION

As illustration, the exponential distribution

$$f(z) := \begin{cases} \beta e^{-z/\beta} & \beta > 0,\ z \ge 0 \\ 0 & \text{elsewhere} \end{cases}$$

has d.f. $\{F(z) = 1 - e^{-z/\beta},\ z \ge 0\}$ and is denoted by $\mathcal{E}(\beta)$. If U is from $\mathcal{U}(0, 1)$, then the solution to $F(Y) = U$,

$$Y = -\beta \ln(1-U),$$

is from $\mathcal{E}(\beta)$, as is $Z = -\beta \ln U$. ■

EXAMPLE 3.3.2

SAMPLING FROM A PIECEWISE CONTINUOUS p.d.f.

As a second illustration, consider the d.f.

$$F(z) = \begin{cases} \frac{\alpha-1}{\alpha\beta} z & 0 \le z \le \beta \\ \frac{\alpha-1}{\alpha} + \frac{1}{\alpha}\left(1-\frac{\beta}{z}\right)^{\alpha-1} & \beta < z < \infty \end{cases} \qquad \alpha > 1, \quad \beta > 0,$$

which arises in Section 5.2.1 in the discussion of ruin probability. Solving $F(Z) = U$ for Z gives

$$Z = \begin{cases} \frac{\alpha\beta}{\alpha-1} U & \text{if } 0 \le U \le 1-\frac{1}{\alpha} \\ \frac{\beta}{[\alpha(1-U)]^{1/(\alpha-1)}} & \text{if } 1-\frac{1}{\alpha} \le U \le 1. \end{cases}$$

■

In sampling from a continuous distribution, the IT method has three appealing properties:

- For a particular d.f., the sample generating time is independent of the values assigned to its parameters.
- The variate Z increases monotonically as U increases. If the preferred form is used, Z decreases monotonically with U. This property is especially valuable when employing antithetic variates (see Section 5.4) to reduce variance.
- The method also applies to sampling from a restricted domain $[a', b'] \subseteq [a, b]$.

For discrete distributions, the sample generating cost for the IT method is also independent of the values of the parameters of F when precomputed tables are available, as in the cutpoint method in Section 3.4.

3.3.1 Restricted Sampling

Suppose sampling is required from the modified distribution

$$\tilde{F}(z) := \frac{F(z) - F(a')}{F(b') - F(a')}, \quad a \le a' \le z \le b' \le b. \tag{3.1}$$

Replacing U by $U' := F(a') + [F(b') - F(a')]U$ as the argument of the inverse d.f. accomplishes this result. That is,

$$Z' = F^{-1}(F(a') + [F(b') - F(a')]U)$$

has the d.f. $\tilde{F}$. Except for the additional computation of $F(a')$, $F(b')$, and U', the times to generate Z from $\{F(z)\}$ and $\{\tilde{F}(z)\}$ are the same.

EXAMPLE 3.3.3 **SAMPLING FROM A RESTRICTED p.d.f.**

As illustration, suppose we want to sample from the restricted exponential distribution

$$\tilde{f}(z) := \begin{cases} \frac{e^{-z/\beta}}{\beta(e^{-a'/\beta} - e^{-b'/\beta})} & a' \leq z \leq b' \\ 0 & \text{elsewhere} \end{cases}$$

with d.f.

$$\tilde{F}(z) = \frac{e^{-a'/\beta} - e^{-z/\beta}}{e^{-a'/\beta} - e^{-b'/\beta}} \qquad a' \leq z \leq b'.$$

For U from $\mathcal{U}(0, 1)$, $\tilde{F}(Z) = U$ is equivalent to $F(Z) = U'$ so that

$$Z = -\beta \ln \left[e^{-a'/\beta} - \left(e^{-a'/\beta} - e^{-b'/\beta}\right) U\right]$$

has the desired distribution. ■

Although the standard normal distribution (2.11) does not have a closed-form inverse, the frequent need for normal samples has motivated the development of highly efficient numerical IT procedures for generating them. In particular, Moro (1995) describes C-language code for normal generation via IT, which can be downloaded from

http://www.or.unc.edu/~gfish/fcmc.html

For other continuous d.f.s that lack closed-form inverses, search methods can be used to solve $F(Z) = U$ for Z (e.g., Devroye 1986, Fishman 1996).

3.3.2 Discrete Distributions

Suppose F is a discrete d.f. on the integers $a, a+1, \ldots, b$ with the property that $F(z) < F(z+1)$ for all $a \leq z < b$. For conciseness let

$$q_z := \begin{cases} F(z) & z = a, a+1, \ldots, b \\ 0 & \text{elsewhere.} \end{cases}$$

Then

$$Z := \min[z : q_{z-1} < U \leq q_z,\ U \sim \mathcal{U}(0, 1)]$$

has d.f. F. Table 3.2 lists several commonly encountered discrete distributions.

TABLE 3.2 | ***Selected discrete distributions*** **($0 < p < 1$)**

	P.m.f.	
binomial distribution $\mathcal{B}(r, p)$	$\binom{r}{z} p^z (1-p)^{r-z}$	$z = 0, 1, \ldots r$
geometric distribution $\mathcal{G}e(p)$	$(1-p)p^z$	$z = 0, 1, \ldots$
hypergeometric distribution $\mathcal{H}(\alpha, \beta, n)$	$\frac{\binom{\alpha}{z}\binom{\beta}{n-z}}{\binom{\alpha+\beta}{n}}$	α, β, n, z integer $\alpha + \beta \geq n > 0$ $\max(0, n-\beta-1) \leq z \leq \min(\alpha, n)$
negative binomial distribution $\mathcal{NB}(r, p)$	$\frac{\Gamma(r+z)}{\Gamma(r)\Gamma(z+1)}(1-p)^r p^z$	$r > 0, \ z = 0, 1, \ldots$
Poisson distribution $\mathcal{P}(\lambda)$	$\frac{\lambda^z e^{-\lambda}}{z!}$	$\lambda > 0, \ z = 0, 1, \ldots$
uniform distribution $\mathcal{U}_d(\alpha, \beta)$	$\frac{1}{\beta - \alpha + 1}$	α and β integer, $\alpha < \beta$ $z = \alpha, \alpha+1, \ldots, \beta$

EXAMPLE 3.3.4

BERNOULLI SAMPLING

Bernoulli sampling provides the most elementary application of the IT method to a discrete distribution. Algorithm ZERONE in Chapter 2 has already made use of the technique. Here we merely restate its essential features. Let Z denote a random variable with p.m.f.

$$q_z := \begin{cases} 1-p & \text{if } z = 0 \\ p & \text{if } z = 1 \end{cases} \qquad 0 \leq p \leq 1,$$

denoted by *Ber(p)*. To generate a random sample, we

1. Randomly generate U from $\mathcal{U}(0, 1)$.

2. If $U \leq p$, $Z = 1$.

3. Otherwise, $Z = 0$. ■

EXAMPLE 3.3.5

SAMPLING FROM THE GEOMETRIC DISTRIBUTION

As an illustration, the geometric distribution with d.f.

$$F(z) = 1 - p^{z+1} \qquad 0 < p < 1 \qquad z = 0, 1, \ldots,$$

denoted by $\mathcal{G}e(p)$, leads to

$$Z = \left\lfloor \frac{\ln U}{\ln p} \right\rfloor,$$

so that sample generating time is independent of p.

Applying the IT method to discrete d.f.s other than the Bernoulli and geometric usually requires a search. The simplest and most direct search executes the steps:

- Randomly generate U from $\mathcal{U}(0, 1)$ and set $Z = a$.
- While $U > q_Z$: $Z = Z + 1$.
- Return Z.

Then N, the number of comparisons until success, has mean

$$\begin{aligned} \mathrm{E}N &= q_a + 2(q_{a+1} - q_a) + 3(q_{a+2} - q_{a+1}) + \cdots + (b - a + 1)(q_b - q_{b-1}) \\ &= 1 + \mathrm{E}Z - a. \end{aligned}$$

■

When $\mathrm{E}Z - a$ is small, this algorithm provides a relatively simple and computationally efficient sampling method. However, since mean sample generating time increases linearly with $\mathrm{E}N$, the algorithm is not competitive with alternative sampling methods when $\mathrm{E}Z - a$ is large. Moreover, it is then subject to non-negligible accumulated round-off error if q_Z is computed recursively in line as $q_z = q_{z-1} + p_z$ for probabilities $p_z > 0$ for $a \le z \le b$ and $p_a + \cdots + p_b = 1$.

Section 3.4 describes an IT search algorithm that eliminates dependence on $\mathrm{E}Z$ when repeatedly drawing samples from a particular distribution with fixed parametric values. Although this table-based bounded-sample-generating-time approach does not apply for distributions whose parameters vary on each draw, it is possible to construct sampling algorithms whose mean sample generating time is bounded above by a constant independent of parametric values. Example 3.7.3 illustrates how this can be done for Gamma generation, and Example 3.7.4 does likewise for Poisson generation.

3.4 SAMPLING FROM A TABLE

The *cutpoint method* (Chen and Asau 1974) allows sampling from a discrete distribution F with finite support $[a, b]$ in time bounded above by a constant independent of $b - a + 1$. Its principal appeal arises in sampling repeatedly from the same distribution and when $b - a + 1$ is large. The present discussion follows Fishman and Moore (1984), who give a detailed analysis of the efficiency of the method. The procedure uses sequential search to determine Z but begins the search at a point in $\{a, a + 1, \ldots, b\}$ considerably closer on average to the solution Z than when starting at a. It achieves this result by computing and storing m *cutpoints* or pointers

$$I_j := \begin{cases} \min\,[i : q_i > (j-1)/m,\ a \le i \le b] & j = 1, \ldots, m \\ b & j = m + 1 \end{cases} \tag{3.2}$$

beforehand. Then each generation randomly selects $L = \lceil mU \rceil$, the smallest integer greater than or equal to mU, and starts the search at I_L. Algorithm CM describes the essential steps.

Because

$$\mathrm{pr}(I_l \le Z \le I_{l+1} | L = l) = 1,$$

the maximal number of comparisons needed to determine Z for given L is $I_{L+1} - I_L + 1$, and the expected maximal number of comparisons is

$$\begin{aligned} c_m &:= \frac{1}{m}(I_2 - I_1 + 1) + \frac{1}{m}(I_3 - I_2 + 1) + \cdots + \frac{1}{m}(I_{m+1} - I_m + 1) \\ &= \frac{1}{m}(I_{m+1} - I_1 + m) . \end{aligned} \tag{3.3}$$

ALGORITHM CM

Purpose: To generate Z randomly from the distribution $\{0 < q_a < q_{a+1} < \cdots < q_b = 1\}$.

Input: Distribution $\{q_a, \ldots, q_b\}$ and pointers $\{I_1, \ldots, I_m\}$ as in (3.2).

Output: Z.

Source: Fishman and Moore (1984).

Method:

Randomly generate U from $\mathcal{U}(0, 1)$.
$L = \lceil mU \rceil$.
$Z = I_L$.
While $U > q_Z$: $Z = Z + 1$.
Return Z. ■

The algorithm requires a total of $b - a + m + 1$ locations, $b - a + 1$ to store $\{q_a, \ldots, q_b\}$ and m to store $I_1, \ldots, I_m$.

If $m \ge b - a$, then $c_m \le 2$, ensuring that no more than two comparisons are required on average to determine Z. Also,

$$J_m := 1 + Z - I_L$$

gives the total number of comparisons, so that

$$\mathrm{E}J_m = 1 + \mathrm{E}Z - \mathrm{E}I_L = 1 + \mathrm{E}Z - \frac{1}{m}\sum_{j=1}^{m} I_j. \tag{3.4}$$

EXAMPLE 3.4.1 TABLE SAMPLING

Table 3.3 lists c_m and $\mathrm{E}J_m$ for an eight-point distribution. As an illustration, $m = 4$ leads to $I_1 = 1$, $I_2 = 4$, $I_3 = 5$, and $I_4 = 7$. Also, $\mathrm{E}Z = 5.31$ and $\mathrm{E}I_L = 17/4 = 4.25$, so that $\mathrm{E}J_4 = 2.06$. If space is not an issue, one may regard $\mathrm{E}J_m - 1$ as the cost of

TABLE 3.3 | *Example of cutpoint method*

i	p_i	q_i	m	$m+b$	c_m	EJ_m
$a=1$	.01	.01	1	9	8.00	5.31
2	.04	.05	2	10	4.50	3.31
3	.07	.12	3	11	3.33	2.31
4	.15	.27	4	12	2.75	2.06
5	.28	.55	5	13	2.40	1.71
6	.19	.74	6	14	2.17	1.64
7	.21	.95	7	15	2.00	1.45
$b=8$	.05	1.00	8	16	1.88	1.44
			16	24	1.44	1.19

sampling. Algorithm CMSET shows how to compute the cutpoints $I_1, \ldots, I_m$ in a presampling step. ■

ALGORITHM

CMSET

Purpose: To compute pointers for use in Algorithm CM.

Input: Distribution $\{q_a, \ldots, q_b\}$ and number of pointers m.

Output: $I_1, \ldots, I_m$ as in (3.2).

Source: Fishman and Moore (1984).

Method:

$i = 0$, $j = a - 1$ and $A = 0$.
While $i < m$:
  While $A \leq i$:
    $j = j + 1$.
    $A = mq_j$.
  $i = i + 1$.
  $I_i = j$.
Return $I_1, \ldots, I_m$. ■

The *alias method* provides an alternative way to sample from a table in mean time independent of size (e.g., Devroye 1986, Fishman 1996). If the p.m.f. of interest changes after generating each sample, then neither the cutpoint nor the alias method can guarantee a constant generating time.[1]

[1] For a commonly encountered form of changing k-cell p.m.f., Fishman and Yarberry (1993) describe a generating procedure that takes time proportional to $\log_2 k$. Also, see Fishman (1996, Section 3.31).

3.4.1 Restricted Sampling

The cutpoint method offers several additional features that make it attractive in practice. Of these, restricted sampling is one of the most important. Suppose that one wants to generate a sample from the modified distribution $\{r_i := (q_i - q_{a'-1})/(q_{b'} - q_{a'-1}) \colon a \leq a' \leq i \leq b' \leq b\}$. Generating U from $\mathcal{U}(q_{a'-1}, q_{b'})$ instead of from $\mathcal{U}(0, 1)$ in Algorithm CM achieves this result while preserving the efficiency of the method. In particular, once the original tables of $\{q_a, \ldots, q_b\}$ and $\{I_1, \ldots, I_m\}$ are computed and stored, sampling from them with varying $a' < b'$ incurs no additional cost. No other method of sampling from tables offers this advantage.

3.5 COMPOSITION METHOD

This sampling approach relies on *mixtures* of d.f.s to achieve its effectiveness. Let Z denote a random variable on $[a, b]$ with d.f. F that admits the decomposition

$$\begin{aligned} F(z) &= \sum_{i=1}^{r} \theta_i F_i(z), \\ 0 < \theta_i &< 1, \quad i = 1, \ldots, r, \\ \sum_{i=1}^{r} \theta_i &= 1, \end{aligned} \tag{3.5}$$

where $\{F_1(z),\ a \leq a_1 \leq z \leq b_1 \leq b\}, \ldots, \{F_r(z),\ a \leq a_r \leq z \leq b_r \leq b\}$ also denote d.f.s. To generate Z, one performs the steps in Algorithm COMPM.

ALGORITHM COMPM

Purpose: To generate Z randomly from the d.f. $\{F(z),\ a \leq z \leq b\}$ by the composition method.

Input: Mixing p.m.f. $\{\theta_i;\ i = 1, \ldots, r\}$ and d.f.s $\{F_i(z),\ a_i \leq z \leq b_i\}$ for $i = 1, \ldots, r$.

Method:

Randomly generate I from $\{\theta_1, \ldots, \theta_r\}$.
Randomly generate Z from $\{F_I(z),\ a_I \leq z \leq b_I\}$.
Return Z. ■

At least two motivations exist for using this method. First, direct generation from F may be difficult and expensive. Second, generation from $\{\theta_1, \ldots, \theta_r\}$ and the chosen F_i may be simple and inexpensive.

EXAMPLE 3.5.1

SAMPLING FROM A MIXTURE OF TWO EXPONENTIALS

Suppose that

$$F(z) = \alpha(1 - e^{-z/\beta_1}) + (1 - \alpha)(1 - e^{-z/\beta_2})$$
$$= 1 - \alpha e^{-z/\beta_1} - (1 - \alpha)e^{-z/\beta_2}, \quad 0 < \alpha < 1, \quad \beta_1, \beta_2 > 0, \quad z \geq 0.$$

Although the IT method does not easily apply here, the composition method, with

$$\theta_1 := \alpha, \qquad \theta_2 := 1 - \alpha,$$
$$F_1(z) := 1 - e^{-z/\beta_1}, F_2(z) := 1 - e^{-z/\beta_2},$$

does. ■

EXAMPLE 3.5.2

SAMPLING POISSON VARIATES FROM A TABLE

Suppose repeated sampling is required from the Poisson distribution

$$p_z = \begin{cases} \frac{\beta^z e^{-\beta}}{z!} & \beta > 0 \qquad z = 0, 1, \ldots \\ 0 & \text{elsewhere} \end{cases}$$

for some fixed β. We denote this distribution by $\mathcal{P}(\beta)$. An application of the composition method, combined with the cutpoint method, can induce computational efficiency. Define z^* as a positive integer and note the identities

$$p_z = q_{z^*} \frac{p_z}{q_{z^*}} I_{[0,z^*]}(z) + (1 - q_{z^*}) \frac{p_z}{1 - q_{z^*}} I_{(z^*,\infty)}(z)$$

and

$$q_z = q_{z^*} \min\left(1, \frac{q_z}{q_{z^*}}\right) + (1 - q_{z^*}) \max\left(0, \frac{q_z - q_{z^*}}{1 - q_{z^*}}\right),$$

where

$$I_{\mathcal{A}}(x) := \begin{cases} 1 & \text{if } x \text{ is in the set } \mathcal{A} \\ 0 & \text{elsewhere} \end{cases} \tag{3.6}$$

defines the *indicator function*. In terms of the parameters in (3.5),

$$\theta_1 = q_{z^*}, \qquad \theta_2 = 1 - q_{z^*},$$
$$F_1(z) = \min\left(1, \frac{q_z}{q_{z^*}}\right), \quad F_2(z) = \max\left(0, \frac{q_z - q_{z^*}}{1 - q_{z^*}}\right) \qquad z = 0, 1, \ldots$$

Given β, p_{z^*}, and q_{z^*} and using the recursion $p_z/p_{z-1} = \frac{\beta}{z-1}$ for $z > 0$, the following steps randomly generate Z from $\mathcal{P}(\beta)$:

- Randomly sample U from $\mathcal{U}(0, 1)$.
- If $U \leq q_{z^*}$, use U/q_{z^*} with the cutpoint method applied to the d.f. $\left\{\frac{q_1}{q_{z^*}}, \ldots, \frac{q_{z^*-1}}{q_{z^*}}, 1\right\}$ with m cutpoints to generate Z.

TABLE 3.4 | *Cutpoint method applied to the Poisson distribution: m = z**

β	z^*	Upper bound (3.7)
5	10	2.13
	12	2.02
10	18	2.13
	20	2.03
15	25	2.15
	28	2.02

- Otherwise:
  $Z = z^*$, $p = p_{z^*}$, and $q = q_{z^*}$.
  Until $U \leq q$:
  $Z = Z + 1$
  $q = q + p$
  $p = \frac{p\beta}{Z-1}$.
- Return Z.

The mean number of comparisons has the upper bound

$$q_{z^*}\left(1 + \frac{z^*}{m}\right) + \sum_{z=z^*+1}^{\infty} z p_z. \tag{3.7}$$

Table 3.4 lists (3.7) for selected β and z^*. Note that the small average number of comparisons controls round-off error.

There remains the issue of how to choose z^* more generally. One approach relies on the Central Limit Theorem. Since Z from $\mathcal{P}(\beta)$ has mean β and variance β, it is known that $\frac{Z-\beta}{\sqrt{\beta}} \xrightarrow{d} \mathcal{N}(0, 1)$ as $\beta \to \infty$. Choosing z^* close to $\beta + 3\sqrt{\beta}$ should keep the upper bound (3.7) close to 2. That is,

$$\text{pr}(Z \geq \beta + 3\sqrt{\beta}) = 1 - \Phi(3) < 10^{-3}.$$ ■

This sampling approach extends to the other discrete distributions with infinite support. For example, the *negative binomial distribution*

$$p_z = \begin{cases} \frac{\Gamma(r+z)}{\Gamma(z+1)\Gamma(r)}(1-p)^r p^z & 0 < p < 1, \quad r > 0, \quad z = 0, 1, \ldots \\ 0 & \text{elsewhere,} \end{cases} \tag{3.8}$$

denoted by $\mathcal{NB}(r, p)$, admits the recursion

$$p_z = \begin{cases} (1-p)^r & z = 0 \\ \frac{r+z-1}{z} p_{z-1} & z = 1, 2, \ldots . \end{cases} \tag{3.9}$$

Also, it has mean $\mu = \frac{rp}{1-p}$ and variance $\sigma^2 = \frac{rp}{(1-p)^2}$. Because it can be shown that $\frac{Z-\mu}{\sigma} \xrightarrow{d} \mathcal{N}(0, 1)$ as $r \to \infty$, z^* in the neighborhood of $\frac{rp}{1-p} + 3\sqrt{\frac{rp}{(1-p)^2}}$ is worth considering.

3.6 ACCEPTANCE-REJECTION METHOD

Sometimes the inverse d.f. of a distribution is unavailable in closed form, and no efficient code exists for a numerical inversion. Then the *acceptance-rejection (AR) method* and the *ratio-of-uniforms (ROU) method* (see Section 3.7) provide alternative generating techniques that are also theoretically exact. However, the sample generating time for each is now itself random, thus providing examples of *probabilistic algorithms*. These methods have many nuances, which we describe here and in Section 3.7.

Theorem 3.6.1

(AR method) Let f and η be distributions on $\mathcal{X}$ with the property that

$$\kappa(f, \eta) := \max_{x \in \mathcal{X}} \frac{f(x)}{\eta(x)} \tag{3.10}$$

exists. Then the iteration:

Repeat:

sample X from η and U from $\mathcal{U}(0, 1)$

until $U \leq \frac{f(X)}{\kappa(f,\eta)\eta(X)}$.

randomly generates X from f. ■

PROOF Let S denote the event $U \leq \frac{f(X)}{\kappa(f,\eta)\eta(X)}$ and $\{k(x \mid \mathrm{S})\}$, the conditional distribution of X given S. Because

$$k(x \mid \mathrm{S}) = \frac{\mathrm{pr}(\mathrm{S} \mid X = x)\eta(x)}{\mathrm{pr}(\mathrm{S})}, \tag{3.11}$$

where

$$\mathrm{pr}(\mathrm{S} \mid X = x) = \frac{f(x)}{\kappa(f, \eta)\eta(x)},$$

then

$$\text{pr(S)} = \frac{1}{\kappa(f, \eta)},$$

implying, upon substitution into (3.11), that $k(x \mid \text{S}) = f(x)$. ■

Let N denote the number of trials until the first success. Then the mean number of required trials to generate X from f is

$$\text{E}N = 1 + \sum_{i=0}^{\infty} i\,[1 - \text{pr(S)}]^i\, \text{pr(S)} = \kappa(f, \eta).$$

Algorithm A-R describes the steps.

ALGORITHM A-R

Purpose: To sample Z randomly from the p.d.f. f on $[a, b]$, using the acceptance-rejection method.

Given: p.d.f.s f and η on $[a, b]$, $\kappa(f, \eta) := \max_z \frac{f(z)}{\eta(z)} < \infty$, and

$$\nu(z) := \frac{f(z)}{\eta(z)\kappa(f, \eta)} \in [0, 1] \qquad a \le z \le b. \tag{3.12}$$

Method:

Repeat:
  randomly sample Y from η
  randomly sample U from $\mathcal{U}(0, 1)$
until $U \le \nu(Y)$.
Return $Z = Y$. ■

Recall that the inverse d.f. of the standard normal distribution (2.11) is not available in closed form.[2] The next example shows how the AR method works for $\mathcal{N}(0, 1)$ and sets the stage for our discussion of the *squeeze method* in Section 3.6.1 and its illustration in Example 3.6.2.

EXAMPLE 3.6.1 A-R NORMAL GENERATION

Let

$$f(z) = \sqrt{\frac{2}{\pi}} e^{-z^2/2} \qquad z \ge 0. \tag{3.13}$$

[2]However, there are computationally efficient C-language implementations of the IT method for generating a sample from $\mathcal{N}(0, 1)$, as in Moro (1995).

Then Algorithm A-R with

$$\begin{aligned}\eta(z) &= \mathrm{e}^{-z}\\ \nu(z) &= \mathrm{e}^{(z-1)^2/2}\\ \kappa(f,\eta) &= \sqrt{\frac{2\mathrm{e}}{\pi}} = 1.3155\end{aligned}$$

generates a sample from the half-normal distribution f on $[0, \infty)$. If we then sample V from $\mathcal{U}(0, 1)$, the quantity $Z\left[2I_{[0,1/2]}(V) - 1\right]$ is from $\mathcal{N}(0, 1)$, where (3.6) defines the indicator function I. ■

3.6.1 Squeeze Method

On average, success in the half-normal example requires 1.3155 exponential function evaluations per trial. Reducing this number can lead to a significant reduction in computing time. The *squeeze method* allows one to do this (e.g., Marsaglia 1977). Suppose there exist simple functions $\{\nu_L(z), a \le z \le b\}$ and $\{\nu_U(z), a \le z \le b\}$ such that

$$\nu_L(z) \le \nu(z) \le \nu_U(z) \quad a \le z \le b, \tag{3.14}$$

where (3.12) defines $\{\nu(z)\}$. Here the modifier "simple" means that the time to evaluate $\nu_L(z)$ and $\nu_U(z)$ is less than the time to evaluate $\nu(z)$. For some z, note that $\nu_L(z)$ may assume negative values and $\nu_U(z)$ may exceed unity. Algorithm ARS shows how to use these bounding functions to reduce computing time. On each trial, the algorithm leads to

$$\text{pr}[\text{evaluating } \nu(Z)] = \text{pr}[\nu_L(Z) \le U \le \nu_U(Z)].$$

The checks, $U \le \nu_L(Z)$ and $U > \nu_U(Z)$, are sometimes called *preliminary tests* or *pretests*.

ALGORITHM ARS

Purpose: To generate Z randomly from the p.d.f. $\{f(z),\ a \le z \le b\}$ using the acceptance-rejection and squeeze methods.

Input: p.d.f. $\{\eta(z),\ a \le z \le b\}$, acceptance function $\{\nu(z),\ a \le z \le b\}$ and bounding functions $\{\nu_L(z),\ a \le z \le b\}$ and $\{\nu_U(z),\ a \le z \le b\}$.

Method:

Repeat:
  randomly generate U from $\mathcal{U}(0, 1)$
  randomly generate Z from $\{\eta(z)\}$
  if $U \le \nu_L(Z)$, accept

otherwise:
    if $U \leq \nu_U(Z)$: if $U \leq \nu(Z)$, accept
    otherwise, reject
until accept.
Return Z.

■

Let S denote the event $U \leq \nu(Z)$, S_L the event $U \leq \nu_L(Z)$, and S_U the event $U \leq \nu_U(Z)$. Then

$$\left.\begin{aligned} \mathrm{pr}(\mathrm{S}_L) &= \int_a^b \max[0, \nu_L(z)]\eta(z)dz \\ \mathrm{pr}(\mathrm{S}_U) &= \int_a^b \min[1, \nu_U(z)]\eta(z)dz, \end{aligned}\right\} \tag{3.15}$$

so that on a single trial,

$$\mathrm{pr}[\nu(Z) \text{ is evaluated}] = \mathrm{pr}(\mathrm{S}_U) - \mathrm{pr}(\mathrm{S}_L). \tag{3.16}$$

If N denotes the number of evaluations of $\nu(Z)$ until success when employing the squeeze method, then

$$\mathrm{E}N = \mathrm{pr}\left(\bar{\mathrm{S}}_L, \mathrm{S}_U|\bar{\mathrm{S}}\right)\left[\frac{1}{\mathrm{pr(S)}} - 1\right] + \mathrm{pr}\left(\bar{\mathrm{S}}_L, \mathrm{S}_U|\mathrm{S}\right), \tag{3.17}$$

where $\bar{\mathrm{S}}_L$ and $\bar{\mathrm{S}}$ denote the events $U > \nu_L(Z)$ and $U > \nu(Z)$, respectively. In the absence of squeezing, the mean number of evaluations of $\nu(Z)$ is

$$[\mathrm{pr(S)}]^{-1} = \left[\int_a^b \nu(z)\eta(z)dz\right]^{-1} = \kappa(f, \eta).$$

Because

$$\left.\begin{aligned} \mathrm{pr}\left(\bar{\mathrm{S}}_L, \mathrm{S}_U|\bar{\mathrm{S}}\right) &= \tfrac{\mathrm{pr}(\mathrm{S}_U)-\mathrm{pr(S)}}{1-\mathrm{pr(S)}} \\ \mathrm{pr}\left(\bar{\mathrm{S}}_L, \mathrm{S}_U|\mathrm{S}\right) &= \tfrac{\mathrm{pr(S)}-\mathrm{pr}(\mathrm{S}_L)}{\mathrm{pr(S)}}, \end{aligned}\right\} \tag{3.18}$$

it follows that

$$\mathrm{E}N = \frac{\mathrm{pr}(\mathrm{S}_U) - \mathrm{pr}(\mathrm{S}_L)}{\mathrm{pr(S)}}. \tag{3.19}$$

Using the inequalities (see Exercise 31)

$$1 - x \leq e^{-x} \leq 1 - x + x^2/2 \qquad x \geq 0 \tag{3.20}$$

reduces the frequency of exponential evaluation. Using the inequalities

$$\left.\begin{aligned} 1 - 1/x &\leq \ln x \leq -1 + x & x &\geq 0 \\ x - 1/x &\leq 2\ln x \leq -3 + 4x - x^2 & 0 &< x \leq 1 \end{aligned}\right\} \tag{3.21}$$

reduces the frequency of logarithmic evaluation.

EXAMPLE 3.6.2

A-R HALF-NORMAL GENERATION WITH SQUEEZING

For the half-normal distribution in Example 3.6.1 with $\nu(z) = e^{-(z-1)^2/2}$, inequalities (3.20) lead to the lower and upper bounds

$$\nu_L(z) = 1 - (z-1)^2/2 \tag{3.22}$$

$$\nu_U(z) = 1 - (z-1)^2/2 + (z-1)^4/8$$

$$= \nu_L(z) + (z-1)^4/8 \tag{3.23}$$

respectively. Moreover,

$$\text{pr}(S_L) = 1/2 + (1+\sqrt{2})e^{-(1+\sqrt{2})} = 0.7159$$

$$\text{pr}(S_U) = 13/8 - 16e^{-3} = 0.8284 \tag{3.24}$$

$$\text{pr[evaluating } \nu(Z) \text{ on a trial]} = \text{pr}(S_U) - \text{pr}(S_L) = 0.1125$$

and the mean number of evaluations of $\nu(Z)$ until success is $EN = 0.1480$. Incorporating the squeeze method into the acceptance-rejection method in this example reduces the mean number of evaluations of $\nu(Z)$ from 1.3155 to 0.1340. Of course, some of these savings are expended in the two additional pretests for S_L and $\bar{S}_U$. ■

In principle, acceptance-rejection sampling also applies to multivariate generation. Section 6.18 returns to this topic in the setting of Hastings-Metropolis sampling.

3.6.2 Adaptive Methods

For a given target distribution f, the sample generating cost of acceptance-rejection sampling depends crucially on the choice of nominating distribution η. It determines $\kappa(f,\eta)$, the mean number of trials until success, and is a principal contributor to the cost of evaluating $\frac{f(Z)}{\eta(Z)}$. Although judiciously chosen squeeze functions can reduce the latter cost, large $\kappa(f,\eta)$ implies that many trials are needed to get a sample from f. Since there is nothing sacred about the choice of η, it is possible on successive rejections to reduce $\kappa(f,\eta)$ by updating η and to reduce the frequency of evaluating $\frac{f(Z)}{\eta(Z)}$ by updating the squeeze functions. This is done by making use of past acceptance-rejection experience. Collectively called *adaptive rejection sampling*, the only requirement is that $\{\log f(z),\ a \le z \le b\}$ be a concave function. That is, $\frac{\partial^2 \log f(z)}{\partial z^2} < 0$ for $a \le z \le b$ (e.g., Gilks and Wild 1992, Wild and Gilks 1993). This property frequently holds.

An adaptive algorithm reduces the time spent prior to any generation in finding suitable η and squeeze functions. From the user's point of view, that is a virtue. It can be especially effective when f is not a commonly encountered distribution and $\kappa(f,\eta)$ is large. However, it is not clear that an adaptive algorithm is computationally more efficient than an approach that devotes a pre-analysis to finding suitable η and squeeze functions to induce low sample generating cost. This uncertainty arises from the fact that although an adaptive algorithm usually hones in on a low mean required number of trials for success, it also incurs a higher cost of updating η and squeeze functions on each rejection.

3.7 RATIO-OF-UNIFORMS METHOD

As with the acceptance-rejection method, the *ratio-of-uniforms method* relies on a random number of independent trials until success to generate a sample from a specified distribution. Algorithms based on this approach are among the most computationally efficient for sampling from the Beta, Gamma, binomial, hypergeometric, negative binomial, and Poisson distributions (e.g., Cheng 1978; Cheng and Feast 1979; Ahrens and Dieter 1991; and Stadlober 1989, 1991; also see Fishman 1996).

The method generates $Z = a_1 + a_2 Y/X$, where (X, Y) is uniformly distributed over a bounded region for which a_1, a_2, and the shape of the region depends on $\{f(z)\}$. In principle, the work to generate (X, Y) is considerably less, thus potentially reducing the computing time required to obtain a sample from $\{f(z)\}$. Theorem 3.7.1 describes a generalization of the original proposal of Kinderman and Monahan (1977).

Theorem 3.7.1

Let $\{r(z), -\infty < z < \infty\}$ denote a non-negative integrable function, and let $\mathcal{Q}$ denote the region

$$\mathcal{Q} := \mathcal{Q}(a_1, a_2) = \{(x, y)\colon 0 \le x \le r^{\frac{1}{2}}(a_1 + a_2 y/x), \quad -\infty < a_1 < \infty,\ a_2 > 0\}. \tag{3.25}$$

Let (X, Y) be uniformly distributed on a bounded region $\mathcal{P} \supseteq \mathcal{Q}$ and set $W = X^2$ and $Z = a_1 + a_2 Y/X$. If $W \le r(Z)$, then Z has the distribution

$$f_Z(z) = r(z)/w, \quad -\infty < z < \infty, \tag{3.26}$$

where

$$w := 2a_2|\mathcal{Q}| = \int_{-\infty}^{\infty} r(z)\mathrm{d}z. \tag{3.27}$$

■

Change of Variables. The proof of Theorem 3.7.1 employs a *change of variables* to facilitate integration. Its many uses in the remainder of this book, especially in the hands-on exercises of Chapter 5, make it worthwhile to describe the formalism in some detail. We do so for two variables but note that the concept readily extends to higher-dimensional problems.

Let k denote a continuous and differentiable function defined on the two-dimensional region $\mathcal{A}$, and suppose it is of interest to evaluate the integral

$$I = \int_{\mathcal{A}} k(x, y)\mathrm{d}x\mathrm{d}y.$$

When this cannot readily be done with respect to the $\mathcal{A}$ space, a change of variables can sometimes lead to evaluation on a transformed space. Suppose there exist functions

$$x = k_1(u, v) \quad \text{and} \quad y = k_2(u, v)$$

that lead to a mapping of a region $\mathcal{A}$ of the x-y plane into a region $\mathcal{B}$ of the u-v plane. Under suitable restrictions on the functions k_1 and k_2, we have

$$\begin{aligned} I &= \int_{\mathcal{A}} k(x, y)\mathrm{d}x\mathrm{d}y \\ &= \int_{\mathcal{B}} k(k_1(u, v), k_2(u, v))\frac{\partial(x, y)}{\partial(u, v)}\mathrm{d}u\mathrm{d}v, \end{aligned}$$

where

$$\frac{\partial(x, y)}{\partial(u, v)} := \left|\frac{\partial x}{\partial u}\frac{\partial y}{\partial v} - \frac{\partial x}{\partial v}\frac{\partial y}{\partial u}\right|$$

is called the *Jacobian* of the transformation.

PROOF If $(X, Y) \in \mathcal{Q}$, then X and Y have the joint p.d.f.

$$f_{X,Y}(x, y) = 1/|\mathcal{Q}|, \qquad (x, y) \in \mathcal{Q}.$$

Then by a standard change of variables, we have

$$f_{W,Z}(w, z) = J\, f_{X,Y}(\sqrt{w}, (z - a_1)\sqrt{w}/a_2), \quad 0 \le w \le r(z), \qquad -\infty < z < \infty,$$

J being the *Jacobian*

$$J := \left|\frac{\partial x}{\partial w}\frac{\partial y}{\partial z} - \frac{\partial x}{\partial z}\frac{\partial y}{\partial w}\right| = \frac{1}{2a_2}.$$

Then

$$f_Z(z)\mathrm{d}z = \int_0^{r(z)} f_{W,Z}(w, z)\mathrm{d}w = \frac{r(z)}{2a_2|\mathcal{Q}|}, \ -\infty < z < \infty,$$

and

$$\int_{-\infty}^{\infty} f_Z(z)\mathrm{d}z = \frac{1}{2a_2|\mathcal{Q}|}\int_{-\infty}^{\infty} r(z)\mathrm{d}z = 1.$$

■

The equations

$$\begin{aligned} x(z) &= r^{\frac{1}{2}}(z) \\ y(z) &= (z - a_1)x(z)/a_2 = (z - a_1)r^{\frac{1}{2}}(z)/a_2 \end{aligned}$$

define the boundary of $\mathcal{Q}$, and for specified a_1 and a_2, every point in $\mathcal{Q}$ lies in the rectangle

$$\mathcal{R} := \{(x, y) : 0 \le x \le x^*,\ y_* \le y \le y^*\}, \tag{3.28}$$

where

$$\left.\begin{aligned} x^* &:= \max_z r^{\frac{1}{2}}(z), \\ y_* &:= \min_z[(z-a_1)r^{\frac{1}{2}}(z)/a_2], \\ y^* &:= \max_z[(z-a_1)r^{\frac{1}{2}}(z)/a_2]. \end{aligned}\right\} \tag{3.29}$$

In order for y^* to be finite, $\{z^2 r(z)\}$ must be integrable.

The motivation behind the ratio-of-uniforms method is now clear. Suppose one sets $r(z) = wf(z)$ and generates (X, Y) uniformly on $\mathcal{R}$. If $X \le r^{\frac{1}{2}}(a_1 + a_2 Y/X)$, then $(X, Y) \in \mathcal{Q}$, and $Z = a_1 + a_2 Y/X$ has the desired p.d.f. $\{f(z)\}$. Algorithm ROU describes the steps.

ALGORITHM **ROU**

Purpose: To generate Z randomly from the p.d.f. $\{f(z),\ -\infty < z < \infty\}$ using the ratio-of-uniforms method.

Input: Function $\{r(z) = wf(z), -\infty < z < \infty\}$, constants $a_1 \in (-\infty, \infty)$ and $a_2 \in (0, \infty)$.

Setup: $x^* = \max_z r^{\frac{1}{2}}(z)$, $y_* = \min_z[(z-a_1)r^{\frac{1}{2}}(z)/a_2]$, and $y^* = \max_z[(z-a_1)r^{\frac{1}{2}}(z)/a_2]$.

Method:

Repeat:
  randomly generate X from $\mathcal{U}(0, x^*)$
  randomly generate Y from $\mathcal{U}(y_*, y^*)$
  $Z = a_1 + a_2 Y/X$
until $X^2 \le r(Z)$.
Return Z.

■

Let T denote the event $X^2 \le r(Z)$. Because pr(T) = pr$[X \le r^{\frac{1}{2}}(a_1 + a_2 Y/X)] =$ pr$[(X, Y) \in \mathcal{Q}] = |\mathcal{Q}|/|\mathcal{R}|$ and because trials are independent, the mean number of trials until success is

$$\frac{|\mathcal{R}|}{|\mathcal{Q}|} = 2w^{-1}a_2 x^*(y^* - y_*). \tag{3.30}$$

Generating (X, Y) from $\mathcal{R}$ usually takes considerably less time than the time generating Z from $\{\eta(z)\}$ using the acceptance-rejection method. Although the inclusive region $\mathcal{R}$ may assume a shape other than rectangular in order to make the ratio $|\mathcal{R}|/|\mathcal{Q}|$ closer to unity, one must balance this reduction against the potentially greater cost of generating (X, Y) from a less simple region. Example 3.7.4 shows how the judicious choice of a_1 and a_2 allows one to sample from a simple region to generate samples from the Poisson distribution with bounded mean computing time.

EXAMPLE 3.7.1

ROU HALF-NORMAL GENERATION

An example based on the half-normal distribution (3.13) illustrates the features of the ratio-of-uniforms method. Here we take $a_1 = 0$ and $a_2 = 1$, so that

$$\begin{aligned} \mathcal{Q} &= \{(x, y) : 0 \le x \le e^{-(y/2x)^2}\}, \\ r(z) &= e^{-z^2/2}, \\ |\mathcal{Q}| &= \frac{1}{2}\int_0^\infty e^{-z^2/2} dz = \frac{1}{2}\sqrt{\pi/2}, \\ x^* &= 1,\ y_* = 0,\ y^* = \sqrt{2/e}, \end{aligned}$$

and

$$\frac{|\mathcal{R}|}{|\mathcal{Q}|} = \frac{4}{\sqrt{\pi e}} = 1.3688.$$

■

The acceptance test in Algorithm ROU takes the form

$$X \le e^{-(Y/2X)^2},$$

which is equivalent to

$$Y^2 \le -4X^2 \ln X. \tag{3.31}$$

The acceptance probability

$$\text{pr}(T) = \frac{|\mathcal{Q}|}{|\mathcal{R}|} = \sqrt{\pi e}/4 = 0.7306 \tag{3.32}$$

is close to the acceptance probability $\text{pr}(S) = \sqrt{\pi/2e} = 0.7602$ in Example 3.6.1.

EXAMPLE 3.7.2

ROU HALF-NORMAL GENERATION WITH SQUEEZING

Although (3.31) eliminates the need for exponential generation and exponential function evaluation on each trial, it does require a logarithmic function evaluation. Pretesting based on the inequalities (3.21) for $0 < z \le 1$ reduces the frequency of this evaluation. Let T_L and T_U denote the respective events

$$Y^2 \le 4X^2(1 - X) \tag{3.33}$$

$$Y^2 \le 4X(1 - X). \tag{3.34}$$

One accepts $Z = Y/X$ if T_L occurs on the first pretest but rejects Z if T_U does not occur on the second pretest, where

$$\begin{aligned}\text{pr}(\text{T}_L) &= \frac{1}{\sqrt{2/\text{e}}}\int_0^1\int_0^{2x\sqrt{1-x}} \text{d}y\,\text{d}x \\ &= \sqrt{2\text{e}}\Gamma(\tfrac{3}{2})/\Gamma(\tfrac{7}{2}) = 4(2\text{e})^{\frac{1}{2}}/15 \\ &= 0.6218\end{aligned} \tag{3.35}$$

and

$$\begin{aligned}\text{pr}(\text{T}_U) &= \frac{1}{\sqrt{2/\text{e}}}\int_0^1\int_0^{\sqrt{\min[2/\text{e},4x(1-x)]}} \text{d}y\text{d}x \\ &= \frac{1}{2}\sqrt{1-2/\text{e}} + \frac{\sin^{-1}\left(-\sqrt{1-2/\text{e}}\right)+\pi/2}{2\sqrt{2/\text{e}}} \\ &= 0.8579.\end{aligned} \tag{3.36}$$

Then

$$\text{pr}[r(Y/X) \text{ is evaluated}] = \text{pr}(\text{T}_U) - \text{pr}(\text{T}_L) = 0.2361.$$

If M denotes the number of evaluations of $r(Y/X)$ until success, then

$$\begin{aligned}\text{E}M &= \frac{\text{pr}(\text{T}_U)-\text{pr}(\text{T})}{1-\text{pr}(\text{T})}\left[\frac{1}{\text{pr}(\text{T})}-1\right] + \frac{\text{pr}(\text{T})-\text{pr}(\text{T}_L)}{\text{pr}(\text{T})} \\ &= \frac{\text{pr}(\text{T}_U)-\text{pr}(\text{T}_L)}{\text{pr}(\text{T})} = 0.3232.\end{aligned} \tag{3.37}$$

This result follows in a manner analogous to that for EN in (3.19) for the acceptance-rejection method.

In summary, incorporating the squeeze method into the ratio-of-uniforms method for generating a sample from the half-normal distribution reduces the mean number of logarithmic function evaluations from 1.3688 to 0.3232. Although the two pretests consume part of this time saving, one should note that $100 \times (1 - 0.3232/1.3688) = 76.39\%$ of the time, no logarithmic or exponential function evaluations are required, whereas the acceptance-rejection method in Section 3.6.1 requires one exponential deviate generation on every trial, regardless of whether the squeeze method is used. ■

EXAMPLE 3.7.3

GAMMA GENERATION WITH BOUNDED MEAN COMPUTING TIME

Although normal and half-normal generation illustrate the most essential features of the ROU and squeeze methods, other normal generation techniques are known to be more computationally efficient. However, the ROU method combined with squeezing is competitive for sampling from many other distributions. This example describes the subtle issues that arise when devising an efficient algorithm for (continuous) Gamma generation, and Example 3.7.4 does the same for (discrete) Poisson generation. For a more complete review of the ROU method combined with squeezing for the Beta, binomial, hypergeometric, and negative binomial, see Stadlober (1991). Also, see Fishman (1996).

Let Z denote a random variable having the Gamma p.d.f.

$$f(z) = \begin{cases} \frac{z^{\alpha-1}}{\beta^{\alpha}\Gamma(\alpha)} e^{-z/\beta} & \alpha, \beta > 0,\ z \geq 0 \\ 0 & \text{elsewhere,} \end{cases} \tag{3.38}$$

denoted by $\mathcal{G}(\alpha, \beta)$. Relevant properties include:

1. If $Z_1, \ldots, Z_n$ are independent from $\mathcal{G}(\alpha_1, \beta), \ldots, \mathcal{G}(\alpha_n, \beta)$, respectively, then $Z = Z_1 + \cdots + Z_n$ is from $\mathcal{G}(\alpha_1 + \cdots + \alpha_n, \beta)$.
2. If $\alpha = 1$, then Z has the exponential distribution $\mathcal{E}(\beta)$.
3. If α is an integer, then Z is the sum of α i.i.d. exponential random variables from $\mathcal{E}(\beta)$, and its distribution is called the *Erlang distribution* with α phases.
4. If α is an integer and $\beta = 2$, then Z has the chi-squared distribution with 2α degrees of freedom.
5. If Y is from $\mathcal{G}(\alpha, 1)$, then $Z = \beta Y$ is from $\mathcal{G}(\alpha, \beta)$.
6. If $\alpha < 1$, then $\{f(z)\}$ is unbounded at $z = 0$.

Without loss of generality, we hereafter take $\beta = 1$. ■

Unlike the distributions in Table 3.1, Gamma generating time can vary with the shape parameter α. For example, if α is an integer, then Z can be viewed as the sum $Z = Z_1 + \cdots + Z_\alpha$, where $Z_1, \ldots, Z_\alpha$, are independent identically distributed random variables from $\mathcal{E}(1)$. This suggests

> $Z = 0$ and $i = 1$.
> Repeat:
>   randomly generate Z_i from $\mathcal{E}(1)$
> $Z = Z + Z_i$
> $i = i + 1$
> until $i > \alpha$.
> Return Z.

This procedure takes time proportional to α and can be costly in an MC sampling experiment with large α. We now describe algorithms for Gamma generation that are known to be computationally efficient. In particular, Algorithms GKM2 and GKM3 lead to bounded mean sample generating times as $\alpha \to \infty$.

$\alpha \leq 1$. If $\alpha < 1$, then f in (3.38) is unbounded at $z = 0$ and the ROU method to be described shortly for $\alpha > 1$ does not apply. Therefore, for completeness we begin by describing a bounded mean sample generating time method that applies for $\alpha < 1$.

In their Algorithm GS, Ahrens and Dieter (1974) generate Z from $\mathcal{G}(\alpha, 1)$ with $0 < \alpha \leq 1$, using a combination of the acceptance-rejection, composition, and IT methods with

$$\begin{aligned} \eta(z) &= \begin{cases} \frac{e\alpha z^{\alpha-1}}{\alpha+e} & \text{if } 0 \leq z \leq 1 \\ \frac{\alpha e^{-z+1}}{\alpha+e} & \text{if } z > 1, \end{cases} \\ \nu(z) &= \begin{cases} e^{-z} & \text{if } 0 \leq z \leq 1 \\ z^{\alpha-1} & \text{if } z > 1, \end{cases} \\ \kappa(f, \eta) &= (\alpha + e)/\alpha e\Gamma(\alpha), \end{aligned} \tag{3.39}$$

where $\{\eta(z)\}$, $\{\nu(z)\}$, and $\kappa(f, \eta)$ are defined in Algorithm A-R. Observe that $f(z) = \kappa(f, \eta)\nu(z)\eta(z)$ gives the Gamma p.d.f. Sampling from the p.d.f. $\{\eta(z)\}$ uses the composition method. The composition method determines whether to generate Z with probability $\frac{e}{\alpha+e}$ from the d.f. $\{z^\alpha, 0 \leq z \leq 1\}$ or with probability $\frac{\alpha}{\alpha+e}$ from the d.f. $\{1 - e^{-z+1}, 1 < z < \infty\}$ by the IT method. Based on this selection, the acceptance-rejection method uses η to determine whether Z is from $\mathcal{G}(\alpha, 1)$. The mean number of trials until success, $\kappa(f, \eta)$, has the upper bound 1.39. Algorithm GS* in Fishman (1996) describes the steps.

$\boldsymbol{\alpha > 1}$**.** We describe Gamma generation via the ratio-of-uniforms method (Cheng and Feast 1979). Based on (3.29), the acceptance region is

$$\mathcal{Q}(\alpha) = \{0 \leq x \leq r^{\frac{1}{2}}(y/x) := [(y/x)^{\alpha-1}e^{-y/x}]^{\frac{1}{2}}\}, \tag{3.40}$$

implying that $\mathcal{Q}(\alpha)$ lies in the rectangle $\mathcal{R}(\alpha) := \{0 \leq x \leq x^* := [(\alpha - 1)/e]^{(\alpha-1)/2}$, $0 \leq y \leq y^* := [(\alpha + 1)/e]^{(\alpha+1)/2}\}$. Sampling (X, Y) uniformly from $\mathcal{R}$ would lead to a mean number of trials until success

$$c(\alpha) := \frac{|\mathcal{R}(\alpha)|}{|\mathcal{Q}(\alpha)|} = 2(\alpha + 1)^{(\alpha+1)/2}(\alpha - 1)^{(\alpha-1)/2}/\Gamma(\alpha)e^\alpha, \tag{3.41}$$

for which

$$\alpha^{-1/2}c(\alpha) \to 2/(2\pi)^{\frac{1}{2}} \text{ as } \alpha \to \infty, \tag{3.42}$$

revealing a mean generating time proportional to $\alpha^{\frac{1}{2}}$. In addition, x^* and y^* need to be computed.

To eliminate the asymptotic dependence on α, Cheng and Feast employ a parallelogram $\mathcal{R}'(\alpha)$, as in Figure 3.1, to enclose $\mathcal{Q}(\alpha)$. It leads to a mean number of trials until success

$$c'(\alpha) := \frac{|\mathcal{R}'(\alpha)|}{|\mathcal{Q}(\alpha)|} = \frac{2x^*y^*(1 + \sqrt{2/e})}{\Gamma(\alpha)\sqrt{\alpha}}, \tag{3.43}$$

where

$$c'(\alpha) \to 1.482 \quad \text{as } \alpha \to \infty.$$

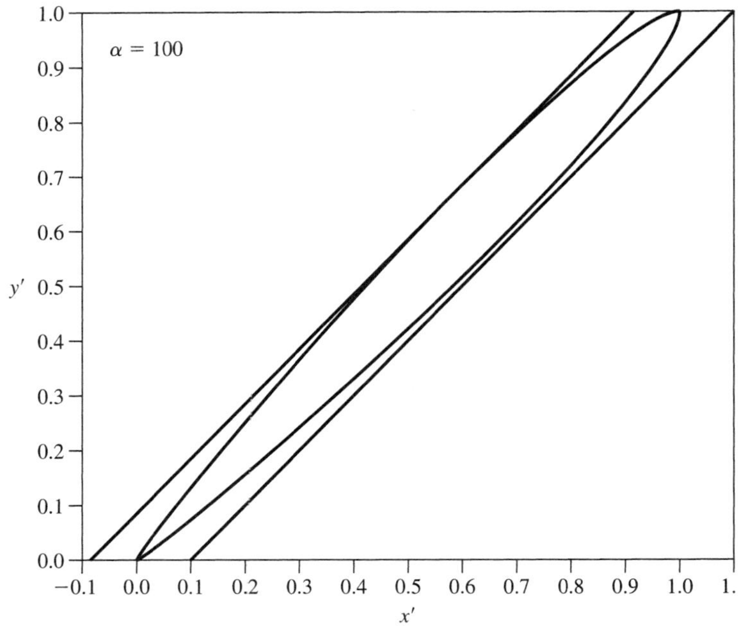

FIGURE 3.1 | *Gamma generation: Ratio-of-uniforms method* $x' = x/x^*$ *and* $y' = y/y^*$

This limit obtains because as α increases, the diagonal parallel sides of the enclosing parallelogram converge to a single diagonal line.

There are additional issues that call for attention in making the ROU method computationally efficient for Gamma generation. Table 3.5 reveals that the square $\mathcal{R}(\alpha)$ has an advantage for $\alpha \leq 3$, whereas for $\alpha \geq 4$, the parallelogram $\mathcal{R}'(\alpha)$ has an ever-increasing advantage as $\alpha \to \infty$. This suggests a procedure that uses the square for $\alpha \leq 3$ and the parallelogram for $\alpha \geq 4$. However, an additional issue plays a role.

Recall that if (X', Y') is uniformly distributed in the unit square $[0, 1]^2$, then $(X = x^*X', Y = y^*Y')$ is uniformly distributed on $\mathcal{R}(\alpha)$. If generation from the same $\mathcal{G}(\alpha, 1)$ is to occur repeatedly, then one can compute x^* and y^* once at the beginning, store them, and use these values as input to the algorithm on each call. If many distinct α's are used, this approach may not be feasible, and without modification one would repeatedly incur the relatively high setup cost of computing x^* and y^*. A modification of the acceptance region substantially reduces this cost at a relatively small increase in the mean number of trials until success for small α.

Let (X', Y') be uniformly distributed on $[0, 1]^2$. Let $X = x^*X'$ and $Y = y^{**}Y'$, where $y^{**} \geq y^*$. Then (X, Y) is uniformly distributed on the rectangle

$$\mathcal{R}_1(\alpha) := \{(x, y) : 0 \leq x \leq x^*, \ 0 \leq y \leq y^{**}\}, \tag{3.44}$$

TABLE 3.5 | *Gamma variate generation: Ratio-of-uniforms method*

	Mean number of trials until success		
α	$c(\alpha)$	$c'(\alpha)$	Ratio
1.5	1.331	2.019	.6592
2	1.406	1.848	.7608
2.5	1.499	1.761	.8512
3	1.593	1.709	.9321
4	1.773	1.647	1.077
5	1.941	1.612	1.204
6	2.096	1.590	1.318
7	2.241	1.574	1.424
8	2.378	1.562	1.522
9	2.507	1.553	1.614
10	2.631	1.545	1.703
15	3.177	1.524	2.085
20	3.643	1.514	2.406
25	4.056	1.507	2.691
30	4.431	1.503	2.948
35	4.777	1.500	3.185
40	5.099	1.498	3.404
45	5.402	1.496	3.611
50	5.689	1.495	3.805
∞	∞	1.482	∞

where $\mathcal{Q}(\alpha) \subseteq \mathcal{R}(\alpha) \subseteq \mathcal{R}_1(\alpha)$. Let $W = X^2$ and $Z = Y/X$. If $W \le r(Z)$, then Z is from $\mathcal{G}(\alpha, 1)$. This last inequality implies

$$\frac{2}{\alpha - 1} \ln X' \le \ln(bY'/X') - bY'/X' + 1, \tag{3.45}$$

where $b := y^{**}/x^*(\alpha - 1)$, suggesting that one merely has to take $b \ge y^*/x^*(\alpha - 1) = (\alpha + 1)/(\alpha - 1)^{(\alpha+1)/2}/\mathrm{e}$ and $Z = (\alpha - 1)bY'/X'$. Cheng and Feast suggest

$$b := (\alpha - 1/6\alpha)/(\alpha - 1), \tag{3.46}$$

which ensures $b \ge y^*/x^*(\alpha - 1)$ and eliminates the need to evaluate x^*, y^*, and y^{**}.

Because W and Z have the joint p.d.f.

$$f_{W,Z}(w, z) = w/b(\alpha - 1), \qquad 0 \le z \le b(\alpha - 1)/w, \qquad 0 \le w \le 1,$$

it follows that

$$\text{pr}[W \le r(Z)] = \frac{|\mathcal{Q}(\alpha)|}{|\mathcal{R}_1(\alpha)|} = \frac{\Gamma(\alpha)e^{\alpha-1}}{2b(\alpha-1)^{\alpha}}$$
$$= \frac{\Gamma(\alpha)e^{\alpha-1}}{2(\alpha - 1/6\alpha)(\alpha-1)^{\alpha-1}},$$

and Z, given $W \le r(Z)$ has the desired Gamma p.d.f. Because X' and Y' are uniformly distributed over $[0, 1]^2$, the mean number of trials is $2(\alpha/2\pi)^{1/2}$ as $\alpha \to \infty$, which agrees with the limit (3.42).

Because $\mathcal{R}'(\alpha) \cap [0, 1]^2 \subset \mathcal{R}(\alpha)$, this alteration also applies to sampling from $\mathcal{R}'(\alpha)$. While preserving the validity of the ratio-of-uniforms approach to Gamma generation for $\alpha > 1$, the change merely scales $c(\alpha)$ and $c'(\alpha)$ by $b(\alpha-1)x^*/y^*$, which decreases from 1.133 at $\alpha = 1$ to unity as $\alpha \to \infty$.

Algorithm GKM1 describes the generating method when (X', Y') are randomly generated on the unit square $\mathcal{R}(\alpha)$. Note the pretest $mX' - d + V + V^{-1} \le 0$, which is based on the inequality $\ln w \le w - 1$ for $w > 0$. This pretest reduces the frequency of logarithmic evaluation (Exercise 32). Whenever possible, evaluating a, b, m, and d beforehand, storing them, and using them as input in place of α on repeated generations for the same α saves time.

ALGORITHM GKM1

Purpose: To sample Z randomly from the Gamma distribution $\mathcal{G}(\alpha, 1)$ with $\alpha > 1$.

Input: α.

Setup: $a = \alpha - 1$, $b = (\alpha - 1/6\alpha)/a$, $m = 2/a$, $d = m + 2$.

Source: Cheng and Feast (1979).

Method:

Repeat:
  randomly generate X' and Y' from $\mathcal{U}(0, 1)$
  $V = bY'/X'$
  if $mX' - d + V + V^{-1} \le 0$, accept (fast acceptance)
  if $m \ln X' - \ln V + V - 1 \le 0$, accept
until accept.
Return $Z = aV$. ■

Algorithm GKM2 describes the steps in sampling (X', Y') from the parallelogram $\mathcal{R}'(\alpha)$. As before, precomputing a, b, m, d and especially c for the same α saves time. Note that $\sqrt{2/e} = 0.857764$. It is left to the reader (Exercise 32) to show that if $0 < X' < 1$, then (X', Y') is uniformly distributed on $\mathcal{R}'(\alpha) \cap [0, 1]^2$ and that on any trial

$$\text{pr}(0 < X' < 1) = 1 - \frac{e+2}{2\sqrt{\alpha e}\left(\sqrt{e}+\sqrt{2}\right)} \ge 0.5328. \quad (3.47)$$

ALGORITHM **GKM2**

Purpose: To generate Z randomly from $\mathcal{G}(\alpha, 1)$ for $\alpha > 1$.

Input: α.

Setup: $a = \alpha - 1$, $b = (\alpha - 1/6\alpha)/a$, $m, = 2/a$, $d = m + 2$, $c = \sqrt{\alpha}$.

Source: Cheng and Feast (1979).

Method:

Repeat:
  Repeat:
    randomly generate X and Y' from $\mathcal{U}(0, 1)$
    $X' = Y' + (1 - 1.857764X)/c$
  until $0 < X' < 1$.
  $V = bY'/X'$
  if $mX' - d + V + V^{-1} \leq 0$, accept (fast acceptance)
  if $m \ln X' - \ln V + V - 1 \leq 0$, accept
until accept.
Return $Z = aV$. ■

Performing the step $c = \sqrt{\alpha}$ in line keeps Algorithm GMK2 from completely dominating Algorithm GMK1 with regard to computing time. However, the associated constant mean evaluation time increasingly favors Algorithm GMK2 when compared to the mean generating time proportional to $\alpha^{\frac{1}{2}}$ for Algorithm GMK1 as $\alpha \to \infty$. Algorithm GKM3 reconciles these two options. Cheng and Feast recommend $\alpha_0 = 2.5$.

ALGORITHM **GKM3**

Purpose: To generate Z randomly from $\mathcal{G}(\alpha, 1)$ for $\alpha > 1$.

Input: α and α_0.

Source: Cheng and Feast (1979).

Method:

If $\alpha \leq \alpha_0$, use Algorithm GKM1.
Otherwise, use Algorithm GKM2. ■

EXAMPLE 3.7.4 **POISSON GENERATION WITH BOUNDED MEAN COMPUTING TIME**

Let Z have the Poisson p.m.f.

$$p_z = \begin{cases} \frac{e^{-\beta}\beta^z}{z!} & \beta > 0, \quad z = 0, 1, \ldots \\ 0 & \text{elsewhere,} \end{cases} \tag{3.48}$$

denoted by $\mathcal{P}(\beta)$. The distribution has the properties:

1. If $X_1, X_2, \ldots$ are i.i.d. random variables from $\mathcal{E}(1)$ and Z is the smallest non-negative integer such that $X_1 + \cdots + X_{Z+1} > \beta$, then Z has the Poisson distribution $\mathcal{P}(\lambda)$.
2. If $U_1, U_2, \ldots$ are i.i.d. random variables from $\mathcal{U}(0, 1)$ and Z is the smallest non-negative integer such that $\Pi_{i=1}^{Z+1} U_i > \beta$, then Z has the Poisson distribution $\mathcal{R}(\beta)$.
3. If Z is from $\mathcal{P}(\beta)$, then $(Z-\beta)/\sqrt{\beta}$ has a distribution that converges to $\mathcal{N}(0, 1)$ as $\beta \to \infty$.

For $\beta < 1$, direct IT search (see Section 3.3.2) is relatively efficient with negligible round-off error. For sufficiently large β, the normal limit is operative. In particular, let Y be from $\mathcal{N}(0, 1)$. Property 3 implies that treating $\max(0, \lfloor \beta+1/2+Y\sqrt{\beta} \rfloor)$ as a Poisson variate from $\mathcal{P}(\beta)$ incurs an error that diminishes as β increases. The $\frac{1}{2}$ term is a continuity correction that reduces the error of approximation for every fixed β. Although this constant-time approach is acceptable in problems in which the error is small relative to errors from other sources, it is unacceptable in problems where results are sensitive to a departure from the Poisson distribution. This may be especially so in applying the MC method to estimate the distributional characteristics of a statistic that is a function of Poisson random variables.

Example 3.5.2 describes a computationally efficient method for repeatedly sampling from $\mathcal{P}(\beta)$ for the same $\beta > 1$. The present example addresses the case in which β varies on successive draws, using the computationally efficient ROU algorithm of Ahrens and Dieter (1991). In particular, it has a finite upper bound on mean sample generating time as β increases without bound.

Let

$$r(z) := \frac{f(z)}{\max_x \; f(x)} \qquad -\infty < z < \infty, \tag{3.49}$$

and

$$\mathcal{R} := \{(x, y) : \; 0 \geq x \geq 1, \; -1 \leq y \leq 1\}.$$

Because the Poisson distribution has its mode at $\lfloor \beta \rfloor$,

$$r(z) = \frac{p_{\lfloor z \rfloor}}{\max_t \; p_{\lfloor t \rfloor}} = \beta^{\lfloor z \rfloor - \lfloor \beta \rfloor} \lfloor \beta \rfloor ! / \lfloor z \rfloor !. \tag{3.50}$$

Its mean number of trials is

$$\frac{|\mathcal{R}|}{|\mathcal{Q}(a_1, a_2)|} = 4a_2 e^{-\beta} \beta^{\lfloor \beta \rfloor} / \lfloor \beta \rfloor !. \tag{3.51}$$

Ahrens and Dieter recommend the assignments

$$a_1 := \beta + \frac{1}{2} \tag{3.52}$$

and

$$a_2 := \sqrt{2a_1/\mathrm{e}} + \frac{3}{2} - \sqrt{3/\mathrm{e}},$$

for which

$$\frac{|\mathcal{R}|}{|\mathcal{Q}(a_1, a_2)|} = 4\left[\sqrt{(2\beta+1)/\mathrm{e}} + \frac{3}{2} - \sqrt{3/\mathrm{e}}\right] \mathrm{e}^{-\beta}\beta^{\lfloor\beta\rfloor}/\lfloor\beta\rfloor! \tag{3.53}$$

and

$$\lim_{\beta\to\infty} \frac{|\mathcal{R}|}{|\mathcal{Q}(a_1, a_2)|} = \frac{4}{\sqrt{\pi \mathrm{e}}} = 1.3688.$$

This is the same limit as would obtain if a_1 and a_2 were chosen to minimize (3.51). ■

Algorithm PRUA* describes a modification of Algorithm PRUA^t in Stadlober (1989), which is a modification of Algorithm PRUA in Ahrens and Dieter (1991). Note that the setup calls for a precomputed table of $\gamma_k = \ln k!$ for $k = 1, \ldots, k'$. Moreover, for $k > k'$, $\ln k!$ can be computed by using a truncated form of the expansion (Hastings 1955):

$$\gamma_k := \ln\sqrt{2\pi} + (k+1/2)\ln\, k - k + 1/12k - 1/360k^3 \\ + 1/1260k^5 - 1/1680k^7 \text{as } k \to \infty. \tag{3.54}$$

Stadlober reports an absolute truncation error less than 7.9×10^{-9} when terms up to and including $1/360k^3$ are retained. The choice of k' is at the user's discretion and, in principle, depends on space availability. Making k' large reduces the need to compute d_7 and γ_k in line and considerably improves the efficiency of the procedure. If Poisson generation with $\lfloor\beta\rfloor \le k'$ is contemplated, one can omit the test in the setup entirely. The rejection test $Z \ge d_8$, based on the precision factor d_0, allows the algorithm to avoid computing insignificant probabilities. As before, if repeated sampling is to occur with the same β, then one can compute the setup prior to any generation, store d_3, d_5, d_6, d_7, γ_{d_7}, d_8, and d_9, and use these as input on successive executions.

ALGORITHM **PRUA***

Purpose: To generate K randomly from $\mathcal{P}(\beta)$ with $\beta \ge 1$.

Input: β and in-line function for computing $\gamma_k = \ln k!$.

Setup: Precomputed table of $\gamma_k = \ln k!$ for $k = 0, 1, \ldots, k'$.
Constants: $d_0 = 7$ (for nine-decimal-digit precision),
$d_1 = 2\sqrt{2/\mathrm{e}} = 1.715527770$, $d_2 = 3 - 2\sqrt{3/\mathrm{e}} = 0.898916162$,
$d_3 = \beta + 1/2$, $d_4 = \sqrt{d_3}$, $d_5 = d_1 d_4 + d_2$, $d_6 = \ln\beta$, $d_7 = \lfloor\beta\rfloor$,
$d_8 = \lfloor d_3 + d_0(d_4 + 1)\rfloor$. If $d_7 > k'$, compute γ_{d_7} in line.
$d_9 = d_6 d_7 - \gamma_{d_7}$.

Method:

Repeat:
randomly generate X and Y from $\mathcal{U}(0, 1)$.
$Z = d_3 + d_5(Y - 1/2)/X$
if $Z < 0$ or $Z \geq d_8$, reject (Preliminary test 1) (Fast rejection)

otherwise:
$K = \lfloor Z \rfloor$; if $K > k'$, compute γ_K in line
$T = Kd_6 - \gamma_K - d_9$
if $T \geq X(4 - X) - 3$, accept (Preliminary test 2) (Fast acceptance)

otherwise:
if $X(X - T) \geq 1$, reject (Preliminary test 3) (Fast rejection)

otherwise:
if $2 \ln X \leq T$, accept

until accept.
Return K.

■

3.8 LESSONS LEARNED

- At least four methods are employed to generate samples from distributions on computers. They are the inverse-transform (IT), composition (CM), acceptance-rejection (AR), and ratio-of-uniforms (ROU) methods.
- Computational efficiency can be realized by using different algorithmic implementations when repeatedly sampling from a distribution whose parameters do not change value and when sampling from one whose parameters do change value on each draw.
- For a continuous distribution, the IT method's sample generating time is essentially independent of the values of the parameters of the distribution.
- For a discrete distribution, an implementation of the IT method called the cutpoint method (see Section 3.4) that relies on tables leads to a sample generating time that is essentially independent of the values of the parameters of the distribution.
- Suppose the form of a d.f. makes it difficult to draw samples from it by the IT, AR, and ROU methods. The composition method facilitates sampling by representing the d.f. as a mixture of d.f.s that are more amenable to sampling by at least one of these methods.
- The AR and ROU methods both have random sample generating times. For each method, there exist algorithms that ensure an upper bound on mean computing time that is independent of the values of the parameters of the distribution.

- Squeeze methods applied within the setting of the AR and ROU methods can reduce sample generating time.

3.9 EXERCISES

These exercises provide an opportunity for students to familiarize themselves with the mechanics of the IT, AR, ROU, and CM methods for generating samples. Although exercises focus on the first three methods, some may need to be used in conjunction with the CM method to facilitate generation. The solutions to the exercises are not necessarily the most computationally efficient, but by applying several methods to the same distribution, a student will be able to discern the solutions' relative costs, and, therefore, their relative desirability.

Inverse-Transform Method

Using, at most, two independent samples U and V from $\mathcal{U}(0, 1)$, apply the IT method to obtain an explicit expression for Z with

1. P.d.f.

$$f(z) = \begin{cases} \frac{5}{16}z + \frac{3}{32}z^2 + \frac{5}{256}z^4 & 0 \le z \le 2 \\ 0 & \text{elsewhere.} \end{cases} \tag{3.55}$$

2. P.d.f.

$$f(z) = \begin{cases} \frac{1}{9}z^{1/2} + \frac{55}{1728}z^{3/2} & 0 \le z \le 4 \\ 0 & \text{elsewhere.} \end{cases} \tag{3.56}$$

3. P.d.f.

$$f(z) = \begin{cases} \frac{1}{2\beta}e^{-|\alpha - z|/\beta} & -\infty < \alpha < \infty,\ \beta > 0, -\infty < z < \infty \\ 0 & \text{elsewhere.} \end{cases} \tag{3.57}$$

4. P.m.f.

$$f(z) = \begin{cases} \frac{20}{9}\left(\frac{3}{8}\right)^{z+1} + \frac{5^{z-1}}{7^z} & z = 1, 2, \ldots \\ 0 & \text{elsewhere.} \end{cases} \tag{3.58}$$

5. Trapezoidal p.d.f.

$$f(z) = \begin{cases} \frac{z-c}{ab} & c \le z \le a + c \\ \frac{1}{b} & a + c \le z \le b + c \\ \frac{a+b+c-z}{ab} & b + c \le z \le a + b + c \\ 0 & \text{elsewhere.} \end{cases} \tag{3.59}$$

6. D.f.

$$F(z) = \begin{cases} \left(1 - \frac{2\cos^{-1} z}{\pi}\right)^2 & 0 \le z \le 1 \\ 0 & \text{elsewhere.} \end{cases} \tag{3.60}$$

7. Triangular p.d.f.

$$f(z) = \begin{cases} \frac{2(z-m_1)}{(m_3-m_1)(m_2-m_1)} & m_1 \le z \le m_2 \\ \frac{2(m_3-z)}{(m_3-m_1)(m_3-m_2)} & m_2 \le z \le m_3 \\ 0 & \text{elsewhere.} \end{cases} \tag{3.61}$$

8. P.d.f.

$$f(z) = \begin{cases} \alpha \frac{(z-a)^{\alpha-1}}{(b-a)^{\alpha}} & \alpha > 0,\ 0 \le a \le z \le b < \infty \\ 0 & \text{elsewhere.} \end{cases} \tag{3.62}$$

Acceptance-Rejection Method

Apply the AR method to generate a sample from each of the following distributions. Specify a candidate-generating distribution $\{\eta(z)\}$ and derive its corresponding $\kappa(f, \eta) :=$ mean number of trials until success in (3.10) for the following distributions:

9. In (3.55).

10. In (3.56).

11. In (3.62).

12.

$$f(z) = \begin{cases} \frac{1}{2\log 2}(\log z + 1) & 1 \le z \le 2 \\ 0 & \text{elsewhere.} \end{cases} \tag{3.63}$$

13.

$$f(z) = \begin{cases} \frac{1}{2}\sin z & 0 \le z \le \pi \\ 0 & \text{elsewhere.} \end{cases} \tag{3.64}$$

14. (Fishman 1976). Using the exponential distribution

$$\eta(z) = \begin{cases} \nu^{-1}e^{-z/\nu} & \nu > 0,\ z \ge 0 \\ 0 & \text{elsewhere} \end{cases} \tag{3.65}$$

to generate candidates for the Gamma distribution in (3.38) with $\alpha \ge 1$.

a. Derive an expression for $\kappa(f, \eta)$ in terms of α, β, and ν.

b. Find the value of ν that minimizes $\kappa(f, \eta)$.

c. How does this minimal $\kappa(f, \eta)$ vary with increasing α?

15. (Cheng 1977). Using the distribution

$$\eta(z) = \begin{cases} \frac{\theta\mu z^{\theta-1}}{(\mu+z^\theta)^2} & z \geq 0 \\ 0 & \text{elsewhere,} \end{cases} \tag{3.66}$$

where

$$\mu := \alpha^\theta$$
$$\theta := (2\alpha - 1)^{1/2},$$

to generate samples from the Gamma distribution in (3.38) with $\alpha \geq 1$.

a. Derive an expression for $\kappa(f, \eta)$ in terms of α.

b. Show that $\kappa(f, \eta) \to \frac{2}{\sqrt{\pi}}$ as $\alpha \to \infty$.

c. Compare this $\kappa(f, \eta)$ with that in (3.43) in Example 3.7.3 as a function of α and discuss why the latter algorithm is usually preferred.

d. Compare this $\kappa(f, \eta)$ with that in Exercise 14.

Ratio-of-Uniforms Method

Apply the ROU method to generate a sample from each of the following distributions. Specify the acceptance region $\mathcal{Q}$ in (3.25) and the sampling region $\mathcal{R}$ in (3.28), and compute the mean number of trials until success. Also, specify an alternative $\mathcal{R}$ that leads to a smaller mean number of trials until success, and comment on the pros and cons of using it in practice.

16. In (3.55).

17. In (3.56).

18. In (3.62).

Order Statistics

Let $Z_1, \ldots, Z_r$ denote i.i.d. random variables with distribution f, and let $Z_{(1)} \leq \ldots \leq Z_{(r)}$ denote the sequence ordered from smallest to largest. Occasionally, it is of interest to generate these *order statistics* or some subset of them, such as the minimum $Z_{(1)}$ or the maximum $Z_{(r)}$. One approach randomly and independently samples $Z_1, \ldots, Z_r$ from f and then sorts them into order. Since the fastest algorithm for sorting $Z_1, \ldots, Z_r$ into $Z_{(1)} \leq \ldots \leq Z_{(r)}$ takes time proportional to $r \log r$, it is of interest to explore other ways of generating the order statistics that eliminate the need to sort. Exercises 19 through 25 describe a basis for doing this.

19. Let $U_1, \ldots, U_r$ be i.i.d. from $\mathcal{U}(0, 1)$, and let $U_{(1)}, \ldots, U_{(r)}$ denote the corresponding order statistics.

a. Show that $U_{(1)}$ has p.d.f.

$$f_1(u) = \begin{cases} r(1-u)^{r-1} & 0 \leq u \leq 1 \\ 0 & \text{elsewhere.} \end{cases}$$

b. For $0 \leq v_j \leq 1$ and $j = 2, \ldots, r$, show that $U_{(j)}$ given $U_{(j-1)} = v_{j-1}$

has p.d.f.

$$f_j(u \mid v_{j-1}) = \begin{cases} (r-j+1)\frac{(1-u)^{r-j}}{(1-v_{j-1})^{r-j+1}} & v_{j-1} \le u \le 1 \\ 0 & \text{elsewhere.} \end{cases}$$

c. Using the properties established in parts a and b, describe an algorithm for randomly generating $U_{(1)}, \ldots, U_{(r)}$ without sorting.

20. Let $Z_1, \ldots, Z_r$ be i.i.d. with p.d.f.

$$f(z) = \begin{cases} \frac{\alpha\beta^\alpha}{(z+\beta)^{\alpha+1}} & \alpha,\ \beta > 0,\ z \ge 0 \\ 0 & \text{elsewhere,} \end{cases} \tag{3.67}$$

and let $Z_{(1)}, \ldots, Z_{(r)}$ denote the corresponding order statistics. Use the knowledge obtained in Exercise 19 to generate $Z_{(1)}, \ldots, Z_{(r)}$ randomly without sorting.

Note: This exercise illustrates how to generate order statistics for *any* distribution without sorting.

Generating the Maximum of r i.i.d. Random Variables. Let $Z_1, \ldots, Z_r$ be i.i.d. with p.d.f. $\{f(z),\ a \le z \le b\}$, d.f. $\{F(z) = \int_a^z f(w)dw,\ a \le z \le b\}$, and

$$Z_{\max} := \max_{1 \le i \le r} Z_i.$$

21. Prove that

$$\text{pr}(Z_{\max} \le z) = F^r(z) \qquad a \le z \le b.$$

22. Using a single random sample U from $\mathcal{U}(0,1)$, describe an algorithm for randomly generating $Z_{\max}$ whose computing time is independent of r when the Z_i are from (3.67).

23. Using a single random sample U from $\mathcal{U}(0,1)$, describe an algorithm for randomly generating $Z_{\max}$ when the Z_i are from the exponential distribution (3.65).

24. Using no more than two independent samples U_1 and U_2 from $\mathcal{U}(0,1)$, describe an algorithm for generating $Z_{\max}$ when the Z_i have p.d.f. (3.56). Except for a one-time setup whose computing time depends on r but whose calculated results can be used repeatedly on successive generations of $Z_{\max}$, the algorithm's computing time should be independent of r. Give formulae for all constants in the setup.

25. Using no more than two independent samples U_1 and U_2 from $\mathcal{U}(0,1)$, describe an algorithm for generating $Z_{\max}$ when the Z_i have p.d.f. (3.55). Give formulae for all constants in the setup.

Special Cases

Sometimes the properties of a distribution suggest a more convenient way to generate a sample than relying solely on the direct application of one of the principal methods. Exercises 26 through 28 provide instructive examples.

26. Show that for appropriately chosen a_1, a_2, and a_3 and independent U and V from $\mathcal{U}(0, 1)$, $Z = a_1U + a_2V + a_3$ has the trapezoidal distribution (3.59).

27. For the special case of $m_2 - m_1 = m_3 - m_2$, devise a generating algorithm that relies on a linear combination of independent U and V from $\mathcal{U}(0, 1)$ to generate a sample from (3.61).

28. Negative Binomial Distribution. Prove that if Y is from $\mathcal{G}(r, 1)$ and Z given Y is from $\mathcal{P}(Yp/(1-p))$ for $0 < p < 1$ and $r > 0$, then Z has the negative binomial distribution

$$f(z) = \begin{cases} \frac{\Gamma(r+z)}{\Gamma(r)\Gamma(z+1)}(1-p)^r p^z & z = 0, 1, \ldots \\ 0 & \text{elsewhere.} \end{cases} \tag{3.68}$$

Derivations and Proofs

29. Show that the recursion

$$p_z = \begin{cases} e^{-\beta} & z = 0 \\ \frac{\beta}{z} p_{z-1} & z = 1, 2, \ldots \end{cases}$$

gives the p.m.f. for $\mathcal{P}(\beta)$.

30. Show that the recursion in (3.9) gives the negative binomial distribution (3.8).

31. Derive the probability of evaluating $\ln V$ in Algorithm GKM1.

32. In Algorithm GKM2, prove that if $0 < X' < 1$, then (X', Y') is uniformly distributed on the intersection of $\mathcal{R}'$ and the unit square $[0, 1]^2$.

3.10 | REFERENCES

Ahrens, J. H., and U. Dieter (1974). Computer methods for sampling from Gamma, Beta, Poisson, and binomial distributions, *Computing,* **12,** 223–246.

Ahrens, J. H., and U. Dieter (1991). A convenient sampling method with bounded computation times for Poisson distributions, *The Frontiers of Statistical Computation, Simulation and Modeling.* P. R. Nelson, E. J. Dudewicz, A. Öztürk, and E. C. van der Meulen, editors, American Science Press, Syracuse, N.Y., 137–149.

Chen, H.-C., and Y. Asau (1974). On generating random variates from an empirical distribution, *AIIE Trans.,* **6,** no. 2, 163–166.

Cheng, R. C. H. (1977). The generation of Gamma variables with nonintegral shape parameter, *Appl. Stat.,* **26,** 71–75.

Cheng, R. C. H. (1978). Generating beta variates with nonintegral shape parameters, *Comm. ACM,* **21,** 317–322.

Cheng, R. C. H., and G. M. Feast (1979). Some simple Gamma variate generators, *Appl. Stat.,* **28,** 290–295.

Devroye, L. (1986). *Non-Uniform Random Variate Generation,* Springer-Verlag, New York.

Fishman, G. S. (1976). Sampling from the gamma distribution on a computer, *Comm. ACM*, **19,** 407–409.

Fishman, G. S. (1996). *Monte Carlo: Concepts, Algorithms, and Applications*, Springer-Verlag, New York.

Fishman, G. S., and L. R. Moore (1984). Sampling from a discrete distribution while preserving monotonicity, *Amer. Stat.*, **38,** 219–223.

Fishman, G. S., and L. S. Yarberry (1993). Generating a sample from a k-cell table with changing probabilities in $O(\log_2 k)$ time, *ACM Trans. Math. Software,* **19,** 257–261.

Gilks, W. R., and P. Wild (1992). Adaptive rejection sampling for Gibbs sampling, *Appl. Stat.*, **41,** 337–348.

Hastings, C., Jr. (1955). *Approximations for Digital Computers*, Princeton University Press, Princeton, N.J.

Kinderman, A. J., and J. F. Monahan (1977). Computer generation of random variables using the ratio of uniform deviates, *ACM Trans. Math. Software,* **3,** 257–260.

Ledold, J., G. Derflinger, G. Tirler, and W. Hörmann (2001). An automatic code generator for nonuniform random variate generation, University of Vienna.

Marsaglia, G. (1977). The squeeze method for generating gamma variates, *Comput. and Math. with Appl.*, **3,** 321–325.

Moro, B. (1995). The full Monte, *Risk,* **8,** 57–58.

Stadlober, E. (1989). Sampling from Poisson, binomial, and hypergeometric distributions: Ratio of uniforms as a simple and fast alternative, *Math. Statist. Sektion 303,* Forschungsgesellschaft Joanneum, Graz, Austria.

Stadlober, E. (1991). Binomial random variate generation: A method based on ratio of uniforms, *The Frontiers of Statistical Computation, Simulation, and Modeling,* **1,** P. R. Nelson, E. J. Dudewicz, A. Öztürk, E. C. van der Meulen, editors, American Sciences Press, Syracuse, N.Y., 93–112.

Wild, P., and W. R. Gilks (1993). Algorithm AS287: Adaptive rejection sampling from log-concave density, *Appl. Stat.*, **42,** 701–709.

CHAPTER 4

Pseudorandom Number Generation

As Chapter 3 indicates, the ability to generate a random sample from a specified distribution depends crucially on the availability of a sequence of i.i.d. random values drawn from $\mathcal{U}(0, 1)$. Although no MC technique can literally produce sequences with exactly this property, *pseudorandom number generators* (PNGs) exist that produce sequences that closely resemble them, the error of approximation depending on the particular method of generation. These include:

- Linear congruential generators (LCGs)
- Multiplicative recursive generators (MRGs)
- Feedback shift registers (FSRs)
- Generalized feedback shift registers (GFSRs)
- Combined multiplicative recursive generators (CMRGs)
- Twisted generalized feedback shift registers (TGFSRs)
- Add-with-carry (AWC) and subtract-with-borrow (SWB) generators
- Inversive generators

LCGs, MRGs, CMRGs, and AWC/SWB and inversive generators produce pseudorandom numbers by operating recursively on previously generated numbers. By contrast, FSRs, GFSRs, and TGFSRs operate recursively on the individual bits of previously generated numbers.

This chapter provides concise accounts of LCGs, CMRGs, TGFSRs, and AWC/SWB generators. The last three generate sequences that approximate the characteristics of i.i.d. $\mathcal{U}(0, 1)$ sequences much better than LCGs do. Moreover, accumulating evidence points to TGFSRs, especially the *Mersenne twister* in Section 4.7.1, as having particularly superior performance. However, LCGs are the most frequently employed PNGs. Why is this so?

To facilitate MC sampling, applications software often provides its own PNGs. LCGs are among the oldest proposals for pseudorandom number generation. Their prevalence in so much of applications software tacitly encourages a prospective MC user to employ them, merely as a convenience. Although it is true that some MC applications can produce meaningful answers even with poor-quality PNGs, some applications, sensitive to a PNG's limitations, generate misleading results. For example, see Example 4.4.1. Also see Ferrenberg et al. (1992).

A more beneficial modus operandi is for an MC user first to become familiar with the properties of the PNG in available applications software in order to determine whether its use will lead to tolerable error. If it does not, resorting to a better-performing PNG can reduce error to an acceptable level. More generally, the Internet site

http://random.mat.sbg.ac.at/links/rando.html

provides considerable information about and links to these better PNGs.

Although they are defective as a class, we first describe LCGs. The motivation for this comes from their availability in so much applications software and because they have been around so long; it is easy to convince oneself that longevity is a sign of acceptability. This chapter shows how they work and both their theoretical and empirically observed deficiencies. It then describes CMRGs in Section 4.5, AWC/SWB generators in Section 4.6, and TGFSRs in Sections 4.7 and 4.7.1, and shows why they compare favorably with LCGs.

Our account relies heavily on the mathematical operation of *modulo reduction*. Every PNG produces a sequence of non-negative integers $Z_1, Z_2, \ldots$ with integer upper bound $M > Z_i$ for all $i \geq 1$ and then uses $U_1, U_2, \ldots$, where $U_i := Z_i/M$, as an approximation to an i.i.d. sequence from $\mathcal{U}(0, 1)$. All generators described in this chapter accomplish this by exploiting the concept of *congruence*. We say that *a is congruent to b modulo M* if M divides $a-b$. In symbols, one writes $a \equiv b \pmod{M}$, where M is called the *modulus* and $\equiv$ is called the congruence symbol. As examples, $7 \equiv 1 \pmod 6$, $11 \equiv 2 \pmod 3$, $338 \equiv 8 \pmod{11}$, and $-5 \equiv 1 \pmod 6$. Also, $X = (B + C) \pmod M$ denotes that X equals the quantity $B+C$ reduced modulo M.

4.1 LINEAR CONGRUENTIAL GENERATORS

To generate $Z_1, Z_2, \ldots$, applications software frequently relies on a *linear congruential pseudorandom number generator* (LCG) of the form

$$Z_i = AZ_{i-1} \pmod M \qquad i \geq 1, \tag{4.1}$$

where Z_0 denotes the *seed* and A is a positive integer called the *multiplier*. Expression (4.1) is algebraically equivalent to

$$Z_i = AZ_{i-1} - MK_i \qquad i \geq 1, \tag{4.2}$$

where

$$\begin{aligned} K_i &:= \text{largest integer less than or equal to } AZ_{i-1}/M \\ &= \lfloor AZ_{i-1}/M \rfloor. \end{aligned} \tag{4.3}$$

Then the sequence of fractions $\{U_i := \frac{Z_i}{M},\ i \geq 1\}$ is used to approximate an i.i.d. sequence from $\mathcal{U}(0, 1)$. Throughout this chapter, we use the notation $\lfloor\ \rfloor$ for conciseness.

Initialization. To generate numbers from (4.1), one needs to *prime* the generator with a seed. For example, to generate Z_1, Z_0 acts as the seed; to generate Z_2, Z_1 acts as the seed; and so on. The simplicity of priming this generator is one of its attractive features.

Reproducibility. In designing an MC experiment, it is not unusual to conduct a preliminary run to identify errors in program logic and previously unexpected sample-path behavior. This identification often prompts modification of the model or program. By recording the seed Z_0 for the first run, one can make a new run with the modified program but using the same seed as input. Provided that the modified program leaves untouched the logic for assigning pseudorandom numbers to each source of variation, the sample-path behavior of the new run reflects only the code change and contains no new variation because of sample generation. Therefore, the ability to reproduce exactly the same sampling variation on successive runs is a valuable tool in judging sample-path changes that result from program modification. All pseudorandom number generators described in this chapter offer this feature.

Discretization Error. Whereas an i.i.d. sequence from $\mathcal{U}(0,1)$ is continuous on the interval $[0, 1)$, PNG-generated $U_1, U_2, \ldots$ form a discrete sequence. In particular, $U_i \in \mathcal{M} \subseteq \mathcal{M}_0 := \{0, \frac{1}{M}, \frac{2}{M}, \ldots, \frac{M-1}{M}\}$ for all i, revealing a *discretization error* that depends on how large M is and on how closely $\mathcal{M}$ resembles $\mathcal{M}_0$. Moreover, the selected PNG determines how well the distribution of any subset of $U_1, U_2, \ldots,$ namely $U_{j_1}, U_{j_2}, \ldots, U_{j_k}$ for arbitrary positive integers $k, j_1, \ldots, j_k$, approximates the joint distribution of k i.i.d. $\mathcal{U}(0, 1)$ random variables. Later sections rely on this and related concepts to evaluate a PNG's performance. Also, Example 4.4.1 describes the important issue of how these discretization and distributional errors in the generation of $U_1, U_2, \ldots$ propagate themselves in sampling from a nonuniform distribution.

Period. Expression (4.1) makes clear that the number of distinct values the Z_i assume cannot exceed M. Just how many distinct values are possible depends on the relationship between A and M. To see how this occurs, first note that the recursion (4.2) induces

$$\begin{aligned}
Z_1 &= AZ_0 - MK_1, \\
Z_2 &= A^2 Z_0 - MK_2 - MK_1 A, \\
Z_3 &= A^3 Z_0 - MK_3 - MK_2 A - MK_1 A^2, \\
&\ \vdots \\
Z_n &= A^n Z_0 - M(K_{n-1} A + \cdots + K_1 A^{n-1}).
\end{aligned} \tag{4.4}$$

Because $1 \le Z_i < M$ for $1 \le i \le n$, the quantity $K_n + K_{n-1}A + \cdots + K_1 A^{n-1}$ must be the largest integer in $A^n Z_0 / M$, so that expression (4.1) has the alternative representation

$$Z_n = A^n Z_0 \pmod{M}. \tag{4.5}$$

Once $Z_n = Z_0$ for some integer $n > 0$, then $Z_{n+i} = Z_i$ for all $i = 1, 2, \ldots$. That is, the sequence repeats itself. We call

$$P(Z_0, A, M) := \min(n \ge 1 : \ Z_n = Z_0) \tag{4.6}$$

the *period* and $Z_1, Z_2, \ldots, Z_{P(Z_0,A,M)}$ a *cycle* of the generator for given Z_0, A, M. Clearly, for given Z_0 and M we would like to choose the multiplier A to maximize the period. That is, $P(Z_0, A, M) = M - 1$. This would make the period identical for all seeds $Z_0 \in \mathcal{M} = \mathcal{M}_0$ and thus independent of the starting seed. Moreover, there would be just one cycle.

Observe that $Z_n = Z_0$ in (4.5) and (4.6) implies

$$(A^n - 1)Z_0 \pmod{M} = 0. \tag{4.7}$$

If $n = M - 1$ is the smallest integer satisfying the congruence (4.7), then it must hold for all $Z_0 \in \mathcal{M}_0$, so that

$$(A^{M-1} - 1) \pmod{M} = 0. \tag{4.8}$$

It also would maximize the density of points $\left\{\frac{1}{M}, \ldots, \frac{M-1}{M}\right\}$ that the U_i can assume.

Observe that Z_0 completely determines the $\{Z_i,\ i \ge 1\}$ integer sequence once A and M are fixed. Therefore, the extent to which $U_1, U_2, \ldots$ resemble a *continuous-valued* i.i.d. $\mathcal{U}(0,1)$ sequence inevitably depends on the choice of A and M.

Independent Experiments. When an MC sampling experiment calls for independent replications, LCGs can easily accommodate the need. In particular, storing the last pseudorandom number generated on replication i and then using it as the seed for replication $i + 1$ ensures that pseudorandom subsequences generated on successive replications do not overlap. That is, none share the same numbers. With the exception of one-step inversive PNGs (e.g., see Fishman 1996, p. 672), no other class of PNGs admits as convenient a method for generating independent pseudorandom subsequences.

For k independent experiments, the formal steps are:

- Select a seed Z_0.
- On experiment 1, use seed Z_0 to generate N_1 pseudorandom numbers $Z_1, \ldots, Z_{N_1}$.
- On experiment 2, use seed Z_{N_1} to generate N_2 pseudorandom numbers $Z_{N_1+1}, \ldots, Z_{N_1+N_2}$.

$\vdots$

- On experiment k, use seed $Z_{N_1+\cdots+N_{k-1}}$ to generate N_k pseudorandom numbers $Z_{N_1+\cdots+N_{k-1}+1}, \ldots, Z_{N_1+\cdots+N_k}$.

Although all PNGs based on recursion conceptually allow access to their last generated numbers, their implementations in applications software sometimes do not. This limitation denies the MC user a valuable form of control in sampling experiments. Section 4.7.1 returns to this topic for the Mersenne twister.

Computational Efficiency. In many MC experiments, pseudorandom number generation accounts for a major proportion of computing cost. When an experiment requires considerable computing time to achieve a specified accuracy, it is naturally desirable to choose a computationally efficient PNG implementation. The one-step generator (4.1) caters to this objective. In particular, selecting a sufficiently small A prevents overflow problems, which can be costly to overcome for larger A. However, theoretical analysis indicates that a small A restricts a sequence's tendency to appear uniformly distributed in $k > 1$ dimensions (L'Ecuyer 1997).

Portability. In principle, a PNG should produce the same sequence of numbers on all computing platforms, most notably on 32-bit and 64-bit machines, which handle overflow differently. Unless special attention is devoted to this issue in software design, this *portability* is not realized in practice. To eliminate the overflow problem and ensure portability, many implementations partition the multiplier A and arithmetic operations in a way that eliminates the overflow problem on a wide range of computing platforms, thereby ensuring portability.

As illustration, let $A_1 := A \pmod{2^{16}}$, $A_2 := A - A_1$, $X_i = A_1 Z_{i-1} \pmod{2^{31}}$, and $Y_i := A_2 Z_{i-1} \pmod{2^{31}}$. Then (4.1) has the equivalent form

$$Z_i = (X_i + Y_i + \lfloor A_1 Z_{i-1}/2^{31} \rfloor + \lfloor A_2 Z_{i-1}/2^{31} \rfloor) \pmod{2^{31} - 1},$$

where $A_1 Z_{i-1} \leq 2^{47} - 1$, and $A_2 Z_{i-1} \leq 2^{62} - 1$. Since the least significant 16 bits of $A_2 Z_{i-1}$ are identically zero, this partition of AZ_{i-1} allows one to compute Z_i with no overflow, provided that a 47-bit mantissa is available. Hörmann and Derflinger (1993) give algorithms for effecting this portability for $M = 2^{31} - 1$ and $A < 2^{30}$.

4.1.1 Prime Modulus

If M is prime, there exist $\phi(M-1)$ integers $A \in \mathcal{M}_0$ for which $n = M - 1$ is the smallest integer satisfying the congruence (4.8), where

$$\phi(n) := n \prod_{p|n} \left(1 - \frac{1}{p}\right), \tag{4.9}$$

and the product is taken over all distinct primes p dividing n. An A that satisfies expression (4.8) for M prime is called a *primitive root* modulo M. Expression (4.9) defines the *Euler totient function.*

As illustration, the largest integer that a 32-bit computer word can hold (except for sign bit) is $M = 2^{31} - 1$, which happens to be prime. Since $M - 1$ has the prime factorization $M - 1 = 2^{31} - 2 = 2 \times 3^2 \times 7 \times 11 \times 31 \times 151 \times 331$, there exist $\phi(M-1) = 534,600,000$ primitive roots of $2^{31} - 1$, each one of which guarantees a full period $P = M - 1$ for (4.1).

Multiplicative Inverse. For every A a primitive root of M prime, there exists a $B \neq A$ that generates the sequence in (4.1) in reverse order. That is,

$$Z_{i-1} = BZ_i \pmod{M} \qquad i \geq 1.$$

By direct substitution, it is seen that

$$AB \equiv 1 \pmod{M}.$$

Here B is called the *multiplicative inverse* of A modulo M. Because of this dualism, the properties of sequences from each generator are the same except for direction, thus reducing the number of primitive roots of interest by one half.

4.2 EVALUATING PNGs

Statistical inference based on MC-generated data assumes that the hypothesis

$$\mathrm{H}_* := \{U_i,\ i \geq 1\} \text{ is an i.i.d. } \mathcal{U}(0, 1) \text{ sequence} \tag{4.10}$$

is true. Because H_* refers to continuous random variables and because PNG-generated sequences $\left\{U_i := \frac{Z_i}{M},\ i \geq 1\right\}$ can assume at most M values $\left\{0, \frac{1}{M}, \ldots, \frac{M-1}{M}\right\}$, H_* cannot be literally true. At best, we can work with a PNG whose error of approximation is sufficiently small so as to have negligible effect on the estimates computed from its pseudorandom numbers.

Two types of approximating errors exist. One characterizes the class of a PNG. That is, every PNG within the class has the property. The other arises from the choice of numerical values assigned to the parameters of the PNG. For example, although an LCG with prime modulus M has many primitive roots that ensure a full period $M - 1$, some induce smaller errors of approximation than others. Naturally, it is desirable to choose the class of PNGs with the best approximating properties, and within that class, to assign parametric values that induce the smallest error. Although this is a tall order to fill in practice, understanding how these errors materialize allows the potential MC user to make an informed choice when several PNGs are available.

Two complementary modalities exist for evaluating how well a PNG's sequence satisfies H_*. Section 4.2.1 describes aspects of theoretical evaluation and Section 4.2.2, aspects of empirical testing. Complementarity arises from the fact that neither analysis alone suffices to give a total evaluation of a PNG's properties.

4.2.1 Theoretical Evaluation

The properties of k-tuples $\{\mathbf{U}_{ik} := (U_i, \ldots, U_{i+k-1}),\ i \geq 1\}$ provide a convenient basis for analyzing distributional behavior. Hypothesis H_* implies that for each $k \geq 1$, $\mathbf{U}_{ik}, \mathbf{U}_{i+1,k}, \ldots$ are continuous-valued random vectors, independently and uniformly distributed in the k-dimensional unit hypercube $[0, 1)^k$. Since a PNG produces U_i's that take values in the countably finite set $\left\{0, \frac{1}{M}, \ldots, \frac{M-1}{M}\right\}$, the best one can hope for is that for each $k \geq 1$, it generates every k-tuple on the k-dimensional *cubic lattice* $\left\{0, \frac{1}{M}, \ldots, \frac{M-1}{M}\right\}^k$ an equal number of times before repeating its cycle of k-tuples. This property is called *k-distributedness*. Figure 4.1 illustrates the concept in $k = 2$ dimensions for a 47×47 array on the interval $[0, 0.001) \times [0, 0.001)$.

We emphasize that although k-distributedness is desirable, it is not sufficient to grade a PNG as acceptable. In particular, it tells us nothing about the sequential order in which the sites are visited during a cycle and, therefore, is uninformative about the independence of k-tuples. However, empirical evaluation as in Section 4.2.2 provides us with a test of this independence. It is in this sense that theoretical evaluation and empirical evaluation complement each other.

An LCG with prime modulus M and multiplier A, a primitive root of M, is 1-distributed on $\left\{\frac{1}{M}, \ldots, \frac{M-1}{M}\right\}$. If the $M-1$ 2-tuples in a cycle were 2-distributed,

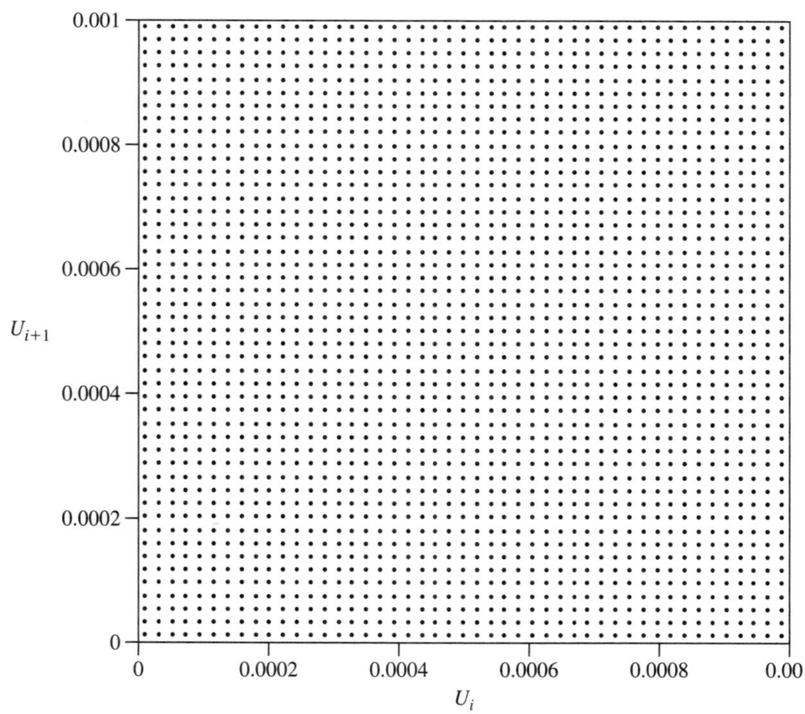

FIGURE 4.1 | 47×47 *array of 2-tuples in* $[0, 0.001) \times [0, 0.001)$

then for $M = 2^{31} - 1$ we would expect them to be arranged roughly as in Figure 4.1, with distance $\frac{1}{M^{1/2}} \approx \frac{.001}{47} = 2.16 \times 10^{-5}$ between successively parallel grid lines. However, Figure 4.2 illustrates what occurs in practice. It displays all the 2-tuples (U_i, U_{i+1}) that fall into the square $[0, 0.001) \times [0, 0.001)$ for two primitive roots[1] of $M = 2^{31} - 1$. Although the 2-tuples are clearly not 2-distributed, they differ in the extent of their non-uniformity. Figure 4.2a, which corresponds to a commonly employed multiplier, $A = 16807$, appears considerably less uniform than Figure 4.2b, whose multiplier $A = 742938295$, among all primitive roots of $2^{31} - 1$, is known to have superior uniformity properties in dimensions 2 through 6 (Fishman and Moore 1986).

Both graphs reveal a regularity that turns out to be a direct consequence of the congruential recursion (4.1) for all primitive roots of $2^{31} - 1$. The 2-tuples are arrayed on grids or *lattices* that are not cubic. In particular, Figure 4.2a shows 2-tuples packed together tightly on 17 parallel lines, whereas Figure 4.2b shows 2-tuples considerably more sparsely (uniformly) distributed on 50 parallel lines. These numbers arise from counting the number of parallel lines within the set of parallel lines that have *maximal distance* between successive lines. Because any square of width and length 0.001 in the unit square exhibits approximately the same appearance, $[0, 0.001) \times [0, 0.001)$ contains about $(0.001)^2 \times M \approx 2147$ 2-tuples. As comparison, Figure 4.1 contains roughly the same number of points, namely, $47^2 = 2209$ 2-tuples. For the unit square $[0, 1)^2$, these numbers suggest that 2-tuples in Figure 4.2a for $A = 16807$ are distributed on about 17,000 parallel lines in the unit square, whereas 2-tuples in Figure 4.2b for $A = 74938285$ are distributed on about 50,000 parallel lines.

Spectral Test. Although the number of parallel lines provides a measure of uniformity, it is not *rotation invariant*. That is, for a given distance between successive parallel lines, the number of lines in $[0, 1)^2$ varies with the angle of orientation in the x-y plane. Because distance between successive parallel lines remains unaffected by rotation of the grid, the spectral test provides a superior measure of uniformity.

To use distance effectively to discriminate among generators, we need to resolve one issue. Figure 4.3 displays 2-tuples for the prime-modulus LCG with $M = 61$ and $A = 6$. The solid lines constitute one family of parallel lines and the dashed lines, a second family. Indeed, there exist 243 different families of parallel lines for this LCG in two dimensions. Among them, at least one has maximal distance between successive parallel lines, and it is this distance that we use to characterize two-dimensional uniformity for this generator. The solid lines in Figure 4.3 provide the basis for computing this maximal distance. The spectral test serves as a worst-case measure of a PNG's performance.

[1]The square $[0, 0.001) \times [0, 0.001)$, which is representative of the one million squares of this size that make up the unit square $[0, 1) \times [0, 1)$, allows us to discern 2-tuple behavior with considerable clarity.

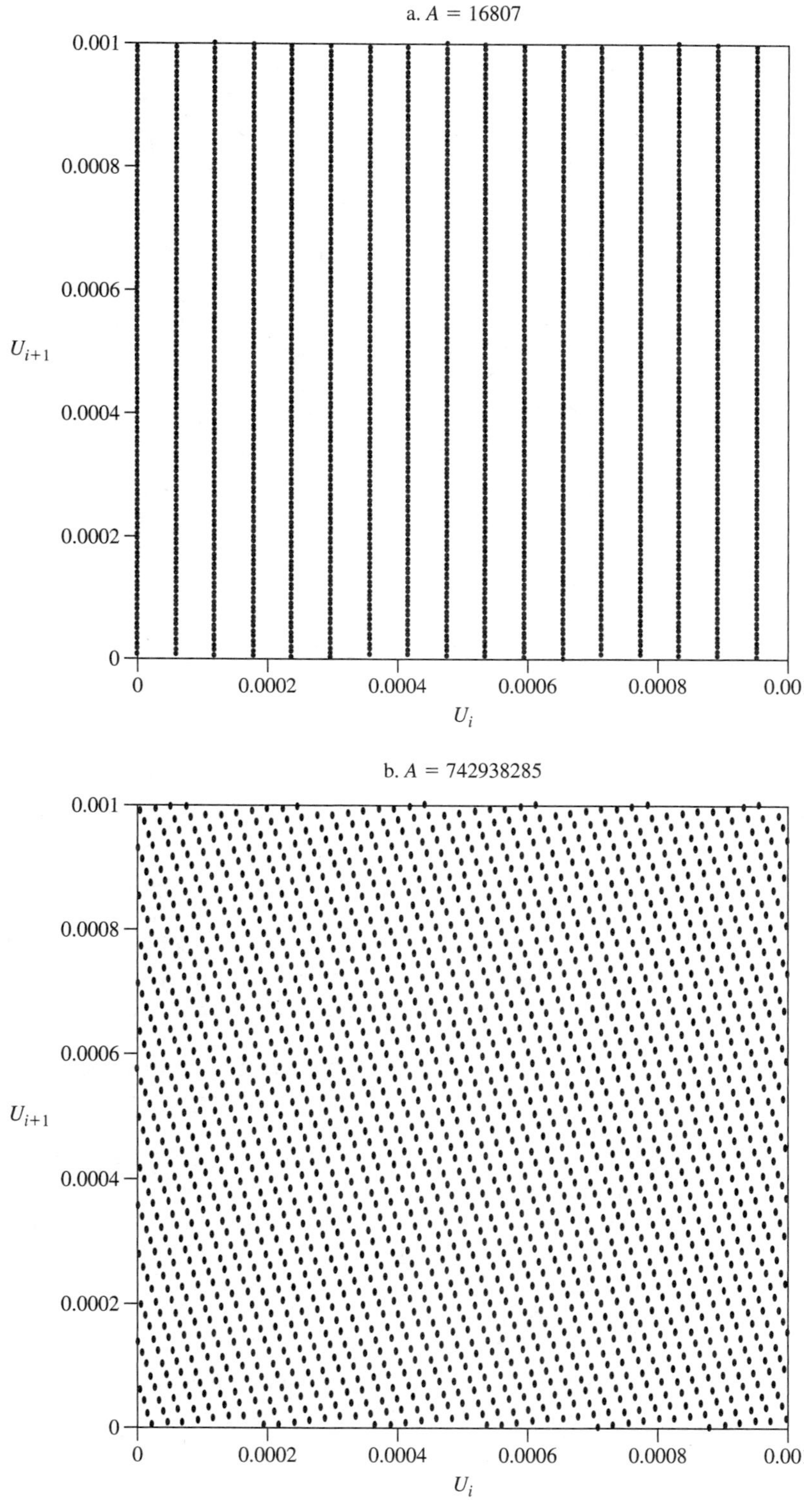

FIGURE 4.2 | *2-tuples* (U_i, U_{i+1}); $U_i := \frac{Z_i}{2^{31}-1}$, $Z_i = AZ_{i-1} (\mathrm{mod}\, 2^{31} - 1), i \geq 1$

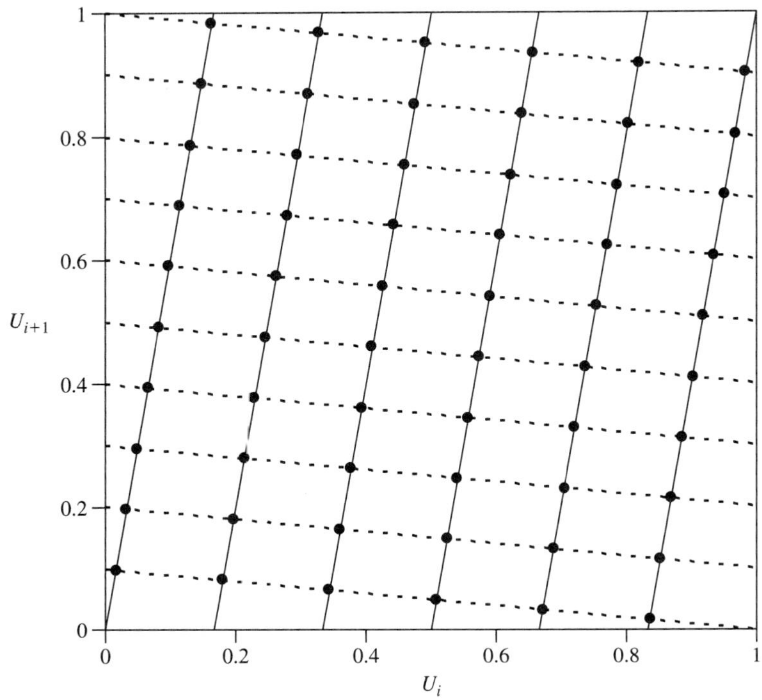

FIGURE 4.3 | *2-tuples* (U_i, U_{i+1}); $U_i := \frac{Z_i}{61}$, $Z_i = 6Z_{i-1} \pmod{61}$, $i \geq 1$
solid line: $\{U_{i+1} = j - 1 + 6U_i;\ j = 1, \ldots, 6\}$
dashed line: $\{U_{i+1} = \frac{j}{10} - \frac{1}{10}U_i;\ j = 1, \ldots, 10\}$

More generally, a prime-modulus LCG as in (4.1) with primitive root A has its k-tuples $\mathbf{U}_{ik}, \mathbf{U}_{i+1,k}, \ldots$ distributed on *parallel hyperplanes* in $[0, 1)^k$. There are $2^k M^{k-1} - 1$ such families of parallel hyperplanes. Let

$$d_k(A, M) := \text{the maximal distance between successive parallel hyperplanes in } k \text{ dimensions.} \tag{4.11}$$

Then the tendency to uniformity in k dimensions increases as $d_k(A, M)$ decreases. For multipliers A and A', A is said to perform better than A' in dimension k if $d_k(A, M) < d_k(A', M)$. As illustration, $A = 16807$ has $d_k(16807, 2^{31} - 1) \approx 5.9 \times 10^{-5}$, whereas $A = 742938285$ has $d_k(742938285, 2^{31} - 1) \approx 2.3 \times 10^{-5}$, implying the latter multiplier has better distributional properties in two dimensions. Evaluating PNGs in this way is called the *spectral test* (e.g., Knuth 1981, Fishman 1996).

Exhaustive search has shown that for $M = 2^{31} - 1$, $A = 742938285, 950706376, 1226874159, 62089911, 1343714438$, and their multiplicative inverses perform better than other multipliers in dimensions $k = 2, \ldots, 6$ (Fishman and Moore 1986).

How good is best? The answer is that $d_k(A, M)$ for all LCGs with modulus M has the lower bound $M^{1/k} d_k(A, M) \geq \gamma_k$ for each $k = 2, 3, \ldots$, where $\{\gamma_k\}$ are constants based on geometric considerations (Cassels 1959, p. 332). In particular, $M = 2^{31} - 1$ induces

$$d_k(A, 2^{31}-1) \geq \begin{cases} 2.008 \times 10^{-5} & k=2 \\ 6.905 \times 10^{-4} & k=3 \\ 3.906 \times 10^{-3} & k=4 \\ 1.105 \times 10^{-2} & k=5 \\ 2.157 \times 10^{-2} & k=6 \\ 3.463 \times 10^{-2} & k=7 \\ 4.819 \times 10^{-2} & k=8, \end{cases} \tag{4.12}$$

revealing the coarseness of the grid in as few as four dimensions. Since this limitation applies to all prime modulus LCGs with $M = 2^{31} - 1$, the reader who uses one of them should be aware of the limited accuracy that results for estimates of λ, no matter how large sample size n is. Accordingly, if a multiplier of the form (4.1) is adopted, the reader is encouraged to use a multiplier for which $d_k(A, M)$ is close to its lower bound $\gamma_k/M^{1/k}$ for $k \geq 2$, thus limiting grid coarseness to the neighborhood of its global minimum.

There exists an additional lower bound that deserves attention. It is known that the form of the LCG in (4.1) imposes a lower bound $d_k(A, M) \geq (1 + A^2)^{-1/2}$, which militates against a small A (L'Ecuyer 1997). As illustration of how severely degraded an LCG can become because of a poor choice of multiplier, $A = 7$, a primitive root of $2^{31}-1$, induces the lower bound $d_k(A, M) \geq (1+7^2)^{-1/2} = 0.1414$. For $A = 742938285$, the lower bound is 1.35×10^{-9}, far below the theoretical bound in (4.12). However, for $A = 16807$, it is only 5.95×10^{-5}, somewhat comparable to the bound for $k = 2$.

4.2.2 Empirical Testing

Testing a simple hypothesis H against all alternatives usually requires specifying a

$$\text{type I error} := \text{pr(rejecting H} \mid \text{H is true)},$$

which determines the *rejection region,* the set of values of the test statistic for which the test rejects H. Although the corresponding

$$\text{type II error} := \text{pr(accepting H} \mid \text{H is false)}$$

is unknown, it usually converges to 0 as the sample size of the data set used in the test becomes large. Sometimes, one refers to the

$$\begin{aligned} \text{power} &:= 1 - \text{type II error} \\ &= \text{probability that the test correctly rejects } \text{H}_* \text{ when it is false,} \end{aligned}$$

which increases with sample size. In testing H_* in (4.10) for a PNG, no one has the illusion that it is true. Therefore, testing sample data sets of successively larger sizes should increase the power of the test, eventually causing H_* to be rejected. The sample size at which rejection occurs can serve as a measure of how close a PNG comes to producing an i.i.d. $\mathcal{U}(0, 1)$ sequence. Presumably, a small sample size implies a poor approximation, whereas a sample size considerably larger than that required for an MC experiment indicates an acceptable working approximation.

Many diverse statistical tests exist for evaluating the performance of PNGs. We concentrate on two, the collision and the birthday spacings tests, because of their focus, in L'Ecuyer (2001), on sample-path length as a measure of performance. Sections 4.2.3 and 4.2.4 describe the tests, revealing the relatively poor results when they are applied to LCGs. Later sections describe how other PNGs perform better.

4.2.3 Collision Test

Suppose we randomly toss t balls into l urns. Then

$$\begin{aligned} q_i &:= \text{pr(an urn contains } i \text{ balls)} \\ &= \binom{t}{i}\left(\frac{1}{l}\right)^i\left(1-\frac{1}{l}\right)^{t-i} \qquad i = 0, 1, \ldots, t. \end{aligned}$$

Moreover, the random variable

$$Y := \text{number of urns containing more than one ball} \tag{4.13}$$

has mean

$$\begin{aligned} \alpha &= l(1 - q_0 - q_1) \\ &\approx \frac{t^2}{l} \qquad \text{for } l \gg t, \end{aligned}$$

and, for fixed ratio t^2/l, its distribution converges to the Poisson with mean t^2/l, denoted by $\mathcal{P}(t^2/l)$, as $t \to \infty$. By fixed ratio, we mean that l increases in direct proportion to t^2.

Let $\{\mathbf{U}_{ki,k};\ i = 1, \ldots, t\}$ denote a PNG-generated sequence of t non-overlapping k-tuples. In the present context, each k-tuple corresponds to a ball, and a partition of the k-dimensional unit hypercube $[0, 1)^k$ into l cubes of equal volume corresponds to the l urns. To test H_*, the *collision test* uses Y in (4.13) calculated for the t k-tuples. Small and large values of Y are of concern. Small values imply that the k-tuples are more evenly distributed in $[0, 1)^k$ than in a truly random sample from an i.i.d. $\mathcal{U}(0, 1)$ sequence. Large values imply substantial non-uniformity in the sample sequence.

Recall that Section 4.2.1 focuses on overlapping k-tuples, whereas we are now focusing on non-overlapping k-tuples. If non-overlapping k-tuples are k-distributed, then so are overlapping k-tuples. To test H_* with its assertion of independence, we require non-overlapping k-tuples.

Given a sample sequence of t non-overlapping k-tuples, a *level 1* version of the collision test consists of the steps:

1. Choose a significance level

   $$\eta := \text{probability of rejecting } H_* \text{ when it is true } (\textit{type I error}).$$

   This is usually set at 0.01 or 0.05.

2. Choose $l :=$ number of cells, which determines each cell's side length l^{-1}.
3. Compute the counts in each cell.
4. Compute the test statistic Y.
5. From $\mathcal{P}(t^2/l)$, compute the following:

$$\begin{aligned} p_{\mathrm{R}}(Y) &:= \text{pr(observing a test statistic larger than } Y|\mathrm{H}_*) \\ p_{\mathrm{L}}(Y) &:= \text{pr(observing a test statistic smaller than } Y|\mathrm{H}_*) \\ \text{p-value} &= \min[p_{\mathrm{R}}(Y), p_{\mathrm{L}}(Y)] \qquad \text{(two-sided p-value)} \end{aligned}$$

6. If p-value $< \eta$, reject H_*; otherwise, no evidence of non-uniformity exists at the η level of significance. The two-sided p-value detects low and high values of Y.

LCGs that perform well on the theoretical spectral test begin failing the collision test at $t \approx 4 \times (\text{period})^{1/2}$, and they fail it decisively for $t \approx 16 \times (\text{period})^{1/2}$ (L'Ecuyer et al. 2000, L'Ecuyer and Simard 2001).

Using the rejection criterion p-value $< \eta = 0.01$, an application of this procedure for $k = 2$ revealed that of six tested LCGs with modulus $M = 2^{31} - 1$, five failed decisively, starting at sample-path lengths of about $4 \times (\text{period})^{1/2} \approx 185{,}364$ (L'Ecuyer et al. 2000). The failed LCGs include $A = 16807$ and $A = 742938285$. The one exception began failing at $8 \times (\text{period})^{1/2} \approx 370{,}728$. Because MC experiments can require considerably large sample sizes, these findings encourage the use of alternative generators such as CMRGs (see Section 4.5) and TGFSRs (see Sections 4.7 and 4.7.1), which have considerably longer periods and perform much better on these tests. Indeed, no rejection occurred on the collision test for the tested combined LCG (Section 4.5) and Mersenne twister (Section 4.7.1) for all sample-path lengths up to and including $t = 2^{20} \approx 10^6$.

4.2.4 Birthday Spacings Test

Let $B_1, \ldots, B_t$ denote a list of *birthdays* in a year of l days. Suppose we:

- Sort the birthdays into increasing order $B_{(1)} \leq B_{(2)} \leq \cdots \leq B_{(t)}$.
- Compute

$$\begin{aligned} Y := {} & \text{number of values that occur more than once among the spacings} \\ & B_{(2)} - B_{(1)}, B_{(3)} - B_{(2)}, \ldots, B_{(t)} - B_{(t-1)} \text{ in the sorted list.} \end{aligned} \tag{4.14}$$

Under the hypothesis

$$\mathrm{H}_{**} := \text{birthdays are uniformly distributed on the days } \{1, \ldots, l\}$$

and for fixed t^3/l, the distribution of Y in (4.14) converges to the Poisson distribution, $\mathcal{P}(t^3/4l)$, as $t \to \infty$ (Knuth 1981, Marsaglia 1985).

Again, consider a partition of the k-dimensional unit hypercube $[0, 1)^k$ into l cubes of equal volume and let the cubes be numbered $0, 1, \ldots, k-1$, for example, in lexicographical order of the coordinates of their centers (L'Ecuyer 2001). Then

$$B_j := \text{number of the cube into which the } j\text{th PNG-generated } k\text{-tuple falls}$$

serves as the birthday for this k-tuple. Again, small values of Y in (4.14) imply a more even distribution of the t k-tuples than a sample sequence of i.i.d. $\mathcal{U}(0, 1)$ random variables would induce, whereas large values imply more clustering than would be expected.

In a level 1 implementation of the test, the steps are essentially the same as those described above for the collision test, but with Y defined as in (4.14) and $\mathcal{P}(t^3/4l)$ replacing $\mathcal{P}(t^2/l)$. As before, the sample-path length t at which rejection occurs provides a measure of how well a particular LCG succeeds in producing sequences that resemble i.i.d. $\mathcal{U}(0, 1)$ sequences. It is known that LCGs that perform well on the spectral test decisively fail the birthday spacings test at $t \approx 16 \times (\text{period})^{1/3}$ (L'Ecuyer et al. 2000, L'Ecuyer and Simard 2001). For $M = 2^{31} - 1$, this would be at $t \approx 20{,}643$. Actually, for $A = 16807$, the test began failing at sample-path lengths as small as $t = 2^{14} = 16{,}384$. In all failed tests, the actual p-values were substantially below the $\eta = 0.01$ rejection criterion. By contrast, no rejection occurred for the tested CMRG in Section 4.5 and the Mersenne twister in Section 4.7.1 for all sample-path lengths up to and including $t = 2^{18} = 262{,}144$.

A *level 2* version of these tests uses the *chi-squared* or *Kolmogorov-Smirnov goodness-of-fit tests* on n non-overlapping sample sequences, each of length t_0, to test how well the distribution of Y fits the Poisson. For $t = nt_0$, empirical evidence suggests that if decisive failure occurs for a sequence of length t, then n non-overlapping subsequences, each of length t_0, usually fail the test for most seeds (L'Ecuyer et al. 2002). The advantage of level 1 over level 2 testing is that it allows for a more complete convergence to the limiting Poisson distribution for its longer single subsequence.

4.3 LCGs WITH MODULUS 2^β

LCGs with $M = 2^\beta$ for $\beta \geq 3$ are also in common use, although less so than prime-modulus generators. The appeal of "powers-of-two" LCGs stems from the greater computational efficiency they allow. All modern-day computers perform computations in base 2, using either integer or floating-point arithmetic. Integer arithmetic takes less computing time, and because of the relatively slow speeds that characterized computers in the early days of the computer age, a natural preference arose for using integer rather than floating-point arithmetic, whenever possible.

To evaluate Z_i in expression (4.1) takes multiplication (AZ_{i-1}), division and truncation ($K_i = \lfloor AZ_{i-1}/M \rfloor$), multiplication ($MK_i$), and subtraction

$(AZ_{i-1}-MK_i)$. Of these, multiplication and division are the more time-consuming operations. By choosing a modulus $M = 2^\beta$, a pseudorandom-number-generating program can be written in the machine language of the chosen computer that would use integer arithmetic and replace all but the first multiplication (AZ_{i-1}) by the faster-executing bit inspection and replacement.

Although faster, integer arithmetic on a computer imposes restrictions, because of finite word size, that induce other design constraints on pseudorandom number generation. Moreover, the speeds of current-day computers are so much greater than in the past as to make the penalty incurred in using floating-point arithmetic considerably less of an issue. As a consequence, many PNG implementations rely on extended-precision floating-point arithmetic to perform all computations.

If A and Z_{i-1} are odd integers, then $\left\{Z_i = AZ_{i-1} \pmod{2^\beta},\ i \ge 1\right\}$ is a sequence of odd integers with a period no greater than $2^\beta - 1$. However, the upper bound on period is not achievable. For $M = 2^\beta$, there exists no multiplier A that admits a full period $P(A, 2^\beta, Z_0) = 2^\beta - 1$. Indeed, the maximal achievable period is $2^{\beta-2}$, which occurs for Z_0 odd and

$$A \equiv 3 \pmod{8}$$

or

$$A \equiv 5 \pmod{8}.$$

That is, A must be in the set of integers $\{3 + 8j; j = 0, 1, \ldots, 2^{\beta-3}\} \cup \{5 + 8j; j = 0, 1, \ldots, 2^{\beta-3}\}$. With regard to the generated sequences (Jannson 1966),

- For $A \equiv 5 \pmod{8}$ and $Z_0 \equiv 1 \pmod{4}$, $\{Z_i, 1 \le i \le P\}$ is a permutation of the integers $4j + 1;\ j = 0, 1, \ldots, 2^{\beta-2} - 1$.
- For $A \equiv 5 \pmod{8}$ and $Z_0 \equiv 3 \pmod{4}$, $\{Z_i, 1 \le i \le P\}$ is a permutation of the integers $4j + 3;\ j = 0, 1, \ldots, 2^{\beta-2} - 1$.
- For $A \equiv 3 \pmod{8}$ and $Z_0 \equiv 1$ or $3 \pmod{4}$, $\{Z_i, 1 \le i \le P\}$ is a permutation of the integers $8j + 1$ and $8j + 3;\ j = 0, 1, \ldots, 2^{\beta-3} - 1$.
- For $A \equiv 3 \pmod{8}$ and $Z_0 \equiv 5$ or $7 \pmod{4}$, $\{Z_i, 1 \le i \le P\}$ is a permutation of the integers $8j + 5$ and $8j + 7;\ j = 0, 1, \ldots, 2^{\beta+3} - 1$.

Here $P := P(A, M, Z_0)$. For this PNG, an odd seed is essential and, for each A, determines which of the two disjoint sequences is generated. Because of the inherent non-uniformity that $A \equiv 3 \pmod{8}$ induces, virtually all PNG implementations with $M = 2^\beta$ focus exclusively on $A \equiv 5 \pmod{8}$. As illustration, the Numerical Algorithms Group (NAG) library contains the PNG

$$Z_i = \left(13^{13} Z_{i-1}\right) \pmod{2^{59}}.$$

4.3.1 $M = 2^{32}$

A commonly encountered implementation takes $\beta = 32$, implying a maximal period of $2^{30} = 1{,}073{,}741{,}824$. Moreover, for $A \equiv 5 \pmod{8}$ and odd Z_0, there are $2^{29} = 536{,}870{,}912$ maximal-period multipliers. For $M = 2^{\beta}$ ($\beta \geq 3$) and $A \equiv 5 \,(\text{mod}\,8)$, all k-tuples again lie on parallel hyperplanes.

Theoretical Evaluation. For $M = 2^{32}$ and $2 \leq k \leq 6$, exhaustive search revealed that multipliers $A = 1099087573, 2396548189, 2824527309, 3934873077, 2304580733$, and their multiplicative inverses perform best on the spectral test (Fishman 1990).

4.3.2 $M = 2^{48}$

Theoretical Evaluation. On 64-bit machines, $M = 2^{48}$ is a commonly employed modulus for LCGs. Evaluation of the spectral test in $2 \leq k \leq 6$ dimensions for $2^{26} = 67,108,864$ maximal-period multipliers revealed that $A = 68909602460261, 33952834046453, 43272750451645, 127107890972165, 55151000561141$, and their multiplicative inverses perform best (Fishman 1990).

Empirical Testing. Application of the level 1 collision test for $M = 2^{48}$ and $k = 2$ to $A = 68909602460261$ and two other multipliers revealed that all three began failing the test decisively at sample-path length $t = 2^{23} = 8{,}388{,}608$ (L'Ecuyer et al. 2002). Also, the LCG with $A = 25214903917$, which is in the SUN standard C library, failed the birthday spacings test at $t = 2^{19} = 524{,}288$ (L'Ecuyer and Simard 2001).

4.4 MIXED LINEAR CONGRUENTIAL GENERATORS

Consider the mixed generator

$$Z_i = (AZ_{i-1} + C) \,(\text{mod}\, M) \qquad i \geq 1, \tag{4.15}$$

which has the additional parameter C. For $M = 2^{\beta}$, C an odd integer, $A \equiv 1 \,(\text{mod}\ 4)$, and $\gcd(M, C) = 1$, this generator has full period M consisting of all integers $\{0, 1, \ldots, M-1\}$. The notation gcd denotes the greatest common divisor of M and C. Note that

- Z_i assumes all values in $\{0, 1, \ldots, M-1\}$ once per cycle.
- Any odd C satisfies $\gcd(M, C) = 1$.
- Bit shifting suffices to implement modulo reduction.

As examples of mixed LCGs,

$$\begin{aligned} Z_i &= (25214903917 Z_{i-1} + 11) \,(\text{mod}\ 2^{48}) \\ U_i &= \left(2^{27} \lfloor Z_{2i}/2^{22} \rfloor + \lfloor Z_{2i+1}/2^{21} \rfloor\right) / 2^{53} \end{aligned} \tag{4.16}$$

is used to implement the "nextdouble" method in the class java.util.Random of the Java standard library (L'Ecuyer 2001). The rand48 generator in the Unix library also employs this recurrence, but with $U_i = Z_i/2^{48}$.

The Ansi-C system PNG is

$$Z_i = (1103515245Z_{i-1} + 12345) \pmod{2^{31}}.$$

Microsoft Visual Basic (VB) employs

$$\begin{aligned} Z_i &= (1140671485Z_{i-1} + 12820163) \pmod{2^{24}} \\ U_i &= Z_i/2^{24}, \end{aligned} \tag{4.17}$$

which, because of its small modulus, can hardly be expected to give a good approximation to an i.i.d. $\mathcal{U}(0, 1)$ sequence.

For Windows 2.0, 3.0, 4.0, 97, 95, 7.0, and 5.0, Microsoft Excel employs a PNG called RAND():

$$U_i = (9821.0U_{i-1} + 0.211327) \pmod{1}. \tag{4.18}$$

This form suggests a direct floating-point implementation. Because of finite word size, this PNG inevitably encounters numerical roundoff error that prevents us from theoretically determining the PNG's period. However, suppose that the PNG were

$$Z_i = (9821Z_{i-1} + 211327) \pmod{10^6}. \tag{4.19}$$

Then it is known (see Jannson 1966, Theorem 3.13) that (4.19) has full period 10^6, far too small to give a good approximation to an i.i.d. $\mathcal{U}(0, 1)$ sequence. Although an empirical investigation of (4.18) that generated 6×10^6 observations has revealed that $U_i \neq U_{i+10^6}$ for $i \geq 1$ (Wagner 1999), the inability to characterize the period of this Excel PNG in (4.18) suffices to make it suspect. To overcome the limitations of this generator, an update to Excel 2003 incorporates a CMRG of the type described in Section 4.5.

Theoretical Evaluation. Although the full-period LCG in (4.15) with $M = 2^\beta$ may appear to be different from the maximal-period LCG in (4.1), elementary analysis reveals the former merely to be of the latter form with its values dilated and shifted. This is called an *affine transformation* (e.g., Overstreet 1974). As a consequence, one can show that an analysis of the randomness properties of maximal-period generators (4.1) with modulus $2^{\beta+2}$ suffices to characterize the randomness properties of full-period generators (4.15) with modulus 2^β. For example, for (4.15) with $M = 2^{48}$, $d_k(A, 2^{50})$ can be easily computed.

Empirical Testing. The Java PNG passed the collision test at sample-path lengths tested up to $t = 2^{20}$ (L'Ecuyer 2001). However, the VB PNG failed the test starting at $t = 2^{15} = 32768$, as did the Excel PNG starting at $t = 2^{18}$. The Java PNG failed the birthday spacings test at $t = 2^{18} = 262{,}144$, whereas the VB and Excel PNGs began failing it at $t = 2^{10}$ and $t = 2^{12}$ respectively.

EXAMPLE 4.4.1

IMPLICATION OF NON-UNIFORMITY

Because Figure 4.2 reveals that $A = 16807$ induces larger empty regions in $[0, 1)^2$ than $A = 742938285$, its use in practice is more suspect. However, some say that even with these large null regions $A = 16807$ continues to be an acceptable generator "to an acceptable level of approximation." Examples based on the acceptance-rejection method (see Section 3.6) illustrate that this is not so (Hörmann and Derflinger 1993).

The objective of the first example was to generate independent samples from $\mathcal{N}(0, 1)$ using a theoretically exact acceptance-rejection algorithm. The algorithm generated proposals from the Cauchy distribution $\{\eta(x) = [\pi(1 + x^2)]^{[-1]}, -\infty < x \leq \infty\}$ by the inverse-transform method. This was done for each of the prime-modulus LCGs in Figure 4.4 for 1,000,000 pseudorandom numbers each. The chi-squared goodness-of-fit test was applied to each sample to test the hypothesis H_0: the 1,000,000 samples were from $\mathcal{N}(0, 1)$. Using intervals of length, 10^{-5} resulted in the chi-squared test statistics $\chi^2_{99,999} = 206{,}056.0$ for $A = 16807$ and $\chi^2_{99,999} = 100,015.8$ for $A = 742938285$. The numbers are scores that need to be interpreted within a statistical context.

Let Y denote a random variable from the chi-squared distribution with 99,999 degrees of freedom. Then under H_0, the probability of observing Y larger than the test statistic for $A = 742938285$ is $\text{pr}(Y > 100{,}015.18) = 0.4853$. For $A = 16807$, it is $\text{pr}(Y > 206{,}056.0) = 10^{-25}$. These are p-values for the one-sided goodness-of-fit

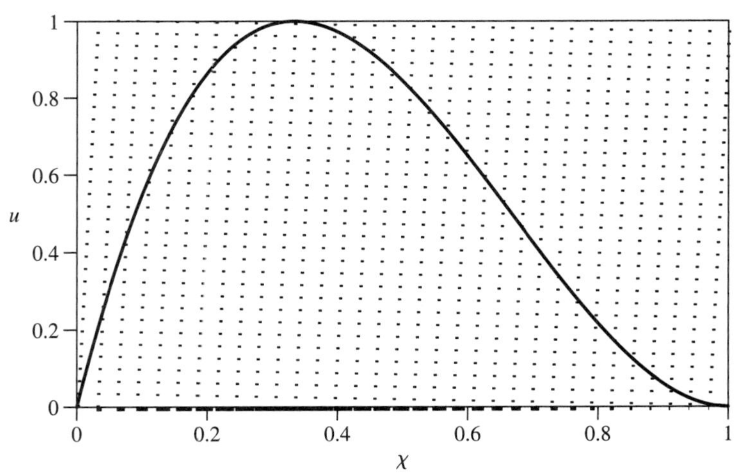

FIGURE 4.4 | *Acceptance pattern for* $\{f(x) = 12x(1-x)^2, 0 \leq x \leq 1\}$ *with* $\{Z_i = (33Z_{i-1} + 1)(\text{mod } 2^{10}),\ i \geq 1\}$

test. The low p-value for $A = 16807$ gives little credibility to H_0. The relatively large p-value for $A = 742938285$ does not contradict H_0.

A second example applied the acceptance-rejection method to the distribution

$$f(x) = 12x(1-x)^2 \qquad 0 \le x \le 1 \tag{4.20}$$

with proposals from the uniform distribution $\{\eta(x) = 1,\ 0 \le x \le 1\}$. Let H_0 now denote the hypothesis that the sample is drawn from (4.20). For the same sample size and intervals, $\chi^2_{99,999} = 367,131.6$ for $A = 16807$ and $\chi^2_{99,999} = 99,404.0$ for $A = 742938285$. Since $\mathrm{pr}(Y > 99,404.0) = 0.9089$ and $\mathrm{pr}(Y > 367,131.6) = 10^{-25}$, the test again rejects H_0 for $A = 16807$ but does not reject $A = 742938285$.

To probe into the cause of this rejection for $A = 16807$, again consider sampling via the acceptance-rejection method from (4.20). Following Section 3.6, we accept U_i as from (4.20) if

$$U_{i+1} \le \frac{9}{16} f(U_i). \tag{4.21}$$

Suppose we take $U_i := \frac{Z_i}{2^{10}}$, where the PNG $\{Z_i = (33Z_{i-1} + 1) \pmod{2^{10}},\ i \ge 1\}$ generates the pseudorandom numbers. Then acceptance occurs only for those points satisfying (4.21), whose projections onto the horizontal axis in Figure 4.4 reveal substantial gaps, especially for U_i near 0 and 1. Perusal of the graph suggests that these gaps arise from the way in which the 2-tuples are oriented in $[0, 1)^2$, and, therefore, would be more pronounced for $A = 16807$ than for $A = 742938285$. The PNG used here allows us to visualize the projection of gaps far more easily than would a generator with a larger modulus. ■

4.5 COMBINED GENERATORS

LCGs are merely a special case of the more general *multiple recursive generator* (MRG)

$$Z_i = (A_1 Z_{i-1} + \cdots + A_p Z_{p-1}) \pmod{M}, \tag{4.22}$$

where M and p are positive integers, $A_i \in \{0, 1, \ldots, M-1\}$ for each $i = 1, \ldots, p-1$, and $A_p \in \{1, \ldots, M-1\}$. This recurrence has period $M^p - 1$ if M is prime and the *characteristic polynomial* $z^p - A_1 z^{p-1} \cdots - A_{p-1}z - A_p$ is primitive.[2] Moreover, k-tuples $\{\mathbf{U}_{ik},\ i \ge 1\}$ are k-distributed on the k-dimensional set of points $\{0, \frac{1}{M}, \ldots, \frac{M-1}{M}\}^k$ for $k = 1, \ldots, p$, with one qualification. Each value occurs M^{p-k} times per cycle with the exception of $(0, \ldots, 0)$, which occurs $M^{p-k} - 1$ per cycle.

Although (4.22) is computationally easier to implement for small $A_1, \ldots, A_p$, $d_k(A_1, \ldots, A_p, M)$, the maximal distance between neighboring parallel hyperplanes in k dimensions has lower bound $(1 + \sum_{i=1}^{p} A_i^2)^{-1/2}$, implying that large $A_1, \ldots, A_p$ are more beneficial with regard to distributional uniformity (Grube

[2]The concept of a primitive polynomial extends the notion of primitivity given in Section 4.1.1. The exact definition need not concern us here.

1973, L'Ecuyer 1997). Although the conflicting goals of computational efficiency and good distributional properties limit direct application of (4.22), both objectives are achievable by combining several MRGs.

Consider J MRGs (L'Ecuyer 1996)

$$Z_{ji} = \left(A_{j1} Z_{j,i-1} + \cdots + A_{jp} Z_{j,i-p}\right) \pmod{M_j} \qquad j = 1, \ldots, J, \tag{4.23}$$

where $M_1, \ldots, M_J$ are distinct primes, and the characteristic polynomial $z^p - A_{j,1} z^{p-1} \cdots - A_{j,p-1} z - A_{j,p}$ is primitive for each $j = 1, \ldots, J$. As illustration, consider the combination of $J = 2$ MRGs with $p = 3$, $M_1 = 2^{32} - 209 = 4294967087$, $M_2 = 2^{32} - 22853 = 4294944443$, and

$$\begin{array}{lll} A_{11} = 0 & A_{12} = 1403580 & A_{13} = -810728 \\ A_{21} = 527612 & A_{22} = 0 & A_{23} = -1370589, \end{array} \tag{4.24}$$

which satisfy primality and primitivity.

Let $\delta_1, \ldots, \delta_J$ denote arbitrary integers such that for each j, δ_j and M_j have no common factors. Consider the combined generator

$$\tilde{Z}_i := \left(\sum_{j=1}^{J} \delta_j Z_{ji}\right) \pmod{M_1} \quad i \geq 1 \tag{4.25}$$

and the normalized sequence $\{\tilde{U}_i := \tilde{Z}_i / M_1,\ i \geq 1\}$. The CMRG with moduli and coefficients as above, $\delta_1 = 1$, and $\delta_2 = -1$ is called mrg32k3a (L'Ecuyer 1999).

What can we say about the properties of CRMGs? Because answering this question directly for (4.25) is difficult, we consider a related generator that is considerably easier to analyze and whose properties are remarkably close to those for (4.22). Let

$$\begin{aligned} U_{ji} &:= Z_{ji}/M_i, \\ V_i &:= \left(\sum_{j=1}^{J} \delta_j U_{ji}\right) \pmod{1} \qquad i \geq 1, \end{aligned} \tag{4.26}$$

and

$$\tilde{M} := \Pi_{j=1}^{J} M_j.$$

Then

P6. $\{V_i,\ i \geq 1\}$ obeys the recurrence

$$Y_i = (\tilde{A}_1\, Y_{i-1} + \cdots + \tilde{A}_p\, Y_{i-p}) \pmod{\tilde{M}}, \tag{4.27}$$

where $Y_i = \tilde{M} V_i$ and $\tilde{A}_1, \ldots, \tilde{A}_p$ are computable and do not depend on $\{\delta_j\}$.

P7. $\{V_i\}$ and, therefore, $\{Y_i\}$ have period

$$\begin{aligned} P &= \operatorname{lcm}\left(M_1^p - 1, \ldots, M_J^p - 1\right), \\ &:= \text{the lowest common multiple of } M_1^p - 1, \ldots, M_J^p - 1, \end{aligned}$$

where

$$P \leq (M_1^p - 1) \times \ldots \times (M_J^p - 1)/2^{J-1}.$$

P8. $\tilde{U}_i = V_i + \epsilon_i$, where tight lower and upper bounds can be computed on ϵ_i.

As illustration, mrg32k3a has

$$\begin{aligned}\tilde{M} &= 18446645023178547541\\ \tilde{A}_1 &= 18169668471252892557\\ \tilde{A}_2 &= 3186860506199273833\\ \tilde{A}_3 &= 8738613264398222622\end{aligned}$$

and induces a very small error bound (L'Ecuyer 2003). Observe that $A_{11},\ A_{12},\ A_{13},\ A_{21},\ A_{22},\ A_{23}$ are small, allowing for computational efficiency in implementation. However, $\tilde{A}_1,\ \tilde{A}_2,\ \tilde{A}_{13}$ are large, making for the lower bound $d(\tilde{A}_1, \tilde{A}_2, \tilde{A}_{13}, \tilde{M}) \geq 4.9 \times 10^{-20}$ for the maximal distance between successive parallel hyperplanes in dimensions $k \geq 4$.

Properties P6, P7, and P8 are important. They imply that the period is at least $\min_{1\leq j\leq J}(M_j^p - 1)$ and that judiciously chosen $M_1, \ldots, M_J$ can induce a period as large as $(M_1^p - 1) \times \ldots \times \left(M_J^p - 1\right)/2^{J-1}$. In particular, the maximum is achieved if $\left(M_1^p - 1\right)/2, \ldots, \left(M_J^p - 1\right)/2$ have no factors in common, as in the case of M_1 and M_2 in mrg32k3a, for which $P \approx 2^{191} \approx 3.1 \times 10^{57}$. This period is substantially larger than those for LCGs with moduli $2^{31} - 1$ and 2^{32}.

Property P6 tells us that the normalized $\{V_i,\ i \geq 1\}$ satisfy a p-step recursion $V_i = (\tilde{A}_1\ V_{i-1} + \cdots + \tilde{A}_p\ V_{i-p}) \pmod 1$. Property P8 reveals that $\tilde{U}_i$, based on the generator (4.25), differs from V_i in (4.26) by a negligible amount. Therefore, it is not unreasonable to characterize the properties of $\{\tilde{U}_i\}$ by those of $\{V_i\}$, say, by the spectral test.

Excel 2003 also uses combined PNGs. It has three one-step LCGs (4.1) with

$$\begin{aligned}A_1 &= 171 \quad M_1 = 30269\\ A_2 &= 172 \quad M_2 = 30307\\ A_3 &= 170 \quad M_3 = 30323,\end{aligned}$$

where $M_1,\ M_2$, and M_3 are prime and A_i is a primitive root of M_i for $i = 1, 2, 3$ (Wichmann and Hill, 1982, 1987). It has period $\tilde{M} \geq 2.78 \times 10^{13}$ and is 2-distributed.

Theoretical Evaluation. The CMRG mrg32k3a is 3-distributed. Recall that in a full period of $M - 1 = 2^{31} - 2$ numbers, the two LCGs in Figures 4.2a and 4.2b each show regular patterns of about 2,147 2-tuples on a skewed grid in $[0, 0.001) \times [0, 0.001)$. As a basis for comparison, Figure 4.5 displays the first 2,147 2-tuples in an mrg32k3a subsequence that fell into $[0, 0.001) \times [0, 0.001)$. The subsequence length of 2,110,829,754 numbers is close to the period of the LCGs. In contrast to the skewed grids in Figures 4.2a and 4.2b, the absence of regularity in Figure 4.5 allows us to conclude that for sequences of approximate

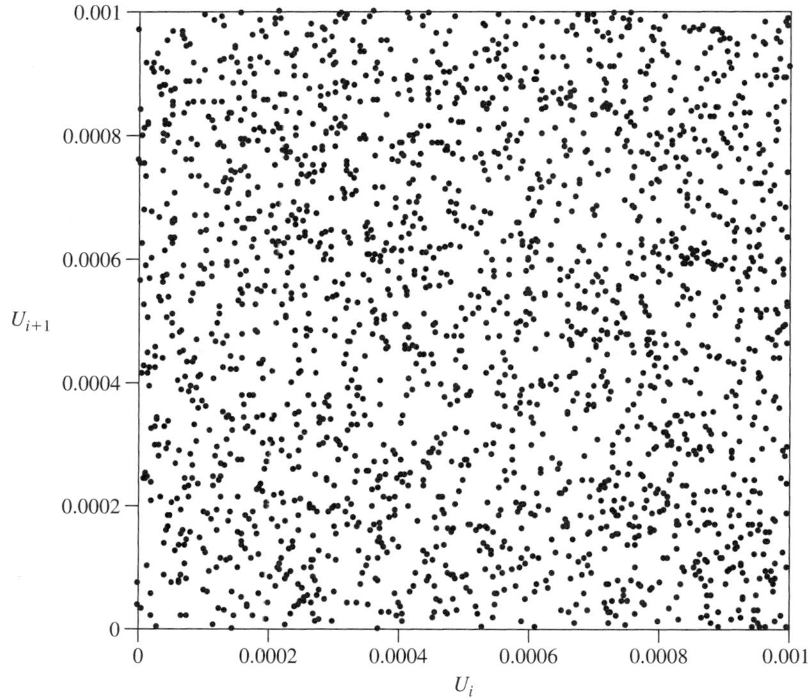

FIGURE 4.5 | *2-tuples* $(\tilde{U}_i, \tilde{U}_{i+1})$: $\tilde{U}_i := \frac{\tilde{Z}_i}{M_1+1}$ *for mrg32k3a generator*

length $2^{31} - 2$, the combined generator eliminates the apparent lattice structure inherent in LCGs.

We emphasize that Figures 4.2a and 4.2b are based on complete cyles of all numbers in the LCGs, whereas Figure 4.5 is based on a subsequence of numbers from mrg32k3a. Because of 2-distributedness, a complete cyle of $P > 2^{191}$ numbers from mrg32k3a would array the 2-tuples in $[0, 0.001), \times [0, 0.001)$ on a square grid with distance $\frac{1}{M_1} \approx 2^{-32}$ between successive parallel lines. For dimensions 4 through 10, Table 4.1 shows that mrg32k3a's performance on the spectral test substantially exceeds that for LCGs with modulus $2^{31} - 1$.

Empirical Evaluation. Application of the collision test to mrg32k3a failed to reject H_* for sample-path lengths up to and including $t = 2^{20} \approx 10^6$. Application of the birthday spacings test failed to reject this generator for sample-path lengths up to and including $t = 2^{18} = 262,144$ (L'Ecuyer 2001).

Initialization, Reproducibility, and Independent Experiments. In contrast to a single seed for LCGs, a CMRG requires Jp seeds $Z_{j,0}, Z_{j,-1}, \ldots, Z_{j,-(p-1)}$ for $j = 1, \ldots, J$ to prime the J recursions. For MRG j, $Z_{j,0}, Z_{j,-1}, \ldots, Z_{j,-(p-1)}$ can each assume any integer value in $\{0, 1, \ldots, M_j - 1\}$, but not all 0. Saving these p seeds for each component generator enables one to

TABLE 4.1 | *Distance between parallel hyperplanes**

Dimension k	LCG lower bound $d_k(A, 2^{31}-1) \geq$	Minimal distance for mrg32k3a $d_k(\tilde{A}, \tilde{M})$
2	2.008×10^{-5}	
3	6.905×10^{-4}	
4	3.906×10^{-3}	3.52×10^{-15}
5	1.105×10^{-2}	3.27×10^{-12}
6	2.157×10^{-2}	2.61×10^{-10}
7	3.463×10^{-2}	5.57×10^{-9}
8	4.819×10^{-2}	6.02×10^{-8}
9	6.495×10^{-2}	3.66×10^{-7}
10	8.713×10^{-2}	1.50×10^{-6}

*Source: Cassels (1959, p. 332) and L'Ecuyer (2003).

reproduce the same subsequences of pseudorandom numbers after correcting or modifying the MC experiment's executable code. Saving the last p numbers for each MRG allows one to use them to prime the recursions on a subsequent run and thereby have independent subsequences.

Implementation. L'Ecuyer (1999) contains concise C-language code for mrg32k3a. It uses floating-point arithmetic and works under the condition that all integers between -2^{53} and 2^{53} are representable exactly as floating-point numbers. To avoid $\tilde{U}_i = 0$, it computes $\tilde{U}_i$ as $\tilde{Z}_i/(M_1+1)$, in place of M_1 in the denominator. Doing so avoids an error in execution when computing $\ln \tilde{U}_i$ or $\ln(1-\tilde{U}_i)$ for exponential variate generation. L'Ecuyer (1998) describes the implementation of other CMRGs for moduli close to 2^{63}, which take advantage of the larger word size available on unix machines.

4.6 AWC AND SWB GENERATORS

Add-with-carry (AWC) generators take the form (Marsaglia and Zaman 1990)

$$Z_i = (Z_{i-s} + Z_{i-r} + C_i) \pmod{M} \qquad i \geq 1, \tag{4.28}$$

where lags r and s and base M are positive integers and

$$C_{i+1} := \begin{cases} 1 & \text{if } Z_{i-s} + Z_{i-r} + C_i \geq M \\ 0 & \text{elsewhere.} \end{cases}$$

Because they require no multiplication, these generators are faster than LCGs and CMRGs. The sequence $\{Z_i,\ i \geq 1\}$ has period $P = M^r + M^s - 2$, which is full period provided that $M^r + M^s - 1$ is prime. As illustration, M about 2^{31} and r

about 20 yield $P \approx 2^{620}$ (Tezuka et al. 1992), a considerably larger period than previously described PNGs offer.

The corresponding fractional sequence takes the form

$$U_i := \sum_{j=1}^{L} \frac{Z_{Li+j+1}}{M^j} \qquad i \geq 1, \tag{4.29}$$

where $L \leq r$. If $M = 2^q$ for some positive integer q, then U_i is an Lq-bit fraction. In particular, for $q = 32$, $L = 1$ suffices to generate 32-bit fractions. If L and $M-1$ have no factors in common other than unity, U_i and Z_i have the same period. For $q \geq 32$, $L = 1$ is a frequent choice, thus ensuring a common period.

Subtract-with-borrow (SWB) generators differ from (4.28) in that

$$Z_i = (Z_{i-s} + -Z_{i-r} - C_i) \pmod{M} \qquad i \geq 1, \tag{4.30}$$

with period $P = M^r - M^s$, which is full provided that $M^r - M^s + 1$ is prime.[3] As illustration, the SWB generator in Marsaglia et al. (1990), hereafter called MNZ, has $M = 2^{32} - 5$, $r = 43$, $s = 22$, and $L = 1$. Since the corresponding $M^r - M^s + 1$ is prime and M is a primitive root of this prime, MNZ has full period $P = 2^{1376} - 2^{704} + 1 \approx 1.6 \times 10^{414}$. Marsaglia and Zaman (1990) provide other parametric values for AWC and SWB generators.

Although these generators appear different than the LCG formulation (4.1), they are in reality closely related.

Theorem 4.6.1 (Tezuka et al. 1992)

Consider the LCG

$$Y_i = \ddot{A} Y_{i-1} \ (mod \ddot{M}) \qquad i \geq 1, \tag{4.31}$$

where

$$\ddot{M} := \begin{cases} M^r + M^s - 1 & \textit{for an AWC generator} \\ M^r - M^s + 1 & \textit{for an SWB generator} \end{cases}$$

is prime and

$$\ddot{A} := \textit{multiplicative inverse of } M (mod \ddot{M})$$

$$V_i := \frac{Y_i}{\ddot{M}} \tag{4.32}$$

$$W_i := V_{Li} = \frac{Y_{Li}}{\ddot{M}}.$$

If Z_0 in (4.28) based on Z_{-r}, Z_{-s} and C_0, and Y_0 in (4.31) correspond, then for all $i \geq r$, the (fractional) digital expansion in base M of U_i in (4.29) and W_i in (4.32) have the same first L digits. In other words, $U_i = M^{-L}(M^L W_i)$. ■

[3]Two additional variants of this generator also exist.

The condition on Z_0 and Y_0 means that the two sequences have the same initial seed. (See Section 4.1.1 for the definition of multiplicative inverse.) As illustration, MNZ has

$$\ddot{A} = \left((2^{32} - 5)^{21} - (2^{32} - 5)^{42}\right) \pmod{\ddot{M}}$$
$$\ddot{M} = 2^{1376} - 2^{704} + 2.$$

Initialization, Reproducibility, and Independent Experiments. An AWC/SWB generator requires r numbers $Z_0, Z_{-1}, \ldots, Z_{-(r-1)}$ to prime the recursion. For example, MNZ requires 43 numbers. Saving the r seeds enables one to reproduce the same subsequence of pseudorandom numbers after correcting or modifying the MC experiment's executable code. Saving the last r numbers at the end of a run allows one to use them to prime the recursion on a subsequent experiment and thereby have independent subsequences.

Theoretical Evaluation. If $M^r + M^s - 1$ is prime for an AWC generator, then its k-tuples based on (4.28) are k-distributed for all $k = 1, 2, \ldots, r$. The same property holds for SWB generators if $M^r - M^s + 1$ is prime. As illustration, MNZ is k-distributed up to $k = 43$ dimensions. For $k > r$, AWC/SWB generators become severely degraded.

Theorem 4.6.2 (Tezuka et al. 1992)

For the LCG (4.32),

$$d_k(\ddot{A}, \ddot{M}) \geq \frac{1}{\sqrt{3}} \qquad k \geq r + 1.$$

■

For MNZ, this implies that $d_k(\ddot{A}, \ddot{M}) \geq \frac{1}{\sqrt{3}}$ for $k \geq 44$ dimensions.

For $M^r + M^s - 1$ (or $M^r - M^s + 1$) prime, Theorem 4.6.1 has merely academic interest. However, if not prime, the k-distributedness would not hold in all dimensions $k \leq r$, and the spectral distance $\left\{d_k(\ddot{A}, \ddot{M})\right\}$ would then provide an assessment of how well the points are distributed. (For details, see Marsaglia and Zaman [1990] and Tezuka et al. [1992].)

Empirical Evaluation. A level 2 application of the birthday spacings test applied to MNZ in L'Ecuyer (1992) yielded a p-value $< 10^{-15}$, far below the $\eta = 0.01$ significance level discussed earlier. The test was based on 100 non-overlapping subsequences, each of length $t = 5{,}000$. The reference also reveals that MNZ failed the *poker test* at the 0.01 significance level.

Implementation. Marsaglia et al. (1990) describe an algorithm for fast MNZ generation.

4.7 TWISTED GFSR GENERATORS

Among the several classes of PNGs listed in the introduction to this chapter, feedback shift registers (FSRs) and generalized feedback shift registers (GFSRs) operate directly on the individual binary bits of pseudorandom numbers. The Mersenne twister, a generalization of GFSRs with demonstrably superior randomness properties compared to other PNGs, especially LCGs, is the product of successive refinements of GFSR procedures.

For integers $q \geq L > 0$, let $\{B_i,\ i \geq 1\}$ denote a $0-1$ sequence of binary bits that form the L-bit integers

$$Z_i := \sum_{l=1}^{L} 2^{L-l} B_{i-q(l-1)} \in \{0, 1, \ldots, 2^L - 1\} \qquad i \geq 1, \tag{4.33}$$

which are then transformed into fractions

$$U_i := \frac{Z_i}{2^L} = \sum_{j=1}^{L} 2^{-l} B_{i-q(l-1)} \in [0, 1) \qquad i \geq 1.$$

The corresponding binary vector representation is

$$\mathbf{W}_i := (B_i, B_{i-q}, \ldots, B_{i-q(L-1)}), \tag{4.34}$$

which consists of an L-bit number whose successive bits are spaced q places apart in the binary sequence. The parameter q is called the *delay*.

For integers $N > L \geq Q > 0$, $B_{i-q(l-1)}$, the lth bit of $\mathbf{W}_i$, satisfies the recursion

$$B_{i-q(l-1)} = \left(B_{i-Q-q(l-1)} + \sum_{j=1}^{L} B_{i-N-q[(j-1)+(l-1)]} C_{jl} \right) \pmod 2 \qquad l = 1, \ldots, L, \tag{4.35}$$

where each coefficient $C_{jl} \in \{0, 1\}$ is either 0 or 1. Then[4]

$$\mathbf{W}_i = \mathbf{W}_{i-Q} \oplus (\mathbf{W}_{i-N}\mathbf{C}) \qquad i \geq 1, \tag{4.36}$$

where $\mathbf{C}$ denotes an $L \times L$ matrix with C_{jl} in row j and column l, $\mathbf{W}_{i-N}\mathbf{C}$ denotes the row vector with $\sum_{j=0}^{L-1} B_{i-N-jq} C_{jl}$ in column l for $l = 1, \ldots, L$. The notation $\oplus$ denotes the *exclusive or* operation, $0 \oplus 0 = 0$, $1 \oplus 0 = 1$, $0 \oplus 1 = 1$, and $1 \oplus 1 = 0$. That is, addition is performed bit-wise modulo 2. Expression (4.36) is merely a concise way of expressing the entries

$$B_{i-(l-1)q} = B_{i-Q-(l-1)q} \oplus \left(\sum_{j=1}^{L} B_{i-N-jq} C_{jl} \right) \qquad l = 1, \ldots, L$$

in the row vector $\mathbf{W}_i$.

[4]Here $\mathbf{C} := \mathbf{T}^{-1}\mathbf{A}\mathbf{T}$ for the $\mathbf{A}$ matrix and invertible *tempering matrix* $\mathbf{T}$ employed in Matsumoto and Kurita (1994) and Matsumoto and Nishimura (1998).

Let $\mathbf{I}$ denote an $L \times L$ *identity matrix* with 1 in each diagonal entry and 0 elsewhere. For $\mathbf{C} = \mathbf{I}$,

$$\mathbf{W}_i = \mathbf{W}_{i-Q} \oplus \mathbf{W}_{i-N} \qquad i \geq 1$$

is called a GFSR. If N and Q are suitably chosen, then $\{\mathbf{W}_i,\ i \geq 1\}$ has period $2^N - 1$ (Lewis and Payne 1973). By "suitably chosen," we mean the characteristic polynomial of the generator (4.36) is primitive over the space of L-bit integers.

Theoretical Evaluation. The concept of *k-distributedness* again provides a way to characterize randomness. Suppose that a sequence $\{U_i \in [0, 1),\ i \geq 1\}$ has period $2^p - 1$. The sequence is said to be k-distributed to L-bit accuracy on $[0, 1)^k$ if and only if the sequence of k-tuples $\{\mathbf{U}_{ik};\ i = 1, \ldots, 2^p - 1\}$ assumes each value in $\left\{\frac{1}{2^L}, \frac{2}{2^L}, \ldots, \frac{2^L-1}{2^L}\right\}^k$ 2^{p-kl} times and assumes the value $\{0, \ldots, 0\}$ $2^{p-kl} - 1$ times. By contrast, it is easily seen that LCGs are at most 1-distributed. In that case, the spectral test (see Section 4.2.1) is about the best one can do to assess an individual LCG's properties in $k > 1$ dimensions.

For GFSRs, we can do much better. For $Q = L$, N suitably chosen, and $\mathbf{W}_1, \ldots, \mathbf{W}_N$ assigned appropriate initial values, the sequence $\{\mathbf{W}_i, i \geq N+1\}$ is k-distributed to L-bit accuracy for all $k \leq \lfloor N/L \rfloor$ (Fushimi and Tezuka 1983). For example, $L = 32$ and $N = 521$ ensure the property for $k \leq 16$ (Fushimi 1990). The initializing L-seed requirement arises from the fact that GFSRs do not realize all NL-bit binary subsequences during a cycle. As a consequence, the period $2^N - 1$ cannot be realized unless the NL bits used to prime the GFSR form a subsequence of the full $L(2^N - 1)$-bit sequence.

Twisted GFSRs are a generalization of GFSRs that eliminate the need for the specialized initializing procedure just described. Moreover, they can substantially increase period and the maximal number of dimensions, $k(L)$, in which k-distributedness to L-bit accuracy applies. For $Q = L$ and suitably chosen N and $\mathbf{C}$, (4.36) has period $2^{NL} - 1$ and upper bound $k(L) \leq N$ on k-distributedness. Moreover, because all NL-bit subsequences occur in a cycle, any NL-bit subsequence, except all zeros, can initialize the generator to achieve these properties (Matsumoto and Kurita 1994). The matrix $\mathbf{C}$ can be chosen so as to bring $k(L)$ close or equal to the upper bound N. For example, TT800 in Matsumoto and Kurita (1994) with $L = 32$, $Q = 7$, $N = 25$ has period $2^{800} - 1 \approx 10^{241}$ and $k(32) = 25$. The reference also lists a C-language program for this generator for 32-bit machines.

4.7.1 Mersenne-Twisted GFSRs

The determination of a suitable N requires the factorization of $2^{NL} - 1$ into its prime integers, which becomes computationally intractable as N gets large. This has limited twisted GFSRs to relatively small N, although $2^{800} - 1$ is a substantial period compared to virtually all previous alternatives. However, the *Mersenne twister* provides a means of overcoming this limitation on N and thus admits larger periods and larger-dimensional k-distributedness.

Let r be in $\{0, 1, \ldots, L-1\}$, let $\mathbf{W}_i^u$ denote $\mathbf{W}_i$ in (4.34), with bits $1, \ldots, L-r$ replaced by zeros, and let $\mathbf{W}_i^l$ denote $\mathbf{W}_i$ with bits $L-r+1, \ldots, L$ replaced by zeros. Let $\mathbf{W}_i^u|\mathbf{W}_j^l$ denote the L-bit vector obtained by adding $\mathbf{W}_i^u$ and $\mathbf{W}_j^l$. This is called the *concatenation* of these quantities. The Mersenne twister takes the form (Matsumoto and Nishimura 1998)

$$\mathbf{W}_i = \mathbf{W}_{i-Q} \oplus (\mathbf{W}_{i-N}^u|\mathbf{W}_{i-N+1}^l)\mathbf{C} \qquad i \geq 1. \tag{4.37}$$

The corresponding bit generator is

$$B_{i-q(l-1)} = \left(B_{i-l(q-1)-Q} + \left[\sum_{j=1}^{L-r} B_{i-N-q(l-1)-q(j-1)} \right.\right. \\ \left.\left. + \sum_{j=L-r+1}^{L} B_{i+1-N-q(l-1)-q(j-1)}\right] C_{jl}\right) \pmod 2. \tag{4.38}$$

If $r = 0$, then we have the twisted GFSR (4.36). However, if N, Q, r, and $\mathbf{C}$ are suitably chosen, then $\{\mathbf{W}_i,\ i \geq 1\}$ in (4.37) and the corresponding sequence $\{U_i, i \geq 1\}$ has period $2^{NL} - 1$ and is $k(L)$-distributed to L-bit accuracy for $k(L) \leq N - 1$.

Theoretical Evaluation. The extra parameter r is the key to increasing the period and the dimension of k-distributedness. For fixed L and N, choosing r so that $2^{NL-r} - 1$ is prime eliminates the computational intractability of determining the prime factors of $2^{NL-r} - 1$ so that a larger N is possible than in the original formulation (4.36). In addition to the checking for primality being far less costly than prime factorization, this approach facilitates the check for the suitability of the proposed Q, N, and $\mathbf{C}$ for full period and k-distributedness. Prime numbers of the form $2^p - 1$ are called *Mersenne numbers,* hence the name Mersenne twister.

Matsumoto and Nishimura (1998) give generator MT19937 with $L = 32$, $N = 624$, $Q = 7$, $r = 31$, and suitable $\mathbf{C}$, yielding period $2^{19937} - 1 \approx 10^{6002}$ and k-distributedness for $k(32) = 623$ dimensions, considerably better attributes for a PNG than those for any other proposed generators. SAS has incorporated MT19937 into its application software.

For a subsequence of MT19937, Figure 4.6 displays the first 2,147 2-tuples that fell into $[0, 0.001) \times [0, 0.001)$. The length of the subsequence was 2,054,820,571, close to $2^{31} - 2$, the period of the LCGs associated with Figures 4.2a and 4.2b. Since this number 2,147 is exactly the same number as in these earlier displays, we can make a revealing comparison with the displays in those figures. Most notably, the absence of any regularity in Figure 4.6 allows us to conclude that for subsequences whose lengths are approximately the period of the LCGs in Figures 4.2a and 4.2b, this Mersenne twister succeeds in eliminating the apparent lattice structure inherent in LCGs. Again, we note that a plot of all 2-tuples in the MT19937 full-period sequence of $2^{19937} - 1$ numbers would array themselves on a square grid in equal numbers except one less for $(0, 0)$, with distance 2^{-32}

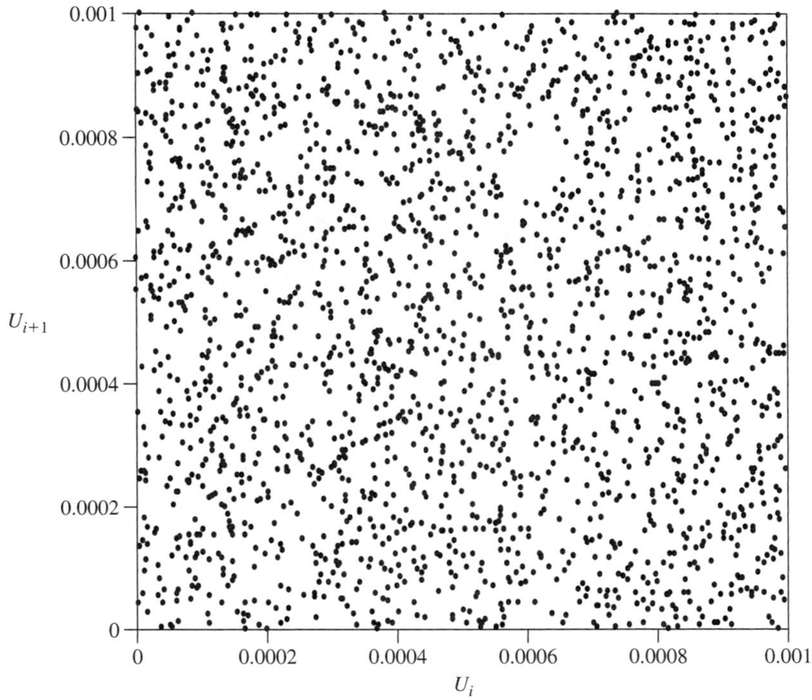

FIGURE 4.6 | *2-tuples* (U_i, U_{i+1}) *for the* $MT19937$ *Mersenne twister*

between successive parallel lines. Indeed because of 623-distributedness, k-tuples in dimensions 2 through 623 array themselves in equal numbers, with one less for $(0, \ldots, 0)$.

Empirical Testing. Level 1 applications of the collision and birthday spacings tests to MT19937 failed to reject H_* for sample-path lengths up to and including $t = 2^{20} \approx 10^6$ and $t = 2^{18} = 262{,}144$ respectively (L'Ecuyer 2001).

Implementation, Initialization and Reproducibility, and Independent Experiments. A C-language implementation mt19937ar-cok.c, available at

http://www.math.sci.hiroshima-u.ac.jp/~m-mat/MT/MT2002/CODES/MTARCOK/mt19937ar-cok.c

is more computationally efficient than the original code in Matsumoto and Nishimura (1998). The new code has options for generating 32-bit numbers in [0, 1] (genrand_real11), in [0, 1) (genrand_real12), and in (0, 1) (genrand_real13), and 53-bit numbers in [0, 1) (genrand_res53). It can also generate 31-bit integers (genrand_int31) and 32-bit integers (genrand_int32).

Whereas an LCG requires a single quantity Z_0 to prime the generator, a twisted GFSR (TGFSR) requires N L-bit integers. In principle, any sequence of N L-bit integers, except all zeros, allows a TGFSR to realize its full period and its k-distributedness properties. In practice, the initial choice can markedly influence the values of the finite sequence of pseudorandom numbers generated in an MC experiment, seriously degrading the quality of final results in an MC experiment. For example, $N-1$ zero numbers followed by unity for the Nth number, which is equivalent to $Z_{-1} = \cdots = Z_{-N+1} = 0$ and $Z_{-N} = 1$, tends to produce small Z_i for $i < 10{,}000$ (Matsumoto 2002). This can be very serious when the sampling experiment uses fewer than this many pseudorandom numbers. In short, TGFSRs primed with an arbitrary sequence of NL bits have a warm-up interval much as Markov-chain sample paths begun from arbitrary starting states do in Chapter 6.

The C-language code in mt19937ar-cok.c contains two options for initializing the PNG each time it is used. They are designed to reduce the severity of the initialization problem. However, their straightforward use admits the possibility that successive MC experiments would use overlapping sequences of pseudorandom numbers and, therefore, not be independent. To overcome the initialization problem and ensure non-overlapping sequences on successive runs, an alternative approach proceeds as follows:

a. On the first use of mt19937ar-cok.c, assign 624 arbitrarily selected 32-bit numbers to state(1), ..., state(624), the array on which mt19937ar-cok.c operates.

b. Generate T uniform deviates using genrand_real12 for some $T \gg 624$.

c. Save the final values of state(1), ..., state(624) and the value of the "left" variable in a *seed file.*

One potential source of the first 624 arbitrary numbers is init_genrand() in mt19937ar-cok.c. Taking $T = 100{,}000$ should reduce the influence of the 624 initial numbers on the final saved numbers to a negligible level. When thereafter executing an MC experiment,

d. Load the 624 numbers from the seed file into state(1), ..., state(624).

e. Execute the MC experiment with successive calls to genrand_real3 each time a uniform deviate is needed.

f. At the end of the MC experiment, save the final state(1), ..., state(624) and the value of the "left" variable in the seed file.

Performing steps d, e, and f ensures that the subsequence of uniform deviates generated on successive MC experiments do not overlap, an essential property when the intention is to make the different experiments statistically independent. Note that steps a, b, and c need to be performed only once.

A C-language program, rng_afm.c, executes steps a through f and is available at

http://www.or.unc.edu/~gfish/fcmc.html

Unix machines. For 64-bit machines, Nishimura (2000) describes two sets of parameters for the Mersenne twister with period $2^{19937} - 1$ and k-distributed up to $k(64) = 312$ dimensions. Nishimura also provides a C-language program.

4.8 LESSONS LEARNED

- Many different algorithms and implementations exist for pseudorandom number generators (PNGs).
- Linear congruential generators (LCGs) are the most commonly implemented (PNGs). However, they exhibit larger errors than other PNGs when generating sequences as approximations to i.i.d. $\mathcal{U}(0, 1)$ sequences.
- Theoretical and empirical tests exist to evaluate the quality of PNGs. They confirm the relatively poor ranking for LCGs when compared with other PNGs. More generally, testing reveals that the Mersenne twister (see Section 4.7.1) performs better on the theoretical tests than other classes of generators and at least as well on empirical tests.
- All PNGs require seeds to prime them. For example, an LCG needs a single number as input, whereas the Mersenne twister mt19937 requires 623 numbers. An appropriate software implementation can automatically store the final pseudorandom numbers on one run and use them on the next run. Doing so ensures that the PNG subsequences are independent in successive runs. Then MC results on different runs are independent.
- Samples generated by the acceptance-rejection and ratio-of-uniforms methods are more sensitive to the errors in their PNG input than is the inverse-transform method.

4.9 REFERENCES

Cassels, J. W. S. (1959). *An Introduction to the Geometry of Numbers,* Springer-Verlag, Berlin.

Ferrenberg, A. M., D. P. Landau, and Y. J. Wong (1992). Monte Carlo simulations: Hidden errors from "good" random number generators, *Physical Review Letters,* **69,** 3382–3384.

Fishman, G. S. (1990). Multiplicative congruential random number generators with modulus 2^β: An exhaustive analysis for $\beta = 32$ and a partial analysis for $\beta = 48$, *Mathematics of Computation,* **54,** 331–334.

Fishman, G. S. (1996). *Monte Carlo: Concepts, Algorithms, and Applications,* Springer-Verlag, New York.

Fishman, G. S., and L. Moore (1986). An exhaustive analysis of multiplicative congruential random number generators with modulus $2^{31} - 1$, *SIAM Journal of Scientific and Statistical Computation,* **7,** 24–45.

Fushimi, M. (1990). Random number generation with the recursion $X_{t-3p} \oplus X_{t-3q}$, *Journal of Computational and Applied Mathematics,* **31,** 105–118.

Fushimi, M., and S. Tezuka (1983). The *k*-distribution of the generalized feedback shift register pseudorandom numbers, *Communications of the ACM,* **26,** 516–523.

Grube, A. (1973). Mehrfach rekursiv-erzeugte Pseudo-Zufallszahlen, *Zeitschrift für Angewandte Mathematik und Mechanik,* **53,** T223–225.

Hörmann, W., and G. Derflinger (1993). A portable random number generator well suited for the rejection method, *ACM Transactions on Mathematical Software,* **19,** 489–495.

Jannson, B. (1966). *Random Number Generators,* Almquist and Wiksell, Stockholm.

Knuth, D. (1981). *The Art of Computer Programming, Volume 2: Seminumerical Algorithms,* second edition, Addison-Wesley.

L'Ecuyer (1992). Testing random number generators, *Proceedings of the 1992 Winter Simulation Conference,* IEEE Press, Piscataway, N.J., 305–313.

L'Ecuyer (1996). Combined multiple recursive random number generators, *Operations Research,* **44,** 816–822.

L'Ecuyer, P. (1997). Bad lattice structures for vectors of non-successive values produced by some linear recurrences, *INFORMS Journal on Computing,* **9,** 57–60.

L'Ecuyer, P. (1998). Good parameters and implementations for combined multiple recursive random number generators, Department of Information and Operations Research, University of Montreal.

L'Ecuyer, P. (1999). Good parameters and implementations for combined multiple recurrence random number generators, *Operations Research,* **47,** 159–165.

L'Ecuyer, P. (2001). Software for uniform random number generation: Distinguishing the good from the bad, *Proceedings of the 2001 Winter Simulation Conference,* IEEE Press, Piscataway, N.J. 95–105.

L'Ecuyer, P. (2003). Personal communication.

L'Ecuyer, P., and R. Simard (2001). On the performance of birthday spacings tests with certain families of random number generators, *Mathematics and Computers in Simulation,* **55,** 131–137.

L'Ecuyer, P., J. F. Cordeau, and R. Simard (2000). Close-point spatial tests and their application to random number generators, *Operations Research,* **48,** 308–317.

L'Ecuyer, P., R. Simard, and S. Wegenkittl (2002). Sparse serial tests of uniformity for random number generators, *Journal on Scientific Computing,* to appear.

Lewis, T. G., and W. H. Payne (1973). Generalized feedback shift register pseudorandom number algorithms, *Journal of the ACM,* **20,** 456–468.

Marsaglia, G. (1985). A current view of random number generators, *Computer Science and Statistics: 16th Symposium on the Interface,* L. Brillard, ed., Elsevier Science Publishers, Amsterdam.

Marsaglia, G., B. Narasimhan, and A. Zaman (1990). A random number generator for PC's, *Computer Physics Communications,* **60,** 345–349.

Marsaglia, G., and A. Zaman (1990). A new class of random number generators, *Ann. Appl. Prob.,* **1,** 482–480.

Matsumoto, M. (2002). Personal communication.

Matsumoto, M., and Y. Kurita (1994). Twisted GFSR generators II, *ACM Transactions on Modeling and Simulation,* **4,** 254–266.

Matsumoto, M., and T. Nishimura (1998). Mersenne twister: A 623-dimensionally equidistributed uniform pseudo-random number generator, *ACM Transactions on Modeling and Simulation,* **8,** 3–30.

Nishimura, T. (2000). Tables of 64-Mersenne twisters, *ACM Transactions on Modeling and Simulation,* **10,** 348–357.

Overstreet, C. L., Jr. (1974). The relationship between the multiplicative and mixed generators modulo 2^b, Computer Science Department, Virginia Polytechnic Institute and State University, Blacksburg.

Tezuka, S., P. L'Ecuyer, and R. Couture (1992). On the lattice structure of the add-with-carry and subtract-with-borrow random number generators, *Transactions on Modeling and Computer Simulation,* **3,** 315–331.

Wagner, H. (1999). Personal communication.

Wichmann, B. A., and I. D. Hill (1982). Algorithm 183: An efficient and portable pseudo-random number generator, *Applied Statistics,* **31,** 188–190.

Wichmann, B. A., and I. D. Hill (1987). Building a random number generator—A Pascal routine for very-long-cycle random-number sequences, *Byte,* **12,** 127–128.

CHAPTER 5

Variance Reduction

Recall that two quantities characterize every comparison of the relative benefits of different MC methods: the cost ratio (2.20) and the variance ratio (2.34). In this chapter, we assume that the representation (2.23) holds and that we want to compare standard MC (see Section 2.4) with alternative IMC methods. Then (2.20) and (2.34) become

$$\mathrm{CR} := \frac{\text{unit generating cost for standard MC}}{\text{unit generating cost for variance-reducing technique}} \tag{5.1}$$

and

$$\mathrm{VR} := \frac{\text{variance for standard MC}}{\text{variance for variance-reducing technique}}. \tag{5.2}$$

Also recall that MC techniques that achieve a specified accuracy with less computing time than standard MC ($\mathrm{CR} \times \mathrm{VR} > 1$) are said to improve efficiency. Those that reduce unit generating time ($\mathrm{CR} > 1$) are said to improve computational efficiency. Chapter 3 described how to keep an often major component of this time, sample generating time, as low as possible. Those procedures that reduce variance ($\mathrm{VR} > 1$) are said to improve statistical efficiency and are called *variance-reducing methods*. They include:

- Importance sampling
- Stratified sampling
- Rao-Blackwellization
- Control variates
- Antithetic variates
- Quasirandom numbers

Section 2.4 described how importance sampling gains its advantage. The current chapter describes the remaining five techniques, illustrating each with an example that reveals how it works and the extent to which it can increase efficiency.

Although we cannot say that among all variance-reducing techniques one induces the greatest increase in efficiency for all problems; the properties of individual problems often point to one or several techniques as likely candidates for the best success. In addition to conceptually describing many of these properties, the examples show the benefits that result from exploiting them. Because experience helps in making an informed choice, the hands-on exercises at the end of the chapter invite the reader to apply several different techniques to the same problems solved by standard MC in Chapter 2. Comparisons of CR $\times$ VR for the several techniques provide a basis for evaluating the relative benefits of each method.

As described in Section 2.3, their product, CR $\times$ VR, gives the relative increase in computing time required to achieve a specified accuracy using standard MC rather than an alternative MC technique. Although CR <1 and VR >1, frequently CR $\times$ VR >1, so that the technique offers an unequivocal advantage. Conversely, CR $\times$ VR <1 reveals the superiority of standard MC. As the importance-sampling problems in Examples 2.7.3 and 2.7.4 indicate, CR ≈ 1 for some MC methods. For others, both CR >1 and VR >1, as can occur for Rao-Blackwellization in Section 5.2.

The effective use of a variance-reducing technique inevitably requires that the potential user devote more time to presampling analysis than standard MC would require. With the current availability of so much low-cost computing power, there is a natural tendency to avoid this additional effort and merely resort to standard MC. This strategy makes considerable sense when the solution requires relatively moderate accuracy and the longer computing time does not impede overall progress. However, a high-accuracy requirement, a need to perform the same MC sampling repeatedly with different parametric values as input, or both makes the time and effort invested in a presampling analysis well spent. One can regard user effort as a fixed cost whose corresponding average cost per replication becomes negligible as the number of replications becomes large.

5.1 STRATIFIED SAMPLING

Let $\mathcal{Y}_1, \ldots, \mathcal{Y}_r$ denote a partition of $\mathcal{X}$ into subsets of $\mathcal{X}$ that are exclusive (i.e., $\mathcal{Y}_i \cap \mathcal{Y}_j = \emptyset,\ \forall i \neq j$) and exhaustive (i.e., $\cup_{i=1}^{r} \mathcal{Y}_i = \mathcal{X}$). As before, our objective is to estimate λ in (2.1), which in the context of the partition has the equivalent additive form

$$\lambda = \lambda_1 + \cdots + \lambda_r,$$

where

$$\lambda_i := \int_{\mathcal{Y}_i} g(\mathbf{x}) d\mathbf{x} \qquad i = 1, \ldots, r. \tag{5.3}$$

Moreover, any chosen sampling distribution ω on $\mathcal{X}$ admits the decomposition

$$\omega(\mathbf{x}) = p_1\omega_1(\mathbf{x}) + \cdots + p_r\omega_r(\mathbf{x}) \qquad \mathbf{x} \in \mathcal{X}, \tag{5.4}$$

where

$$\begin{aligned} p_i &:= \int_{\mathcal{Y}_i} \omega(\mathbf{x})d\mathbf{x} \\ p_1 + \cdots + p_r &= 1 \\ \omega_i(\mathbf{x}) &:= \begin{cases} p_i^{-1}\omega(\mathbf{x}) & \text{if } \mathbf{x} \in \mathcal{Y}_i \\ 0 & \text{otherwise.} \end{cases} \end{aligned} \tag{5.5}$$

For each stratum $i = 1, \ldots, r$, *stratified sampling* randomly generates data from $\{\omega_i(\mathbf{x}),\ \mathbf{x} \in \mathcal{Y}_i\}$ to compute an estimate of λ_i. The sum of these estimates provides an estimate of λ. We now show that if $\mathcal{Y}_1, \ldots, \mathcal{Y}_r$ are properly chosen, the resulting estimate of λ, based on n independent replications within each of the r strata, has smaller variance than a standard MC estimate based on nr independent replications using $\{\omega(\mathbf{x}),\ \mathbf{x} \in \mathcal{X}\}$.

Recall that standard MC randomly generates $\mathbf{X}$ from ω so that $\frac{g(\mathbf{X})}{\omega(\mathbf{X})}$ has expectation

$$\mathrm{E}_\omega\left[\frac{g(\mathbf{X})}{\omega(\mathbf{X})}\right] = \int_{\mathcal{X}} \left[\frac{g(\mathbf{x})}{\omega(\mathbf{x})}\right]\omega(\mathbf{x})d\mathbf{x} = \lambda$$

and variance

$$\begin{aligned} \mathrm{var}_\omega\left[\frac{g(\mathbf{X})}{\omega(\mathbf{X})}\right] &= \mathrm{E}_\omega\left[\frac{g(\mathbf{X})}{\omega(\mathbf{X})} - \lambda\right]^2 \\ &= \int_{\mathcal{X}} \omega(\mathbf{x})\left[\frac{g(\mathbf{x})}{\omega(\mathbf{x})} - \lambda\right]^2 d\mathbf{x}. \end{aligned} \tag{5.6}$$

The decomposition (5.4) allows us to express (5.6) in the alternative but equivalent form

$$\begin{aligned} \mathrm{var}_\omega\left[\frac{g(\mathbf{X})}{\omega(\mathbf{X})}\right] &= \sum_{i=1}^r p_i \int_{\mathcal{Y}_i} \omega_i(\mathbf{x})\left[\frac{g(\mathbf{x})}{p_i\omega_i(\mathbf{x})} - \frac{\lambda_i}{p_i} + \frac{\lambda_i}{p_i} - \lambda\right]^2 d\mathbf{x} \\ &= \sum_{i=1}^r \frac{1}{p_i}\int_{\mathcal{Y}_i} \omega_i(\mathbf{x})\left[\frac{g(\mathbf{x})}{\omega_i(\mathbf{x})} - \lambda_i\right]^2 d\mathbf{x} + \sum_{i=1}^r p_i\left(\frac{\lambda_i}{p_i} - \lambda\right)^2. \end{aligned} \tag{5.7}$$

Because randomly sampling $\mathbf{X}_i$ from $\{\omega_i(\mathbf{x}),\ \mathbf{x} \in \mathcal{Y}_i\}$ gives

$$\mathrm{E}_{\omega_i}\left[\frac{g(\mathbf{X})}{\omega_i(\mathbf{X})}\right] = \int_{\mathcal{Y}_i}\left[\frac{g(\mathbf{x})}{\omega_i(\mathbf{x})}\right]\omega_i(\mathbf{x})d\mathbf{x} = \lambda_i$$

and

$$\mathrm{var}_{\omega_i}\left[\frac{g(\mathbf{X})}{\omega_i(\mathbf{X})}\right] = \int_{\mathcal{Y}_i} \omega_i(\mathbf{x})\left[\frac{g(\mathbf{x})}{\omega_i(\mathbf{x})} - \lambda_i\right]^2 \mathrm{d}\mathbf{x},$$

we have

$$\mathrm{var}_{\omega}\left[\frac{g(\mathbf{X})}{\omega(\mathbf{X})}\right] = \sum_{i=1}^{r}\frac{1}{p_i}\mathrm{var}_{\omega_i}\left[\frac{g(\mathbf{X})}{\omega_i(\mathbf{X})}\right] + \sum_{i=1}^{r}\frac{\lambda_i^2}{p_i} - \lambda^2. \tag{5.8}$$

Let Ω denote the collection of distributions $\{\omega_1, \ldots, \omega_r\}$. Then the pooled estimator

$$\widetilde{\lambda} := \sum_{i=1}^{r}\frac{g(\mathbf{X}_i)}{\omega_i(\mathbf{X}_i)} \tag{5.9}$$

has expectation

$$\begin{aligned} \mathrm{E}_{\Omega}\widetilde{\lambda} &= \sum_{i=1}^{r}\mathrm{E}_{\omega_i}\left[\frac{g(\mathbf{X}_i)}{\omega_i(\mathbf{X}_i)}\right] \\ &= \lambda \end{aligned}$$

and variance

$$\mathrm{var}_{\Omega}\widetilde{\lambda} = \sum_{i=1}^{r}\mathrm{var}_{\omega_i}\left[\frac{g(\mathbf{X})}{\omega_i(\mathbf{X})}\right]. \tag{5.10}$$

If $\mathcal{Y}_1, \ldots, \mathcal{Y}_r$ are chosen so that $p_1 = \cdots = p_r = \frac{1}{r}$, then

$$\mathrm{var}_{\omega}\left[\frac{g(\mathbf{X})}{\omega(\mathbf{X})}\right] = r\mathrm{var}_{\Omega}\widetilde{\lambda} + r\sum_{i=1}^{r}\lambda_i^2 - \lambda^2.$$

Most importantly,

$$\begin{aligned} \mathrm{var}_{\Omega}\widetilde{\lambda} &\le \tfrac{1}{r}\mathrm{var}_{\omega}\left[\frac{g(\mathbf{X})}{\omega(\mathbf{X})}\right] \\ &= \mathrm{var}_{\omega}\bar{\lambda}_r, \end{aligned} \tag{5.11}$$

where $\bar{\lambda}_r$ denotes the standard MC estimator (2.9) based on r independent samples from ω. That is, $\widetilde{\lambda}$, based on independently sampling one observation in each of the r strata $\mathcal{Y}_1, \ldots, \mathcal{Y}_r$, has smaller variance than $\bar{\lambda}_r$, based on r independent samples from the original ω on $\mathcal{X}$. This is the principal benefit of stratified sampling.

If sampling from each ω_i takes about the same computing time as sampling from ω, then stratified MC sampling offers an unequivocal advantage over standard MC sampling. Moreover (see Exercise 1),

$$\mathrm{var}_{\Omega}\widetilde{\lambda} = \mathrm{var}_{\omega}\bar{\lambda}_r - \sum_{i=1}^{r}\lambda_i^2 + \frac{\lambda^2}{r} \tag{5.12}$$

implies that for any given ω, selecting the partition $\mathcal{Y}_1, \ldots, \mathcal{Y}_r$ that maximizes $\sum_{i=1}^{r}\lambda_i^2$ minimizes $\mathrm{var}_{\Omega}\widetilde{\lambda}$, subject to the r constraints $\int_{\mathcal{Y}_i}\omega(\mathbf{x})\mathrm{d}\mathbf{x} = \frac{1}{r}$ for $i = 1, \ldots, r$.

Let the superscript $^{(j)}$ denote replication j. For n independent replications per stratum, stratified sampling gives

$$\ddot{\lambda}_{in} := \frac{1}{n} \sum_{j=1}^{n} \frac{g\left(\mathbf{X}_i^{(j)}\right)}{\omega\left(\mathbf{X}_i^{(j)}\right)}$$

as an estimate of λ_i,

$$\widetilde{\lambda}_n := \ddot{\lambda}_{1n} + \cdots + \ddot{\lambda}_{rn} \tag{5.13}$$

as an estimate of λ,

$$\ddot{s}_{in}^2 := \frac{1}{n-1} \sum_{j=1}^{n} \left[\frac{g\left(\mathbf{X}_i^{(j)}\right)}{\omega_i\left(\mathbf{X}_i^{(j)}\right)} - \widetilde{\lambda}_n\right]^2$$

as an estimate of $\mathrm{var}_{\omega_i}\left[\frac{g\left(\mathbf{X}_i^{(j)}\right)}{\omega_i\left(\mathbf{X}_i^{(j)}\right)}\right]$, and

$$\widetilde{s}_n^2 := \frac{1}{n}\left(\ddot{s}_{1n}^2 + \cdots + \ddot{s}_{rn}^2\right)$$

as an estimate of $\mathrm{var}\,\ddot{\lambda}_{in}$. Moreover, $\widetilde{\lambda}_n$ has standard error s.e.$(\widetilde{\lambda}_n) = \sqrt{\frac{\widetilde{s}_n^2}{n}}$, relative error r.e.$(\widetilde{\lambda}_n) = \frac{\text{s.e.}(\widetilde{\lambda}_n)}{\widetilde{\lambda}_n}$, and for fixed r

P9. $\widetilde{\lambda}_n \to \lambda$ w.p.1 as $n \to \infty$.

P10. $\frac{\widetilde{\lambda}_n - \lambda}{\sqrt{\widetilde{s}_n^2/n}} \xrightarrow{d} \mathcal{N}(0,1)$ as $n \to \infty$.

P11. An asymptotically valid $100 \times (1-\delta)$ percent confidence interval is provided for λ by

$$\left[\widetilde{\lambda}_n - \Phi^{-1}(1-\delta/2)\sqrt{\widetilde{s}_n^2/n},\ \widetilde{\lambda}_n + \Phi^{-1}(1-\delta/2)\sqrt{\widetilde{s}_n^2/n}\right].$$

Although (5.11) ensures a variance reduction, stratified sampling incurs limitations for multivariable problems. Suppose that $\mathcal{X} = (-\infty, \infty)^m$. If $\mathcal{X}_i = (-\infty, \infty)$, the set of values that x_i can assume, is partitioned into N exclusive and exhaustive subsets for each variable $i = 1, \ldots, m$, then the number of possible ways of combining strata for all m variables is $r = N^m$, which grows rapidly with m and N. As illustration, Example 2.7.4, on pricing Asian options, with $N = 2$ strata per variable and $m = 16$ intervals has $2^{16} = 65{,}536$ strata. For $m = 64$ intervals, it has $2^{64} \approx 1.8 \times 10^{19}$, clearly not a practicable choice.

Because stratified sampling cost is proportional to nr, a large number of strata, r, may restrict the number of independent replications n, in order to remain within budget. Since properties P10 and P11 rely on n being large, this restriction may invalidate normal theory as a basis for error assessment. Moreover, contrary to assumption, unit generating cost within strata may be substantially greater than for standard MC.

Whereas these issues can impede the direct application of stratified sampling, exploiting special structure frequently ameliorates their seriousness and occasionally eliminates them. Example 5.4.4 demonstrates this for the Asian-options pricing problem described earlier in Example 2.7.4.

EXAMPLE 5.1.1 ONE-DIMENSIONAL INTEGRATION II

This univariate example illustrates how stratified sampling reduces variance by sampling within subsets of $\mathcal{X}$ with smaller ranges of variation than that for standard MC. Table 5.1 displays MC results for the one-dimensional integration problem in Example 2.1.1. They are based on equal sample sizes, n/r, in each stratum generated from the p.d.f.s

$$\omega_i(x) = \begin{cases} \frac{1}{r} & \frac{i-1}{r} \le x < \frac{i}{r} \\ 0 & \text{elsewhere} \end{cases} \qquad i = 1, \ldots, r. \tag{5.14}$$

For each r, the total sample size is $nr = 327680$. The cost ratio is $\text{CR} \approx 1$, and the variance ratio is

$$\text{VR}_r = \frac{\text{variance for standard MC}}{\text{variance for } r \text{ strata}}.$$

To see why stratified sampling induces an increasing VR_r with increasing r, we study the variation of the integrand $g(x) = \mathrm{e}^{-x\cos(\pi x)}$ within strata. In particular, Table 5.2 reveals that the range of variation within each stratum decreases as r increases, thus allowing stratified sampling to reduce variance by successively increasing amounts. ■

TABLE 5.1 | *One-dimensional integration with stratified sampling*

No. of strata r	No. of replications per stratum n	$\tilde{\lambda}_n \pm \Phi^{-1}(.995)\text{s.e.}(\tilde{\lambda}_n)$	$\text{s.e.}(\tilde{\lambda}_n)$	$\text{r.e.}(\tilde{\lambda}_n)$	VR_r
1	327680	$1.3280 \pm 2.6016 \times 10^{-3}$	1.0100×10^{-3}	7.6054×10^{-4}	1.000
2	163840	$1.3289 \pm 1.7096 \times 10^{-3}$	6.6373×10^{-4}	4.9944×10^{-4}	2.314
4	81920	$1.3286 \pm 8.4017 \times 10^{-4}$	3.2617×10^{-4}	2.4551×10^{-4}	9.588
8	40960	$1.3286 \pm 4.1722 \times 10^{-4}$	1.6197×10^{-4}	1.2192×10^{-4}	38.88
16	20480	$1.3287 \pm 2.0822 \times 10^{-4}$	8.0836×10^{-5}	6.0837×10^{-5}	156.1
32	10240	$1.3286 \pm 1.0416 \times 10^{-4}$	4.0438×10^{-5}	3.0438×10^{-5}	623.8

TABLE 5.2 | *Within-strata bounds on variation*

r	Stratum	Within stratum $\min_x g(x)$	$\max_x g(x)$	$\max_x g(x) - \min_x g(x)$
1	[0, 1]	0.8364	e	1.882
2	[0, 1/2]	0.8364	1	0.1636
	[1/2, 1]	1	e	1.718
4	[0, 1/4]	$e^{-\sqrt{2}/8}$	1	0.1620
	[1/4, 1/2]	0.8364	1	0.1636
	[1/2, 3/4]	1	$e^{3\sqrt{2}/8}$	0.6995
	[3/4, 1]	$e^{3\sqrt{2}/8}$	e	1.019

EXAMPLE 5.1.2

PRICING AN ASIAN OPTION II

In contrast to Example 5.1.1, which minimally exploits the structure of the one-dimensional integration problem, Example 2.7.4 showed how a more elaborate exploitation of a multivariable problem can lead once again to a stratified sampling plan in one dimension that reduces variance. This problem has as its objective the estimation of the mean discounted Asian options price

$$\lambda = \int_{\Re^m} g(\mathbf{x})d\mathbf{x}$$
$$= \int_{\Re^m} \pi(\mathbf{x})h(\mathbf{x})d\mathbf{x},$$

where (2.54) and (2.56) define $h(\mathbf{x})$ and $\pi(\mathbf{x})$ respectively. Straightforward application of stratified sampling conceptually partitions $\Re^m$ into r exclusive and exhaustive sets $\mathcal{Y}_1, \ldots, \mathcal{Y}_r$ so that for each i, (5.4) yields the strata sampling distributions (5.5) as

$$\omega_i(\mathbf{x}) = p_i^{-1}\pi(\mathbf{x})$$
$$p_i = \int_{\mathcal{Y}_i} \pi(\mathbf{x})d\mathbf{x}.$$

To cope with the problem of determining the partition $\mathcal{Y}_1, \ldots, \mathcal{Y}_r$ so that $p_1 = \cdots = p_r = \frac{1}{r}$, we rely on the composition method in Section 3.5. It exploits properties of the *multinormal distribution*. ■

P12. Let $u_1, \ldots, u_m$ be constants satisfying $\sum_{i=1}^m u_i^2 = 1$. If $X_1, \ldots, X_m$ are i.i.d. random variables from $\mathcal{N}(0, 1)$ with density ϕ as in (2.11), then for fixed

$y \in (-\infty, \infty)$, $X_1, \ldots, X_m$ given $\sum_{i=1}^m u_i X_i = y$ have the multinormal distribution with

$$\begin{aligned} \mathrm{E}(X_i \mid y) &= y u_i \\ \mathrm{var}(X_i \mid y) &= 1 - u_i^2 \\ \mathrm{cov}(X_i, X_l \mid y) &= -u_i u_l \qquad \forall i \neq l. \end{aligned}$$

P13. If Y is from $\mathcal{N}(0, 1)$ with density ϕ as in (2.11), then

$$\begin{aligned} \int_{-\infty}^{\infty} \phi(\mathbf{x} \mid y)\phi(y)\mathrm{d}y &= \prod_{i=1}^{m} \phi(x_i) \qquad \mathbf{x} \in \Re^m, \\ &= \pi(\mathbf{x}) \end{aligned}$$

where $\{\phi(\mathbf{x} \mid y)\}$ denotes the p.d.f. of the multinormal distribution of $X_1, \ldots, X_m$ given $\sum_{i=1}^m u_i X_i = y$.

Property P13 has immediate relevance for stratified sampling. In particular, let $y_0 = -\infty < y_1 < \cdots < y_{r-1} < y_r = \infty$ be defined by

$$\Phi^{-1}(y_l) = \frac{l}{r} \qquad l = 1, \ldots, r, \tag{5.15}$$

where (2.16) defines the inverse d.f. Φ^{-1}. That is, y_l is the quantile of the standard normal distribution, $\mathcal{N}(0, 1)$ at probability $\frac{l}{r}$. Then $\{\pi(\mathbf{x})\}$ can be written in the form

$$\pi(\mathbf{x}) = \sum_{i=1}^{r} p_i \omega_i(\mathbf{x}) \qquad \mathbf{x} \in \Re^m,$$

where

$$\begin{aligned} p_i &= \int_{y_{i-1}}^{y_i} \left[\int_{\mathcal{Y}(y)} \phi(\mathbf{x} \mid y)\mathrm{d}\mathbf{x} \right] \phi(y)\mathrm{d}y = \frac{1}{r} \\ \omega_i(\mathbf{x}) &= r \int_{y_{i-1}}^{y_i} \phi(\mathbf{x} \mid y)\phi(y)\mathrm{d}y \qquad \mathbf{x} \in \mathcal{Y}(y) \\ \mathcal{Y}(y) :&= \{\mathbf{x} \in \Re^m : \ u_1 x_1 + \cdots + u_m x_m = y\}. \end{aligned}$$

Moreover, $\mathcal{Y}(y) \subset \mathcal{Y}_i$ for $y_{i-1} \leq y \leq y_i$, where

$$\mathcal{Y}_i := \{\mathbf{x} \in \Re^m : \ y_{i-1} < u_1 x_1 + \cdots + u_m x_m < y_i\}.$$

The appeal of property P12 now becomes evident. Whereas $\{\omega_i(\mathbf{x})\}$ is neither multinormal nor is easily employed to generate random samples, $\{r\phi(\mathbf{x} \mid y)\}$ for each $y \in [y_{l-1}, y_l]$ is a multinormal distribution, for which well-established techniques exist for generating random samples (e.g., Devroye 1986, Fishman 1996). Moreover, each Y can be randomly generated from its stratified distribution $\{r\phi(y),\ y_{l-1} < y \leq y_l\}$ by means of the inverse d.f. for the normal distribution.[1] For example, if U is from $\mathcal{U}(0, 1)$, then $Y = \Phi^{-1}\left(\frac{l-1}{r} + \frac{U}{r}\right)$ has distribution

[1]See the discussion at the end of Section 3.3.1.

$\{r\phi(\mathbf{x} \mid y),\ y_{l-1} < y \le y_l\}$. Algorithm ASSTRAT describes the application of this form of stratified sampling to the Asian-options pricing problem.

In the present problem, it is also possible to combine importance sampling and stratified sampling. For constants $\nu_1, \ldots, \nu_m$, let $X_1 - \nu_1, \ldots, X_m - \nu_m$ denote i.i.d. $\mathcal{N}(0, 1)$ random variables. Then property P12 implies that $X_1, \ldots, X_m$, given $u_1X_1 + \cdots + u_mX_m = y + u_1\nu_1 + \cdots + u_m\nu_m$ jointly have the multinormal distribution with means

$$\mathrm{E}\left(X_i \mid y + u_1\nu_1 + \cdots + u_m\nu_m\right) - \nu_i + yu_i \qquad i = 1, \ldots, m$$

and the same variance-covariance structure as in property P9. Because these nonzero means add negligible computing time to multinormal generation, it is relatively straightforward to combine importance sampling with stratified sampling by starting with the overall importance sampling distribution

$$\omega(\mathbf{x}) = (2\pi)^{-m/2} \prod_{i=1}^{m} \mathrm{e}^{-(x_i - \nu_i)^2/2} \qquad \mathbf{x} \in \Re^m,$$

where $\nu_1, \ldots, \nu_m$ are the constants determined when applying importance sampling in Example 2.7.4. Using a near-optimal assignment for $u_1, \ldots, u_r$, Glasserman et al. (1999) combine these techniques.

ALGORITHM

ASSTRAT

Purpose: To estimate λ in (2.55) by stratified sampling.

Given: $\xi_1,\ \xi_2,\ K,\ T,\ m$ for Asian options problem, $r :=$ number of strata, $\{y_l := \frac{l}{r}\text{th quantile of } \mathcal{N}(0, 1),\ i = 1, \ldots, r\}$, coefficients $u_1, \ldots, u_m$, and n = number of independent replications.

Method:

$i = 1$.
While $i \le r$:
  $j = 1,\ V_{i1} = 0$, and $V_{i2} = 0$
  While $j \le n$:
    randomly sample Y from $\{r\phi(y),\ y_{i-1} \le y \le y_i\}$
    randomly sample $X_1, \ldots, X_m$ jointly from $\{\phi(\mathbf{x} \mid Y),\ \mathbf{x} \in \mathcal{Y}(Y)\}$
    compute S_m as in (2.53)
    $Z = \mathrm{e}^{-\xi_1 T}(S_m - K)^+$
    $V_{i1} = V_{i1} + Z$, and $V_{i2} = V_{i12} + Z^2$
    $j = j + 1$.
  $\ddot{\lambda}_{in} = \frac{V_{i1}}{n}$ and $\widetilde{s}^2_{in} = \frac{1}{n-1}\left(V_{i2} - n\ddot{\lambda}^2_{in}\right)$
  $i = i + 1$.
$\ddot{\lambda}_n = \ddot{\lambda}_{1n} + \cdots + \ddot{\lambda}_{rn}$.
$\mathrm{s.e.}(\ddot{\lambda}_n) = \sqrt{\frac{1}{n}\left(\widetilde{s}^2_{1n} + \cdots + \widetilde{s}^2_{rn}\right)}$ and $\mathrm{r.e.}(\ddot{\lambda}_n) = \frac{\mathrm{s.e.}(\ddot{\lambda}_n)}{\ddot{\lambda}_n}$. ■

TABLE 5.3 | *Variance ratio (VR) for Asian options pricing MC
$S(0) = 50,\ \xi_1 = 0.05,\ T = 1.0,\ r = 100$ *strata*, $n = 10^6$

				Variance ratio (VR)	
m	ξ_2	K	Price	Importance sampling	Importance and stratified sampling
16	0.10	45	6.05	11.0	1,246
		50	1.92	7.0	5,710
		55	0.20	21.0	17,026
16	0.30	45	7.15	8.3	1,664
		50	4.17	9.2	1,899
		55	2.20	12.0	2,296
64	0.10	45	6.00	11.0	1,022
		50	1.85	7.3	5,665
		55	0.17	23.0	17,841
64	0.30	45	7.02	8.3	1,694
		50	4.02	9.2	1,971
		55	2.08	12.0	2,402

*Source: Glasserman et al. (1999).

Table 5.3 shows their resulting variance ratios. Those for importance sampling are reproduced from Table 2.7. Those for stratified sampling are based on $\frac{n}{r}$ independent replications per stratum. The ratios for combined sampling reveal substantial variance-reducing benefits. As illustration, stratified IMC with $m = 64,\ \xi_2 = 0.10$, and $K = 45$ would require 1,022 times as many replications to achieve the same standard error, and with $m = 64,\ \xi_2 = 0.10$, and $K = 55$ it would require 17,841 times as many replications.

Cost Considerations. The cost ratio (CR) in (5.1) depends principally on unit generating time within strata for stratified sampling and across strata for standard MC. In the present example, the former is larger, merely because it requires generating m dependent, rather than independent, normal variates plus the need to generate Y within each strata. However, this increase rarely is more than a small multiple of standard MC unit generating time.

5.1.1 Unequal Sample Sizes

Suppose an observation for stratum j has variance σ_j^2 so that

$$\mathrm{var}\tilde{\lambda}_n = \frac{r}{n}\left(\sigma_1^2 + \cdots + \sigma_r^2\right).$$

If σ_j^2 is much larger than the variances for the other $r-1$ strata, then its contribution is substantially larger to $\mathrm{var}\tilde{\lambda}_n$. This observation suggests that we execute a greater number of replications for stratum j than for any other stratum. We describe a rule that formalizes this concept:

Let n_j denote the number of replications in stratum j. Then stratified sampling based on these sample sizes produces an estimate $\tilde{\lambda}_{n_1,\ldots,n_r}$ with variance

$$\mathrm{var}\tilde{\lambda}_{n_1,\ldots,n_r} = \frac{\sigma_1^2}{n_1} + \cdots + \frac{\sigma_r^2}{n_r},$$

which is minimized subject to $n = n_1 + \cdots + n_r$, by the assignment

$$n_j = \frac{n\sigma_j}{\sum_{l=1}^{r}\sigma_l} \qquad j = 1, \ldots, r. \tag{5.16}$$

That is,

$$\begin{aligned}\mathrm{var}\tilde{\lambda}_{n_1,\ldots,n_r} &= \frac{1}{n}\left(\sum_{l=1}^{r}\sigma_l\right)^2 \\ &\leq \mathrm{var}\tilde{\lambda}_n.\end{aligned}$$

Although $\sigma_1, \ldots, \sigma_r$ are unknown in practice, perusal of the integrand often reveals the relative variations in different strata. This observation can guide the choice of a near-optimal assignment for $n_1, \ldots, n_r$. The next example and Exercise H.5 illustrate the technique.

EXAMPLE 5.1.3

ONE-DIMENSIONAL INTEGRATION III

Exploiting the bounds in Table 5.2 for the integrand in Example 2.1.1 can further reduce variance. In particular, it is easily seen that for each stratum $i = 1, \ldots, r$,

$$\sigma_i \leq \max_x g(x) - \min_x g(x),$$

where the maximum and minimum are taken over stratum i. Suppose we partition $[0, 1)$ into $r = 4$ strata and substitute the upper bound for $\sigma_1, \ldots, \sigma_r$ in (5.16). This leads to

i	Stratum i	$\frac{n_i}{n}$
1	$[0, 0.25)$	0.07926
2	$[0.25, 0.50)$	0.08004
3	$[0.50, 0.75)$	0.3422
4	$[0.75, 1)$	0.4985

Then stratified sampling from the uniform p.d.f.s (5.14) for a total of 327,680 independent replications gives

$$\tilde{\lambda}_n = 1.3229$$
$$\text{s.e.}(\tilde{\lambda}_n) = 2.6429 \times 10^{-4}$$
$$\text{r.e.}(\tilde{\lambda}_n) = 1.9887 \times 10^{-4}.$$

When compared to stratified sampling for $r = 4$ and equal sample sizes per stratum in Table 5.1, the variance ratio,

$$\text{VR}_r = \left(\frac{3.2617 \times 10^{-4}}{2.6429 \times 10^{-4}} \right)^2 = 1.52,$$

reveals about a 50% increase in statistical efficiency and CR ≈ 1. In comparison to standard MC in Example 2.1.1, the variance reduction is greater than 10. This example shows the advantage to be gained by choosing $n_1, \ldots, n_r$ via (5.16) with bounds in place of σ_i.

Adjusting sample sizes within strata provides one way to increase the potency of stratified sampling. An alternative approach adjusts the sizes or volumes of the strata so as to reduce the range of variation of g within each. Since the bounds in Table 5.2 for $r = 4$ are substantially greater for strata 3 and 4 than for strata 1 and 2, we consider a small increase in the number of strata to $r = 6$ and partition [0, 1) according to the following:

i	Stratum i	$\text{Max}_x g(x) - \text{min}_x g(x)$	$\frac{n_i}{n}$
1	[0, 1/4)	0.1620	0.07926
2	[1/4, 1/2)	0.1636	0.08004
3	[1/2, 5/8)	0.2702	0.1322
4	[5/8, 3/4)	0.4293	0.2100
5	[3/4, 7/8)	0.5441	0.2666
6	[7/8, 1)	0.4740	0.2320

Then sampling a total of $nr = 327{,}680$ independent replications from the uniform p.d.f.s

$$\omega_i(x) = \begin{cases} \frac{1}{4} & \frac{i-1}{4} \le x < \frac{i}{4} \\ 0 & \text{elsewhere} \end{cases} \qquad i = 1 \text{ and } 2$$

and

$$\omega_i(x) = \begin{cases} \frac{1}{8} & \frac{i+1}{8} \le x < \frac{i+2}{8} \\ 0 & \text{elsewhere} \end{cases} \qquad i = 3, \ldots, 6$$

$$\tilde{\lambda}_n = 1.3285$$
$$\text{s.e.}\left(\tilde{\lambda}_n\right) = 1.5915 \times 10^{-4}$$
$$\text{r.e.}\left(\tilde{\lambda}_n\right) = 1.1981 \times 10^{-4}.$$

When compared to standard MC in Table 2.1, the variance ratio was VR = 40.30. More notably, this ratio is more than four times the VR for $r = 8$ strata in that table. The last comparison reveals the benefit to be gained by a smaller number of strata applied to subsets of $\mathcal{X}$ with tighter bounds on their ranges of variation for g. ■

5.2 RAO-BLACKWELLIZATION

Stratified sampling exemplifies how sampling from conditional distributions, induced by stratification, can reduce variance. For sample generation, the composition method in Section 3.5 shows how to reduce unit generating cost by exploiting conditional probability. This section shows how exploiting conditional probability occasionally can lead to an estimator with reduced variance, regardless of the adopted MC sampling plan and sample generating method.

For $\mathbf{x} := (x_1, \ldots, x_m) \in \Re^m$, write $\mathbf{x} = (\mathbf{y}, \mathbf{z})$ where $\mathbf{y} := (x_1, \ldots, x_l) \in \Re^l$ and $\mathbf{z} := (x_{l+1}, \ldots, x_m) \in \Re^{m-l}$. Suppose that for each $\mathbf{x} \in \Re^m$, g in (2.1) admits the factorization

$$g(\mathbf{x}) = g_a(\mathbf{y}, \mathbf{z}) g_b(\mathbf{y}) \qquad l < m, \tag{5.17}$$

where the function

$$\eta(\mathbf{y}) := \int_{\Re^{m-l}} g_a(\mathbf{y}, \mathbf{z}) d\mathbf{z} \qquad \mathbf{y} \in \Re^l \tag{5.18}$$

is known analytically. As before, the objective is to estimate the unknown λ in (2.1), which we can now write as the l-dimensional integral

$$\lambda = \int_{\Re^l} \eta(\mathbf{y}) g_b(\mathbf{y}) d\mathbf{y}.$$

Let $\mathbf{Y} := (X_1, \ldots, X_l)$ denote an l-dimensional random point with distribution $\{\omega_{\mathbf{Y}}(\mathbf{y}),\ \mathbf{y} \in \Re^l\}$. Then $\frac{\eta(\mathbf{Y}) g_b(\mathbf{Y})}{\omega_{\mathbf{Y}}(\mathbf{Y})}$, based solely on the variates 1 through l in $\mathbf{Y}$, has mean

$$\mathrm{E}_{\omega_{\mathbf{Y}}} \left[\frac{\eta(\mathbf{Y}) g_b(\mathbf{Y})}{\omega_{\mathbf{Y}}(\mathbf{Y})} \right] = \lambda$$

and

$$\mathrm{var}_{\omega_{\mathbf{Y}}} \left[\frac{\eta(\mathbf{Y}) g_b(\mathbf{Y})}{\omega_{\mathbf{Y}}(\mathbf{Y})} \right] = \int_{\Re^l} \frac{g_b(\mathbf{y})}{\omega_{\mathbf{Y}}(\mathbf{y})} \left[\int_{\Re^{m-l}} g_a(\mathbf{y}, \mathbf{z}) d\mathbf{z} \right]^2 d\mathbf{y} - \lambda^2.$$

To compare this method of estimating λ with standard MC, consider any joint distribution $\{\omega(\mathbf{y}, \mathbf{z}),\ (\mathbf{y}, \mathbf{z}) \in \Re^m\}$ with the property that

$$\omega_{\mathbf{Y}}(\mathbf{y}) = \int_{\Re^{m-l}} \omega(\mathbf{y}, \mathbf{z}) d\mathbf{z} \qquad \mathbf{y} \in \Re^l.$$

The standard MC estimator $\frac{g(\mathbf{Y},\mathbf{Z})}{\omega(\mathbf{Y},\mathbf{Z})}$ has mean λ and

$$\begin{aligned}\operatorname{var}_{\omega}\left[\frac{g(\mathbf{Y},\mathbf{Z})}{\omega(\mathbf{Y},\mathbf{Z})}\right] &= \int_{\Re^l} g_b^2(\mathbf{y}) \int_{\Re^{m-l}} \frac{g_a^2(\mathbf{y},\mathbf{z})}{\omega(\mathbf{y},\mathbf{z})}\mathbf{dzdy} - \lambda^2 \\ &= \int_{\Re^l} \frac{g_b^2(\mathbf{y})}{\omega_{\mathbf{Y}}(\mathbf{y})} \int_{\Re^{m-l}} \frac{g_a^2(\mathbf{y},\mathbf{z})}{\omega_{\mathbf{Z}|\mathbf{Y}}(\mathbf{z})}\mathbf{dzdy} - \lambda^2,\end{aligned}$$

where

$$\omega_{\mathbf{Z}|\mathbf{Y}}(\mathbf{z}) := \frac{\omega(\mathbf{y},\mathbf{z})}{\omega_{\mathbf{Y}}(\mathbf{y})}.$$

Because

$$\begin{aligned}&\operatorname{var}_{\omega}\left[\frac{g(\mathbf{Y},\mathbf{Z})}{\omega(\mathbf{Y},\mathbf{Z})}\right] - \operatorname{var}_{\omega_{\mathbf{Y}}}\left[\frac{\eta(\mathbf{Y})g_b(\mathbf{Y})}{\omega_{\mathbf{Y}}(\mathbf{Y})}\right] \\ &= \int_{\Re^l} \frac{g_b^2(\mathbf{y})}{\omega_{\mathbf{Y}}(\mathbf{y})}\left\{\int_{\Re^{m-l}} \frac{g_a^2(\mathbf{y},\mathbf{z})}{\omega_{\mathbf{Z}|\mathbf{Y}}(\mathbf{z})}\mathbf{dz} - \left[\int_{\Re^{m-l}} g_a(\mathbf{y},\mathbf{z})\mathbf{dz}\right]^2\right\}\mathbf{dy} \\ &= \int_{\Re^l} \frac{g_b^2(\mathbf{y})}{\omega_{\mathbf{Y}}(\mathbf{Y})} \mathrm{E}_{\omega_{\mathbf{Z}|\mathbf{Y}}}\left[\frac{g_a(\mathbf{y},\mathbf{Z})}{\omega_{\mathbf{Z}|\mathbf{Y}}(\mathbf{Z})} - \int_{\Re^{m-l}} g_a(\mathbf{y},\mathbf{z})\mathbf{dz}\right]^2\mathbf{dy} \\ &= \int_{\Re^l} \frac{g_b^2(\mathbf{y})}{\omega_{\mathbf{Y}}(\mathbf{y})} \operatorname{var}_{\omega_{\mathbf{Z}|\mathbf{Y}}}\left[\frac{g_a(\mathbf{y},\mathbf{Z})}{\omega_{\mathbf{Z}|\mathbf{Y}}(\mathbf{Z})}\right]\mathbf{dy} \\ &\geq 0,\end{aligned}$$

a variance reduction results when estimating λ by $\frac{\eta(\mathbf{Y})g_b(\mathbf{Y})}{\omega_{\mathbf{Y}}(\mathbf{Y})}$.

If g takes the form (5.17) and standard MC samples from $\{\omega(\mathbf{x}) = \pi(\mathbf{x})\}$, this is called the *Rao-Blackwell estimator* to acknowledge its intrinsic tie to the Rao-Blackwell theorem (e.g., Lehman 1983 or Mood et al. 1974). Some writers call this variance-reducing technique *conditional sampling*. Since that term actually refers to an MC method for estimating a conditional mean (e.g., Fishman 1996, pp. 321–324), we adopt the Rao-Blackwellization appellation here.

For $\mathbf{Y}^{(1)}, \ldots, \mathbf{Y}^{(n)}$ randomly sampled from $\omega_{\mathbf{Y}}$, on n independent replications, the Rao-Blackwell estimator of λ is

$$\dot{\lambda}_n := \frac{1}{n}\sum_{i=1}^{n}\left[\frac{\eta(\mathbf{Y}^{(i)})g_b(\mathbf{Y}^{(i)})}{\omega_{\mathbf{Y}}(\mathbf{Y}^{(i)})}\right], \tag{5.19}$$

and

$$\dot{s}_n^2 = \frac{1}{n-1}\sum_{i=1}^{n}\left[\frac{\eta(\mathbf{Y}^{(i)})g_b(\mathbf{Y}^{(i)})}{\omega_{\mathbf{Y}}(\mathbf{Y}^{(i)})} - \dot{\lambda}_n\right]^2$$

is an unbiased estimator of $\operatorname{var}\eta\left(\mathbf{Y}_l^{(i)}\right)$. Then

P14. $\dot{\lambda}_n \to \lambda$ w.p.1 as $n \to \infty$.

P15. $\frac{\dot{\lambda}_n-\lambda}{\sqrt{\dot{s}_n^2}} \xrightarrow{d} \mathcal{N}(0,1)$ as $n \to \infty$.

P16. $\text{s.e.}(\dot{\lambda}_n) = \sqrt{\frac{\dot{s}_n^2}{n}}$ and $\text{r.e.}(\dot{\lambda}_n) = \frac{\text{s.e.}(\dot{\lambda}_n)}{\dot{\lambda}_n}$.

P17. $\left[\dot{\lambda}_n - \Phi^{-1}(1-\delta/2)\text{s.e.}(\dot{\lambda}_n),\ \dot{\lambda}_n + \Phi^{-1}(1-\delta/2)\text{s.e.}(\dot{\lambda}_n)\right]$ (5.20)

includes λ with probability that converges to $1-\delta$ as $n \to \infty$.

Algorithm RB describes an implementation of Rao-Blackwell estimation. Example 5.2.1 illustrates how the technique can reduce variance in estimating the probability of a rarely occurring event.

ALGORITHM RB

Purpose: To estimate λ in (2.1) using Rao-Blackwellization.

Given: $\{g_b(\mathbf{y}),\ \mathbf{y} \in \Re^{m-l}\}$, $\{\eta(\mathbf{y}),\ \mathbf{y} \in \Re^{m-l}\}$, $\{\omega_{\mathbf{Y}}(\mathbf{y}),\ \mathbf{y} \in \Re^{m-l}\}$, and $n :=$ number of independent replications.

Method:

$i = 1$, $V_1 = 0$, and $V_2 = 0$.
While $i \le n$:
  randomly sample $\mathbf{Y}$ from $\omega_{\mathbf{Y}}$
  $Z = \frac{\eta(\mathbf{Y})g_b(\mathbf{Y})}{\omega_{\mathbf{Y}}(\mathbf{Y})}$
  $V_1 = V_1 + Z$ and $V_2 = V_2 + Z^2$
  $i = i + 1$.
$\dot{\lambda}_n = \frac{V_1}{n}$.
$\dot{s}_n^2(\dot{\lambda}_n) = \frac{1}{n-1}\left(V_2 - n\dot{\lambda}_n^2\right)$.
$\text{s.e.}(\dot{\lambda}_n) = \sqrt{\frac{\dot{s}_n^2(\dot{\lambda}_n)}{n}}$ and $\text{r.e.}(\dot{\lambda}_n) = \frac{\text{s.e.}(\dot{\lambda}_n)}{(\dot{\lambda}_n)}$. ■

Cost Considerations. Because l ($< m$) variates are generated per replication, the sampling cost under Rao-Blackwellization is generally less than for standard MC. Occasionally, this leads to a reduction in total computing time per replication, implying a cost ratio (CR) greater than 1. This attribute gives the technique additional appeal.

5.2.1 Exceedance Probabilities for Rare Events

Estimating exceedance probabilities is a required exercise in many areas in which risk plays a role, especially in estimating the probability of a *rare event*. These areas include claims analysis, network reliability, warranty analysis, and investment analysis. Let $X_1, \ldots, X_m$ denote i.i.d. non-negative random variables with continuous p.d.f. π and d.f. $\{F(x) = \int_0^x \pi(y)dy,\ 0 \le x < \infty\}$, and let $S_l := X_1 + \cdots + X_l$

for $l = 1, \ldots, m$. It is of interest to estimate the *exceedance probability*

$$\begin{aligned} \mathrm{pr}(S_m > u) &= \psi_m(u) \\ &:= 1 - F^{(m)}(u) \end{aligned} \tag{5.21}$$

for selected $u > 0$, where

$$F^{(m)}(u) := \int_0^u \pi^{(m)}(z)\mathrm{d}z$$

and $\pi^{(m)}$ denotes the p.d.f. of the sum of m i.i.d. random variables each with p.d.f. π, called the m-fold convolution of π with itself.

Some p.d.f.s, including the exponential, Gamma, and normal, have analytically closed forms for $\psi_m(u)$. Since problems encountered in claims analysis, for example, rarely allow the derivation of a closed form for $\psi_m(u)$, MC is a prime candidate for estimating it.

Let

$$\zeta_{1m}(u) := \begin{cases} 1 - F(u) & \text{if } m = 1 \\ \int_{\Re_+^m} h_{1m}(\mathbf{x}, u)\, f_m(\mathbf{x})\mathrm{d}\mathbf{x} & \text{if } m \geq 2, \end{cases} \tag{5.22}$$

where $\Re_+^m := (0, \infty)^m$,

$$f_m(\mathbf{x}) := \prod_{i=1}^{m} \pi(x_i) \tag{5.23}$$

and

$$h_{1m}(\mathbf{x}, u) := \begin{cases} 1 & \text{if } x_1 + \cdots + x_m > u \\ 0 & \text{elsewhere.} \end{cases} \tag{5.24}$$

It is easily seen that $\psi_m(u) = \zeta_{1m}(u)$, suggesting one way to estimate $\psi_m(u)$. On each independent replication for $m \geq 2$, standard MC executes the steps:

Plan 5.1

1. Randomly sample X_i from π for $i = 1, \ldots, m$.

2. Set $Z = h_{1m}(\mathbf{X}, u)$ as in (5.24).

Then Z has mean $\psi_m(u)$, variance $\psi_m(u)[1 - \psi_m(u)]$, and coefficient of variation $\sqrt{1/\psi_m(u) - 1}$, which increases without bound as u increases. Because the objective often is to estimate the probability of a rare event, this sampling method requires an inordinately large sample size n to obtain an accurate estimate of $\psi_m(u)$ when u is large.

Rao-Blackwellization provides at least two ways to improve statistical efficiency. Because

$$\begin{aligned}\psi_m(u) &= \mathrm{pr}(S_m > u)\\ &= \mathrm{pr}(X_m > u - S_{m-1})\\ &= \mathrm{E}_{X_1,\ldots,X_{m-1}}\{\mathrm{pr}[X_m > (u - S_{m-1})^+ | X_1, \ldots, X_{m-1}]\},\end{aligned}$$

it is easily seen that

$$\psi_m(u) = \mathrm{E}_{X_1,\ldots,X_{m-1}}[1 - F(0 \vee (u - S_{m-1}))],$$

where $S_0 := 0$ and $a \vee b := \max(a, b)$. One approach relies on

$$\zeta_{2m}(u) = \begin{cases} 1 - F(u) & \text{if } m = 1\\ \int_{\Re_+^{m-1}} h_{2,m-1}(\mathbf{x}, u)\, f_{m-1}(\mathbf{x})d\mathbf{x} & \text{if } m \geq 2,\end{cases}$$

where

$$h_{2,m-1}(\mathbf{x}, u) := 1 - F(0, \vee(u - x_1 - \cdots - x_{m-1})) \tag{5.25}$$

Because it is easily seen that $\psi_m(u) = \zeta_{2m}(u)$, an elementary Rao-Blackwell estimator Z for $m > 1$ follows from executing the steps:

Plan 5.2

1. Randomly sample X_i from π for $i = 1, \ldots, m-1$.
2. Set $Z = h_{2,m-1}(\mathbf{X}, u)$ as in (5.25).

Then Z again has mean $\psi_m(u)$, but $\mathrm{var}Z \leq \psi_m(u)[1-\psi_m(u)]$. A second Rao-Blackwell estimator relies on an equivalence in Asmussen and Binswanger (1997) and Asmussen et al. (2000).

Theorem 5.2.1

Let $X_{(1)} < \cdots < X_{(m)}$ denote the order statistics corresponding to $X_1, \ldots, X_m$ and $S_{(l)} := X_{(1)} + \cdots + X_{(l)}$ for $l = 1, \ldots, m$. Then for $m \geq 2$,

$$\psi_m(u) = \mathrm{E}_{X_{(1)},\ldots,X_{(m-1)}}\left[\frac{1 - F(X_{(m-1)} \vee (u - S_{(m-1)}))}{1 - F(X_{(m-1)})}\right]. \tag{5.26}$$

■

PROOF Observe that

$$\mathrm{pr}(X_{(m)} > y \mid X_{(1)}, \ldots, X_{(m-1)}) = \mathrm{pr}(X_{(m)} > y \mid X_{(m-1)})$$

and that $X_{(m-1)}$ and $X_{(m)}$ have the joint p.d.f.

$$f_{X_{(m-1)},X_{(m)}}(x, y) = \begin{cases} m(m-1)F^{(m-2)}(x)\pi(x)\pi(y) & y \geq x \geq 0 \\ 0 & \text{otherwise,} \end{cases}$$

so that

$$\text{pr}(X_{(m)} > y \mid X_{(m-1)} = x) = \begin{cases} \frac{1-F(x)}{1-F(y)} & y \geq x \geq 0 \\ 1 & \text{otherwise.} \end{cases}$$

Because $S_m = S_{(m)}$,

$$\begin{aligned} \text{pr}(S_m > u) &= \text{E}_{X_{(1)},\ldots,X_{(m-1)}}\text{pr}(S_{(m)} > u \mid X_{(1)}, \ldots, X_{(m-1)}) \\ &= \text{E}_{X_{(1)},\ldots,X_{(m-1)}}\text{pr}(X_{(m)} > u - S_{(m-1)} \mid X_{(m-1)}) \\ &= \zeta_{3m}(u), \end{aligned}$$

where

$$\zeta_{3m}(u) := \begin{cases} 1 - F(u) & \text{if } m = 1 \\ \int_{\Re_+^{m-1}} h_{3,m-1}(\mathbf{x}, u) f_{m-1}(\mathbf{x})d\mathbf{x} & \text{if } m \geq 2, \end{cases}$$

$$h_{3,m-1}(\mathbf{x}, u) := \frac{1 - F(x_{(m-1)} \vee (u - x_{(1)} - \cdots - x_{(m-1)}))}{1 - F(x_{(m-1)})}, \tag{5.27}$$

and $x_{(1)} < \cdots < x_{(m)}$ are the values of $x_1, \ldots, x_m$ ordered from smallest to largest. ■

To implement this form of Rao-Blackwellization for $m \geq 2$, one executes the steps:

Plan 5.3

1. Randomly sample X_i from π for $i = 1, \ldots, m$.
2. Set $Z = h_{3,m-1}(\mathbf{X}, u)$ as in (5.27).

Then Z again has mean $\psi_m(u)$ and $\text{var}Z \leq \psi_m(u)[1 - \psi_m(u)]$. Most importantly, if $1 - F(u) = L(u)/u^a$ for some $a > 1$ and L, a *slowly varying function* of u, then (Asmussen et al. 2000, Theorem 3.1)

$$\frac{\sqrt{\text{var}h_{3,m-1}(\mathbf{X}, u)}}{\psi_m(u)} \leq [\psi_m(u)]^c$$

for some $c > 1$.

Definition. The function L is slowly varying if $\frac{L(sx)}{L(s)} \to 1$ as $s \to \infty$ for every $x > 0$ (e.g., see Feller 1971).

Because $\psi_m(u) \to 0$ as $u \to \infty$, $h_{3,m-1}(\mathbf{X}, u)$ has a coefficient of variation that remains bounded and actually decreases as $u \to \infty$, if F has this property. This implies that estimation becomes more efficient as $u \to \infty$. Estimators with this property are sometimes called *asymptotically efficient* (see Heidelberger 1995). Example 5.2.1 illustrates the efficacy of these techniques for claims analysis.

EXAMPLE 5.2.1

CLAIMS ANALYSIS

An example in Asmussen and Binswanger (1997) illustrates Rao-Blackwellization. Consider an insurance company that receives claims whose interarrival times are i.i.d. random variables with the exponential distribution

$$k(x) = \begin{cases} \theta e^{-\theta x} & \theta > 0, \ x \geq 0 \\ 0 & \text{elsewhere.} \end{cases} \tag{5.28}$$

If $M(t)$ denotes the number of claims received in the time interval $[0, t)$, then $\{M(\tau), \ \tau \geq 0\}$ is a Poisson process with rate θ. Let X_i denote the size of the ith claim and assume that $X_1, X_2, \ldots$ are i.i.d. random variables with d.f. $\{B(x), \ x \geq 0\}$, mean μ_B, and independent of claim interarrival times. Suppose customers pay their premiums at a constant rate c, where prudence dictates $c > \theta\mu_B$. Then the insurance surplus at time τ is

$$V(\tau) := v + c\tau - X_1 - \cdots - X_{M(\tau)}, \tag{5.29}$$

where v denotes the cash reserve established by the insurance company. If $V(\tau) > 0$, there is a net surplus. If $V(\tau) < 0$, claims exceed funds and ruin occurs.

The *ruin probability* is

$$\begin{aligned} \psi(v) &= \text{pr}\,[\min_{\tau \geq 0} V(\tau) < 0] \\ &= \text{probability that at some time } \tau \geq 0, \ V(\tau) < 0. \end{aligned} \tag{5.30}$$

Because $\psi(v)$ is monotone decreasing in v, it is of interest to estimate the risk $\psi(v)$ that is incurred for selected values of the reserve level v. By the Pollaczeck-Khinchine theorem (e.g., Asmussen 1987),

$$\psi(v) = (1 - \rho) \sum_{l=0}^{\infty} \rho^l \psi_l(v), \tag{5.31}$$

where (5.21) defines $\psi_l(u)$,

$$\pi(x) = \frac{1 - B(x)}{\mu_B} \qquad x \geq 0, \tag{5.32}$$

and

$$\rho := \frac{\theta\mu_B}{c}.$$

Because (5.31) reveals the ruin probability to be a *mixture* of an infinite number of integrals, directly estimating this summation would require sampling from an infinite number of distributions on each replication. Fortunately, we can exploit the structure of the mixture to make the conditional expectation on each replication a finite-dimensional integral.

Let M denote a discrete random variable from the geometric distribution $\mathcal{G}e(\rho)$. Algorithm RUIN incorporates Plan 5.3 on each replication. ■

ALGORITHM RUIN

Randomly sample M from $\mathcal{G}e(\rho)$ (see Section 3.3.2).
If $M = 0$, set $Z = 0$.
If $M = 1$, set $Z = 1 - F(v)$.
If $M > 1$:
  randomly sample X_i from π, as defined in (5.32), for $i = 1, \ldots, M$
  compute $S_M = X_1 + \cdots + X_M$, $X_{(M-1)}$, and $X_{(M)}$
  $Z = h_{3,M-1}(\mathbf{X}, v)$ as in (5.27).
■

Determining $X_{(M-1)}$ and $X_{(M)}$ takes mean computing time proportional to $\mathrm{E}M = 1/(1-\rho)$. For Plan 5.1 with (5.24) and Plan 5.2 with (5.25), only S_M needs to be computed.

As illustration, suppose the claim size has the Pareto distribution with d.f.

$$B(z) = \begin{cases} 0 & \text{if } 0 \le z \le \beta \\ 1 - \left(\frac{\beta}{z}\right)^{\alpha} & \text{if } \beta < z < \infty \end{cases} \tag{5.33}$$

and mean $\mu_B = \alpha\beta/(\alpha - 1)$ and is denoted by $\mathcal{P}a(\alpha, \beta)$. Then $\pi(x)$ in (5.32) takes the form

$$\pi(x) = \begin{cases} \frac{\alpha-1}{\alpha\beta} & \text{if } 0 \le x \le \beta \\ \frac{\alpha-1}{\alpha\beta}\left(\frac{\beta}{x}\right)^{\alpha} & \text{if } \beta < x < \infty \end{cases} \tag{5.34}$$

with d.f.

$$F(x) = \frac{\alpha - 1}{\alpha\beta}\min(x, \beta) + \frac{1}{\alpha}\max\left[0, 1 - \left(\frac{\beta}{x}\right)^{\alpha-1}\right] I_{(\beta,\infty)}(x). \tag{5.35}$$

Figure 5.1 displays this d.f. for $\alpha = 2$ and $\beta = 1$.

Table 5.4 lists results for Plans 5.1, 5.2, and 5.3 for $\mathcal{P}a(2, 1)$ based on $n = 1{,}000$ independent replications. We leave the proof that Z in Algorithm RUIN has a bounded coefficient of variation as an exercise. Note that for F, as in (5.35), Z does not have a bounded coefficient of variation (see Exercise 2).

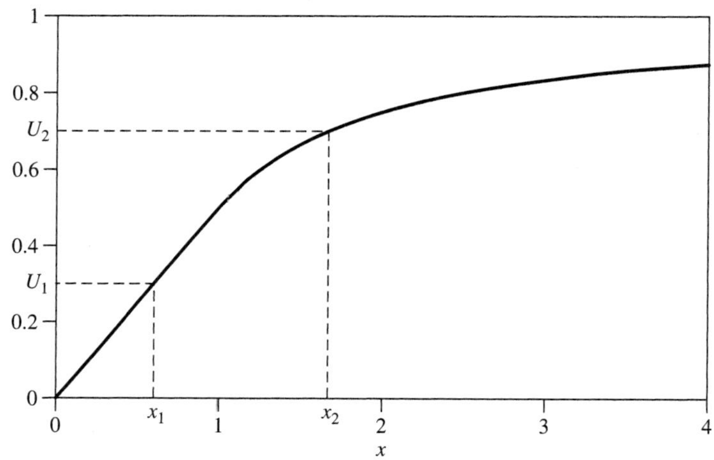

FIGURE 5.1 | $\{F(x) = \frac{1}{2}\min(1, x) + \frac{1}{2}\min(0, 1 - \frac{1}{x}), 0 < x < \infty\}$

TABLE 5.4 | ***Variance ratio (VR) for ruin probability****
$\{B(x)\}$ *is* $\mathcal{P}a(2, 1),\ \rho = 1/1.1, n = 10{,}000$

	$\hat{\psi}(v)$			Variance Ratio		
v	Plan 5.1	Plan 5.2	Plan 5.3	Plan 5.1 / Plan 5.2	Plan 5.1 / Plan 5.3	$\psi(v)$
10	.564	.610	.564	1.56	1.23	.548
100	.069	.0956	.0858	1.00	2.09	.0853
1,000	.00630	.00618	.00529	1.10	23.5	.00540

*Source: Ratios were computed from results in Binswanger (1997, Table 3.2). $\psi(v)$ is based on a numerical approximation in the reference.

5.3 | CONTROL VARIATES

The method of control variates reduces variance by exploiting correlation. Suppose g admits the representation

$$g(\mathbf{x}) = g_a(\mathbf{x}) + g_b(\mathbf{x}) \qquad \mathbf{x} \in \mathcal{X},$$

where

$$\lambda(g_a) := \int_{\mathcal{X}} g_a(\mathbf{x})d\mathbf{x}$$

is known exactly, but

$$\lambda(g_b) := \int_{\mathcal{X}} g_b(\mathbf{x})d\mathbf{x}$$

is not. If $\mathbf{X}$ has distribution ω, then

$$Z := \lambda(g_a) + \frac{g_b(\mathbf{X})}{\omega(\mathbf{X})} \tag{5.36}$$

has mean λ and

$$\text{var}Z = \text{var}\left[\frac{g_b(\mathbf{X})}{\omega(\mathbf{X})}\right].$$

In principle, one prefers Z to the standard MC estimator $\bar{\lambda}_1$ in (2.9) for $n = 1$ if

$$\text{var}Z < \text{var}\bar{\lambda}_1. \tag{5.37}$$

For $k_a(\mathbf{x}) := g_a(\mathbf{x})/\omega(\mathbf{x})$ and $k_b(\mathbf{x}) := g_b(\mathbf{x})/\omega(\mathbf{x})$, this obtains if and only if (see Exercise 3)

$$\frac{\text{cov}\,[k_a(\mathbf{X})k_b(\mathbf{X})]}{\text{var}k_a(\mathbf{X})} > -\frac{1}{2}. \tag{5.38}$$

If positive correlation prevails, Z offers a clear advantage over $\bar{\lambda}_1$. However, strong negative correlation ($< -\frac{1}{2}$) makes standard MC more accurate. Although knowing $\lambda(g_a)$ may not be beneficial when using (5.36), a closely related formulation reduces variance, regardless of the sign of the correlation.

Let

$$k(\mathbf{x}, \beta) := \frac{g(\mathbf{x})}{\omega(\mathbf{x})} + \beta\left[\frac{g_a(\mathbf{x})}{\omega(\mathbf{x})} - \lambda(g_a)\right] \quad -\infty < \beta < \infty, \quad \mathbf{x} \in \mathcal{X}. \tag{5.39}$$

For $\mathbf{X}$ from ω, $k(\mathbf{X}, \beta)$ has mean λ and

$$\begin{aligned}\text{var}k(\mathbf{X}, \beta) &= (1+\beta)^2\text{var}k_a(\mathbf{X}) + 2(1+\beta)\text{cov}\,[k_a(\mathbf{X}), k_b(\mathbf{X})] + \text{var}k_b(\mathbf{X}) \\ &\geq \left\{1 - \text{corr}^2\,[k_a(\mathbf{X}), k_b(\mathbf{X})]\right\}\text{var}k_b(\mathbf{X}).\end{aligned}$$

Minimizing $\text{var}k(\mathbf{X}, \beta)$ with respect to β achieves the lower bound at $\beta = \beta_{\text{opt}}$, where

$$\beta_{\text{opt}} = -\left\{1 + \frac{\text{cov}\,[k_a(\mathbf{X}), k_b(\mathbf{X})]}{\text{var}k_a(\mathbf{X})}\right\}.$$

Most importantly, it is easily seen that

$$\text{var}k(\mathbf{X}, \beta_{\text{opt}}) \leq \text{var}\bar{\lambda}_1,$$

implying a reduced variance for any nonzero covariance, and

$$\text{var}k(\mathbf{X}, \beta) \to 0 \quad \text{as}\ |\text{corr}\,[g_a(\mathbf{X}), g_b(\mathbf{X})]| \to 1,$$

regardless of sign.

A more general version of this technique, called the *control variate method,* takes the form

$$\tau(\mathbf{x}, \beta) = \frac{g(\mathbf{x})}{\omega(\mathbf{x})} + \beta\left[\frac{q(\mathbf{x})}{\omega(\mathbf{x})} - \theta\right], \tag{5.40}$$

where q is a function defined on $\mathcal{X}$, $\theta := \int_{\mathcal{X}} q(\mathbf{x})d\mathbf{x}$ is known, and $\int_{\mathcal{X}} \frac{q^2(\mathbf{x})}{\omega(\mathbf{x})}d\mathbf{x} < \infty$. Then

$$\beta_{\text{opt}} = -\frac{\text{cov}\left[\frac{g(\mathbf{X})}{\omega(\mathbf{X})}, \frac{q(\mathbf{X})}{\omega(\mathbf{X})}\right]}{\text{var}\left[\frac{q(\mathbf{X})}{\omega(\mathbf{X})}\right]}$$

induces the minimal variance

$$\text{var}\tau(\mathbf{X}, \beta_{\text{opt}}) = \left(1 - \rho^2\right) \text{var}\left[\frac{g(\mathbf{X})}{\omega(\mathbf{X})}\right],$$

where

$$\rho := \text{corr}\left[\frac{g(\mathbf{X})}{\omega(\mathbf{X})}, \frac{q(\mathbf{X})}{\omega(\mathbf{X})}\right].$$

The resulting maximal variance reduction is $\frac{1}{1-\rho^2}$, revealing the advantage of choosing a function q that has strong positive or negative correlation with g.

In practice, β_{opt} is unknown. Therefore, it is of interest to see how errors in approximating β_{opt} affect variance. For (5.40), it can be shown that

$$\text{var}\tau(\mathbf{X}, \beta) \leq \text{var}\tau(\mathbf{X}, 0) = \text{var}\left[\frac{g(\mathbf{X})}{\omega(\mathbf{X})}\right] \tag{5.41}$$

for all β satisfying $\beta(\beta + 2\beta_{\text{opt}}) \leq 0$. In particular, a variance reduction occurs for

$$0 \leq \beta \leq 2\beta_{\text{opt}} \qquad \text{if } \beta_{\text{opt}} > 0$$

and

$$2\beta_{\text{opt}} \leq \beta \leq 0 \qquad \text{if } \beta_{\text{opt}} < 0,$$

where $\beta_{\text{opt}} > 0$ if and only if $\rho < 0$ (see Exercise 4). It can also be shown that these same inequalities hold for $k(\mathbf{x}, \beta)$ in (5.39).

With regard to sensitivity to errors of approximation, the variance ratio

$$\begin{aligned}\text{VR}(\beta) &:= \frac{\text{var}\tau(\mathbf{X}, 0)}{\text{var}\tau(\mathbf{X}, \beta)} \\ &= \frac{1}{1 - \rho^2 + \left(\beta - \beta_{\text{opt}}\right)^2 R},\end{aligned}$$

where

$$R := \frac{\text{var}\frac{q(\mathbf{X})}{\omega(\mathbf{X})}}{\text{var}\frac{g(\mathbf{X})}{\omega(\mathbf{X})}}.$$

These expressions reveal that variance reduction depends on

- The extent of correlation, ρ, between $g(\mathbf{X})$ and $q(\mathbf{X})$.
- The ratio, R, of the variances of $\frac{g(\mathbf{X})}{\omega(\mathbf{X})}$ and $\frac{q(\mathbf{X})}{\omega(\mathbf{X})}$.
- How close the chosen β is to β_{opt}.

If $\beta = \beta_{\text{opt}}$, then ρ is the sole determinant of variance. As illustration,

ρ	VR
.50	1.33
.80	2.78
.90	5.26
.95	10.3
.99	50.3

reveal that a relatively high squared correlation is required to realize a substantial variance reduction.

If $\beta \neq \beta_{\text{opt}}$, as usually occurs in practice, $\text{VR}(\beta) \le \text{VR}(\beta_{\text{opt}})$, but

$$\text{VR}(\beta) \to \text{VR}(\beta_{\text{opt}}) \qquad \text{as } R \to \infty.$$

This analysis reveals that it is most desirable to choose a control variate q for which $\frac{q(\mathbf{X})}{\omega(\mathbf{X})}$ has considerably smaller variance than $\frac{g(\mathbf{X})}{\omega(\mathbf{X})}$ as well as strong correlation between $\frac{q(\mathbf{X})}{\omega(\mathbf{X})}$ and $\frac{g(\mathbf{X})}{\omega(\mathbf{X})}$.

Suppose that β_{opt} is estimated from sample data within the MC experiment. Then for $\mathbf{X}^{(1)}, \ldots, \mathbf{X}^{(n)}$ randomly sampled from ω on n independent replications,

$$\breve{\lambda}_n := \frac{1}{n}\sum_{i=1}^{n} \tau\left(\mathbf{X}^{(i)}, \breve{\beta}_n\right) \tag{5.42}$$

is the control-variate estimator of λ, where the least-squares method gives

$$\breve{\beta}_n := -\frac{\sum_{i=1}^{n}\left\{\left[\frac{g(\mathbf{X}^{(i)})}{\omega(\mathbf{X}^{(i)})} - \bar{\lambda}_n\right]\left[\frac{q(\mathbf{X}^{(i)})}{\omega(\mathbf{X}^{(i)})} - \breve{\theta}_n\right]\right\}}{\sum_{i=1}^{n}\left[\frac{q(\mathbf{X}^{(i)})}{\omega(\mathbf{X}^{(i)})} - \breve{\theta}_n\right]^2} \tag{5.43}$$

as an estimate of β_{opt} and

$$\breve{\theta}_n := \frac{1}{n}\sum_{i=1}^{n}\left[\frac{q(\mathbf{X}^{(i)})}{\omega\left(\mathbf{X}^{(i)}\right)}\right].$$

The control-variate estimator (5.42) has the properties

P18. $\breve{\beta}_n \to \beta_{\text{opt}}$ w.p.1 as $n \to \infty$.

P19. $\breve{\lambda}_n \to \lambda$ w.p.1 as $n \to \infty$.

P20. $\frac{\breve{\lambda}_n - \lambda}{\sqrt{\breve{s}_n^2/n}} \xrightarrow{d} \mathcal{N}(0, 1)$ as $n \to \infty$, where

$$\breve{s}_n^2 := \frac{1}{n-1}\sum_{i=1}^{n}\left[\tau\left(\mathbf{X}^{(i)}, \breve{\beta}_n\right) - \breve{\lambda}_n\right]^2. \tag{5.44}$$

P21. The estimator $\breve{\lambda}_n$ has standard error and relative error $\text{s.e.}(\breve{\lambda}_n) = \sqrt{\frac{\breve{s}_n^2}{n}}$ and $\text{r.e.}(\breve{\lambda}_n) = \frac{\text{s.e.}(\breve{\lambda}_n)}{\breve{\lambda}_n}$ respectively.

P22. The interval

$$\left[\breve{\lambda}_n - \Phi^{-1}(1-\delta/2)\text{s.e.}(\breve{\lambda}_n),\ \breve{\lambda}_n + \Phi^{-1}(1-\delta/2)\text{s.e.}(\breve{\lambda}_n)\right]$$

includes λ with probability that converges to $1-\delta$ as n increases.

Expression (5.42) uses data from the same n independent replications to compute both $\breve{\lambda}_n$ and $\breve{\beta}_n$. Although this procedure introduces a bias in $\breve{\lambda}_n$ as an estimator of λ, the bias decreases at a rate proportional to n^{-1} as n becomes large. An alternative procedure computes $\breve{\beta}_n$ as in (5.43) from N ($\ll n$) preliminary replications. It then uses this estimate together with data from n subsequent replications to compute $\breve{\lambda}_n$. This procedure reduces the bias in $\breve{\lambda}_n$ for any fixed n. In the remainder of this chapter, we assume that the first procedure is followed with n sufficiently large to make the bias negligible (see property P18). (For more on the estimation of β_{opt}, see Fishman 1996, Section 4.2.)

Simple to apply, the control-variate method merely calls for the collection of supplementary data during IMC sampling. Algorithm CV describes the steps. The expression for $\breve{s}_n^2$ follows from substitution of definitions into (5.44).

ALGORITHM CV

Purpose: To estimate λ in (2.1) using the control-variate method.

Given: θ, $\{g(\mathbf{x}),\ \mathbf{x} \in \mathcal{X}\}$, $\{q(\mathbf{x}),\ \mathbf{x} \in \mathcal{X}\}$, and $\{\omega(\mathbf{x}),\ \mathbf{x} \in \mathcal{X}\}$.

Method:

$i = 1,\ V_1 = \cdots = V_5 = 0.$

While $i \le n$:

  randomly sample $\mathbf{X}$ from ω

  $W_1 = \frac{g(\mathbf{X})}{\omega(\mathbf{X})}$ and $W_2 = \frac{q(\mathbf{X})}{\omega(\mathbf{X})}$

  $V_1 = V_1 + W_1$ and $V_2 = V_2 + W_1^2$

  $V_3 = V_3 + W_2$ and $V_4 = V_4 + W_2^2$

  $V_5 = V_5 + W_1 W_2$

  $i = i + 1.$

$\bar{\lambda}_n = \frac{V_1}{n}$ and $\bar{\theta}_n = \frac{V_3}{n}$.

$\breve{\beta}_n = -\frac{V_5 - n\bar{\lambda}_n\bar{\theta}_n}{V_4 - n\bar{\theta}_n^2}$.

$\breve{\lambda}_n = \bar{\lambda}_n + \breve{\beta}_n(\bar{\theta}_n - \theta)$.

$\breve{s}_n^2 = \left[V_2 + 2\breve{\beta}_n(V_5 - \theta V_1) + \breve{\beta}_n^2(V_4 - 2\theta V_3 + n\theta^2) - n\breve{\lambda}_n^2\right]$.

$\text{s.e.}(\breve{\lambda}_n) = \sqrt{\frac{\breve{s}_n^2}{n}}$ and $\text{r.e.}(\breve{\lambda}_n) = \frac{\text{s.e.}(\breve{\lambda}_n)}{\breve{\lambda}_n}$. ■

Cost Considerations. The need to evaluate $\frac{q(\mathbf{x})}{\omega(\mathbf{x})}$ on each replication and to evaluate $\breve{\beta}_n$ from the n samples at the end of the experiment makes the time expenditure for the control-variate method greater than for standard MC. In a large-scale problem, this additional expenditure is usually negligible.

EXAMPLE 5.3.1

ONE-DIMENSIONAL INTEGRATION IV

As a control variate for the problem in Example 2.1.1, let

$$q(x) := \mathrm{e}^{-x} \qquad 0 \le x \le 1, \tag{5.45}$$

with

$$\begin{aligned} \theta &= \int_0^1 q(x)\mathrm{d}x \\ &= 1 - \mathrm{e}^{-1} = 0.6321. \end{aligned}$$

For $n = 2^{15} \times 10 = 327{,}680$ independent replications with sampling from $\mathcal{U}(0, 1)$, this control variate gives

$$\begin{aligned} \breve{\lambda}_n &= 1.330 \\ \text{s.e.}\left(\breve{\lambda}_n\right) &= 5.8955 \times 10^{-4} \\ \text{r.e.}\left(\breve{\lambda}_n\right) &= 4.4343 \times 10^{-4} \\ \breve{\beta}_n &= 2.5986 \\ \breve{\rho} &= 0.8128 \\ \text{VR} &= 2.947, \end{aligned}$$

where the VR comparison is with standard MC and $\breve{\rho}$ denotes the sample correlation coefficient. Also, CR is slightly greater than 1.

Although a worthwhile reduction in variance for a negligible increase in cost, this q mimics g over the relatively small subinterval [0, 0.2739). Thereafter, g increases with x, but q continues to decrease. An alternative control function more closely follows the shape of g. Let

$$q(x) := (ax - 1)^2 \qquad 0 \le x \le 1,$$

where

$$a := \mathrm{e}^{1/2} + 1.$$

Then

$$\begin{aligned} q(0) &= 1 \\ q(1) &= \mathrm{e}, \end{aligned}$$

which are equal to $g(0)$ and $g(1)$ respectively.

Also,

$$\theta = \int_0^1 q(x)\mathrm{d}x$$
$$= \frac{\mathrm{e}^{3/2}+1}{3(\mathrm{e}^{1/2}+1)} = 0.6899.$$

Although q has its minimum at $x = 1/(\mathrm{e}^{1/2}+1) = 0.3775$ whereas g has its minimum at $x = 0.2739$, both g and q decrease over $[0, 0.2739)$ and increase over $[0.3775, 1)$. That is, there is a considerably smaller interval, $[0.2739, 0.3775]$, where the two functions move in opposite directions.

For the same n as before,

$$\breve{\lambda}_n = 1.330$$
$$\text{s.e.}\left(\breve{\lambda}_n\right) = 3.2021 \times 10^{-4}$$
$$\text{r.e.}\left(\breve{\lambda}_n\right) = 2.4061 \times 10^{-4}$$
$$\breve{\beta}_n = -0.76188$$
$$\breve{\rho} = 0.9486$$
$$\text{VR} = 10.01,$$

where the VR comparison is with standard MC.

For essentially the same cost ratio of about CR ≈ 1, this second control function clearly has a reduction in variance more than three times larger than that corresponding to (5.45). This outcome reflects the advantages of selecting a control function that has the same general shape as g. ■

EXAMPLE 5.3.2 PRICING AN ASIAN OPTION III

This example again shows the advantage of pre-analyzing the integrand before choosing a control variate. Recall from Example 2.7.4 that the objective is to estimate the mean discounted Asian-option price $\lambda = \mathrm{e}^{-\xi_1 T}\mathrm{E}_\pi\left(\overline{S}_m - K\right)^+$ where (2.53) defines the sample *arithmetic average*, $\overline{S}_m$, of $S_1, \ldots, S_m$. Although computing λ exactly is an intractable problem, the discounted expectation

$$\theta := \mathrm{e}^{-\xi_1 T}\mathrm{E}_\pi\left(\widetilde{S}_m - K\right)^+ \tag{5.46}$$

is known in closed form (Turnbull and Wakeman 1991), where

$$\widetilde{S}_m := \left(\prod_{j=1}^m S_j\right)^{1/m} \tag{5.47}$$

TABLE 5.5 | *Variance ratio (VR) for Asian options pricing Monte Carlo based on a control variate**
$m = 50,\ \xi_1 = .10,\ K = 100,\ T = .20,\ n = 10^4$

ξ_2	$\frac{K}{S_0}$	VR = $\frac{\text{variance using standard MC}}{\text{variance using control variate}}$
0.20	0.90	312.1
	1.0	7396
	1.1	8899
0.40	0.90	4840
	1.0	1667
	1.1	2345
0.60	0.90	390.0
	1.0	766.4
	1.1	942.6

*Source: Boyle et al. (1997).

defines the sample *geometric average* of $S_1, \ldots, S_m$. Moreover, it is known that $\overline{S}_m$ and $\widetilde{S}_m$ are highly correlated, making $q(\mathbf{X}) := \mathrm{e}^{-\xi_1 T}(\widetilde{S}_m - K)^+ - \theta$ a control variate worth studying. Occasionally, we refer hereafter to λ as the discounted arithmetic average price and to θ as the discounted geometric average price.

Table 5.5 displays variance ratios computed from results in Boyle et al. (1997) using this control variate. They reveal a substantial benefit. For example, for the case of $\xi_2 = 0.20$ and $\frac{K}{S_0} = 1.0$, standard MC requires an estimated 7,396 times more replications than the corresponding control-variate MC to obtain the same variance. Because $\mathrm{VR} = \frac{1}{1-\rho^2}$, the table estimates $\rho^2 \geq 1 - \frac{1}{312} = 0.9968$. Although this level of association is considerably higher than what is to be expected in a typical application of control variates, it does demonstrate the substantial reward that careful control-variate selection prior to MC sampling can sometimes produce. ■

5.4 | ANTITHETIC VARIATES

Suppose that g admits the factorization (2.23) and that $\mathbf{X} := (X_1, \ldots, X_m)$ and $\mathbf{Y} := (Y_1, \ldots, Y_m)$ each have the p.d.f. π. Then

$$Z := \frac{1}{2}\left[h(\mathbf{X}) + h(\mathbf{Y})\right] \tag{5.48}$$

has mean λ and variance

$$\begin{aligned}\mathrm{var}Z &= \frac{1}{4}\{\mathrm{var}h(\mathbf{X}) + 2\mathrm{cov}\,[h(\mathbf{X}), h(\mathbf{Y})] + \mathrm{var}h(\mathbf{Y})\}\\ &= \frac{1}{2}\mathrm{var}h(\mathbf{X})\{1 + \mathrm{corr}\,[h(\mathbf{X}), h(\mathbf{Y})]\},\end{aligned}$$

where $\mathrm{var}h(\mathbf{Y}) = \mathrm{var}h(\mathbf{X})$ and

$$\mathrm{corr}\,[h(\mathbf{X}), h(\mathbf{Y})] = \frac{\mathrm{cov}\,[h(\mathbf{X}), h(\mathbf{Y})]}{\sqrt{\mathrm{var}h(\mathbf{X})\mathrm{var}h(\mathbf{Y})}}.$$

Recall that all preceding MC sampling plans rely on independent replications, for which $\mathrm{cov}\,[h(\mathbf{X}), h(\mathbf{Y})] = 0$ and, therefore, $\mathrm{corr}\,[h(\mathbf{X}), h(\mathbf{Y})] = 0$. However, the *antithetic-variate method* gains its advantage by inducing negative correlation between X_i and Y_i for each coordinate $i = 1, \ldots, m$. Provided that h has a certain monotone property, which we describe shortly, this correlation between $h(\mathbf{X})$ and $h(\mathbf{Y})$ is negative. This implies

$$\mathrm{var}Z \le \mathrm{var}\bar{\lambda}_2,$$

where $\bar{\lambda}_2$ is the standard MC estimator of λ for $n = 2$ independent replications. Moreover, it achieves this variance reduction at a cost per replication no greater, and sometimes less, than the cost per replication for standard MC.

If U is from $\mathcal{U}(0,1)$, then it easily seen that $1 - U$ is also from $\mathcal{U}(0,1)$. However, U and $1 - U$ are not independent. In particular,

$$\begin{aligned}\mathrm{cov}(U, 1-U) &= \int_0^1 u(1-u)\mathrm{d}u - \int_0^1 u\mathrm{d}u \int_0^1 (1-u)\mathrm{d}u\\ &= -\frac{1}{12}.\end{aligned}$$

Because $\mathrm{var}U = \mathrm{var}(1-U) = \frac{1}{12}$, we have

$$\mathrm{corr}(U, 1-U) = -1$$

so that U and $1 - U$ have perfect negative correlation.

An analogous property holds more generally. Suppose X and Y have p.d.f. $\pi := \{\pi(z),\ a \le z \le b\}$ and d.f. $\{F(z) := \int_a^z \pi(w)\mathrm{d}w\}$. It is known that there exist many different joint p.d.f.s $\{G(x,y),\ a \le x \le b,\ a \le y \le b\}$ for X and Y with marginal p.d.f.s π. Among these, it also known that those that satisfy the constraint $F(X) + F(Y) = 1$ induce the most negative correlation (Hoeffding 1940). In reality, there exists only one joint p.d.f. that satisfies this constraint. Let U be from $\mathcal{U}(0,1)$ and recall from Theorem 3.3.1 that solving the equation $F(X) = U$ for X generates X via the inverse-transform method. To satisfy the constraint $F(X) + F(Y) = 1$, it must be that solving $F(Y) = 1 - U$ for Y induces the most negative correlation between X and Y. Several examples illustrate the concept.

EXAMPLE 5.4.1

ANTITHETIC EXPONENTIAL VARIATES

The exponential distribution $\{F(x) = 1 - e^{-x/\beta},\ 0 \le x < \infty\}$ with $\beta > 0$ has $X = -\beta \ln(1-U)$ for $\beta > 0$, $Y = -\beta \ln U$, and

$$\begin{aligned}\operatorname{cov}(X, Y) &= \beta^2 \int_0^1 \ln[u(1-u)]du - \int_0^1 \ln u du \int_0^1 \ln(1-u)du \\ &= \beta^2 \left(1 - \frac{\pi^2}{6}\right).\end{aligned}$$

Because $\operatorname{var} X = \operatorname{var} Y = \beta^2$, the most negative achievable correlation for exponential variates is

$$\begin{aligned}\operatorname{corr}(X, Y) &= 1 - \frac{\pi^2}{6} \\ &\approx -0.645.\end{aligned}$$

■

EXAMPLE 5.4.2

ANTITHETIC VARIATES WITH SYMMETRIC DISTRIBUTIONS

Suppose that $\Phi(X) = U$ and $\Phi(Y) = 1 - U$, where (2.17) defines the standard normal d.f. Φ. Since the standard normal p.d.f. (2.11) is symmetric around 0, in other words, $\phi(z) = \phi(-z)$, it follows that $\Phi(z) = 1 - \Phi(-z)$, implying $X = -Y$ and, therefore, $\operatorname{corr}(X, Y) = -1$. More generally, if X and Y have a p.d.f. symmetric around some point c and d.f. F, then $F(z-c) = 1 - F(c-z)$. Therefore, solving $F(X-c) = U$ for X and $F(Y-c) = 1 - U$ for Y gives $Y = 2c - X$ and $\operatorname{corr}(X, Y) = -1$. ■

EXAMPLE 5.4.3

DISCRETE ANTITHETIC VARIATES

The antithetic-variate method also applies to discrete distributions. Suppose that X and Y have the discrete d.f. $\{q_{a-1} = 0 < q_a < \cdots < q_{b-1} < q_b = 1\}$. Then for U from $\mathcal{U}(0, 1)$, the inverse-transform method (see Section 3.3) gives

$$X = \min(z :\ q_{z-1} < U \le q_z)$$

and

$$Y = \min(z :\ q_{z-1} < 1 - U \le q_z).$$

Figure 5.2 and Table 5.6 illustrate the concept for the d.f. in Table 3.3. The figure graphically depicts how the antithetic-variate method determines X and Y using this d.f. The diagonal line is $\{(u, 1-u),\ 0 \le u \le 1\}$, and the dashed lines correspond to $\{q_1 < \ldots < q_8\}$ in the table. Whereas X and Y can assume values from 1 to 8, $\frac{1}{2}(X + Y)$ is restricted to 4.5, 5, and 5.5, hinting at the variance-reducing

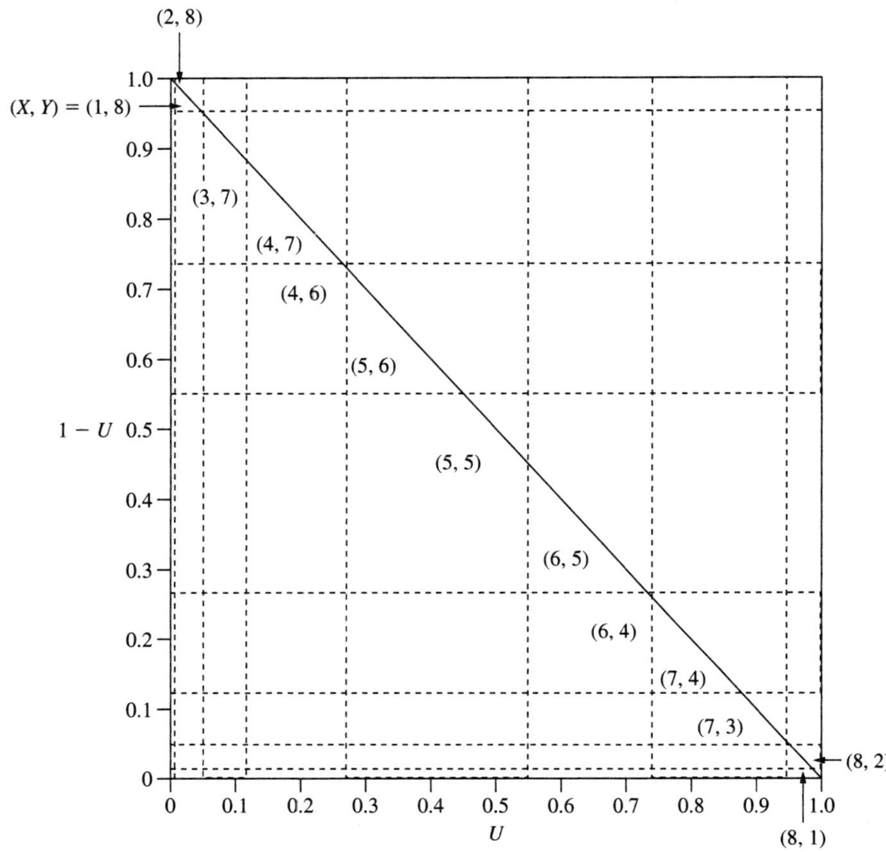

FIGURE 5.2 | *Joint (X, Y) antithetic distribution with marginal d.f. in Table 3.3*

potential of the antithetic-variate method. Table 5.6 also shows the achievable variance ratios for selected functions h. They range from 1.80 for $\{h(x) := e^x\}$ to 20.5 for $\{h(x) = x^2\}$, suggesting that the form of h significantly affects the variance ratio. Indeed, this is a critical issue.

If $\{h(x)\}$ is either monotone nondecreasing or monotone nonincreasing in x and $F(X) + F(Y) = 1$, then $\text{corr}[h(X), h(Y)] \le 0$, as desired. In the absence of monotonicity, negative correlation cannot be guaranteed. Perusal of Table 5.6 reveals that all functions are monotone. ■

With qualification, negative correlation also holds in $m \ge 1$ dimensions:

- Let $\{h(\mathbf{z}),\ \mathbf{z} := (z_1, \ldots, z_m) \in \mathcal{X}\}$ in (2.23) be either monotone nondecreasing or monotone nonincreasing in each variable z_i for $i = 1, \ldots, m$.
- Let the p.d.f. π in (2.23) have the form

$$\pi(\mathbf{z}) := \prod_{i=1}^{m} \pi_i(z_i). \tag{5.49}$$

TABLE 5.6 | *Antithetic joint distribution for marginal d.f. in Table 3.3*

$q_x + q_y = 1$

x	y	$\text{pr}(X = x, Y = y)$	$z = \frac{1}{2}(x + y)$	$\text{pr}\left[\frac{1}{2}(X + Y) = z\right]$
1	8	0.01	4.5	0.02
2	8	0.04	5.0	0.34
3	7	0.07	5.5	0.64
4	7	0.14		
4	6	0.01		
5	6	0.18		
5	5	0.10		
6	5	0.18		
6	4	0.01		
7	4	0.14		
7	3	0.07		
8	2	0.04		
8	1	0.01		

$h(x)$	λ	$\text{var}h(X)$	$\min\left(\frac{x+y}{2}\right)$	$\max\left(\frac{x+y}{2}\right)$	$\text{var}\left[\frac{h(X)+h(Y)}{2}\right]$	VR	$\text{corr}(X, Y)$
x	5.31	2.33	4.5	5.5	0.0689	16.9	−0.941
x^2	30.5	243	25	32.5	5.92	20.5	−0.951
e^x	507	476,899	229.0	1494	132,674	1.80	−0.444
$\log x$	0.702	0.0208	0.4515	0.739	0.00290	4.10	−0.756

- The random variables $X_1, \ldots, X_m$ are independent, and X_i has p.d.f. π_i and d.f. $\{F_i(z) := \int_{-\infty}^{z} \pi_i(w)dw\}$ for $i = 1, \ldots, m$.
- The random variables $Y_1, \ldots, Y_m$ are independent, and Y_i has p.d.f. π_i and d.f. $\{F_i(z) := \int_{-\infty}^{z} \pi_i(w)dw\}$ for $i = 1, \ldots, m$.
- If for each $i = 1, \ldots, m$, (X_i, Y_i) satisfies $F_i(X_i) + F_i(Y_i) = 1$, then for $\mathbf{X} := (X_1, \ldots, X_m)$ and $\mathbf{Y} := (Y_1, \ldots, Y_m)$

$$\text{cov}\,[h(\mathbf{X}), h(\mathbf{Y})] \le 0,$$

and Z in (5.48) has

$$\text{var}Z \le \frac{1}{2}\text{var}h(\mathbf{X}).$$

If IMC were to rely on a sampling distribution $\{\omega(\mathbf{z}) \neq \pi(\mathbf{z})\}$ of the form

$$\omega(\mathbf{z}) := \prod_{i=1}^{m} \omega_i(\mathbf{z}_i),$$

and (X_i, Y_i) satisfies $\int_{-\infty}^{X_i} \omega_i(\mathbf{z})d\mathbf{z} + \int_{-\infty}^{Y_i} \omega_i(\mathbf{z})d\mathbf{z} = 1$, then a variance reduction again accrues if $\left\{\frac{g(\mathbf{z})}{\omega(\mathbf{z})}\right\}$ is either monotone nondecreasing or monotone nonincreasing in each variable.

ALGORITHM **ANTVAR**

Purpose: To estimate λ in (2.23) using the antithetic-variate method where π satisfies (5.49).

Given: D.f.s $F_1, \ldots, F_m$, where $\{F_i(z) := \int_{-\infty}^{z} \pi_i(w)dw, -\infty < z < \infty\}$ for $i = 1, \ldots, m$, $n =$ number of independent replications.

Method:

$i = 0$, $V_1 = 0$, and $V_2 = 0$.
While $i \le n$:
  $j = 1$
  while $j \le m$:
    randomly sample U from $\mathcal{U}(0, 1)$
    solve $F_i(X_i) = U$ for X_i and $F_i(Y_i) = 1 - U$ for Y_i
    $j = j + 1$
  $Z = \frac{1}{2}[h(X_1, \ldots, X_m) + h(Y_1, \ldots, Y_m)]$
  $V_1 = V_1 + Z$ and $V_2 = V_2 + Z^2$
  $i = i + 1$
$\grave{\lambda}_n = \frac{V_1}{n}$ and $\grave{s}_n^2 = \frac{1}{n-1}\left(V_2 - n\grave{\lambda}_n^2\right)$.
$\text{s.e.}(\grave{\lambda}_n) = \sqrt{\grave{s}_n^2 n}$ and $\text{r.e.}(\grave{\lambda}_n) = \frac{\text{s.e.}(\grave{\lambda}_n)}{\grave{\lambda}_n}$.

Algorithm ANTVAR describes one way to implement antithetic variates. With regard to error assessment: ■

P23. $\grave{\lambda}_n \to \lambda$ w.p.1 as $n \to \infty$.

P24. $\frac{\grave{\lambda}_n - \lambda}{\sqrt{\grave{s}_n^2/n}} \xrightarrow{d} \mathcal{N}(0, 1)$ as $n \to \infty$,

where

$$\grave{s}_n^2 := \frac{1}{n-1}\sum_{i=1}^{n}\left\{\frac{1}{2}[h(\mathbf{X}^{(i)}) + h(\mathbf{Y}^{(i)})] - \grave{\lambda}_n\right\}^2 \tag{5.50}$$

and $\mathbf{X}^{(i)}$ and $\mathbf{Y}^{(i)}$ denote $X_1, \ldots, X_m$ and $Y_1, \ldots, Y_m$, respectively, on replication i.

P25. The estimator $\grave{\lambda}_n$ has standard error and relative error $\text{s.e.}(\grave{\lambda}_n) = \sqrt{\frac{\grave{s}_n^2}{n}}$ and $\text{r.e.}(\grave{\lambda}_n) = \frac{\text{s.e.}(\grave{\lambda}_n)}{\grave{\lambda}_n}$ respectively.

P26. The interval

$$\left[\grave{\lambda}_n - \Phi^{-1}(1-\delta/2)\text{s.e.}(\grave{\lambda}_n),\ \grave{\lambda}_n + \Phi^{-1}(1-\delta/2)\text{s.e.}(\grave{\lambda}_n)\right]$$

includes λ with probability that converges to $1-\delta$ as n increases.

EXAMPLE 5.4.4

PRICING AN ASIAN OPTION IV

Recall from Example 2.7.4 that the objective is to estimate the discounted payoff $\lambda := \mathrm{E}_\pi h(\mathbf{X})$, where (2.54) defines h and $X_1, \ldots, X_m$ are i.i.d. from $\mathcal{N}(0, 1)$. Because h is monotone increasing in each variable, the problem is a candidate for variance reduction using antithetic variates. In particular,

$$\text{var}\left\{\frac{1}{2}\left[h(\mathbf{X}) + h(\mathbf{Y})\right]\right\} \leq \frac{1}{2}\text{var}h(\mathbf{X}).$$

Again, it is possible to glimpse the potential for variance reduction before sampling. We focus on $S(\tau_j) := S_j$ in (2.52) and $\overline{S}_m$ in (2.53).

Lognormal Distribution. Let Z denote a random variable from $\mathcal{N}(0, 1)$. Then for constants $\alpha \in (-\infty, \infty)$ and $\beta \in (0, \infty)$, $W := \mathrm{e}^{\alpha+\beta Z}$ has the lognormal distribution

$$f(w) = \begin{cases} \dfrac{1}{\sqrt{2\pi}\beta w}\mathrm{e}^{-(\ln w-\alpha)^2/2\beta^2} & 0 \leq w < \infty \\ 0 & \text{elsewhere,} \end{cases}$$

denoted by $\mathcal{LN}(\alpha, \beta)$, with

$$\begin{aligned} \mathrm{E}W &= (2\pi)^{-1/2}\int_{-\infty}^{\infty} \mathrm{e}^{\alpha+\beta z-z^2/2}\mathrm{d}z \\ &= (2\pi)^{-1/2}\int_{-\infty}^{\infty} \mathrm{e}^{\alpha+\beta^2/2-(z-\beta)^2/2}\mathrm{d}z \\ &= \mathrm{e}^{\alpha+\beta^2/2} \\ \mathrm{E}W^2 &= (2\pi)^{-1/2}\int_{-\infty}^{\infty} \mathrm{e}^{2\alpha+2\beta z-z^2/2}\mathrm{d}z \\ &= (2\pi)^{-1/2}\int_{-\infty}^{\infty} \mathrm{e}^{2\alpha+2\beta^2-(z-2\beta)^2/2}\mathrm{d}z \\ &= \mathrm{e}^{(2\alpha+\beta^2)} \\ \text{var}W &= \mathrm{E}W^2 - \mathrm{E}^2 W \\ &= \mathrm{e}^{2\alpha+\beta^2}(\mathrm{e}^{\beta^2}-1). \end{aligned}$$

In addition to having the same lognormal distribution $\mathcal{LN}(\alpha, \beta)$ and, therefore, the same mean and variance as W, the antithetic variate $W^* := e^{\alpha+\beta(-Z)}$ has

$$\begin{aligned}\text{cov}(W, W^*) &= \text{E}(W, W^*) - \text{E}W\text{E}W^* \\ &= e^{2\alpha}(1 - e^{\beta^2})\end{aligned}$$

and

$$\begin{aligned}\text{corr}(W, W^*) &= -e^{-\beta^2} \\ &< 0.\end{aligned} \tag{5.51}$$

Expression (5.51) reveals that negative correlation increases with decreasing β.

Let

$$\overline{S}^*_m := \frac{1}{m}\sum_{j=1}^{m} S^*_j,$$

where

$$S^*_j := S_0 e^{(\xi_1-\xi_2^2/2)\Delta-\xi_2\sqrt{\Delta}X_j} \qquad j = 1, \ldots, m,$$

so that

$$h(-\mathbf{X}) = e^{-\xi_1 T}\left(\overline{S}^*_m - K\right)^+.$$

Although evaluating $\text{cov}[h(\mathbf{X}), h(-\mathbf{X})]$ is not possible, we can use the previous results for W and W^* to derive

$$\begin{aligned}\text{cov}(\overline{S}_m, \overline{S}^*_m) &= \frac{1}{m^2}\sum_{j=1}^{m}\sum_{l=1}^{m}\text{cov}(S_j, S^*_l) \\ &= \left(\frac{S_0}{m}\right)^2\sum_{j=1}^{m}\sum_{l=1}^{m} e^{(j+l)(\alpha+\beta^2/2)}\left(e^{-\min(j,l)\beta^2} - 1\right) \\ &\le 0\end{aligned}$$

and

$$\begin{aligned}\text{cov}(S_j, S^*_l) &= e^{-j\beta^2}\left(\frac{e^{j\beta^2} - 1}{e^{l\beta^2} - 1}\right) \qquad 1 \le j \le l \le m, \\ &\le 0\end{aligned} \tag{5.52}$$

where $\alpha := (\xi_1 - \xi_2^2/2)\Delta$ and $\beta := \xi_2\sqrt{\Delta} = \xi_2\sqrt{T/m}$.

Expression (5.52) shows that negative correlation increases with decreasing volatility ξ_2, an increasing number of grid points m, and a decreasing time interval T. Table 5.7 supports this assertion. In particular, the variance ratios are greatest for $\xi_2 = 0.10$, $m = 16$ and $\xi_2 = 0.10$, $m = 64$. Moreover, the cost ratio CR $= 1.41$ confirms our earlier claim that computing time may decrease when employing antithetic variates. In the present case, this arises from the convenience of merely generating $X_1, \ldots, X_m$ once, using them to evaluate $h(X_1, \ldots, X_m)$ and then using $-X_1, \ldots, -X_m$ to evaluate $h(-X_1, \ldots, -X_m)$. The products CR $\times$ VR in

TABLE 5.7 | *Variance ratio (VR) for Asian options pricing Monte Carlo based on antithetic variates*
$S(0) = 50,\ \xi_1 = 0.05,\ T = 1.0,\ n = 500{,}000,\ CR = 1.41$

m	ξ_2	K	$\hat{\lambda}_n$	VR	CR × VR	corr$[h(\mathbf{X}), h(-\mathbf{X})]$
16	0.1	45	6.055	80.45	113.4	−0.988
16	0.1	50	1.919	3.213	4.530	−0.689
16	0.1	55	0.2024	1.080	1.523	−0.074
16	0.3	45	7.153	3.578	5.045	−0.721
16	0.3	50	4.17	1.773	2.450	−0.436
16	0.3	55	2.211	1.272	1.794	−0.214
64	0.1	45	5.995	89.35	126.0	−0.989
64	0.1	50	1.846	3.181	4.485	−0.686
64	0.1	55	0.1744	1.072	1.512	−0.067
64	0.3	45	7.025	3.701	5.218	−0.73
64	0.3	50	4.022	1.772	2.499	−0.436
64	0.3	55	2.080	1.263	1.781	−0.208

the table show the reductions in computing time to obtain a specified variance with antithetic variates as compared with standard MC.

As this example confirms, the variance ratio for antithetic variates is sensitive to the values assigned to the parameters of the problem. Nevertheless, its ease of implementation and its frequently high cost ratio (CR > 1) have encouraged its use, merely to take advantage of any potential variance reduction. However, this should be done only after a pre-analysis confirming monotonicity.

As confirmation of this last point, an analysis prior to combining antithetic and importance sampling for the Asian options pricing problem in Example 2.7.4 quickly reveals that for ω in (2.57),

$$\frac{g(\mathbf{x})}{\omega(\mathbf{x})} = h(\mathbf{x})e^{\sum_{i=1}^{m}(-\nu_i x_i + \frac{1}{2}\nu_i^2)},$$

which is not unqualifiedly monotone in each variable $x_1, \ldots, x_m$. ■

5.5 | QUASIRANDOM NUMBERS

Recall that standard MC based on n independent replications induces a standard error $n^{-1/2}\sigma$. Employing one or more of the variance-reducing techniques described in earlier sections reduces the standard error to $n^{-1/2}a$ for some $a < \sigma$. Of course, this benefit has value only if the relative increase in computing time does not exceed $\frac{\sigma^2}{a^2} - 1$. *Quasirandom numbers* provide an alternative way to improve accuracy. Instead of reducing variance σ^2, application of this technique replaces $n^{-1/2}$ by a more rapidly decreasing function of n for moderate and large n. Moreover, it does this with a relatively small increase in computing time compared to that for standard MC.[2] Many different kinds of quasirandom numbers exist. Collectively, they are called *quasi–Monte Carlo* (QMC).

Given QMC's accelerated error-reducing potential, a reader may wonder why its description has not appeared earlier and more prominently in this book. Two properties account for the delay and limited emphasis:

- Theory has established this accelerated convergence for integrands, $\{g(\mathbf{x})\}$, that exhibit specified boundedness and continuity properties. This restricts the justification for applying QMC to a considerably smaller class of functions than satisfy the relatively mild integrality condition (2.6) for standard MC. Interestingly, empirical evidence indicates that QMC sometimes induces accelerated convergence when applied to integrands that do not satisfy the continuity properties, but contraindications also exist.
- Whereas accelerated convergence holds for fixed dimension m as the sample size n grows large, QMC theory asserts that the rate tends to slow down as m increases. That is, the convergence rate depends on m as well as on n. This again contrasts with most IMC for which standard errors are often insensitive to dimension. However, there exist cases in which no slowdown in the convergence rate occurs as m increases, again a difference between empirical and theoretical properties.

Quasirandom numbers are a special case of *equidistributed sequences*. Consider a sequence of points $\mathcal{Q} = \{\mathbf{x}^{(1)}, \mathbf{x}^{(n)}, \ldots\}$, where $\mathbf{x}_i := (x_{i1}, \ldots, x_{im})$ and $0 \le x_{ij} < 1$ for all $j = 1, \ldots, m$ and $i = 1, \ldots, n$. Let $\mathcal{R}$ denote a region in the m-dimensional unit hypercube $\mathcal{I}^m := [0, 1)^m$, and let

$$W_n(\mathcal{R}, \mathcal{Q}) := \text{number of points } \mathbf{x}_1, \ldots, \mathbf{x}_n \text{ in } \mathcal{R}.$$

The sequence is said to be *equidistributed* if

$$D_n(\mathcal{Q}) := \max_{\mathcal{R} \subseteq \mathcal{I}^m} \left| \frac{W_n(\mathcal{R}, \mathcal{Q})}{n} - |\mathcal{R}| \right| \to 0 \qquad \text{as } n \to \infty. \tag{5.53}$$

The quantity $D_n(\mathcal{Q})$ is called the *discrepancy* with respect to the sequence $\mathcal{Q}$.

The global significance of quasirandom numbers becomes apparent when we turn to the convergence rate for (5.53). Among the many different equidistributed sequences that exist, some exhibit the convergence bound (Halton 1960)

$$D_n^* := \max_{\mathcal{R}} \left| \frac{W_n(\mathcal{R}, \mathcal{Q})}{n} - |\mathcal{R}| \right| = \mathrm{O}(n^{-1}(\log n)^m) \tag{5.54}$$

[2]For proofs of the following properties, see Kuipers and Niederreiter (1974) and Niederreiter (1978, 1992).

over all regions of the form $\mathcal{R} = \prod_{i=1}^{m} [0, u_i) \subseteq \mathcal{I}^m$, where $0 < u_i < 1$ for $i = 1, \ldots, m$. These *low-discrepancy sequences* include the Halton, Sobol$'$, Faure, and Niederreiter sequences (e.g., see Niederreiter 1992). Given an initial value $\mathbf{x}_0$, these sequences can be generated recursively, giving them considerable computational appeal. The quantity D_n^*, called the *star discrepancy,* is related to $D_n(\mathcal{Q})$ by the inequalities

$$D_n^*(\mathcal{Q}) \leq D_n(\mathcal{Q}) \leq 2^m D_n^*(\mathcal{Q}),$$

revealing the same convergence rate for $D_n(\mathcal{Q})$, but with a larger coefficient. Techniques that employ low-discrepancy sequences to estimate λ are called *quasi–Monte Carlo* (QMC) *methods.*

For every equidistributed sequence $\mathcal{Q}$, consider all functions g in (2.1) on some region $\mathcal{R} \subseteq \mathcal{I}^m$ that satisfy the property

$$\left|\lambda_n^* - \lambda\right| \to 0 \qquad \text{as } n \to \infty, \tag{5.55}$$

where

$$\lambda_n^* := \frac{1}{n} \sum_{i=1}^{n} g(\mathbf{x}_i).$$

Functions with this property are called *Riemann integrable.*[3] With regard to the convergence rate for (5.55), the importance of low-discrepancy sequences is apparent from the Koksma-Hlawka bound

$$\left|\lambda_n^* - \lambda\right| \leq V_m(g) D_n^*(\mathcal{Q}) \tag{5.56}$$

for all functions g of bounded variation $V_m(g)$ in the sense of Hardy and Krause (e.g., see Niederreiter 1992). This collection of functions includes all Riemann integrable functions.

Expressions (5.54) and (5.56) are worst-case (L^∞) discrepancies. The L^2 discrepancy

$$T_n^*(\mathcal{Q}) := \left\{ \int_{\mathcal{I}^m} \left| \frac{W(\prod_{i=1}^{m}[0, u_i), \mathcal{Q})}{n} - |\mathcal{R}| \right|^2 d\mathbf{u} \right\}^{1/2},$$

where $\mathbf{u} := (u_1, \ldots, u_m)$, gives an average measure of convergence, much like standard deviation for random sequences. It is known that

$$c_m(\mathcal{Q}) D_n^{(m+2)/2}(\mathcal{Q}) \leq T_n^*(\mathcal{Q}) \leq D_n^*(\mathcal{Q}),$$

where the constant $c_m(\mathcal{Q})$ depends only on m and the sequence $\mathcal{Q}$. Also, for any n points

$$D_n^*(\mathcal{Q}) \geq T_n^*(\mathcal{Q}) \geq c_m' n^{-1} (\log n)^{(m-1)/2},$$

[3]The class of Riemann integrable functions includes all bounded continuous functions and bounded discontinuous functions for which the Lebesgue measure is zero on the points of discontinuity.

where $c'_m = s^{-4m}[(m-1)\log 2]^{(1-m)/2}$. Apart from c'_m, this lower bound is the best possible.

Quasirandom sequences are in no way random. Indeed, it is their nonrandomness that gives them advantages for reducing error. Every low-discrepancy sequence of n points appears spatially more uniform in $\mathcal{I}^m$ than a sample sequence of independent random vectors $\mathbf{U}_1, \ldots, \mathbf{U}_n$ where each $\mathbf{U}_j := (U_{1j}, \ldots, U_{im})$ has the uniform distribution on $\mathcal{I}^m$. This property is responsible for the more favorable convergence rate of low-discrepancy sequences for each fixed m.

Because quasirandom sequences make no pretense to randomness, an asymptotic distributional theory based on normality does not exist for estimating how well λ_n^* approximates λ, as in the IMC case. The following procedure provides an alternative approach for making this error assessment:

1. Perform L independent *macro-replications*, each based on n *micro-replications* using the quasirandom sequence $\{\mathbf{x}_1^{(l)}, \ldots, \mathbf{x}_n^{(l)}\}$ for $l = 1, \ldots, L$.
2. Compute $\lambda_n^{*(l)} := \frac{1}{n}\sum_{i=1}^n g\left(\mathbf{x}_i^{(l)}\right)$ for $l = 1, \ldots, L$.
3. Compute $\lambda_{nL}^* := \frac{1}{L}\sum_{l=1}^L \lambda_n^{*(l)}$.
4. Compute the *root-mean-square error*

$$\text{rms}_n := \left\{\frac{1}{L}\sum_{l=1}^{L}\left[\lambda_n^{*(l)} - \lambda_{nL}^*\right]^2\right\}^{1/2}. \tag{5.57}$$

If the the quantities $\mathbf{x}_0^{(1)}, \ldots, \mathbf{x}_0^{(L)}$ used to initialize the quasirandom number generators are randomly and independently chosen, then $\text{rms}_n/(L-1)^{1/2}$ is the standard error of λ_{nL}^*. Because of this randomized initialization, λ_{nL}^* is the average of L i.i.d. random averages $\lambda_n^{*(1)}, \ldots, \lambda_n^{*(L)}$ so that

$$\frac{\lambda_{nL}^* - \lambda}{\sqrt{\text{rms}_n/L}} \xrightarrow{\text{d}} \mathcal{N}(0,1) \qquad \text{as } L \to \infty,$$

thus providing a basis for assessing accuracy. However, for a total fixed number of quasirandom points nL, the number of macro-replications L is usually chosen to be small relative to n to allow the fullest benefit of quasirandom numbers to materialize. This small L limits the application of asymptotic normal theory to assess accuracy.

Past empirical experience provides useful guidance to the potential QMC user. In particular,

- **Continuous g.** For small dimension m, the $\mathrm{O}(n^{-1}(\log n)^m)$ convergence rate tends to be realized for moderate n. As m increases, theory implies that the rate is realized at larger n. Empirical studies of high-dimensional problems revealed that *dimension-reducing techniques* can effect a more favorable lower-dimensional convergence rate for moderate n (Moskowitz and Caflisch 1996).
- **Effective Dimensions.** Other empirical studies, especially in computational finance, suggest that an increasing m has little effect on the convergence rate. Possibly, the common practice of discounting over time in these problems

causes most of the variation in the integrands g to take place in m^* *effective dimensions*, where m^* is less than and independent of m and controls the convergence rate (Boyle et al. 1997).

- **Nonsmooth and Discontinuous g.** For integrands that vary widely over small intervals and are possibly discontinuous, the $O(n^{-1}(\log n)^m)$ convergence rate is not attained. Nevertheless, convergence occurs faster than $O(n^{-1/2})$. Applying function-smoothing transformations increases the convergence rate (Moskowitz and Caflisch 1996, Spanier and Maize 1994). Importance sampling (see Section 2.4) provides one way to reduce the variation $V_m(g)$ in (5.56) by smoothing.
- **Variation.** The rms_n error showed little dependence on function variation, implying that it provides a more representative error assessment than $V_m(g)D_n^*(\mathcal{Q})$ (Morokoff and Caflisch 1995).
- **Choosing a Sequence.** For dimension $m \leq 6$, the Halton sequence generally gave better results than the Sobol′ and Faure sequences. In higher dimensions, the Sobol′ sequence dominated (Morokoff and Caflisch 1995).

For a continuous function g, more can be said about effective dimension and smoothness, which may explain why convergence occurs more rapidly than at the $O(n^{-1}(\log n)^m)$ rate (Owen 1995). Let $\mathcal{M} := \{1, \ldots, m\}$ denote the coordinate index set, let $\mathcal{M}_1, \ldots, \mathcal{M}_r$ denote subsets of $\mathcal{M}$ such that $\mathcal{M}_i \neq \mathcal{M}_j$ for all $i \neq j$ and for $i = 1, \ldots, r-1$, and let $m_i := |\mathcal{M}_i|$. Then there exists a decomposition

$$g(\mathbf{x}) = \lambda + \sum_{i=1}^{r} g_i(\mathbf{x}) \qquad \mathbf{x} \in \mathcal{X}, \tag{5.58}$$

where g_i denotes a function only of the coordinates in $\mathcal{M}_i$ and

$$\int_{\mathcal{I}^m} g_i(\mathbf{x}) g_j(\mathbf{x}) d\mathbf{x} = 0 \qquad \forall i \neq j. \tag{5.59}$$

Functions satisfying this last property are called *orthogonal* and lead to

$$\int_{\mathcal{I}^m} [g(\mathbf{x}) - \lambda]^2 d\mathbf{x} = \sum_{i=1}^{r} \alpha_i, \tag{5.60}$$

where

$$\alpha_i := \int_{\mathcal{I}^{m_i}} g_i^2(\mathbf{x}) d\mathbf{x} \qquad i = 1, \ldots, r.$$

Expression (5.60) reveals a decomposition of variance into additive components. In the present setting, its significance arises from the relative importance of each α_i. In particular, α_i tends to decrease with increasing dimension m_i. The rate of this decrease increases with the smoothness of g. Because D_n^*, the star discrepancy for estimating $\int_{\mathcal{I}^m} g_i(\mathbf{x}) d\mathbf{x}$ using a sequence of quasirandom numbers is $O(n^{-1}(\log n)^{m_i})$ rather than $O(n^{-1}(\log n)^m)$, a more rapid convergence rate than $O(n^{-1}(\log n)^m)$ is anticipated for high-dimensional smooth functions. That is, m_i ($\leq m$) is the effective dimension.

The decomposition (5.58) is a well-known formulation in classical mathematics. It plays a central role in the *analysis of variance* (ANOVA), a standard tool of statistical inference. Note that some coordinates may be in more than one subset.

EXAMPLE 5.5.1

PRICING AN ASIAN OPTION V

To trade advantageously in the stock market, financial investment firms need highly accurate estimates of option prices for many different securities, each of which has its own yield and volatility. Because prices are continually fluctuating, these estimates need to be computed rapidly and frequently throughout the trading day. Because of their more rapid convergence, the demands of statistical accuracy and computational speed have spurred interest in QMC. This example in Boyle et al. (1997) evaluates the performance of QMC for estimating Asian-options prices over L securities with different volatilities ξ_2, strike prices K, and initial prices S_0.

Recall from Example 2.7.4 that $S(\tau_j)$, the price of an individual security at time τ_j, has yield ξ_1 (interest rate) and volatility ξ_2. For conciseness we again write $S_j := S(\tau_j)$ and $\Delta := \tau_j - \tau_{j-1} = \frac{T}{m}$. At the end of interval j on quasirandom replication i, the sampling experiment takes

$$S_{ij} = S_{i,j-1}\text{exp}\left[\left(\xi_1 - \frac{\xi_2^2}{2}\right)\Delta + \xi_2\sqrt{\Delta}\Phi^{-1}(x_{ij})\right] \tag{5.61}$$

as the price of the security for $j = 1, \ldots, m$ and $i = 1, \ldots, n$, where the inverse normal d.f., Φ^{-1}, defined in (2.16) converts the quasirandom number x_{ij} into a *quasinormal variate.*

Whereas previous examples on this topic have described estimation of the discounted payoff $\text{E}_\pi(\overline{S}_m - K)^+$ in (2.55) based on the arithmetic average $\overline{S}_m$ in (2.53), the current example focuses on the estimation of the discounted payoff $\text{E}_\pi(\widetilde{S}_m - K)^+$ in (9.31), based on the geometric average $\widetilde{S}_m$ in (5.47). In the present context, QMC generates

$$g(\mathbf{x}_i) := \text{e}^{-\xi_1 T}(\widetilde{S}_i - K)^+ \prod_{j=1}^{m} \Phi^{-1}(\mathbf{x}_{ij}), \tag{5.62}$$

where ϕ denotes the standard normal density (2.11) and $\mathbf{x} := (x_{i1}, \ldots, x_{im})$. Analogous to the geometric mean in (5.47) is the result

$$\begin{aligned} \widetilde{S}_{im} &:= \prod_{j=1}^{m} (S_{ij})^{1/m} \\ S_{ij} &= S_0\text{exp}\left[\left(\xi_1 - \frac{\xi_2^2}{2}\right)\Delta + \xi_2\sqrt{\Delta}\frac{1}{m}\sum_{i=1}^{m}\sum_{j=1}^{i} \Phi^{-1}(x_{ij})\right]. \end{aligned} \tag{5.63}$$

Because $E_\pi(\widetilde{S}_m - K)^+$ is available in closed form, as already noted in Section 5.3, focusing on the discounted geometric average in the Boyle et al. (1997) study facilitated the assessment of error rates.

The study considered the estimation of option prices for L securities. Let the superscript $[l]$ denote the lth security. Then

$$\bar{\lambda}_n^{[l]} := \frac{1}{n}\sum_{i=1}^{n} g^{[l]}\left(\mathbf{x}_i^{[l]}\right)$$

is an estimate of the discounted geometric average options price $\lambda^{[l]}$ based on n quasirandom points $\mathbf{x}_1^{[l]}, \ldots, \mathbf{x}_n^{[l]}$, and (5.62) defines $g^{[l]}$ for security l. The L quasirandom point sequences are independent. Because $\lambda^{[l]} := E_\pi(\widetilde{S}_m^{[l]} - K)^+$ is known exactly, the error of approximation, $\bar{\lambda}_n^{[l]} - \lambda^{[l]}$, and the relative error across all L securities can be approximated by

$$\overline{\text{rmsr}}(n) := \left[\sum_{l=1}^{L}\left(\frac{\bar{\lambda}_n^{[l]} - \lambda^{[l]}}{\lambda^{[l]}}\right)^2\right]^{1/2}. \qquad (5.64)$$

That is, rather than estimating error for a single security, the approach of Boyle et al. estimates the average relative rms across L different securities.

Table 5.8 shows results for standard MC and for Faure and Sobol′ sequences for dimensions $m = 10$, 50, and 100 and sample sizes $n = 50{,}000$ and 200,000. Recall that the dimension m denotes the number of intervals used to partition the time horizon T into discrete units. The table reveals the standard MC $\overline{\text{rmsr}}$ to be relatively independent of m. Moreover, the Faure and Sobol′ sequences both show an advantage over standard MC for $n = 200{,}000$ for each m, but the Faure sequence loses its superiority for $m = 100$ and $n = 50{,}000$. The Faure sequence also decreases and then increases in $\overline{\text{rmsr}}$ with increasing m, revealing that (5.54) is a

TABLE 5.8 | *Asian-option pricing using quasirandom numbers**

		$\overline{\text{rmsr}}(n)$		
m	n	IMC	Faure	Sobol′
10	50,000	0.81	0.40	0.07
	200,000	0.39	0.15	0.02
50	50,000	0.75	0.21	0.27
	200,000	0.41	0.10	0.21
100	50,000	0.83	1.00	0.43
	200,000	0.37	0.25	0.23

*Source: Boyle et al. (1997).

bound but not necessarily useful predictor of $\overline{\text{rmsr}}$ for each (m, n). Corresponding results for the Halton sequence were not competitive with those for the Faure and Sobol′ sequences.

The benefit of QMC becomes most apparent in estimating the sample size n^* that IMC would require to obtain the same $\overline{\text{rmsr}}$ that QMC produces. Let $\overline{\text{rmsr}}_{\text{IMC}}(n)$ and $\overline{\text{rmsr}}_{\text{SOBOL}'}(n)$ denote the rms relative errors based on sample size n for standard MC and for a Sobol′ sequence respectively. It is of interest to determine the sample size $n^* > n$ such that

$$\overline{\text{rmsr}}_{\text{IMC}}(n^*) = \overline{\text{rmsr}}_{\text{SOBOL}'}(n).$$

The ratio n^*/n reveals how much larger the sample size is that standard MC needs to achieve the same accuracy as the Sobol′ sequence of n points. Because $\overline{\text{rmsr}}_{\text{IMC}}(n)$ is roughly proportional to $n^{-1/2}$, apart from sampling variation, we have

$$\overline{\text{rmsr}}_{\text{IMC}}(n^*) \approx \overline{\text{rmsr}}_{\text{IMC}}(n)\sqrt{\frac{n}{n^*}},$$

so that

$$n^* \approx n \left[\frac{\overline{\text{rmsr}}_{\text{IMC}}(n)}{\overline{\text{rmsr}}_{\text{SOBOL}'}(n)}\right]^2.$$

At $m = 10$ and $n = 50,000$, standard MC requires $n^* = 50,000 \times \left(\frac{.81}{.07}\right)^2 =$ 6,694,898 sample points. At $n = 200{,}000$, $n^* = 200{,}000 \times \left(\frac{.39}{.02}\right)^2 = 76{,}050{,}000$ are needed. By contrast, at $m = 100$ and $n = 50{,}000$, IMC requires $n^* = 50{,}000 \times \left(\frac{.83}{.43}\right)^2 = 186{,}290$ sample points, and for $n = 200{,}000$ it requires $n^* = 200{,}000 \times \left(\frac{.37}{.23}\right)^2 = 517{,}581$. Collectively these calculations reveal that efficiency of QMC increases with n and decreases with m, as expected. More importantly, the calculations show the substantial increases in efficiency that QMC offers, when it applies. ■

Cost Considerations. Generating a quasirandom number takes more computing time than generating a pseudorandom number. This leads to a CR less than 1. However, the accelerated convergence proportional to $\frac{(\log n)^m}{n}$ inevitably shows QMC to be superior to standard MC as n becomes large. Because the constants of proportionality, corresponding to different QMC sequences, tend to be large, dominance does not necessarily prevail for small and moderate n.

It is also known that the design of software for generating QMC numbers substantially influences unit generating time. In particular, Bratley and Fox (1988) show that attention to this issue results in software for generating Sobol′ sequences with considerably smaller unit generating time than for Faure sequences. (For implementing Sobol′ generation, see Joe and Kuo [2003].)

5.6 LESSONS LEARNED

- For strata of equal volume or size, stratified sampling with equal sample sizes per stratum ($n_1 = \cdots = n_r$) increases statistical efficiency (VR > 1) over standard MC with sample size nr.
- Adjusting strata sample sizes to incorporate bounds on their ranges of variation can further improve statistical efficiency.
- Statistical efficiency in stratified sampling can also be enhanced by choosing strata with roughly equal bounds on the ranges of variation within strata.
- Rao-Blackwellization increases statistical efficiency and occasionally computational efficiency (CR > 1).
- The statistical efficiency of the control variate technique increases as the strength of the correlation increases between the integrand, $g(\mathbf{X})$, and the control variate, $q(\mathbf{X})$; that is, as $|\text{corr}\,[g(\mathbf{X}), q(\mathbf{X})]\,|$ increases.
- The antithetic variate method requires monotonicity for g to ensure an increase in statistical efficiency.
- Provided g has specified properties, the error when using the quasi-Monte Carlo method decreases faster than $n^{-1/2}$, the rate for standard MC.

5.7 EXERCISES

The exercises in this section focus on derivations and proofs.

Stratified Sampling

1. Derive (5.8) and then (5.12).

Rao-Blackwellization

2. For F in (5.35), show that Z in Algorithm RUIN does not have a bounded coefficient of variation.

Control Variates

3. Show that (5.37) holds if and only if (5.38) holds.

4. Show that the inequality (5.41) holds if and only if $\beta(\beta + \beta_{\mathbf{opt}}) \leq 0$. Also show that $\beta_{\mathbf{opt}} > 0$ if and only if $\rho < 0$.

Antithetic Variates

The antithetic-variate method often allows a potential user to predict the method that will have a positive or negative effect on efficiency prior to performing the sampling experiment. Exercises 5 and 6 illustrate the benefit that this pre-execution analysis offers. Exercises H.22 through H.26 also illustrate the benefit.

5. Consider k arcs in series-connecting nodes s and t. Let $x_i = 1$ if arc i operates and $x_i = 0$ if it fails, so that the network has the structure function

$$h(\mathbf{x}) := \prod_{i=1}^{k} x_i = \begin{cases} 1 & \text{if } s \text{ and } t \text{ are connected} \\ 0 & \text{otherwise} \end{cases} \qquad \mathbf{x} := (x_1, \ldots, x_k).$$

Suppose that each arc operates with probability $p \in [0.50, 1]$ and fails with probability $1 - p$. Let $U_1, \ldots, U_k$ denote i.i.d. random variables from $\mathcal{U}(0, 1)$, and let

$$X_i := \begin{cases} 1 & \text{if } U_i \le p \\ 0 & \text{otherwise.} \end{cases}$$

Then s and t are connected if $h(\mathbf{X}) = 1$ and are not connected if $h(\mathbf{X}) = 0$, where $\mathbf{X} := (X_1, \ldots, X_k)$.
For the same sample $U_1, \ldots, U_k$, let

$$Z_i := \begin{cases} 1 & \text{if } 1 - U_i \le p \\ 0 & \text{otherwise.} \end{cases}$$

a. Derive an expression for $\lambda := \mathrm{E}h(\mathbf{X})$, the probability that s and t are connected, in terms of p and k and show that $\mathrm{E}h(\mathbf{X}) = \mathrm{E}h(\mathbf{Z})$.

b. Derive expressions for $\mathrm{var}h(\mathbf{X})$ and $\mathrm{var}h(\mathbf{Z})$ in terms of p and k.

c. As an estimate of λ, show that this antithetic sampling scheme induces a variance reduction for $\frac{1}{2}[h(\mathbf{X}) + h(\mathbf{Z})]$

$$\mathrm{VR}(k, p) = \frac{1}{1 + \frac{(2-1/p)^k - p^k}{1-p^k}} \qquad \frac{1}{2} \le p \le 1,\ k = 1, 2 \ldots.$$

d. How does $\mathrm{VR}(k, p)$ behave as $p \to \frac{1}{2}$ and $p \to 1$ for fixed k and as $k \to \infty$ for fixed p?

e. Derive $\mathrm{VR}(k, p)$ for $p \in [0, 0.50]$ and describe its limiting behavior as $p \to 0$ and $p \to \frac{1}{2}$ for fixed k and as $k \to \infty$ for fixed p.

6. Suppose the arcs in Exercise 5 were in parallel, leading to the structure function $\{h(\mathbf{x}) := 1 - \prod_{i=1}^{k}(1 - x_i),\ x_i \in \{0, 1\};\ i = 1, \ldots, k\}$.

a. Derive an expression for $\lambda := \mathrm{E}h(\mathbf{X})$, the probability that s and t are connected, in terms of p and k and show that $\mathrm{E}h(\mathbf{X}) = \mathrm{E}h(\mathbf{Z})$.

b. Derive expressions for $\mathrm{var}h(\mathbf{X})$ and $\mathrm{var}h(\mathbf{Z})$ in terms of p and k.

c. As an estimate of λ, show that this antithetic sampling scheme induces a variance reduction for $\frac{1}{2}[h(\mathbf{X}) + h(\mathbf{Z})]$:

$$\mathrm{VR}(k, p) = \frac{\lambda}{2\lambda - 1} \qquad \frac{1}{2} \le p \le 1,\ k = 1, \ldots.$$

d. How does $\mathrm{VR}(k, p)$ behave as $p \to \frac{1}{2}$ and $p \to 1$ for fixed k, and as $k \to \infty$ for fixed p?

e. Derive VR(k, p) for $p \in [0, 0.50]$ and describe its limiting behavior as $p \to 0$ and $p \to \frac{1}{2}$ for fixed k and as $k \to \infty$ for fixed p.

5.8 HANDS-ON EXERCISES

The hands-on exercises demonstrate how alternative variance-reducing techniques perform. Each problem differently affects the ratio of computing times, CR, and the ratio of variances, VR. Conversely, for a particular problem different methods affect CR and VR differently. All exercises have solutions with VR ≥ 1, but not necessarily with CR ≥ 1. However, careful attention to computer programming often leads to CR sufficiently close to 1, and for some techniques CR > 1. Although applying a particular variance-reducing technique in some exercises may be problematic, a change of variables readily allows the method to be applied. (See the discussion above the proof of Theorem 3.7.1.)

In addition to including sample quantities listed as part of the solution, students are encouraged to include additional sample quantities in support of their conclusions.

Stratified Sampling

Exercises H.1 through H.5 illustrate how stratified sampling increases efficiency. In practice, MC sampling is performed for a specified number of strata and one sample size per stratum. By calling for several independent experiments at successively increasing numbers of strata and successively decreasing sample sizes per stratum, these exercises show how statistical accuracy varies with stratum size and sample size.

H.1. With g defined as in (2.82) and stratifying on variable x, implement stratified sampling experiments to estimate λ in (2.1) for each number of strata $r \in \{4^l;\ l = 1, \ldots, 4\}$ and $n = 2^{20}/r$ independent replications per stratum. Randomly sample Y from

$$\omega_Y(y) = \begin{cases} \frac{4}{\pi} & \frac{\pi}{2} \leq y \leq \frac{3\pi}{4} \\ 0 & \text{elsewhere.} \end{cases}$$

a. For $j = 1, \ldots, 4^l$, show that

$$\omega_{jX}(x) = \begin{cases} \frac{4^{l+1}}{\pi} & \frac{(j-1)\pi}{4^{l+1}} \leq x < \frac{j\pi}{4^{l+1}} \\ 0 & \text{elsewhere} \end{cases}$$

partitions the x axis, $\left[0, \frac{\pi}{4}\right]$, into intervals of equal volume. Also show that

$$\omega(x, y) = \frac{1}{r} \sum_{j=1}^{r} \omega_{jX}(x)\omega_Y(y),$$

where (2.83) defines $\omega(x, y)$.

b. For each experiment $l = 1, \ldots, 4$, compute $\tilde{\lambda}_n$, s.e.$(\tilde{\lambda}_n)$, r.e.$(\tilde{\lambda}_n)$, and the corresponding computing time. Also compute CR, VR, and CR $\times$ VR when comparing $\tilde{\lambda}_n$ with $\bar{\lambda}_n$ in Exercise 2.H.1.

c. Comment on the desirability of this form of stratified sampling compared to that in Exercise 2.H.1, based on standard MC.

H.2. With g defined as in (2.82) and stratifying on both variables x and y over $\mathcal{X} = \left[0, \frac{\pi}{4}\right] \times \left[\frac{\pi}{2}, \frac{3\pi}{4}\right]$, implement stratified sampling experiments to estimate λ in (2.1) for each number of strata $r \in \{4^l;\ l = 1, \ldots, 4\}$ and $n = 2^{20}/r$ independent replications per stratum.

a. For $j = 1, \ldots, 2^l$ and $k = 1, \ldots, 2^l$, show that

$$\omega_{jX}(x) = \begin{cases} \frac{2^{l+2}}{\pi} & \frac{(j-1)\pi}{2^{l+2}} \leq x \leq \frac{j\pi}{2^{l+2}} \\ 0 & \text{elsewhere} \end{cases}$$

and

$$\omega_{kY}(y) = \begin{cases} \frac{2^{l+2}}{\pi} & \frac{\pi}{2} + \frac{(k-1)\pi}{2^{l+2}} \leq y \leq \frac{\pi}{2} + \frac{k\pi}{2^{l+2}} \\ 0 & \text{elsewhere} \end{cases}$$

partition $\mathcal{X}$ into cells of equal volume. Also show that

$$\omega(x, y) = \frac{1}{r} \sum_{j=1}^{\sqrt{r}} \sum_{k=1}^{\sqrt{r}} \omega_{jX}(x)\omega_{kY}(y),$$

where (2.83) defines $\omega(x, y)$.

b. For each experiment $l = 1, \ldots, 4$, compute $\tilde{\lambda}_n$, s.e. $(\tilde{\lambda}_n)$, r.e. $(\tilde{\lambda}_n)$, and the corresponding computing time. Also compute CR, VR, and CR $\times$ VR when comparing $\tilde{\lambda}_n$ with $\bar{\lambda}_n$ in Exercise 2.H.1.

c. For each experiment $l = 1, \ldots, 4$, compute CR, VR, and CR $\times$ VR when comparing $\tilde{\lambda}_n$ here with $\tilde{\lambda}_n$ in Exercise H.1.

d. Comment on the desirability of this form of stratified sampling compared to that in Exercise H.1, which stratifies on a single variate, and to that in Exercise 2.H.1, based on standard MC.

H.3. With g defined as in (2.84) and stratifying on variable x, implement stratified sampling experiments to estimate λ in (2.1) for each number of strata $r \in \{4^l;\ l = 1, \ldots, 4\}$, $n = 2^{20}/r$ independent replications per stratum, and a random sampling of Y from $\mathcal{U}(0, 1)$.

a. For $j = 1, \ldots, 4^l$, show that

$$\omega_{jX}(x) = \begin{cases} \frac{2^{2l+1}}{\pi} & \frac{(j-1)\pi}{2^{2l+1}} \leq x \leq \frac{j\pi}{2^{2l+1}} \\ 0 & \text{elsewhere} \end{cases}$$

partitions the x axis, $\left[0, \frac{\pi}{2}\right]$, into intervals of equal volume. Also show that

$$\omega(x, y) = \frac{1}{r}\sum_{j=1}^{r} \omega_{jX}(x)\omega_Y(y),$$

where (2.85) defines $\omega(x, y)$.

b. For each experiment $l = 1, \ldots, 4$, compute $\tilde{\lambda}_n$, s.e.$(\tilde{\lambda}_n)$, r.e.$(\tilde{\lambda}_n)$, and the corresponding computing time. Also compute CR, VR, and CR × VR when comparing $\tilde{\lambda}_n$ with $\overline{\lambda}_n$ in Exercise 2.H.3.

c. Comment on the desirability of this form of stratified sampling compared to that in Exercise 2.H.3, based on standard MC.

H.4. With g defined as in (2.84) and stratifying on both variables x and y over $\mathcal{X} = \left[0, \frac{\pi}{2}\right] \times [0, 1]$, implement stratified sampling experiments to estimate λ in (2.1) for each number of strata $r \in \{4^l;\ l = 1, \ldots, 4\}$ and $n = 2^{20}/r$ independent replications per stratum.

a. For $j = 1, \ldots, 2^l$ and $k = 1, \ldots, 2^l$, show that

$$\omega_{jX}(x) = \begin{cases} \frac{2^{l+1}}{\pi} & \frac{(j-1)\pi}{2^{l+1}} \le x \le \frac{j\pi}{2^{l+1}} \\ 0 & \text{elsewhere} \end{cases}$$

and

$$\omega_{kY}(y) = \begin{cases} \frac{2^l}{\pi} & \frac{(k-1)}{2^l} \le y \le \frac{k}{2^l} \\ 0 & \text{elsewhere} \end{cases}$$

partition $\mathcal{X}$ into cells of equal volume. Also show that

$$\omega(x, y) = \frac{1}{r}\sum_{j=1}^{\sqrt{r}}\sum_{k=1}^{\sqrt{r}} \omega_{jX}(x)\omega_{kY}(y),$$

where (2.85) defines $\omega(x, y)$.

b. For each experiment $l = 1, \ldots, 4$, compute $\tilde{\lambda}_n$, s.e.$(\tilde{\lambda}_n)$, r.e.$(\tilde{\lambda}_n)$, and the corresponding computing time. Also compute CR, VR, and CR × VR when comparing $\tilde{\lambda}_n$ with $\overline{\lambda}_n$ in Exercise 2.H.3.

c. For each experiment $l = 1, \ldots, 4$, compute CR, VR, and CR × VR when comparing $\tilde{\lambda}_n$ here with $\tilde{\lambda}_n$ in Exercise H.3.

d. Comment on the desirability of this form of stratified sampling compared to that in Exercise H.3, based on single variate stratification; and to that in Exercise 2.H.3, based on standard MC.

H.5. With g defined as in (2.87) and stratifying on the x, y, z variables over $\mathcal{X} = [0, \infty) \times [0, \infty) \times \{1, \ldots, 9\}$, implement stratified sampling experiments to estimate λ in (2.86) for each number of strata $r \in \{9 \times 4^l;\ l = 0, 1, 2, 3\}$.

a. For $j = 1, \ldots, 2^l$; $k = 1, \ldots, 2^l$; and $q = 1, \ldots, 9$; show that

$$\omega_{jX}(x) = \begin{cases} 2^l e^{-x} & -\ln\left(1 - \frac{j-1}{2^l}\right) \le x \le -\ln\left(1 - \frac{j}{2^l}\right) \\ 0 & \text{elsewhere,} \end{cases}$$

$$\omega_{kY}(y) = \begin{cases} \frac{2^{l+1}}{\pi(1+y^2)} & \text{if } \tan\left(\frac{(j-1)\pi}{2^{l+1}}\right) \le y \le \tan\left(\frac{j\pi}{2^{l+1}}\right) \\ 0 & \text{elsewhere,} \end{cases}$$

and

$$\omega_{qZ}(z) = \begin{cases} 1 & z = q \\ 0 & \text{elsewhere} \end{cases}$$

partition $\mathcal{X}$ into cells of equal volume. Also show that

$$\omega_X(x)\omega_Y(y)\omega_Z(z) = \frac{1}{9r}\sum_{j=1}^{\sqrt{r}}\sum_{k=1}^{\sqrt{r}}\sum_{q=1}^{9}\omega_{jX}(x)\omega_{kY}(y)\omega_{qZ}(z),$$

where (2.88), (2.89), and (2.90) define $\omega_X(x)$, $\omega_Y(y)$, and $\omega_Z(z)$ respectively.

Note: For $l = 0$, stratification occurs only on variable z.

b. Take $n = 10^6$ replications. Inspection of (2.87) reveals that $z^{3/2}$ varies from 1 for $z = 1$ to 27 for $z = 9$, thereby dominating the variation in g. To control this variation, this observation encourages us to make the number of replications per stratum increase as z increases. Let n_k denote the total number of replications for $z = k$. The assignment

$$n_z = \left\lceil \frac{nz^{3/2}}{\sum_{k=1}^{9} k^{3/2}} \right\rceil \qquad z = 1, \ldots, 9$$

does just this in a manner reminiscent of the optimal choice of sample size per stratum when the stratum variances are known (see Section 5.1). If x has 2^i strata and y, 2^j strata, take $n/2^{i+l}$ observations in the strata for which $z = k$. Because $n_1, \ldots, n_9$ need to be integers, their summation will be slightly higher than n. For each each experiment $l = 0, 1, 2, 3$, compute $\tilde{\lambda}_n$, s.e.$(\tilde{\lambda}_n)$, r.e.$(\tilde{\lambda}_n)$, and the corresponding computing time. Also, compute CR, VR, and CR × VR when comparing $\tilde{\lambda}_n$ with $\bar{\lambda}_n$ in Exercise 2.H.5.

c. Compare this form of stratified sampling ($l \ge 1$) with standard MC sampling ($l = 0$).

Rao-Blackwellization

H.6. Show how a change of variables, $v = x$ and $w = x + y$, in Exercise 2.H.1 facilitates Rao-Blackwell sampling.

a. Use $n = 10^6$ independent replications to evaluate the efficacy of the method. Explain your results in detail.

b. What are the prospects for an alternative form of Rao-Blackwell sampling? Explain your answer in detail.

H.7. Derive a Rao-Blackwell estimator for λ in (2.1), based on the integrand g in (2.84) in Exercise 2.H.3.

a. Perform $n = 10^6$ independent replications and compute $\dot{\lambda}_n$ in (5.19), s.e.$(\dot{\lambda}_n)$, r.e.$(\dot{\lambda}_n)$, and the computing time.

b. Compute CR and VR when comparing these results with those in Exercise 2.H.3.

c. How well does Rao-Blackwellization perform relative to standard MC sampling?

H.8. Derive a Rao-Blackwell estimator for λ in (2.86), based on the integrand g in (2.87) in Exercise 2.H.5.

a. Perform $n = 10^6$ independent replications and compute $\dot{\lambda}_n$ in (5.19), s.e.$(\dot{\lambda}_n)$, r.e.$(\dot{\lambda}_n)$, and the computing time.

b. Compute CR and VR when comparing these results with those in Exercise 2.H.5.

c. How well does Rao-Blackwellization perform relative to standard MC sampling?

H.9. Derive a Rao-Blackwell estimator for λ in (2.1), based on the integrand g in (2.91) in Exercise 2.H.7.

a. Perform $n = 10^6$ independent replications and compute $\dot{\lambda}_n$ in (5.19), s.e.$(\dot{\lambda}_n)$, r.e.$(\dot{\lambda}_n)$, and the computing time.

b. Compute CR and VR when comparing these results with those in Exercise 2.H.7.

c. How well does Rao-Blackwellization perform relative to standard MC sampling?

H.10. Derive a Rao-Blackwell estimator for λ in (2.1), based on the integrand g in (2.92) in Exercise 2.H.8.

a. Perform $n = 10^6$ independent replications and compute $\dot{\lambda}_n$ in (5.19), s.e.$(\dot{\lambda}_n)$, r.e.$(\dot{\lambda}_n)$, and the computing time.

b. Compute CR and VR when comparing these results with those in Exercise 2.H.8.

c. How well does Rao-Blackwellization perform relative to standard MC sampling?

H.11. Derive a Rao-Blackwell estimator for λ in (2.1), based on the integrand g in (2.93) in Exercise 2.H.9.

a. Perform $n = 10^6$ independent replications and compute $\dot{\lambda}_n$ in (5.19), s.e.$(\dot{\lambda}_n)$, r.e.$(\dot{\lambda}_n)$, and the computing time.

b. Compute CR and VR when comparing these results with those in Exercise 2.H.9.

c. How well does Rao-Blackwellization perform relative to standard MC sampling?

H.12. Derive a Rao-Blackwell estimator for the volume of the polytope (2.94) in Exercise 2.H.10.

a. Perform $n = 10^6$ independent replications and compute $\dot{\lambda}_n$ in (5.19), s.e.$(\dot{\lambda}_n)$, r.e.$(\dot{\lambda}_n)$, and the computing time.

b. Compute CR and VR when comparing these results with those in Exercise 2.H.10.

c. How well does Rao-Blackwellization perform relative to standard MC sampling?

H.13. Power Generation. From (2.99) we know supply exceeds demand if for some station l in Exercise 2.H.16,

$$a_{lj}c_l Z_l \geq D_{lj},$$

where

$$D_{lj} := d_j - \sum_{\substack{i=1\\ i\neq l}}^{r} a_{ij}c_i Z_i.$$

For $j = 1, \ldots, s$, let

$$w_j(\mathbf{Z}_{-l}) := \begin{cases} \frac{D_{lj}}{a_{lj}c_l} & \text{if } a_{lj} > 0 \\ \infty & \text{if } a_{lj} = 0 \text{ and } D_{lj} > 0 \\ 0 & \text{if } a_{lj} = 0 \text{ and } D_{lj} \leq 0, \end{cases}$$

where

$$\mathbf{Z}_{-l} := (Z_1, \ldots, Z_{1-1}, Z_{1+1}, \ldots, Z_1).$$

Then

$$\begin{aligned} \lambda &:= \text{pr(demand exceeds supply)} \\ &= 1 - \mathrm{E}\left\{\text{pr}\left[Z_l \geq \max_{1\leq j\leq s} w_j(\mathbf{Z}_{-l})|\mathbf{Z}_{-l}\right]\right\} \\ &= \mathrm{E}_{\mathbf{Z}_{-l}} F\left(\max_{1\leq j\leq s} w_j(\mathbf{Z}_{-l})\right), \end{aligned} \tag{5.65}$$

where F denotes the d.f. of Z_l from $\mathcal{B}(m_l, p_l)$.

a. For $n = 10^6$ independent replications, apply Rao-Blackwellization to derive a point estimate of λ in (5.65) and its standard error, its relative error, and the variance reduction it induces. Choose a station l that is most likely to reduce variation and justify your choice.

b. Evaluate the desirability of Rao-Blackwellization when compared to the results in Exercises 2.H.16b and 2.H.17b.

Suggestion: Before executing the IMC experiment, compute and store $\{F(z);\ z = 0, 1, \ldots, m_l\}$ for station l. The p.m.f. for $\mathcal{B}(m, p)$ is

$$p_0 = (1-p)^m$$

$$p_z = p_{z-1}\frac{(n-z+1)}{z} \cdot \frac{p}{1-p} \qquad z = 1, \ldots, m.$$

Also, $F(z) = F(z-1) + p_z$ and $F_0 = p_0$. The p.m.f. for $\mathcal{B}(m, 1-p)$ merely replaces p by $1-p$ everywhere.

Control Variates

H.14. Derive a control-variate estimator for λ in (2.1) based on the integrand g in (2.82) in Exercise 2.H.1.

a. Perform $n = 10^6$ independent replications and compute $\breve{\lambda}_n$ in (5.19), s.e.($\breve{\lambda}_n$), r.e.($\breve{\lambda}_n$), and the computing time.

b. Compute CR and VR when comparing these results with those in Exercise 2.H.1.

c. How well does the control-variate estimator perform relative to standard MC sampling?

H.15. Derive a control-variate estimator for λ in (2.1) based on the integrand g in (2.84) in Exercise 2.H.3.

a. Perform $n = 10^6$ independent replications and compute $\breve{\lambda}_n$ in (5.42), s.e.($\breve{\lambda}_n$), r.e.($\breve{\lambda}_n$), and the computing time.

b. Compute CR and VR when comparing these results with those in Exercise 2.H.3.

c. How well does the control-variate estimator perform relative to standard MC sampling?

H.16. Derive a control-variate estimator for λ in (2.86) based on the integrand g in (2.87) in Exercise 2.H.5.

a. Perform $n = 10^6$ independent replications and compute $\breve{\lambda}_n$ in (5.42), s.e.($\breve{\lambda}_n$), r.e.($\breve{\lambda}_n$), and the computing time.

b. Compute CR and VR when comparing these results with those in Exercise 2.H.5.

c. How well does the control-variate estimator perform relative to standard MC sampling?

H.17. Derive a control-variate estimator for λ in (2.1) based on the integrand g in (2.91) in Exercise 2.H.7.

a. Perform $n = 10^6$ independent replications and compute $\breve{\lambda}_n$ in (5.42), s.e.($\breve{\lambda}_n$), r.e.($\breve{\lambda}_n$), and the computing time.

b. Compute CR and VR when comparing these results with those in Exercise 2.H.7.

c. How well does the control-variate estimator perform relative to standard MC sampling?

H.18. Derive a control-variate estimator for λ in (2.1) based on the integrand g in (2.92) in Exercise 2.H.8.

a. Perform $n = 10^6$ independent replications and compute $\breve{\lambda}_n$ in (5.42), s.e.$(\breve{\lambda}_n)$, r.e.$(\breve{\lambda}_n)$, and the computing time.

b. Compute CR and VR when comparing these results with those in Exercise 2.H.8.

c. How well does the control-variate estimator perform relative to standard MC sampling?

H.19. Derive a control-variate estimator for λ in (2.1) based on the integrand g in (2.93) in Exercise 2.H.9.

a. Perform $n = 10^6$ independent replications and compute $\breve{\lambda}_n$ in (5.42), s.e.$(\breve{\lambda}_n)$, r.e.$(\breve{\lambda}_n)$, and the computing time.

b. Compute CR and VR when comparing these results with those in Exercise 2.H.9.

c. How well does the control-variate estimator perform relative to standard MC sampling?

H.20. Derive a control-variate estimator for the volume of the polytope (2.94) in Exercise 2.H.10.

a. Perform $n = 10^6$ independent replications and compute $\breve{\lambda}_n$ in (5.42), s.e.$(\breve{\lambda}_n)$, r.e.$(\breve{\lambda}_n)$, and the computing time.

b. Compute CR and VR when comparing these results with those in Exercise 2.H.10.

c. How well does the control-variate estimator perform relative to standard MC sampling?

H.21. Pricing an Asian Option. Example 2.7.4 formulates the problem of estimating $\mathrm{E}h(X_1, \ldots, X_m)$, the price of an Asian option on one share of a stock, where h in (2.54) is a function of the arithmetic average $\overline{S}_m$ in (2.53). Although this expectation is not available in closed form, a slight modification does admit a closed-form expectation. The quantity

$$\tilde{S}_m := \left[\prod_{j=0}^{m} S(\tau_j)\right]^{1/(m+1)} \qquad \tau_0 = 0 < \tau_1 < \cdots < \tau_m = T \tag{5.66}$$

denotes the *geometric mean* of $S(\tau_0), S(\tau_1), \ldots, S(\tau_m)$, for which

$$h'(X_1, \ldots, X_m) := \mathrm{e}^{-\xi_1 T}(\tilde{S}_m - K)^+ \tag{5.67}$$

has expectation (Kemna and Vorst 1990)

$$\mathrm{E}h'(X_1, \ldots, X_m) = \mathrm{e}^{-\xi_1 T}\left[\mathrm{e}^{\left(\xi_1 - \xi_2^2/6\right)T/2} S(0)\Phi(d) - K\Phi(d - \xi_2\sqrt{T/3})\right], \tag{5.68}$$

where

$$d := \frac{\ln(S(0)/K) + \frac{1}{2}\left(\xi_1 + \frac{1}{6}\xi_2^2\right)T}{\xi_2\sqrt{T/3}}$$

and Φ denotes the d.f. of the standard normal distribution. Because it is known that the geometric average of non-negative numbers never exceeds the corresponding arithmetic average, we have

$$\mathrm{E}h'(X_1, \ldots, X_m) \le \mathrm{E}h(X_1, \ldots, X_m),$$

thereby providing a lower bound on the price of an Asian option. In an MC setting, the geometric average also provides a basis for applying the method of control variates in Section 5.3 to reduce the variance in estimating $\mathrm{E}h(X_1, \ldots, X_m)$.

a. Using the information about the geometric mean, formulate a control variate estimator of $\lambda := \mathrm{E}h(X_1, \ldots, X_m)$, the price of an Asian option on one share of stock, based on n independent replications.

b. Execute 27 IMC sampling experiments, each for $n = 10,000$ independent replications, for annual interest rates $\xi_1 \in \{0.03, 0.05, 0.07\}$, volatilities $\xi_2 \in \{0.20, 0.30, 0.40\}$, prices $K \in \{35, 40, 45\}$, and initial stock price $S(0) = 40$. Assume that a year has 22×12 trading days and run each experiment for $m = 88$ days so that $T = \frac{88}{22\times 12} = \frac{1}{3}$, $\tau_0 = 0$, and $\tau_j = \tau_{j-1} + T/88$ for $j = 2, \ldots, m$.

For each set of parametric values, compute and display the control variate estimate of λ, its standard error, its relative standard error, $1-$ corr $[h(X_1, \ldots, X_m), h'(X_1, \ldots, X_m)]$, and the variance reduction that the control variate method induces.

c. How do variations in ξ_1, ξ_2, and K affect variance reduction? Use the output to explain why these variations occur.

Antithetic Variates

On $n = 500{,}000$ independent pairs of replications in Exercises H.22 through H.25, choose one variable to serve as the antithetic variate, and independently sample the remaining variables on every replication. As mentioned in the previous section, the antithetic-variate method has the desirable property of often allowing a potential user to predict its effect, positive or negative, on efficiency prior to performing the sampling experiment. Exercise H.26 illustrates the benefit.

H.22. Do you expect the antithetic-variate method to improve efficiency when estimating λ in (2.86) for the integrand in (2.87)?

a. Justify your answer analytically.

b. Execute the experiment and compute $\grave{\lambda}_n$, s.e.$(\grave{\lambda}_n)$, and r.e.$(\grave{\lambda}_n)$ in Algorithm ANTVAR, and CR and VR using the results in Exercise 2.H.5 for comparison.

c. Are your analytical and empirical results consistent?

H.23. Do you expect the antithetic-variate method to improve efficiency when estimating λ in (2.1) for the integrand in (2.91)?

a. Justify your answer analytically.

b. Execute the experiment and compute $\grave{\lambda}_n$, s.e.$(\grave{\lambda}_n)$, and r.e.$(\grave{\lambda}_n)$ in Algorithm ANTVAR, and CR and VR using the results in Exercise 2.H.7 for comparison.

c. Are your analytical and empirical results consistent?

H.24. Do you expect the antithetic-variate method to improve efficiency when estimating λ in (2.1) for the integrand in (2.92)?

a. Justify your answer analytically.

b. Execute the experiment and compute $\grave{\lambda}_n$, s.e.$(\grave{\lambda}_n)$, and r.e.$(\grave{\lambda}_n)$ in Algorithm ANTVAR, and CR and VR using the results in Exercise 2.H.8 for comparison.

c. Are your analytical and empirical results consistent?

H.25. Do you expect the antithetic-variate method to improve efficiency when estimating λ in (2.1) for the integrand in (2.93)?

a. Justify your answer analytically.

b. Execute the experiment and compute $\grave{\lambda}_n$, s.e.$(\grave{\lambda}_n)$, and r.e.$(\grave{\lambda}_n)$ in Algorithm ANTVAR, and CR and VR using the results in Exercise 2.H.9 for comparison.

c. Are your analytical and empirical results consistent?

H.26. **Pricing a European Option.** Example 2.7.4 describes the setup for estimating the price of an Asian option. In practice, several different types of stock options exist. For example, the price of a *European option* on one share of a nondividend stock, based on the Black-Scholes model, is

$$\lambda := \mathrm{E}e^{-\xi_1 T}[0, S(T) - K]^+, \tag{5.69}$$

using the definitions in Section 2.7.4. Whereas the price of an Asian option is based on an arithmetic average, the price of a European option is a function only of the final price

$$S(T) = S(0)e^{\left[(\xi_1 - \frac{1}{2}\xi_2^2)T + \xi_2\sqrt{T}X\right]} \tag{5.70}$$

at time T, where X denotes a random variable from $\mathcal{N}(0, 1)$. Then

$$\lambda = e^{-\xi_1 T}\int_\theta^\infty e^{-x^2/2}\left\{S(0)e^{\left[(\xi_1 - \frac{1}{2}\xi_2^2)T + \xi_2\sqrt{T}x\right]} - K\right\}dx,$$

where

$$\theta := \frac{\ln K - \ln S(0) - \left(\xi_1 - \frac{1}{2}\xi_2^2\right)T}{\xi_2\sqrt{T}}.$$

Although the univariate integral can be evaluated by numerical methods, estimating λ by IMC sampling provides a good illustration of the effectiveness of the antithetic variate method.

Let $X = \Phi^{-1}(U)$ and $X_a = \Phi^{-1}(1-U)$, where U is from $\mathcal{U}(0,1)$ and Φ^{-1} is the inverse d.f. of the standard normal d.f. Let

$$h(X) := e^{-\xi_1 T}[0, S(T) - K]^+,$$

where (5.70) defines $S(T)$.

a. Prove that

$$\text{var}\left\{\frac{1}{2}[h(X) + h(X_a)]\right\} \le \frac{1}{2}\text{var } h(X).$$

b. Perform an IMC sampling experiment using antithetic variates to estimate (5.69) for interest rate $\xi_1 = 0.10$, all volatilities $\xi_2 \in \{0.20, 0.40, 0.60\}$, strike price $K = 100$, all initial-strike price ratios $S(0)/K \in \{0.90, 1.0, 1.10\}$, and time horizons $T \in \{0.20, 0.40\}$. Use $n = 10{,}000$ pairs of antithetic variates for each set of parametric values. On replication i, take

$$h^{(i)} := \frac{1}{2}\left[h(X^{(i)}) + h(X_a^{(i)})\right]$$

as the sample. Compute and display the sample average

$$\bar{\lambda}_n := \frac{1}{n}\sum_{i=1}^{n} h^{(i)},$$

its standard error, and its relative standard error.

c. Show the estimated variance reduction for each set of parametric values.

d. Characterize the variance reduction as a function of volatility ξ_2, initial price $S(0)$, and time horizon T.

5.9 REFERENCES

Asmussen, S. (1987). *Applied Probability and Queues,* Wiley, New York.

Asmussen, S., and K. Binswanger (1997). Simulation of ruin probabilities for subexponential claims, *ASTIN Bulletin,* **27,** 297–318.

Asmussen, S., K. Binswanger, and B. Højgaard (2000). Rare event simulation for heavy-tailed distributions, *Bernoulli,* **6,** 303–322.

Binswanger, K. (1997). *Rare Events and Insurance,* Ph.D. dissertation, Swiss Federal Institute of Technology, Zurich.

Boyle, P., M. Broadie, and P. Glasserman (1997). Monte Carlo methods for security pricing, *Journal of Economic Dynamics and Control,* **21,** 1267–1321.

Bratley, P., and B. L. Fox (1988). Algorithm 659: Implementing Sobol's quasirandom sequence generator, *ACM Transactions on Mathematical Software,* **14,** 88–100.

Devroye, L. (1986). *Non-Uniform Random Variate Generation,* Springer-Verlag, New York.

Efron, B., and R. J. Tibshirani (1993). *An Introduction to the Bootstrap Method,* Chapman and Hall, New York and London.

Feller, W. (1971). *An Introduction to Probability Theory and Its Applications: Volume 2,* Wiley, New York.

Fishman, G. S. (1996). *Monte Carlo: Concepts, Algorithms, and Applications,* Springer-Verlag, New York.

Glasserman, P., P. Heidelberger, and P. Shahabuddin (1999). Asymptotically optimal importance sampling and stratification for pricing path-dependent options, *Mathematical Finance,* **9,** 117–152.

Halton, J. H. (1960). On the efficiency of certain quasi-random sequences of points in evaluating multi-dimensional integrals, *Numerical Mathemathics,* **2,** 84–90.

Heidelberger, P. (1995). Fast simulation of rare events in queueing and reliability models, *ACM Trans. Model. Comput. Simulation,* **6,** 43–85.

Hoeffding, W. (1940). Mass tab vin variante Korrelation theorie, *Schriften des Mathematischen Instituts für Angewandte Mathematik der Universität Berlin,* **5,** 179–233.

Joe, S., and F. Y. Kuo (2003). Algorithm 659: Implementing Sobol's quasirandom sequence generator, *ACM Transactions on Mathematical Software,* **29,** 49–57.

Johnson, N. L., and S. Kotz (1972). *Distributions in Statistics: Continuous Multivariate Distributions,* Wiley, New York.

Kemna, A. G. Z., and A. C. F. Vorst (1990). A pricing method for options based on average asset values, *Journal of Banking and Finance,* **14,** 113–129.

Kuipers, L., and H. Niederreiter (1974). *Uniform Distribution of Sequences,* Wiley, New York.

Lehman, E. L. (1983). *Theory of Point Estimation,* Wadsworth, Pacific Grove, Calif.

Mood, A. M., F. A. Graybill, and D. C. Boes (1974). *Introduction to the Theory of Statistics,* McGraw-Hill, New York.

Morokoff, W. J., and R. E. Caflisch (1995). Quasi–Monte Carlo integration, *Journal of Computational Physics,* **122,** 218–230.

Moskowitz, B., and R. E. Caflisch (1996). Smoothness and dimension reduction in quasi–Monte Carlo methods, *Mathematical and Computer Modelling,* **23,** 37–54.

Niederreiter, H. (1978). Quasi–Monte Carlo methods and pseudorandom numbers, *Bulletin of the American Mathematical Society,* **84,** 957–1041.

Niederreiter, H. (1992). *Random Number Generation and Quasi–Monte Carlo Methods,* SIAM, Philadelphia.

Owen, A. B. (1995). Randomly permuted (t, m, s)–nets and (t, s) sequences, in *Monte Carlo and Quasi–Monte Carlo Methods in Scientific Computing,* H. Niederreiter and P. J. S. Shiue editors, Lecture Notes in Statistics 106, Springer-Verlag, New York.

Spanier, J., and E. H. Maize (1994). Quasi-random methods for estimating integrals using relatively small samples, *SIAM Review,* **36,** 18–44.

Turnbull, S. M., and L. M. Wakeman (1991). A quick algorithm for pricing a European option, *Journal of Financial and Quantitative Analysis,* **26,** 377–389.

CHAPTER 6

Markov Chain Monte Carlo

For an m-dimensional random point $\mathbf{X} = (X_1, \ldots, X_m)$, statistical independence for $X_1, \ldots, X_m$, when it holds, allows straightforward application of the algorithms in Chapter 3 to generate random samples for each coordinate. However, many problems specify that $X_1, \ldots, X_m$ are dependent, and sample generation needs to take account of this property. Sometimes this is easily done; for example, when $X_1, \ldots, X_m$ have the multinormal or multinomial distribution.

When direct sampling from π is computationally intractable, Markov chain Monte Carlo (MCMC) overcomes this limitation. It provides techniques for generating a sample sequence or *sample path* $\mathbf{X}_1, \ldots, \mathbf{X}_t$ where each $\mathbf{X}_j := (X_{1j}, \ldots, X_{mj})$ has the desired distribution π. It does this by randomly generating $\mathbf{X}_j$ conditionally on the previous sample $\mathbf{X}_{j-1}$. Although this conditioning often induces relatively low unit generating cost, there is a penalty. The elements $\mathbf{X}_1, \ldots, \mathbf{X}_t$ tend to be correlated rather than independent, so the statistical properties of a sample path of length t differ from those of an IMC-generated sequence of independent $\mathbf{X}^{(1)}, \ldots, \mathbf{X}^{(t)}$ from the same distribution π. Because the resulting dependence often manifests itself as positive correlation between $\mathbf{X}_j$ and $\mathbf{X}_l$ for $j \neq l$, it frequently induces larger variances for sample averages and, therefore, less statistical accuracy.

Although the dependence dissipates as $|j - l| \to \infty$, its presence for small and moderate $|j - l|$ cannot be ignored. Indeed, to keep the error of approximation within a specified bound, MCMC may require a considerably larger number of observations than IMC requires to achieve the same bound. Nevertheless, there exist problems for which a comparison of total MCMC and IMC costs of estimating λ in (2.1) to within a specified error frequently favors MCMC.

This chapter describes the principal features of MCMC. They include the procedural concepts of Hastings-Metropolis sampling and Gibbs sampling and the more abstract concepts of global balance, detailed balance, irreducibility, recurrence, aperiodicity, and ergodicity. Familiarity with these concepts is essential for understanding how MCMC produces approximations to λ in (2.1) with desirable statistical properties. Among these, the most important is the way in which the error of approximation decreases as sample-path length, t, increases without bound. Chapter 9 describes more advanced MCMC concepts.

MCMC applies to both continuous- and discrete-state spaces. We first focus on the discrete version of (2.23)

$$g(\mathbf{x}) = h(\mathbf{x})\pi(\mathbf{x}) \qquad \forall \mathbf{x} \in \mathcal{X}, \tag{6.1}$$

where $\mathcal{X}$ is a countable set of points, $\pi(\mathbf{x}) > 0$ for all $\mathbf{x} \in \mathcal{X}$, and

$$\sum_{\mathbf{x}\in\mathcal{X}} \pi(\mathbf{x}) = \varphi < \infty. \tag{6.2}$$

That is, $\{\pi(\mathbf{x})/\varphi\}$ is a p.m.f. One major appeal of MCMC is that it can generate samples from $\{\pi(\mathbf{x})/\varphi\}$ when different amounts of information are known. These include

- π and φ are known.
- π is known but φ is unknown.
- π and φ are unknown.

The last of these is most intriguing, and we describe how this occurs in Section 6.14 for discrete-event systems. Because Section 6.1 shows that unknown φ (but known π) is no impediment to MCMC, we hereafter take $\varphi = 1$ so that π is a p.m.f. However, we emphasize that the concepts described here also apply when $\varphi \neq 1$. Section 6.12 extends these concepts to general-state spaces.

For each $\mathbf{x} \in \mathcal{X}$, let

$$K(\mathbf{x}, \cdot) := \{K(\mathbf{x}, \mathbf{y}) \geq 0, \ \mathbf{y} \in \mathcal{X}\}$$

denote a conditional probability distribution satisfying

$$\sum_{\mathbf{x}\in\mathcal{X}} \pi(\mathbf{x})K(\mathbf{x}, \mathbf{y}) = \pi(\mathbf{y}) \qquad \forall \mathbf{y} \in \mathcal{X}. \tag{6.3}$$

Expression (6.3) denotes *global balance* with respect to π. Any π that satisfies (6.3) is called an *invariant distribution* with respect to the *one-step transition kernel*

$$K := \{K(\mathbf{x}, \mathbf{y}); \ \mathbf{y} \in \mathcal{X}, \ \mathbf{x} \in \mathcal{X}\}.$$

A sequence of random variables $\{\mathbf{X}_i := (X_{1i}, \ldots, X_{mi}), \ i \geq 0\}$ each taking values in $\mathcal{X}$ with invariant distribution π and transition kernel K is called a time-homogeneous Markov chain, or, more concisely, a *Markov chain*.

The significance of global balance cannot be overestimated. If $\mathbf{X}_0$ is a random point with distribution π, then global balance implies that randomly sampling $\mathbf{X}_1$ from the conditional distribution $\{K(\mathbf{X}_0, \mathbf{y}), \ \mathbf{y} \in \text{supp}[K(\mathbf{X}_0, \cdot)]\}$ endows $\mathbf{X}_1$ with the same distribution π. Moreover, it is easily shown by induction on (6.3) that for each $i \geq 1$, randomly sampling $\mathbf{X}_i$ from the conditional distribution $\{K(\mathbf{X}_{i-1}, \mathbf{y}), \ \mathbf{y} \in \text{supp}[K(\mathbf{X}_{i-1}, \cdot)]\}$ endows each $\mathbf{X}_i$ with the invariant distribution π, sometimes called the *stationary distribution*. In this context, some writers refer to a step on a sample path as an *iteration*. The motivation for writing the support, $\text{supp}[K(\mathbf{x}, \cdot)]$, instead of $\mathcal{X}$ is to emphasize that not all states in $\mathcal{X}$ are reachable from state $\mathbf{x}$ in one iteration.

Algorithm MCMC describes the principal steps to generate a sample path by MCMC sampling, and

$$\widehat{\lambda}_t := \frac{1}{t}\sum_{i=1}^{t} h(\mathbf{X}_i) \tag{6.4}$$

provides an estimate of λ. Although $\mathbf{X}_0$ rarely has distribution π in practice, we continue to assume it does until Section 6.7, which describes how the properties of the sample path $\mathbf{X}_1, \ldots, \mathbf{X}_t$ are influenced by drawing $\mathbf{X}_0$ from a more easily sampled distribution π_0.

ALGORITHM **MCMC**

Purpose: To generate $\mathbf{X}_1, \ldots, \mathbf{X}_t$ from K via MCMC.

Given: Initial state $\mathbf{X}_0$, transition kernel K, and sample-path length t.

Method:

$j = 1.$
Until $j > t$:
  randomly sample $\mathbf{X}_j$ from $\{K(\mathbf{X}_{j-1}, \mathbf{y}),\ \mathbf{y} \in \text{supp}[K(\mathbf{X}_{j-1}, \cdot)]\}$
  $j = j + 1.$ ■

6.1 HASTINGS-METROPOLIS METHOD

How difficult is it to devise a transition kernel that satisfies global balance? Not very, as we now show.[1]

Theorem 6.1.1

Let $\mathbf{Z}$ *have distribution* $\{\pi(\mathbf{z}),\ \mathbf{z} \in \mathcal{X}\}$, $\mathbf{Y}$ *given* $\mathbf{Z} = \mathbf{z}$ *have distribution* $\{q(\mathbf{z}, \mathbf{y}), \mathbf{y} \in \mathcal{X}\}$, *and* U *have distribution* $\mathcal{U}(0, 1)$. *Let*

$$\mathbf{X} := \begin{cases} \mathbf{Y} & \text{if } U \le \alpha(\mathbf{Z}, \mathbf{Y}) \\ \mathbf{Z} & \text{otherwise,} \end{cases}$$

where $\alpha(\mathbf{x}, \mathbf{y})$ *satisfies*

$$0 \le \alpha(\mathbf{x}, \mathbf{y}) \le 1 \tag{6.5}$$

[1] The concept for symmetric q originated in Metropolis et al. (1953). The current, more general description follows Hastings (1970).

and

$$\alpha(\mathbf{x}, \mathbf{y})\pi(\mathbf{x})q(\mathbf{x}, \mathbf{y}) = \alpha(\mathbf{y}, \mathbf{x})\pi(\mathbf{y})q(\mathbf{y}, \mathbf{x}) \qquad (\mathbf{x}, \mathbf{y}) \in \mathcal{X}^2. \tag{6.6}$$

Then $\mathbf{X}$ *has distribution* π. ■

PROOF Let C denote the event $U \leq \alpha(\mathbf{Z}, \mathbf{Y})$, and $\overline{\mathrm{C}}$, its complement. Then $\mathbf{X}$, $\mathbf{Y}$, $\mathbf{Z}$, and C have the joint distribution

$$\begin{aligned}\pi(\mathbf{x}, \mathbf{y}, \mathbf{z}, \mathrm{C}) &= \pi(\mathbf{z})q(\mathbf{z}, \mathbf{y})\alpha(\mathbf{z}, \mathbf{y})\delta(\mathbf{x} - \mathbf{y}) \\ &= \pi(\mathbf{z})q(\mathbf{y}, \mathbf{z})\alpha(\mathbf{y}, \mathbf{z})\delta(\mathbf{x} - \mathbf{y}),\end{aligned}$$

where

$$\delta(\mathbf{z}) := \begin{cases} 1 & \text{if } \mathbf{z} = \mathbf{0} \\ 0 & \text{elsewhere.} \end{cases} \tag{6.7}$$

Therefore,

$$\pi(\mathbf{x}, \mathbf{y}, \mathrm{C}) = \pi(\mathbf{z})A(\mathbf{y})\delta(\mathbf{x} - \mathbf{y}),$$

where

$$A(\mathbf{y}) := \sum_{\mathbf{z} \in \mathcal{X}} q(\mathbf{y}, \mathbf{z})\alpha(\mathbf{y}, \mathbf{z}),$$

so that

$$\pi(\mathbf{x}, \mathrm{C}) = \pi(\mathbf{x})A(\mathbf{x}).$$

In like manner,

$$\begin{aligned}\pi(\mathbf{x}, \mathbf{y}, \mathbf{z}, \overline{\mathrm{C}}) &= \pi(\mathbf{z})[1 - \alpha(\mathbf{z}, \mathbf{y})]q(\mathbf{z}, \mathbf{y})\delta(\mathbf{z} - \mathbf{x}) \\ \pi(\mathbf{x}, \mathbf{z}, \overline{\mathrm{C}}) &= \pi(\mathbf{z})[1 - A(\mathbf{z})]\delta(\mathbf{z} - \mathbf{x}) \\ \pi(\mathbf{x}, \overline{\mathrm{C}}) &= \pi(\mathbf{x})[1 - A(\mathbf{x})].\end{aligned}$$

Therefore, $\mathbf{X}$ has the distribution

$$\pi(\mathbf{x}, \mathrm{C}) + \pi(\mathbf{x}, \overline{\mathrm{C}}) = \pi(\mathbf{x}) \qquad \forall \mathbf{x} \in \mathcal{X}.$$ ■

Hereafter, we call $q := \{q(\mathbf{x}, \mathbf{y}),\ (\mathbf{x}, \mathbf{y}) \in \mathcal{X}^2\}$ the *nominating kernel*. Because for all $\mathbf{y}$ such that $q(\mathbf{x}, \mathbf{y}) > 0$,

$$\begin{aligned}\alpha(\mathbf{x}, \mathbf{y}) &= \alpha(\mathbf{y}, \mathbf{x})\frac{\pi(\mathbf{y})q(\mathbf{y}, \mathbf{x})}{\pi(\mathbf{x})q(\mathbf{x}, \mathbf{y})} \\ &\leq \min\left[1, \frac{\pi(\mathbf{y})q(\mathbf{y}, \mathbf{x})}{\pi(\mathbf{x})q(\mathbf{x}, \mathbf{y})}\right],\end{aligned}$$

the assignment

$$\alpha(\mathbf{x}, \mathbf{y}) := \min \left[1, \frac{\pi(\mathbf{y}) q(\mathbf{y}, \mathbf{x})}{\pi(\mathbf{x}) q(\mathbf{x}, \mathbf{y})} \right], \tag{6.8}$$

which satisfies (6.6), maximizes the acceptance probability.

Algorithm HM describes a frequently employed form of the *Hastings-Metropolis (HM) method* for randomly generating a sample path. Because successful implementation relies only on the ratio $\pi(\mathbf{y})/\pi(\mathbf{x})$ and not on the distribution itself, the approach applies to a wide range of problems in which π need only be positive, as in (6.1) and (6.2), and not necessarily to a probability distribution. If the initially assigned state $\mathbf{X}_0$ is randomly drawn from π, then each subsequent state $\mathbf{X}_i$ in the generated sample path has the distribution π. The *nominating kernel* q is chosen to facilitate sample generation. However, the choice also affects $\text{var}\widehat{\lambda}_t$, as Section 6.10.1 shows.

ALGORITHM

HM

Purpose: To generate $\mathbf{X}_1, \ldots, \mathbf{X}_t$ randomly by the HM method.

Given: Initial state $\mathbf{X}_0$, nominating kernel q, and sample-path length t.

Method:

$j = 1$.
Until $j > t$:
randomly sample $\mathbf{Y}$ from $\{q(\mathbf{X}_{j-1}, \mathbf{y}),\ \mathbf{y} \in \text{supp}[q(\mathbf{X}_{j-1}, \cdot)]\}$
randomly sample U from $\mathcal{U}(0, 1)$
if $U \le \min \left[1, \frac{\pi(\mathbf{Y}) q(\mathbf{Y}, \mathbf{X}_{j-1})}{\pi(\mathbf{X}_{j-1}) q(\mathbf{X}_{j-1}, \mathbf{Y})} \right]$, $\mathbf{X}_j = \mathbf{Y}$
otherwise, $\mathbf{X}_j = \mathbf{X}_{j-1}$
$j = j + 1$. ■

Algorithm HM induces the one-step transition kernel

$$K(\mathbf{x}, \mathbf{y}) = \begin{cases} \min \left[1, \frac{\pi(\mathbf{y}) q(\mathbf{y}, \mathbf{x})}{\pi(\mathbf{x}) q(\mathbf{x}, \mathbf{y})} \right] q(\mathbf{x}, \mathbf{y}) & \text{if } \mathbf{x} \neq \mathbf{y} \text{ and } q(\mathbf{x}, \mathbf{y}) > 0 \\ 0 & \text{if } \mathbf{x} \neq \mathbf{y} \text{ and } q(\mathbf{x}, \mathbf{y}) = 0 \\ 1 - \sum_{\mathbf{z} \in \mathcal{X} \setminus \{\mathbf{x}\}} q(\mathbf{z}, \mathbf{z}) & \text{if } \mathbf{x} = \mathbf{y}, \end{cases} \tag{6.9}$$

where $\mathcal{X} \setminus \{\mathbf{x}\}$ denotes the state space $\mathcal{X}$ with the point $\mathbf{x}$ omitted. It is easily seen that $K(\mathbf{x}, \mathbf{y}) \ge 0$ for all $(\mathbf{x}, \mathbf{y}) \in \mathcal{X}^2$ and

$$\sum_{\mathbf{y} \in \text{supp}[K(\mathbf{x}, .)]} K(\mathbf{x}, \mathbf{y}) = 1 \qquad \forall \mathbf{x} \in \mathcal{X},$$

so $K(\mathbf{x}, \cdot)$ is a probability distribution. Moreover,

$$\begin{aligned}\pi(\mathbf{x})K(\mathbf{x}, \mathbf{y}) &= \min\left[\pi(\mathbf{x})q(\mathbf{x}, \mathbf{y}), \pi(\mathbf{y})q(\mathbf{y}, \mathbf{x})\right] \\ &= \pi(\mathbf{y})K(\mathbf{y}, \mathbf{x}) \qquad \forall(\mathbf{x}, \mathbf{y}) \in \mathcal{X}^2,\end{aligned} \tag{6.10}$$

which provides an easy way to show that global balance holds. More generally, if

$$\pi(\mathbf{x})K(\mathbf{x}, \mathbf{y}) = \pi(\mathbf{y})K(\mathbf{y}, \mathbf{x}) \qquad \forall(\mathbf{x}, \mathbf{y}) \in \mathcal{X}^2, \tag{6.11}$$

then K is said to obey *detailed balance,* and the Markov chain is said to be *reversible.*

6.2 REVERSIBILITY

The term *reversible* comes from the observation that (6.11) ensures

$$\text{pr}(\mathbf{X}_j = \mathbf{y}, \mathbf{X}_{j-1} = \mathbf{x}) = \text{pr}(\mathbf{X}_{j-1} = \mathbf{y}, \mathbf{X}_j = \mathbf{x}) \tag{6.12}$$

so that

$$\begin{aligned}\text{pr}(\mathbf{X}_j = \mathbf{y} \mid \mathbf{X}_{j-1} = \mathbf{x}) &= \text{pr}(\mathbf{X}_{j-1} = \mathbf{y} \mid \mathbf{X}_j = \mathbf{x}) \\ &= \text{pr}(\mathbf{X}_j = \mathbf{y} \mid \mathbf{X}_{j+1} = \mathbf{x}).\end{aligned} \tag{6.13}$$

This implies that the dynamics of the chain are independent of direction in the sense that the same probability law applies to a sample path $\mathbf{X}_j, \mathbf{X}_{j+1}, \mathbf{X}_{j+2}, \ldots$ as to a sample path $\mathbf{X}_j, \mathbf{X}_{j-1}, \mathbf{X}_{j-2}, \ldots$ evolving in the opposite direction. Later sections make repeated use of this property. Note that detailed balance implies global balance, perhaps one of the most convenient properties of reversibility. Section 9.1 describes how reversibility provides a basis for characterizing convergence for finite-state Markov chains.

6.3 COORDINATE UPDATING

Updating a single coordinate on each iteration is the most commonly encountered form of MCMC sampling. Simple in concept, it allows many of the computationally efficient techniques of Chapter 3 to be used for this univariate sampling. However, single-coordinate updating frequently leads to highly dependent sample sequences $\mathbf{X}_1, \ldots, \mathbf{X}_t$ on iterations 1 through t, which tends to make $\text{var}\widehat{\lambda}_t$ much larger than $t^{-1}\text{var}_\pi h(\mathbf{X}_j)$, the variance in the i.i.d. case (see Section 6.8). This is especially true for Gibbs sampling (Section 6.6) when the equilibrium distribution π indicates substantial dependence among the coordinate values $X_{1j}, \ldots, X_{mj}$. The inflated variance is often so substantial as to require an inordinately large number of iterations t to achieve a specified accuracy for $\widehat{\lambda}_t$.

In contrast to the difficulties of sampling directly from π, HM sampling can facilitate multiple-coordinate updating per iteration by choosing nominating distributions to be multivariate distributions for which computationally efficient generating procedures exist. These include the multinormal and multivariate Student t distributions in the continuous case and the multinomial distribution in the discrete case. In practice, this approach frequently results in low acceptance rates, implying long subsequences in $\mathbf{X}_1, \ldots, \mathbf{X}_t$, which do not change value. This reluctance to change value results in long sample paths to explore the sample space adequately. Section 9.7.5 returns to this topic of multivariate nominating distributions within the context of HM sampling and describes how they can be combined with a modified acceptance-rejection sampling procedure to generate asymptotically (as $t \to \infty$) independent $\mathbf{X}_1, \ldots, \mathbf{X}_t$ from π.

Sometimes, attention to special structure allows computationally efficient multiple-coordinate updating while inducing relatively moderate dependence among $\mathbf{X}_1, \ldots, \mathbf{X}_t$. Examples 6.6.1, 6.11.1, and 6.11.2 illustrate the benefits of exploiting special structure, and Section 9.3 describes a class of models for which $\mathbf{X}_j$ is dependent on only $l < m$ of the variables, $X_{1,j-1}, \ldots, X_{m,j-1}$, on the previous iteration $j-1$, where l is often substantially smaller than m.

6.3.1 Single-Coordinate Updating

As just mentioned, updating at most one coordinate per iteration offers the conceptually simplest and most easily implementable HM generating procedure. We describe several ways that single-coordinate updating can be implemented. Let

$$\begin{aligned} \mathbf{x}_i(y) &:= (x_1, \ldots, x_{i-1}, y, x_{i+1}, \ldots, x_m) \\ &= \mathbf{x} \text{ with } y \text{ replacing } x_i \qquad (6.14) \\ \mathbf{x}_{-i} &:= (x_1, \ldots, x_{i-1}, x_{i+1}, \ldots, x_m) \\ &= \mathbf{x} \text{ with } x_i \text{ omitted} \\ \mathcal{X}_i(\mathbf{x}_{-i}) &:= \text{set of all values } y \text{ such that } \mathbf{x}_i(y) \in \mathcal{X}. \end{aligned}$$

Then for each $\mathbf{x} \in \mathcal{X}$, a univariate distribution $\{q(\mathbf{x}, \mathbf{x}_i(y)),\ y \in \mathcal{X}_i(\mathbf{x}_{-i})\}$ is specified with corresponding acceptance function

$$\alpha_i(\mathbf{x}, \mathbf{x}_i(y)) = \min\left[1, \frac{\pi(\mathbf{x}_i(y))}{\pi(\mathbf{x})} \frac{q_i(\mathbf{x}_i(y), \mathbf{x})}{q_i(\mathbf{x}, \mathbf{x}_i(y))}\right]. \qquad (6.15)$$

In addition to choosing a nominating kernel, a user needs to adopt a rule that determines which coordinate is to be updated on each iteration $j \geq 1$. For single coordinate updating, two options are:

- **Deterministic Coordinate Selection (DCS) Rule**

  Select coordinate $I = (j-1) \pmod m + 1$.

- **Random Coordinate Selection (RCS) Rule**

  Randomly select coordinate $I = i$ with probability w_i, where $w_i > 0$ for all i and $\sum_{l=1}^{m} w_l = 1$.

The DCS rule repeatedly cycles through coordinates $1, \ldots, m$ in that order. The RCS rule randomly and independently selects a coordinate on each step, most commonly with equal weights $w_i = \frac{1}{m}$ for $i = 1, \ldots, m$. Unless otherwise stated, we hereafter assume equal weights.

As our MCMC description unfolds, the reader will note considerably more frequent reference to the RCS than to the DCS rule. This arises because the former is considerably more amenable to theoretical analysis, and therefore its implications for the sampling properties of Markov chains are easier to characterize. However, this more frequent mention should not be construed to mean that RCS induces better sampling properties in $\widehat{\lambda}_t$ than DCS.

Let

$$\mathbf{X}_{ij}(y) := \mathbf{X}_j \text{ with } y \text{ replacing its } i\text{th coordinate value } X_{ij}.$$

Algorithm SCHM describes single-coordinate updating via HM sampling.

ALGORITHM SCHM

Given: Initial state $\mathbf{X}_0$, sample-path length t, and either the DCS or RSC rule for coordinate selection.

Method:

$j = 1.$
Until $j > t$:
  select coordinate I
  randomly select Y from $\{q(\mathbf{X}_{j-1}, \mathbf{X}_{i,j-1}(y))\}$
  randomly sample U from $\mathcal{U}(0, 1)$
  if $U \le \alpha(\mathbf{X}_{j-1}, \mathbf{X}_{i,j-1}(Y)),\ \mathbf{X}_j = \mathbf{X}_{i,j-1}(Y)$
  otherwise, $\mathbf{X}_j = \mathbf{X}_{j-1}$
  $j = j + 1.$ ■

Single-coordinate updating also reduces the amount of information needed to evaluate the acceptance function. Let

$$\{\pi_i(y \mid \mathbf{x}_{-i})\} := \text{conditional distribution of coordinate } i, \text{ given the values of coordinates } 1, \ldots, i-1, i+1, \ldots, m, \tag{6.16}$$

and

$$\{\pi_i(\mathbf{x}_{-i})\} := \text{joint distribution of coordinates } 1, \ldots, i-1, i+1, \ldots, m.$$

Then writing π in the form

$$\pi(\mathbf{x}) = \pi_i(x_i \mid \mathbf{x}_{-i})\pi_i(\mathbf{x}_{-i})$$

immediately reveals that

$$\frac{\pi(\mathbf{x}_i(y))}{\pi(\mathbf{x})} = \frac{\pi_i(y \mid \mathbf{x}_{-i})}{\pi_i(x_i \mid \mathbf{x}_{-i})}.$$

Therefore, availability of the *full conditional distributions* $\{\pi_i(y \mid \mathbf{x}_{-i})\}$ for each coordinate $i = 1, \ldots, m$ suffices to implement HM sampling with univariate updating. This property broadens the range of application to many problems for which the full joint distribution $\{\pi(\mathbf{x})\}$ is not completely specified but the full conditional distributions are.

As to the choice of nominating kernel, one option assigns

$$q(\mathbf{x}, \mathbf{x}_i(y)) = \begin{cases} \frac{\pi_i(y|\mathbf{x}_{-i})}{1-\pi_i(x_i|\mathbf{x}_{-i})} & \text{if } y \neq x_i \\ 0 & \text{if } y = x_i, \end{cases} \tag{6.17}$$

which simplifies the acceptance function to

$$\alpha(\mathbf{x}, \mathbf{x}_i(y)) = \min\left[1, \frac{1 - \pi_i(x_i \mid \mathbf{x}_{-i})}{1 - \pi_i(y \mid \mathbf{x}_{-i})}\right]. \tag{6.18}$$

This choice makes use of the available full conditional distributions, thus relieving the user of the burden of choosing a nominating kernel from other sources.

EXAMPLE 6.3.1 ILLUSTRATING THE HASTINGS-METROPOLIS METHOD

Suppose the objective is to sample from the eight-point distribution in Table 3.3 by the HM method. Consider the nominating kernel

$$\left.\begin{array}{rcl} q(x, x+1) = & \frac{1}{2} & x = 1, \ldots, 7 \\ q(x, x-1) = & \frac{1}{2} & x = 2, \ldots, 8 \\ q(x, 1) = & \frac{1}{2} & x = 8 \\ q(x, 8) = & \frac{1}{2} & x = 1 \\ q(x, y) = & 0 & \text{elsewhere.} \end{array}\right\} \tag{6.19}$$

Figure 6.1 conceptualizes the process where at any site x transitions are possible only to *nearest neighbors*. The nominating kernel (6.19) induces the acceptance

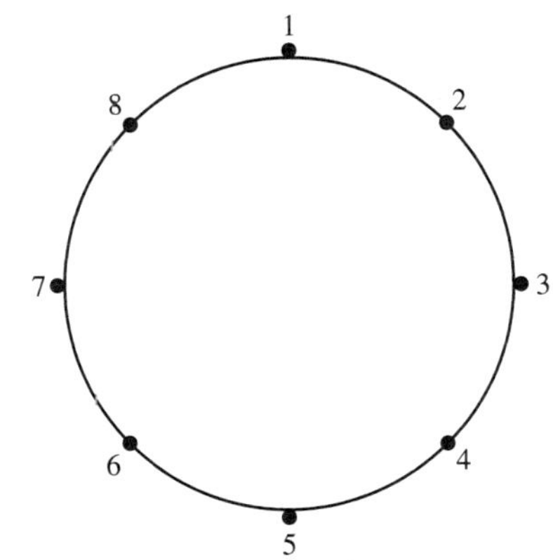

FIGURE 6.1 | *Nearest-neighbor states*

function

$$\alpha(x, y) = \begin{pmatrix} & 1 & & & & & & 1 \\ 0.25 & & 1 & & & & & \\ & 1 & & 1 & & & & \\ & & 0.467 & & 1 & & & \\ & & & 0.536 & & 0.679 & & \\ & & & & 1 & & 1 & \\ & & & & & 0.905 & & 0.238 \\ 0.20 & & & & & & 1 & \end{pmatrix} \tag{6.20}$$

and transition kernel

$$K(x, y) = \begin{pmatrix} & 0.50 & & & & & & 0.50 \\ 0.125 & 0.375 & 0.50 & & & & & \\ & 0.50 & & 0.50 & & & & \\ & & 0.234 & 0.266 & 0.50 & & & \\ & & & 0.268 & 0.392 & 0.340 & & \\ & & & & 0.50 & & 0.50 & \\ & & & & & 0.453 & 0.428 & 0.119 \\ 0.10 & & & & & & 0.50 & 0.40 \end{pmatrix}, \tag{6.21}$$

with entries in row x and column y for $x, y = 1, \ldots, 8$ and where blank fields denote 0 entries. Expression (6.21) reveals that for some states x, the next state must be a nearest neighbor, whereas other states x allow a third option of the next state remaining the current state. Figure 6.2a shows a sample path for $t = 100$ iterations, where X_0 was randomly sampled from the stationary distribution in Table 3.3. Horizontal segments (i.e., $X_j = X_{j+1} = \cdots$) indicate where rejection has occurred.

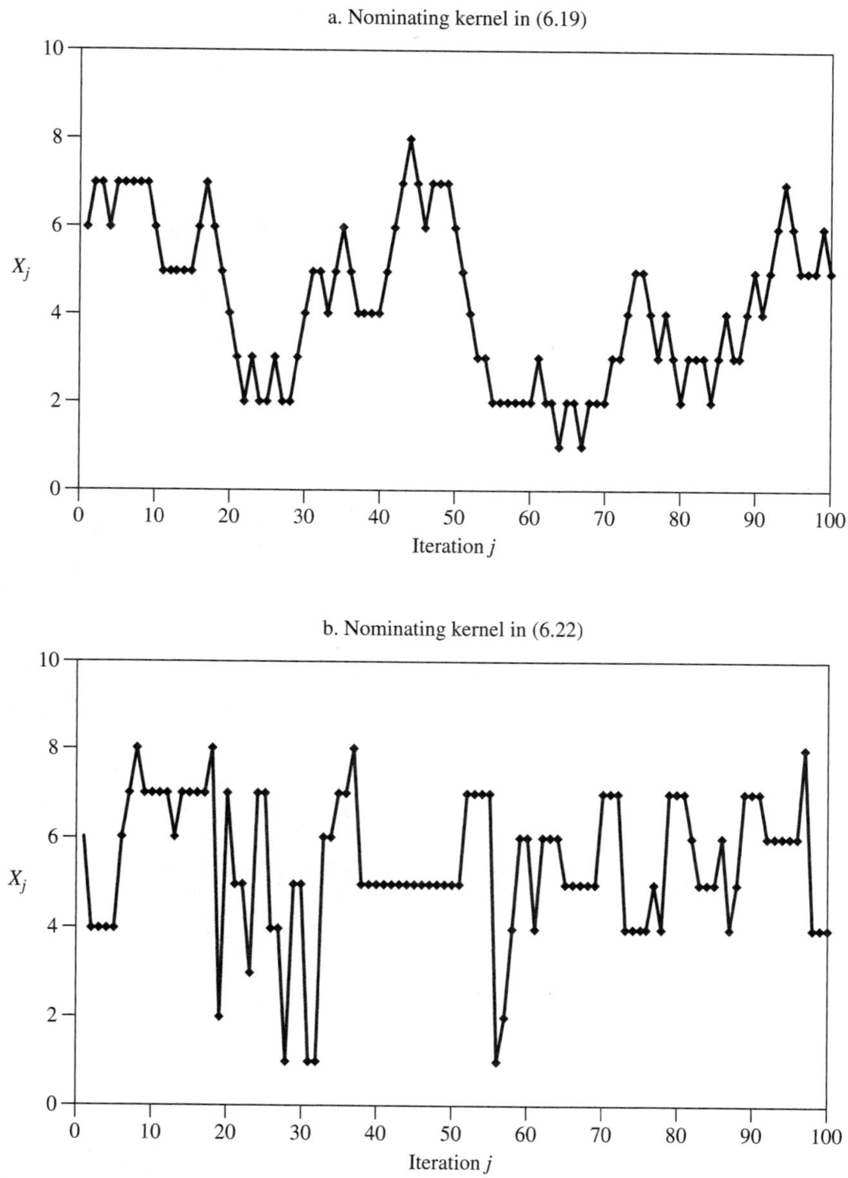

FIGURE 6.2 | *Sample paths*

The freedom to choose a nominating kernel also allows us to select distributions independent of the current state x. For example, consider the nominating kernel

$$q(x, y) = \begin{cases} \frac{1}{8} & y = 1, \ldots, 8 \\ 0 & \text{elsewhere.} \end{cases} \tag{6.22}$$

The corresponding acceptance function and transition kernel are

$$\alpha(x, y) = \begin{pmatrix} 1 & 1 & 1 & 1 & 1 & 1 & 1 & 1 \\ 0.250 & 1 & 1 & 1 & 1 & 1 & 1 & 1 \\ 0.143 & 1 & 1 & 1 & 1 & 1 & 1 & 0.714 \\ 0.0670 & 0.467 & 0.467 & 1 & 1 & 1 & 1 & 0.333 \\ 0.0360 & 0.250 & 0.250 & 0.536 & 1 & 0.679 & 0.750 & 0.179 \\ 0.0530 & 0.368 & 0.368 & 0.789 & 1 & 1 & 1 & 0.263 \\ 0.0480 & 0.333 & 0.333 & 0.714 & 1 & 0.905 & 1 & 0.238 \\ 0.200 & 1 & 1 & 1 & 1 & 1 & 1 & 1 \end{pmatrix} \tag{6.23}$$

and

$$K(x, y) = \begin{pmatrix} 0.125 & 0.125 & 0.125 & 0.125 & 0.125 & 0.125 & 0.125 & 0.125 \\ 0.0310 & 0.219 & 0.125 & 0.125 & 0.125 & 0.125 & 0.125 & 0.125 \\ 0.0180 & 0.125 & 0.268 & 0.125 & 0.125 & 0.125 & 0.125 & 0.0893 \\ 0.0080 & 0.0583 & 0.0583 & 0.458 & 0.125 & 0.125 & 0.125 & 0.0417 \\ 0.0040 & 0.0313 & 0.0313 & 0.0670 & 0.665 & 0.0848 & 0.0938 & 0.0223 \\ 0.0070 & 0.0461 & 0.0461 & 0.0987 & 0.125 & 0.520 & 0.125 & 0.0329 \\ 0.0060 & 0.0417 & 0.0417 & 0.0893 & 0.125 & 0.113 & 0.643 & 0.0298 \\ 0.0250 & 0.125 & 0.125 & 0.125 & 0.125 & 0.125 & 0.125 & 0.225 \end{pmatrix}. \tag{6.24}$$

Figure 6.2b displays a sample path for this choice of nominating kernel. Since the sample paths in each graph were generated independently, they naturally differ. However, Figure 6.2b displays much larger fluctuations than Figure 6.2a, a direct consequence of the choice of q. ■

6.4 BAYESIAN MCMC

Let $\mathbf{Z} := (Z_1, \ldots, Z_r)$ denote a data set of r points with distribution $\{f(\mathbf{z} \mid \beta)\}$, where $\beta := (\beta_1, \ldots, \beta_m)$ denote m parameters whose values are unknown. Classical statistics regards $\beta_1, \ldots, \beta_m$ as constants to be estimated. For example, the maximum likelihood method estimates them by the m-dimensional point, $\widehat{\beta}$, that maximizes $\{f(\mathbf{Z} \mid \beta)\}$ with respect to β. Although inducing many desirable statistical properties, this and other methodologies have limitations. In particular, a corresponding error assessment is frequently unavailable for small and moderate sample sizes r. Also, there is often no provision for incorporating additional information about β from other sources into the estimating procedure.

Bayesian statistical methods provide one way of reducing the severity and sometimes eliminating these limitations. Most notably, they treat $\beta_1, \ldots, \beta_m$ as random with user-specified *prior distribution* $\{f_{\text{prior}}(\beta)\}$, usually based on past experience and expert judgment. They also make use of the well-known probabilistic relationship that if Q and R are two random events, then

$$\mathrm{pr}(\mathrm{Q} \mid \mathrm{R})\mathrm{pr}(\mathrm{R}) = \mathrm{pr}(\mathrm{R} \mid \mathrm{Q})\mathrm{pr}(\mathrm{Q}). \tag{6.25}$$

Letting $\mathbf{Z}$ denote the event Q and β the event R results in the *posterior distribution*

$$f_{\mathrm{posterior}}(\beta \mid \mathbf{Z}) = \frac{f(\mathbf{Z} \mid \beta)\, f_{\mathrm{prior}}(\beta)}{f(\mathbf{Z})}, \tag{6.26}$$

where

$$f(\mathbf{Z}) := \int f(\mathbf{Z} \mid \beta)\, f_{\mathrm{prior}}(\beta) \mathrm{d}\beta.$$

The Bayesian method uses the posterior distribution (6.26) to characterize β. Sometimes a single parameter, say, β_m, is of interest, the remainder $\beta_1, \ldots, \beta_{m-1}$ being nuisance parameters. Then its posterior distribution

$$f_{\mathrm{posterior}}(\beta_m \mid \mathbf{Z}) = \frac{1}{f(\mathbf{Z})} \int f(\mathbf{Z} \mid \beta_1, \ldots, \beta_m)\, f_{\mathrm{prior}}(\beta_1, \ldots, \beta_m) \mathrm{d}\beta_1 \cdots \mathrm{d}\beta_{m-1} \tag{6.27}$$

and, in particular, its mean, mode, and quantiles characterize β_m. Moreover, a posterior distribution concentrated in a small interval in $(-\infty, \infty)$ indicates a most likely region for β_m.

For (6.26) to be completely specified, the integral $f(\mathbf{Z})$ needs to be evaluated. For (6.27), complete specification calls for the evaluation of two integrals. At least two solutions exist to these computational problems:

1. Choose prior distributions that lead to closed-form analytical solutions for the integrals or that facilitate numerical evaluation.
2. Perform an MC sampling experiment that generates data from which the posterior distribution of interest and its mean, mode, and so on can be estimated.

Because the HM as well as other MC methods allow sampling from a distribution π in (6.2) without specification of the normalizing constant φ, we see that it is possible to sample from (6.26) without knowing its normalizing constant $f(\mathbf{Z})$. The next example illustrates how this is done.

EXAMPLE 6.4.1

CAPTURE-RECAPTURE[2]

One method of estimating the size of an animal population relies on a sequence of r capture-recapture surveys. For $l = 1, \ldots, r$, survey l captures a sample from the target population, counts the number of previously marked animals in the sample, counts and marks those that have not been caught in previous surveys, and returns the sample to the target population. For present purposes, assume that

[2] This problem is taken from Castledine (1981) and George and Robert (1992).

i. The target population remains constant throughout the time that the r surveys are performed.

ii. No animals are marked prior to the first survey.

iii. In a particular survey, all animals have the same propensity of being caught regardless of their previous catch history.

iv. Prior to sampling, there is no reason to suppose that the surveyor expends more sampling effort on some surveys than on others.

On survey l, let

$$\begin{aligned} p_l &:= \text{probability of capture} \\ N_l &:= \text{number captured} \\ M_l &:= \text{number captured that have been previously marked (captured)} \\ S_{l+1} &:= \text{number marked on surveys 1 through } l \\ &= \begin{cases} 0 & \text{if } l = 0 \\ S_l + N_l - M_l & \text{if } l \geq 1 \end{cases} \\ \mathbf{p} &:= (p_1, \ldots, p_r) \\ \mathbf{N} &:= (N_1, \ldots, N_r) \\ \mathbf{M} &:= (M_1, \ldots, M_r), \end{aligned}$$

and let

$$\theta := \text{unknown population size to be estimated.}$$

The quantity S_{r+1} denotes the total number of different animals captured on all r surveys. Given θ and p_l, the probability of capturing N_l animals on survey l is

$$\text{pr}(N_l \mid p_l, \theta) = \binom{\theta}{N_l} p_l^{N_l}(1 - p_l)^{\theta - N_l} \qquad N_l = 0, 1, \ldots, \theta,$$

and given N_l and S_{l+1}, the probability that M_l of the N_l animals have not been previously marked is

$$\text{pr}(M_l \mid N_l, S_{l+1}) = \prod_{l=1}^{r} \frac{\binom{\theta - S_{l+1}}{N_l - M_l}\binom{S_{l+1}}{M_l}}{\binom{\theta}{N_l}} \qquad M_l = 0, 1, \ldots \min(N_l, S_{l+1}).$$

These lead to the joint distribution

$$\text{pr}(\mathbf{N}, \mathbf{M} \mid \mathbf{p}, \theta) = \text{pr}(\mathbf{M} \mid \mathbf{N}, \mathbf{p}, \theta)\text{pr}(\mathbf{N} \mid \mathbf{p}, \theta), \tag{6.28}$$

where

$$\text{pr}(\mathbf{N} \mid \mathbf{p}, \theta) = \prod_{l=1}^{r} \binom{\theta}{N_l} p_l^{N_l}(1 - p_l)^{\theta - N_l}$$

and

$$\text{pr}(\mathbf{M} \mid \mathbf{N}, \mathbf{p}, \theta) = \prod_{l=1}^{r} \frac{\binom{\theta - S_{l+1}}{N_l - M_l}\binom{S_{l+1}}{M_l}}{\binom{\theta}{N_l}}.$$

TABLE 6.1 | *Capture-recapture counts for Gordy Lake sunfish[*]

i	1	2	3	4	5	6	7	8	9	10	11	12	13	14
N_i	10	27	17	7	1	5	6	15	9	18	16	5	7	19
M_i	0	0	0	0	0	0	2	1	5	5	4	2	2	3
S_{i+1}	10	37	54	61	62	67	71	85	89	102	114	117	122	138

*Source: Castledine (1981).

Expression (6.28) reduces to

$$\mathrm{pr}(\mathbf{N}, \mathbf{M} \mid \mathbf{p}, \theta) := C(\mathbf{N}, \mathbf{M}) \frac{\theta!}{(\theta - S_{r+1})!} \left[\prod_{l=1}^{r} p_l^{N_l} (1 - p_l)^{\theta - N_l} \right], \tag{6.29}$$

where

$$C(\mathbf{N}, \mathbf{M}) := \prod_{l=1}^{r} \frac{1}{(N_l - M_l)!} \binom{S_{l+1}}{M_l}.$$

Table 6.1 displays data collected on sunfish in Gordy Lake in Michigan. For an estimate of population size, the *maximum likelihood method* chooses the integer-valued θ that maximizes (6.29) with respect to θ and $p_1, \ldots, p_l$. Under the assumptions that $p_1 = \ldots = p_r = p$ and θ is an integer, numerical iteration on θ yields $\widehat{\theta} = 457$ and $\widehat{p} = 0.02532$. Although optimal in the sense of maximizing (6.29), these estimates have the limitations previously described. For a small or moderate number of surveys, an error analysis is not readily available to assess how well $\widehat{\theta}$ approximates the unknown population size θ. Also, $\widehat{\theta}$ and $\widehat{p}$ are functions of **N** and **M** exclusively, thus incorporating no additional information that may be available from other sources. For example, suppose previous unrelated studies of this population reveal that past capture rates have estimated average value A and modal value B.

Bayesian statistical methods overcome these limitations by incorporating this and like information into prior distributions on θ and **p** and then employing the corresponding induced joint posterior distribution on θ and **p** to estimate population size. In the present case, assume

v. Consistent with Assumption iv, $p_1, \ldots, p_r$ are i.i.d. random variables, each with prior distribution $\{\nu(p),\ 0 \le p \le 1\}$.

vi. The random variable θ has the prior distribution $\{ f_{\text{prior}}(z);\ z = S_{r+1}, S_{r+1} + 1, \ldots\}$.

vii. The random variables θ and $(p_1, \ldots, p_r)$ are independent.

If Q in (6.25) denotes the event $(\mathbf{p}, \theta)$ and R the event $(\mathbf{N}, \mathbf{M})$, then **p** and θ jointly have the posterior distribution

$$f_{\text{posterior}}(\mathbf{p}, \theta \mid \mathbf{N}, \mathbf{M}) = \frac{1}{\varphi} \mathrm{pr}(\mathbf{N}, \mathbf{M} \mid \mathbf{p}, \theta)\, f_{\text{prior}}(\theta) \prod_{l=1}^{r} \nu(p_l), \tag{6.30}$$

where

$$\varphi := \sum_{z=1}^{\infty} \varphi(z)\, f_{\text{prior}}(z)$$

and

$$\varphi(z) := \int_{\mathbf{p}\in[0,1]^r} \text{pr}(\mathbf{N}, \mathbf{M} \mid \mathbf{p}, z) \left[\prod_{l=S_{r+1}}^{r} \nu(p_l)\right] d\mathbf{p}.$$

Here φ corresponds to the normalizing $f(\mathbf{Z})$ in (6.26).

The objective is to approximate the posterior distribution of population size

$$f_{\text{posterior}}(\theta \mid \mathbf{N}, \mathbf{M}) = \int_{[0,1]^r} f_{\text{posterior}}(\mathbf{p}, \theta \mid \mathbf{N}, \mathbf{M}) d\mathbf{p}. \tag{6.31}$$

We do this by randomly generating samples of $\mathbf{p}$ and θ from (6.30) and then using the data to estimate (6.31). Here θ can be no less than S_{l+1}, the number of different animals marked in all r surveys. Also, θ is discrete but $p_1, \ldots, p_r$ are continuous. Although previous sections in this chapter have focused exclusively on discrete-state spaces, ignoring the distinction here has negligible consequences.

Suppose that $p_1, \ldots, p_r$ are each assumed to have the Beta prior distribution

$$\nu(p) = \begin{cases} \frac{\Gamma(a+b)}{\Gamma(a)\Gamma(b)} p^{a-1}(1-p)^{b-1} & a, b > 1 \qquad 0 \le p \le 1 \\ 0 & \text{elsewhere,} \end{cases} \tag{6.32}$$

denoted by $\mathcal{B}e(a, b)$, with mean $\frac{a}{a+b}$ and mode $\max\left(0, \frac{a-1}{a+b}\right)$. If, for example, previous studies on catch rates reveal an estimated mean $A = 0.03098$ and mode $B = 0.01509$, then the assignments

$$a = \frac{A}{A-B} = 1.95$$

$$b = \frac{1-A}{A-B} = 61$$

incorporate this information into the priors for $p_1, \ldots, p_r$. Independent samples generated from (6.32) for each $p_1, \ldots, p_r$ reflect differences in climatic conditions at the time of each survey.

Because no additional information is available about population size θ, consider the *noninformative prior distribution* $\{f_{\text{prior}}(\theta) = 1,\ \theta = S_{r+1}, S_{r+1}+1, \ldots\}$ (e.g., Berger 1985, Section 3.3), giving equal weight to each population size no less than S_{r+1}. Although this prior is an *improper distribution* in the sense that it has infinite probability mass, it nevertheless leads to a posterior distribution (6.31) that sums to unity with respect to $\mathbf{p}$ and θ and is thus a *proper probability distribution.*

To employ HM sampling, we first need to identify convenient nominating distributions. Note that (6.30) can be written in the form

$$f_{\text{posterior}}(\mathbf{p}, \theta \mid \mathbf{N}, \mathbf{M}) = \frac{S_{r+1}!}{\varphi\,[1 - w(\mathbf{p})]^{S_{r+1}+1}} \prod_{i=1}^{r} \left[\left(\frac{p_i}{1-p_i}\right)^{N_i}\right] \times \left[\prod_{i=1}^{r} \nu(p_i)\right] \times \eta(\theta - S_{r+1} \mid S_{r+1}, \mathbf{p}), \tag{6.33}$$

where (6.32) defines ν and

$$\eta(l \mid S_{r+1}, \mathbf{p}) := \binom{l + S_{r+1}}{l} [w(\mathbf{p})]^l \, [1 - w(\mathbf{p})]^{S_{r+1}+1} \qquad l = 0, 1, \ldots \tag{6.34}$$

is the probability mass function of the negative binomial distribution $\mathcal{NB}(S_{r+1} + 1, w(\mathbf{p}))$ with

$$w(\mathbf{p}) := \prod_{i=1}^{r} (1 - p_i).$$

Cyclically updating the $r + 1$ coordinates one at a time (according to the DCS rule) offers one approach. On iteration j, let $\mathbf{X}_j := (X_{1j}, \ldots, X_{rj}, X_{r+1,j})$ be an $(r + 1)$-dimensional random point where X_{ij} denotes the capture rate on survey i and $X_{r+1,j}$ denotes population size. Then MCMC provides several options for generating a sample path $\mathbf{X}_1, \ldots, \mathbf{X}_t$ and then estimating $\{f_{\text{posterior}}(\theta \mid \mathbf{N}, \mathbf{M})\}$ from $X_{r+1,1}, \ldots, X_{r+1,t}$. One HM option attempts to update coordinate $i := (j - 1) \pmod{r + 1} + 1$ on iteration j by sampling X_{ij} from

$$q(\mathbf{x}, \mathbf{x}_i(y_i)) = \begin{cases} \nu(y_i) & 0 \le y_i \le 1 \quad \text{if } i \ne r + 1 \\ \eta(y_i \mid S_{r+1}, w(\mathbf{x}_{-(r+1)})) & y_i = 0, 1, \ldots \quad \text{if } i = r + 1 \\ 0 & \text{elsewhere.} \end{cases}$$

The corresponding acceptance function is

$$\alpha(\mathbf{x}, \mathbf{x}_i(y_i)) = \begin{cases} \min\left\{1, \left[\frac{y_i(1-x_i)}{x_i(1-y_i)}\right]^{N_i} \frac{(1-y_i)^{y_{r+1}}}{(1-x_i)^{x_{r+1}}}\right\} & \text{if } 1 \le i \le r \\ 1 & \text{if } i = r + 1, \end{cases}$$

implying that updating population-size coordinate $r + 1$ occurs only once per $r + 1$ iterations, severely limiting the opportunity for updating on each of the t iterations.

Because catch rates are independent of population size (Assumption vii), a second sampling plan is possible. It allows size to be updated more frequently. On each iteration $j > 0$, it nominates new values for coordinate $r + 1$ and for coordinate $i = (j - 1)(\bmod r) + 1 \; (< r + 1)$ using

$$q_i(\mathbf{x}, \mathbf{x}_{i,r+1}(y_i, y_{r+1})) = \begin{cases} \eta(y_{r+1} \mid S_{r+1}, \mathbf{x}_{-(r+1)})\nu(y_i) & 0 \le y_i \le 1, \; y_{r+1} = 0, 1, \ldots \\ 0 & \text{elsewhere,} \end{cases}$$

where

$$\mathbf{x}_{i,r+1}(y_i, y_{r+1}) := (x_1, \ldots, x_{i-1}, y_i, x_{i+1}, \ldots, x_r, y_{r+1}).$$

The corresponding acceptance function is

$$\alpha(\mathbf{x}, \mathbf{x}_{i,r+1}(y_i, y_{r+1})) = \min[1, \zeta(\mathbf{x}, \mathbf{x}_{i,r+1}(y_i, y_{r+1})],$$

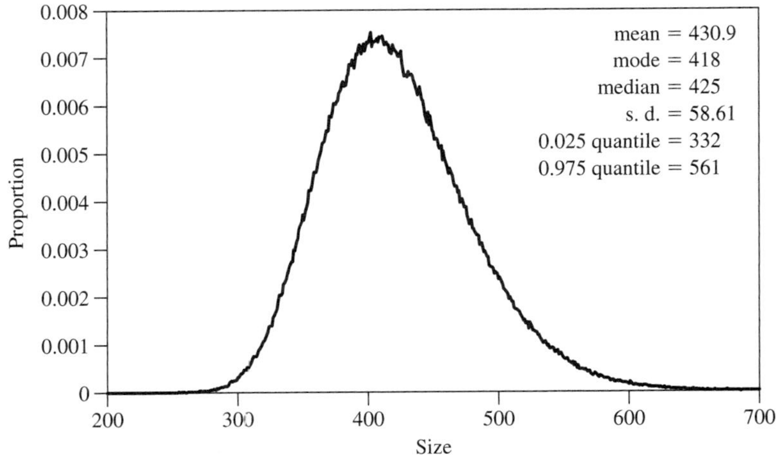

FIGURE 6.3 | *Sunfish population size posterior distribution: Hastings-Metropolis sampling, two coordinates updated per iteration* $t = 10^6$

where

$$\begin{aligned}\zeta(\mathbf{x}, \mathbf{x}_{i,r+1}(y_i, y_{r+1})) := & \left(\frac{y_i}{x_i}\right)^{N_i} \left(\frac{1-x_i}{1-y_i}\right)^{S_{r+1}+N_i} \left(\frac{1-x_i}{1-y_i}\right)^{x_{r+1}+y_{r+1}} \\ & \times \left[\frac{1-w(\mathbf{x})(1-y_i)/(1-x_i)}{1-w(\mathbf{x})}\right]^{S_{r+1}+1}.\end{aligned} \tag{6.35}$$

This option induces more frequent successful updating of population size (coordinate $r+1$) if its overall acceptance rate exceeds the $1/(r+1)$ acceptance rate of single-coordinate updating in the first option.

Figure 6.3 displays the histogram

$$J_l := \frac{\text{number of observations } X_{r+1,1}, \ldots, X_{r+1,t} \text{ equal to } l}{\text{number of observations } t} \qquad l = S_{r+1}, S_{r+1}+1, \ldots, \tag{6.36}$$

computed from a sample path generated by the second option for $a = 1.95$ and $b = 61$, the parameters of the prior distribution in (6.32), and $t = 10^6$. The results estimate a modal size (most likely) of 418, and a probability of 0.95 that size lies in the interval [332, 561]. Later, we compare these estimates with those for the Gibbs sampling approach to this problem in Example 6.6.1.

A third option for estimating population size simultaneously nominates all capture rates $Y_{1j}, \ldots, Y_{rj}$ and size $Y_{r+1,j}$ on iteration j and then tests for acceptance. The nominating distribution is

$$q(\mathbf{x}, \mathbf{y}) = \begin{cases} \eta(y_{r+1} \mid S_{r+1}, \mathbf{x}) \prod_{i=1}^{r} \nu(y_i) & 0 \le y_i \le 1, \quad i = 1, \ldots, r \\ & y_{r+1} = 0, 1, \ldots \\ 0 & \text{elsewhere.} \end{cases}$$

The resulting acceptance function is $\alpha(\mathbf{x}, \mathbf{y}) = \min[1, \zeta'(\mathbf{x}, \mathbf{y})]$, where

$$\zeta'(\mathbf{x}, \mathbf{y}) := \left[\frac{w(\mathbf{y})}{w(\mathbf{x})}\right]^{x_{r+1}+y_{r+1}-S_{r+1}} \left[\frac{1 - w(\mathbf{y})}{1 - w(\mathbf{x})}\right]^{S_{r+1}+1} \prod_{i=1}^{r} \left[\frac{y_i(1 - x_i)}{x_i(1 - y_i)}\right]^{N_i}. \tag{6.37}$$

Unfortunately, the inclusion of all $r + 1$ nominations in one test tends to make $\zeta'(\mathbf{x}, \mathbf{y})$ in (6.37) vary considerably more widely than $\zeta(\mathbf{x}, \mathbf{x}_{i,r+1}(y_i, y_{r+1}))$ in (6.35), and this results in a smaller overall acceptance rate. A sampling experiment verifies this assertion, implying that the simultaneous HM updating of all coordinates is impractical for this problem.

The posterior distribution in Figure 6.3 characterizes population size based on both the survey data $(\mathbf{N}, \mathbf{M})$ and on the specified prior distributions. Inevitably, the characterization varies with the choice of priors and changes in their parametric values. A more comprehensive study than that presented here would also explore the sensitivity of the characterization to these variations in prior specifications (e.g., see George and Robert 1992). ■

6.5 JOINT AND FULL CONDITIONAL DISTRIBUTIONS

The reader will have noticed that the second sampling option in Example 6.4.1 attempts to update two coordinates per iteration, and the third option attempts to update all $m = r + 1$ coordinates per iteration. If $\mathbf{x}$ and $\mathbf{y}$ are in $\mathcal{X}$, then it can be shown (e.g., Besag 1974) that

$$\frac{\pi(\mathbf{y})}{\pi(\mathbf{x})} = \prod_{i=1}^{m} \frac{\pi_i(y_i \mid x_1, \ldots, x_{i-1}, y_{i+1}, \ldots, y_m)}{\pi_i(x_i \mid x_1, \ldots, x_{i-1}, y_{i+1}, \ldots, y_m)}, \tag{6.38}$$

where $\{\pi_i(\cdot \mid \cdot)\}$ denotes the full conditional distribution of coordinate i. This expression reveals that it suffices to know only the ratios of the full conditional distributions in order to evaluate the acceptance probability function $\{\alpha(\mathbf{x}, \mathbf{y}), \mathbf{x}, \mathbf{y} \in \mathcal{X}\}$, thus further broadening the scope of application of HM sampling.[3]

Positivity Condition. It is easily seen that (6.38) holds when $\mathcal{X}$ is the product space $\mathcal{X} = \mathcal{X}_1 \times \cdots \times \mathcal{X}_m$, or equivalently π satisfies the *positivity condition:*

For each $i = 1, \ldots, m$ and $x_i \in \mathcal{X}_i$, $\pi_i(x_i) > 0$ so that
$\pi(x_1, \ldots, x_m) > 0$ for all $\mathbf{x} := (x_1, \ldots, x_m) \in \mathcal{X}_1 \times \cdots \times \mathcal{X}_m$.

If, as is often the case, $\mathcal{X}$ is not a product space, verifying (6.38) can become a computational burden. Sometimes a transformation of variables converts the original space into a product space, but that is more the exception than the rule. Of equal

[3]Section 9.3.1 contains a more detailed description of full conditional distributions in the setting of Markov random fields and Gibbs distributions.

concern, using (6.38) to evaluate the ratio may lead to excessive computing time. For example, if evaluating $\pi_i(x_i \mid \mathbf{x}_{-i})$ takes time proportional to m, then evaluation of the ratio takes time proportional to m^2.

6.6 GIBBS SAMPLING

The availability of the full conditional distributions also provides an alternative to HM sampling for randomly generating a sample path. Called *Gibbs sampling*, this technique generates a coordinate value directly from its full conditional distribution. In particular, consider the one-step transition kernel

$$\begin{aligned} K(\mathbf{x}, \mathbf{x}_i(y)) &:= \pi_i(y \mid \mathbf{x}_{-i}) \\ &= \text{probability of moving from } \mathbf{x} \text{ to } \mathbf{x}_i(y). \end{aligned} \tag{6.39}$$

Because it is easily seen that

$$\sum_{x_i \in \mathcal{X}_i(\mathbf{x}_{-i})} \pi(\mathbf{x}) K(\mathbf{x}, \mathbf{x}_i(y)) = \pi(\mathbf{x}_i(y)),$$

global balance holds. If $\mathbf{X}_0$ has distribution π, Algorithm OGIBBS randomly generates a sample path $\mathbf{X}_1, \ldots, \mathbf{X}_t$ on which each $\mathbf{X}_j$ has the distribution π.

ALGORITHM OGIBBS

Purpose: To generate a sample path $\mathbf{X}_1, \ldots, \mathbf{X}_t$ using Gibbs sampling with the DCS rule.

Given: Initial state $\mathbf{X}_0$, sample-path length t where $t/m =$ integer.

Method:

$j = 1$.
$i = 1$.
Until $j > t$:
  $i = (j-1) \pmod m + 1$
  randomly sample Y_i from
    $\{\pi_i(y \mid Y_1, \ldots, Y_{i-1}, X_{i+1,j-1}, \ldots, X_{m,j-1})\}$.
$\mathbf{X}_j = \mathbf{X}_{i,j-1}(Y_i)$
$j = j + 1$. ■

This form of Gibbs sampling repeatedly cycles through the m coordinates in a fixed order. Algorithm RGIBBS provides an alternative procedure based on the RCS rule. It randomly selects one coordinate on each step and randomly updates its value. For simplicity of exposition, all coordinates are equally likely to be selected on an iteration.

ALGORITHM **RGIBBS**

Purpose: To generate a sample path $\mathbf{X}_1, \ldots, \mathbf{X}_t$ using Gibbs sampling with the RCS rule.

Given: Initial state $\mathbf{X}_0$, sample-path length t.

Method:

$j = 1$.
Until $j > t$:
  randomly sample U from $\mathcal{U}(0, 1)$
  I = largest integer less than or equal to $mU + 0.50$
  randomly sample Y_I from $\{\pi_I(y \mid \mathbf{X}_{-I,j-1}), y \in \mathcal{X}_I\}$
  $\mathbf{X}_j = \mathbf{X}_{j-1}(Y_I)$
  $j = j + 1$.

■

The corresponding one-step transition kernel takes the form

$$K(\mathbf{x}, \mathbf{y}) = \frac{1}{m}\sum_{i=1}^{m} \pi_i(y_i \mid \mathbf{x}_i)\delta(\mathbf{y} - \mathbf{x}_i(y_i)), \tag{6.40}$$

where

$$\delta(\mathbf{z}) := \begin{cases} 1 & \text{if } \mathbf{z} = \mathbf{0} \\ 0 & \text{elsewhere.} \end{cases}$$

Again, global balance holds (see Exercise 1). Although K in (6.39) for Algorithm OGIBBS does not satisfy detailed balanced, K in (6.40) for Algorithm RGIBBS does.

Given Algorithm RGIBBS, the reader may rightly ask why one would employ the more indirect Algorithm SCHM (see Section 6.3.1) with the RCS rule and use (6.17) as the nominating kernel. In particular, if one takes $q(\mathbf{x}, \mathbf{x}_i(y)) = \pi_i(y \mid \mathbf{x}_{-i})$ in place of (6.17) in Algorithm SCHM, $\alpha(\mathbf{x}, \mathbf{x}_i(y)) = 1$ for all $\mathbf{x}$ and $\mathbf{x}_i(y)$ in $\mathcal{X}$ so that random-coordinate Gibbs sampling turns out to be a special case of HM sampling. A partial answer to this question comes in Section 6.8, which shows that for sample-path length t, a sample average based on Algorithm SCHM has variance no greater than the variance for the corresponding sample average based on Algorithm RGIBBS, as t becomes large.

While overcoming the limitation of not knowing π or π/φ, updating one coordinate at a time can induce substantial dependence among $\mathbf{X}_1, \ldots, \mathbf{X}_t$ along a sample path, especially as the number of dimensions m increases. In turn, this dependence frequently increases the sample-path length t needed to approximate λ by $\widehat{\lambda}_t$ to within a specified accuracy. This reality encourages a user to consider algorithms that update more than one coordinate at a time. Section 6.10.1 returns to this subject for HM sampling.

EXAMPLE 6.6.1

CAPTURE-RECAPTURE WITH GIBBS SAMPLING

Recall the population-estimating problem in Example 6.4.1, where the posterior distribution (6.30) gives

$$f_{\text{posterior}}(\theta \mid \mathbf{p}, \mathbf{N}, \mathbf{M}) = \frac{f_{\text{posterior}}(\mathbf{p}, \theta \mid \mathbf{N}, \mathbf{M})}{\sum_{l=S_{r+1}}^{\infty} f_{\text{posterior}}(\mathbf{p}, l \mid \mathbf{N}, \mathbf{M})} \tag{6.41}$$

and

$$f_{\text{posterior}}(\mathbf{p} \mid \theta, \mathbf{N}, \mathbf{M}) = \frac{f_{\text{posterior}}(\mathbf{p}, \theta \mid \mathbf{N}, \mathbf{M})}{\int_{[0,1]^r} f_{\text{posterior}}(\zeta, \theta \mid \mathbf{N}, \mathbf{M}) d\zeta}, \tag{6.42}$$

with $\zeta := (\zeta_1, \ldots, \zeta_r)$. For the non-informative prior $\{f_{\text{prior}}(l) = 1; l = S_{r+1}, S_{r+1} + 1 \ldots\}$ for size, the full conditional distribution for θ is the negative binomial distribution (see Exercise 2)

$$f_{\text{posterior}}(\theta \mid \mathbf{p}, \mathbf{N}, \mathbf{M}) = \begin{cases} \eta(\theta - S_{r+1} \mid S_{r+1}, \mathbf{p}) & \theta = S_{r+1}, S_{r+1} + 1 \ldots \\ 0 & \text{elsewhere,} \end{cases} \tag{6.43}$$

where (6.34) defines η. For the Beta prior $\mathcal{B}e(a, b)$ in (6.32), the full conditional distribution for each p_i is $\mathcal{B}e(N_i + a, \theta - N_i + b)$ (see Exercise 3).

Although Gibbs sampling offers the option of cyclically updating the $r + 1$ coordinates one at a time, this would again restrict the number of updated θ samples to $t/(r + 1)$. Because of the structure of the full conditional distributions, we can again sequentially update size θ and a different capture rate p_i on each of r iterations in a cycle, thus generating t size samples. On iteration j, one

1. Randomly samples Z from $\mathcal{NB}(S_{r+1} + 1, \mathbf{X}_{-(r+1), j-1})$.
2. Computes $X_{r+1,j} = S_{r+1} + Z$.
3. For $i := (j - 1) \pmod{r} + 1$, randomly samples X_{ij} from $\mathcal{B}e(N_i + a, X_{r+1,j-1} - N_i + b)$.
4. Increments j by 1.

Figure 6.4 shows the histogram based on a sample path $X_{r+1,1}, \ldots, X_{r+1,t}$ generated by this form of Gibbs sampling. A comparison with Figure 6.3 shows that the two different methods of sample generation produce similar histograms. Example 6.11.1 returns to Gibbs sampling for this problem where all coordinates are updated on each iteration. ■

6.7 CONVERGENCE FOR $\mathbf{X}_j$

Recall that we began our discussion of MCMC sampling by assuming the availability of an $\mathbf{X}_0$ randomly sampled from π. Because computational intractability often prevents us from generating samples in this way, common practice samples $\mathbf{X}_0$ from a conveniently chosen, and computationally tractable, distribution π_0 on some subset $\mathcal{X}_0 \subseteq \mathcal{X}$ and thereafter samples $\mathbf{X}_j$ from the chosen

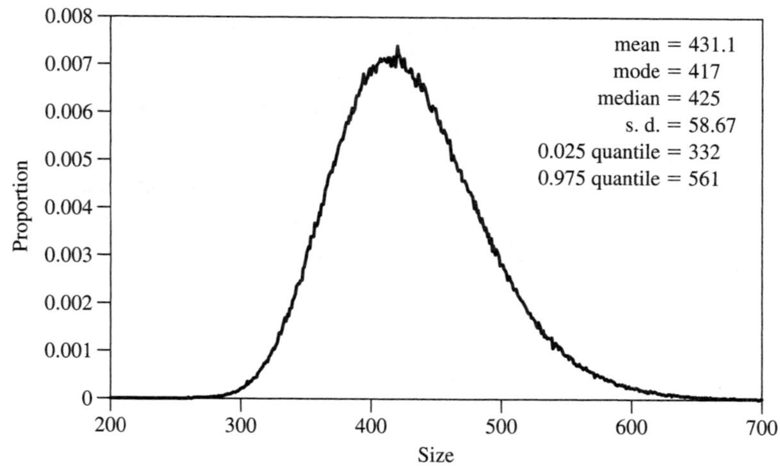

FIGURE 6.4 | *Sunfish population size posterior distribution: Gibbs sampling, two coordinates updated per iteration $t = 10^6$*

$\{K(\mathbf{X}_{j-1}, \mathbf{y}),\ \mathbf{y} \in \mathcal{X}\}$ for $j \geq 1$. For example, one approach arbitrarily assigns $\mathbf{X}_0 = \mathbf{x}_0$, which is equivalent to sampling from a *degenerate distribution* with all its mass on this point.

Because for $\pi_0 \neq \pi$,

$$\sum_{\mathbf{x} \in \mathcal{X}} \pi_0(\mathbf{x}) K(\mathbf{x}, \mathbf{y}) \neq \pi(\mathbf{y}), \tag{6.44}$$

global balance (6.1) is not satisfied, and $\mathbf{X}_1, \mathbf{X}_2, \ldots$ are not from π. Nevertheless, it is possible to restrict K to a class of transition kernels endowed with properties that ensure that $\mathbf{X}_j$ on iteration j has a distribution that converges to π with probability 1 as $j \to \infty$. This implies that for sufficiently large j, $\mathbf{X}_j$ can be regarded as having the distribution π with negligible error. These properties are irreducibility, positive recurrence, and aperiodicity. This section describes all three properties and shows how they lead to the desired convergence.

We define the *j-step transition kernel*

$$\begin{aligned} K^{(j)}(\mathbf{x}, \mathbf{y}) &:= \begin{cases} K(\mathbf{x}, \mathbf{y}) & \text{if } j = 1 \\ \sum_{\mathbf{z} \in \mathcal{X}} K^{(j-1)}(\mathbf{x}, \mathbf{z}) K(\mathbf{z}, \mathbf{y}) & \text{if} j > 1 \end{cases} \qquad \forall (\mathbf{x}, \mathbf{y}) \in \mathcal{X}^2 \\ &= \text{probability of being in state } \mathbf{y} \text{ on step } j + l \text{ given that} \\ &\quad\ \text{the sample path begins in state } \mathbf{x} \text{ on step } l \text{ for } l > 0. \end{aligned} \tag{6.45}$$

Irreducibility. We say that a Markov chain is *irreducible* if there exists an integer $j \geq 1$ such that

$$K^{(j)}(\mathbf{x}, \mathbf{y}) > 0 \qquad \forall (\mathbf{x}, \mathbf{y}) \in \mathcal{X}^2. \tag{6.46}$$

This property ensures a positive probability of moving from any state in $\mathcal{X}$ to any other state in $\mathcal{X}$ in a finite number of steps.

Positive Recurrence. The property of recurrence characterizes the frequency with which a chain returns to individual states. Let

$$T_{\mathbf{x}} := \text{time between successive visits to state } \mathbf{x}$$

and

$$N_{\mathbf{x}j} := \text{number of visits to state } \mathbf{x} \text{ in } j \text{ successive steps.}$$

We say that state $\mathbf{x}$ is *recurrent* if

$$\text{pr}(T_{\mathbf{x}} < \infty \mid \mathbf{X}_0 = \mathbf{x}) = 1. \tag{6.47}$$

That is, the time between successive returns to state $\mathbf{x}$ is finite with probability 1. It can be shown that (6.47) is equivalent to the statement

$$\lim_{j\to\infty} \text{E}(N_{\mathbf{x}j} \mid \mathbf{X}_0 = \mathbf{x}) = \infty, \tag{6.48}$$

implying that returns to state $\mathbf{x}$ occur infinitely often.

For present purposes, we need a stronger property. If

$$\text{E}(T_{\mathbf{x}} \mid \mathbf{X}_0 = \mathbf{x}) < \infty, \tag{6.49}$$

we say that state $\mathbf{x}$ is *positive recurrent*. It can be shown that (6.49) implies

$$\lim_{j\to\infty} \frac{1}{j+1} \text{E}(N_{\mathbf{x}j} \mid \mathbf{X}_0 = \mathbf{x}) > 0, \tag{6.50}$$

so that on average the number of returns to $\mathbf{x}$ must grow at least linearly with i (e.g., Kulkarni 1995, Theorem 3.4).

If every state is positive recurrent, we say that the chain is positive recurrent. This implies that with probability 1 a sample path returns infinitely often to each state as its path length, t, grows without bound. An equivalent statement is that the mean time between successive returns to each state is finite.

Our first use of these properties addresses the inequality in (6.44).

Theorem 6.7.1

If K is irreducible and satisfies global balance with respect to π, then the chain is positive recurrent, and π is the only solution that satisfies

$$\sum_{\mathbf{x}\in\mathcal{X}} \pi(\mathbf{x})K(\mathbf{x},\mathbf{y}) = \pi(\mathbf{y}) \qquad \forall(\mathbf{x},\mathbf{y}) \in \mathcal{X}^2. \tag{6.51}$$

■

That is, if for example $\mathcal{X}$ is countably finite and irreducible, then the chain is also positive recurrent, and π is the only probability distribution that satisfies global balance. If $\varphi > 1$ in (6.2), then π is unique up to the normalizing constant φ.

Aperiodicity. Because sampling the initial state $\mathbf{X}_0$ from π is often computationally intractable, common practice samples $\mathbf{X}_0$ from a distribution π_0 with low unit-generating cost. As a consequence, $\mathbf{X}_1$ does not have distribution π, nor do

$\mathbf{X}_2, \ldots, \mathbf{X}_t$ for finite t. Collectively, global balance, irreducibility, and aperiodicity provide the wherewithal for overcoming this limitation.

A Markov chain is periodic if at least one of its states can be entered only at regularly spaced steps. Otherwise, it is aperiodic. Formally, we say that a chain is aperiodic if the *greatest common divisor* (gcd) of the integers $\{l > 0 : K^{(l)}(\mathbf{x}, \mathbf{x}) > 0\}$ is unity for at least one $\mathbf{x} \in \mathcal{X}$. Note that the property $K(\mathbf{x}, \mathbf{x}) > 0$ for some $\mathbf{x} \in \mathcal{X}$ suffices to ensure aperiodicity.

If a Markov chain $\{\mathbf{X}_j, j \geq 0\}$ has the properties of irreducibility, positive recurrence, and aperiodicity, we say that it is *ergodic*. This property has important implications.

Theorem 6.7.2

Let $\mathcal{F}_0$ denote the collection of all possible initial distributions π_0 on $\mathcal{X}$. If the Markov chain is ergodic, then for any initial distribution $\pi_0 \in \mathcal{F}_0$ and all $\mathbf{y} \in \mathcal{X}$,

$$\sum_{\mathbf{x} \in \mathcal{X}} \pi_0(\mathbf{x}) K^{(j)}(\mathbf{x}, \mathbf{y}) \to \pi(\mathbf{y}) \qquad \text{as } j \to \infty. \tag{6.52}$$

In particular, for the degenerate distribution

$$\pi_0(\mathbf{z}) = \begin{cases} 1 & \text{if } \mathbf{z} = \mathbf{x}_0 \in \mathcal{X} \\ 0 & \text{elsewhere,} \end{cases}$$

$$K^{(j)}(\mathbf{x}, \mathbf{y}) \to \pi(\mathbf{y}) \qquad \text{as } j \to \infty. \tag{6.53}$$

■

Sometimes we equivalently refer to the irreducibility, positive recurrence, and aperiodicity of the transition kernel K and say that it is ergodic.

The significance of Theorem 6.7.2 cannot be overestimated. It assures us that, provided $\mathbf{X}_0$ is assigned a value $\mathbf{x}_0$ in $\mathcal{X}$, the distribution of $\mathbf{X}_j$ converges to π as $j \to \infty$. By contrast, if a chain were periodic rather than aperiodic, then for each $\mathbf{x} \in \mathcal{X}$ and for each $\pi_0 \in \mathcal{F}_0$, $\{K^{(j)}(\mathbf{x}, \mathbf{y}),\ \mathbf{y} \in \mathcal{X}\}$ may converge to a different distribution, a particularly undesirable result.

When (6.52) holds, we say that in addition to being the unique invariant distribution, π is the *equilibrium distribution* for the Markov chain. Note that

$$\left.\begin{array}{l} \left.\begin{array}{l} \text{Detailed} \Longrightarrow \text{Global} \\ \text{balance} \qquad\quad \text{balance} \\ \qquad\qquad\quad\ \text{Irreducibility} \end{array}\right\} \Longrightarrow \begin{array}{l}\text{Positive} \\ \text{recurrence}\end{array} \Longrightarrow \begin{array}{l}\text{Unique} \\ \text{invariant} \\ \text{distribution}\end{array} \\ \qquad\qquad\qquad\qquad\qquad\qquad\qquad\qquad\quad \text{Aperiodicity} \end{array}\right\} \Longrightarrow \text{Ergodicity.}$$

Whenever possible, establishing ergodicity should precede the use of MCMC sampling. Satisfying either of two sufficient conditions ensures the desired property for finite-state, countably infinite-state, and continuous-state spaces.

The conditions are as follows:

a. If $K(\mathbf{x}, \mathbf{y}) > 0$ for all $(\mathbf{x}, \mathbf{y}) \in \mathcal{X}^2$, and it obeys global balance, the chain is ergodic.

b. For single-coordinate updating, for each $i = 1, \ldots, m$, if $K(\mathbf{x}, \mathbf{x}_i(y)) > 0$ for all $\mathbf{x}_i(y) \in \mathcal{X}$ and $\mathbf{x} \in \mathcal{X}$, and it obeys global balance, the chain is ergodic.

When neither condition holds, other conditions specialized to particular MCMC sampling plans may hold (see Section 6.12).

For HM sampling, as in Algorithm HM, to induce an ergodic Markov chain on a discrete-state space, it suffices for the nominating kernel q to be positive everywhere on $\mathcal{X}^2$. Sections 6.10.1 and 6.10.2 describe two kernels that have this property. For Gibbs sampling to induce an ergodic Markov chain using Algorithm OGIBBS or RGIBBS, it suffices for $\mathcal{X}$ to be a product space so that the transition kernels (6.39) and (6.40) are everywhere positive.

EXAMPLE 6.7.1 INITIAL CONDITIONS FOR CAPTURE-RECAPTURE

Recall the capture-recapture Examples 6.4.1 and 6.6.1, where a sample path $X_{r+1,1}, \ldots, X_{r+1,t}$ provides MCMC data for estimating the posterior distribution of sunfish population size for Gordy Lake. Although we cannot conveniently monitor $K^{(j)}(\mathbf{x}, \mathbf{y}) - \pi(\mathbf{y})$ for $j \geq 1$ for each $\mathbf{y}$ in this problem, we can observe $\{X_{r+1,j}, j \geq 1\}$ and, in particular, its behavior in response to starting MCMC in different states. Figure 6.5a shows these size data for the first 100 iterations for HM sampling that updates a capture rate and size on each iteration. The three paths

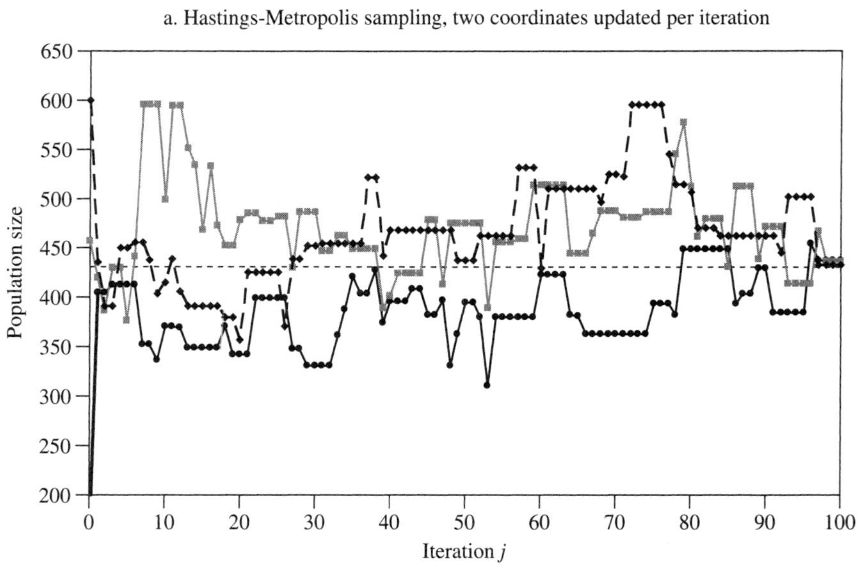

FIGURE 6.5 | *Initial sample paths for sunfish population size*

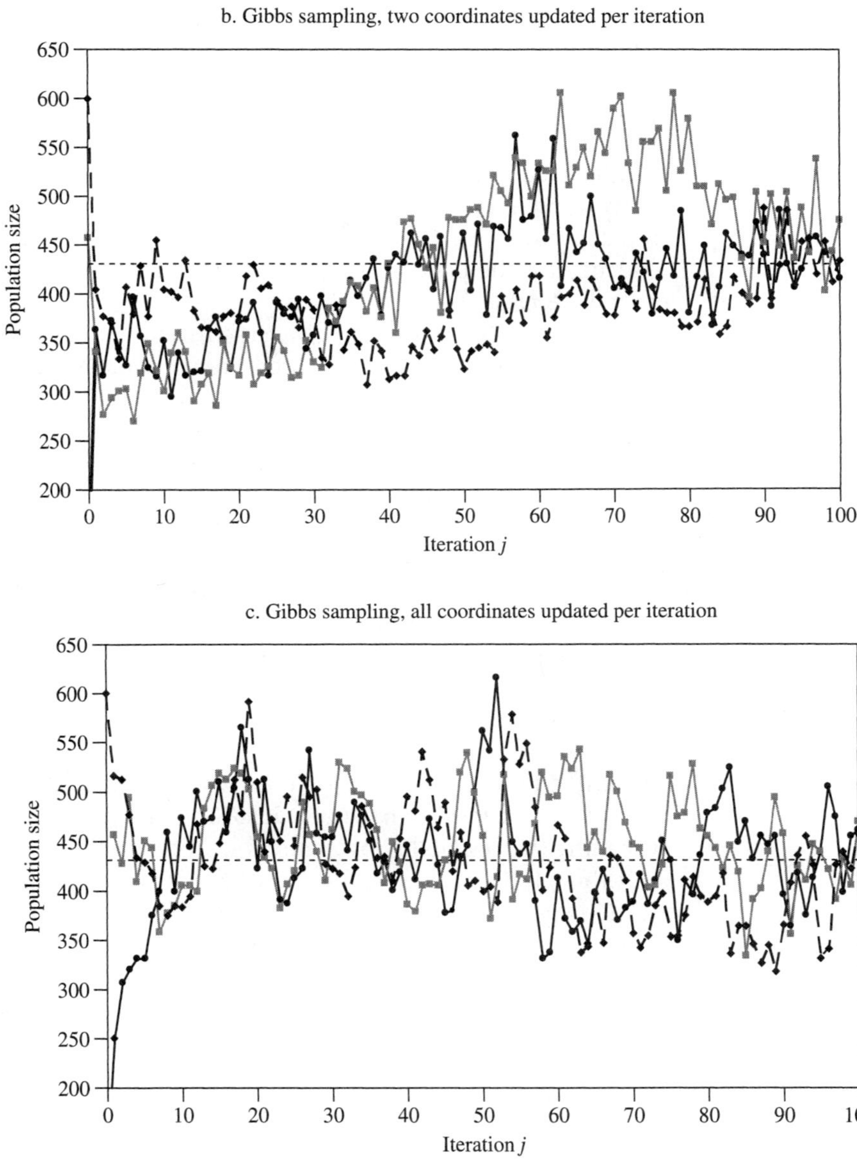

FIGURE 6.5 | *Continued.*

correspond to initial sizes $X_{r+1,0} = 138$ (black line), the minimal size dictated by the data in Table 6.1, $X_{r+1,0} = 457$ (gray line), the maximum likelihood estimator of size θ, and $X_{r+1,0} = 600$ (dashed line), a value thought to exceed population size. The long-run sample average 431.1 (dotted line) for a sample-path length of $t = 10^6$ provides a convenient benchmark for judging when the influence of initial conditions wears off. The horizontal lines on each path are characteristic of HM sampling where rejection on iteration j implies $X_{r+1,j} = X_{r+1,j-1}$.

Suppose that a graphical comparison of sample paths begun in widely different states reveals a positive integer k such that for iterations $j \geq k$, the paths converge to a common region and repeatedly fluctuate about some constant (but unknown) level. This behavior supports the assertion that for $j \geq k$, the sample paths are relatively free of the influence of initial conditions and, therefore, can be regarded as stationary with negligible error. Thus $k - 1$ is an adequate *warm-up interval*. In the present case, the long-run average serves as a convenient reference. In practice, this is not available at the beginning of an MCMC. Nevertheless, a tendency for the paths to intersect and to fluctuate around a constant level is often discernible.

Perusal of Figure 6.5a for HM sampling shows no intersecting behavior of this kind within the first 100 iterations of the sample paths. Figure 6.5b displays corresponding graphs for Gibbs sampling with the same two coordinates being updated as in the HM sample paths. Although one may argue that the three paths intersect for $j \geq k = 90$, longer sample paths would be needed to support an assertion that they are in equilibrium.

Recall that the structure of the capture-recapture problem also allows Gibbs sampling to update all capture rates and sizes on a single iteration. Figure 6.5c displays the sample paths that this alternative option generates for the same three initial states. Most notably, the paths now exhibit fluctuations around the long-run sample average by iteration 10, suggesting that this alternative induces convergence much faster than the other two sampling methods. However, if one assumes that variates all have comparable unit generating costs, this sample path with its $r + 1 = 15$ updates per iteration requires about $15/2 = 7.5$ as much computing time per iteration as do each of the other methods. This observation makes clear the need to balance statistical properties and computing cost when evaluating the desirability of a sampling procedure.

Although these graphs illustrate what happens at the beginning of different MCMC experiments, they alone do not provide an adequate basis for determining a warm-up interval $k - 1$. Most notably, the long-run average is not available as a reference point. Example 7.3.1 describes a considerably more comprehensive warm-up analysis that relies on averages over the independent replications to make inferences about the warm-up interval. ■

6.7.1 Nonconnected State Spaces and Mixtures

Suppose that π has the form

$$\pi(\mathbf{x}) = \alpha\pi_A(\mathbf{x})I_{\mathcal{X}_A}(\mathbf{x}) + (1-\alpha)\pi_B(\mathbf{x})I_{\mathcal{X}_B}(\mathbf{x}) \qquad \mathbf{x} \in \mathcal{X},$$

where

$$I_{\mathcal{R}}(x) = \begin{cases} 1 & \text{if } x \in \mathcal{R} \\ 0 & \text{elsewhere,} \end{cases}$$

$\mathcal{X}_A$ and $\mathcal{X}_B$ denote exclusive (i.e., $\mathcal{X}_A \cap \mathcal{X}_B = \emptyset$) and exhaustive (i.e., $\mathcal{X}_A \cup \mathcal{X}_B = \mathcal{X}$) subsets of $\mathcal{X}$, and $\alpha \in (0, 1)$. That is, $\mathcal{X}$, the support of π, has two nonconnected exhaustive regions $\mathcal{X}_A$ and $\mathcal{X}_B$. Then the choice of a nominating kernel for HM sampling needs to allow for transitions from each region to the other. That is, for some $(\mathbf{x}_1, \mathbf{y}_1) \in \mathcal{X}_A \times \mathcal{X}_B$ and $(\mathbf{x}_2, \mathbf{y}_2) \in \mathcal{X}_A \times \mathcal{X}_B$, $q(\mathbf{x}_1, \mathbf{y}_1)$ and $q(\mathbf{x}_2, \mathbf{y}_2)$ must be positive. Otherwise, the induced Markov chain is said to be *reducible*, and ergodicity does not hold.

Because Gibbs sampling relies on the full conditional distributions (6.16), it poses a more severe problem for nonconnected regions. If the objective is to estimate

$$\lambda = \alpha \sum_{\mathbf{x} \in \mathcal{X}_A} h(\mathbf{x})\pi_A(\mathbf{x}) + (1 - \alpha) \sum_{\mathbf{x} \in \mathcal{X}_B} h(\mathbf{x})\pi_B(\mathbf{x}),$$

one can, in principle, overcome the limitation of nonconnectedness by partitioning the problem into two independent MCMC sampling experiments, one for each region. However, identifying suitable partitions for more than one dimension can be a challenge.

6.8 CONVERGENCE FOR $\widehat{\lambda}_t$

Although ergodicity ensures that $\mathbf{X}_j$ converges in distribution to π as $j \to \infty$, there remains the important issue of how well $\widehat{\lambda}_t$ in (6.4) approximates λ. Three theorems based on the properties of h, π, and K help us answer this question. The first relates to the statistical property of consistency.[4]

Theorem 6.8.1

If the Markov chain is ergodic with equilibrium distribution π, then for any initial $\pi_0 \in \mathcal{F}_0$ and $\sum_{\mathbf{x} \in \mathcal{X}} \pi(\mathbf{x})|h(\mathbf{x})| < \infty$,

$$\widehat{\lambda}_t \to \lambda \text{ w.p.1} \qquad \text{as } t \to \infty. \tag{6.54}$$

■

The second theorem provides a distribution theory for characterizing the error of approximation. It relies on the concept of geometric ergodicity.

Geometric Ergodicity. If there exist a positive function $\{M(\mathbf{x}), \mathbf{x} \in \mathcal{X}\}$ such that

$$\sum_{\mathbf{x} \in \mathcal{X}} M(\mathbf{x})\pi(\mathbf{x}) < \infty$$

and a constant $\rho \in [0, 1)$ such that for all $\mathbf{x} \in \mathcal{X}$ and $\mathbf{y} \in \mathcal{X}$

$$\left|K^{(j)}(\mathbf{x}, \mathbf{y}) - \pi(\mathbf{y})\right| \le M(\mathbf{x})\rho^j \qquad \forall j \ge 1, \tag{6.55}$$

we say that the Markov chain is *geometrically ergodic.*

[4]For a discussion of the proofs of Theorems 6.8.1, 6.8.2, and 6.8.3, see Tierney (1994, p. 1717).

Theorem 6.8.2

If a Markov chain is geometrically ergodic with equilibrium distribution π, and h is finite, there exists a positive constant σ_∞ such that for any initial $\pi_0 \in \mathcal{F}_0$,

$$\sqrt{t}(\widehat{\lambda}_t - \lambda) \xrightarrow{\text{d}} \mathcal{N}\left(0, \sigma_\infty^2\right) \qquad \text{as } t \to \infty. \tag{6.56}$$

■

The third theorem also provides a distribution theory when uniform ergodicity holds.

Uniform Ergodicity. If there exist a finite upper bound M^* on $\{M(\mathbf{x})\}$ and a $\rho \in [0, 1)$ such that for all $\mathbf{x} \in \mathcal{X}$ and $\mathbf{y} \in \mathcal{X}$

$$\left|K^{(j)}(\mathbf{x}, \mathbf{y}) - \pi(\mathbf{y})\right| \le M^* \rho^j \qquad \forall j \ge 1, \tag{6.57}$$

we say that the Markov chain is *uniformly ergodic*. Note that

uniform ergodicity $\Rightarrow$ geometric ergodicity $\Rightarrow$ ergodicity.

Theorem 6.8.3

If the Markov chain is uniformly ergodic with equilibrium distribution π, then for any initial $\pi_0 \in \mathcal{F}_0$, (6.56) holds for all h satisfying $\sum_{\mathbf{x} \in \mathcal{X}} h^2(\mathbf{x})\pi(\mathbf{x}) < \infty$. ■

Theorem 6.8.1 unequivocally establishes the necessity for ergodicity. Without it, one cannot expect $\widehat{\lambda}_t$ to tend to λ as the sample-path length t increases. Moreover, geometric or uniform ergodicity, when it holds, allows us to characterize the behavior of $\widehat{\lambda}_t$ as the sample-path length increases. When geometric ergodicity holds, Theorem 6.8.2 facilitates error assessment by providing an asymptotic distribution theory for bounded functions h. When the stronger property of uniform ergodicity holds, Theorem 6.8.3 provides the corresponding theory for the wider class of mean-square integrable functions h.

Knowing that geometric (uniform) ergodicity holds justifies Theorem 6.8.2 (see Theorem 6.8.3) as the basis for error assessment, whereas knowing that the property does not hold militates against reliance on the theorem. As a first illustration, a finite-state ergodic chain whose one-step transition kernel is *reversible* is uniformly ergodic. Section 6.2 describes the details. As a second illustration, an ergodic Markov chain with a uniform equilibrium distribution on a continuous- and bounded-state space $\mathcal{X}$ whose boundary is sufficiently smooth is uniformly ergodic (Roberts and Rosenthal 1997). As Section 6.13 indicates, not all Markov chains are geometrically or uniformly ergodic. That section describes a considerably more problematic property called polynomial convergence.

6.8.1 Local and Global Moves

When geometric or uniform ergodicity holds, it is of interest to know how the *convergence factor*, ρ, behaves as a function of the shapes of π and, for each $\mathbf{x} \in \mathcal{X}$, $K(\mathbf{x}, \cdot) := \{K(\mathbf{x}, \mathbf{y}),\ \mathbf{y} \in \mathcal{X}\}$. We say that π is *multimodal* if it has more than one maximum on $\mathcal{X}$. With regard to $K(\mathbf{x}, \cdot)$, we distinguish between those kernels that allow only *local moves* in the neighborhood of $\mathbf{x}$ and those that allow more *global moves*.

If a multimodal π concentrates substantial probability mass in the neighborhoods around its peaks, and $K(\mathbf{x}, \cdot)$ allows only local moves, then convergence tends to be slow (ρ is close to 1). Moreover, the more the masses for π and $K(\mathbf{x}, \cdot)$ are concentrated, the slower is convergence (ρ tends toward 1). Although a $K(\mathbf{x}, \cdot)$ that admits global moves may lead to faster convergence, sampling from it is often more costly, especially if $\mathcal{X}$ is a nonproduct space (see Section 2.7). Then each sample, $\mathbf{X}_j$, must satisfy the constraints that define $\mathcal{X}$. Section 6.10 returns to the choice of K.

6.8.2 Problem Size

The convergence factor, ρ, can also be characterized in terms of problem size, defined as the number of coordinates m. Most notably, ρ tends to increase toward 1 as m increases, thereby slowing convergence to equilibrium. We say that the chain is *rapid mixing* if ρ converges polynomially fast in m. The chain is *slow mixing* if it converges exponentially fast or faster in m. Section 9.2 discusses this topic in more detail and extends the definition of size. Section 9.5 describes how a particular class of models is affected by problem size.

6.9 VARIANCE OF ESTIMATE

Although Theorems 6.8.2 and 6.8.3 provide an asymptotic distribution theory for error assessment, the sample-path length t inevitably is finite in practice. This raises at least two procedural issues:

- How does an MCMC user adjust $\widehat{\lambda}_t$ to account for the influence of the initial state $\mathbf{X}_0$ generated from an arbitrary distribution π_0?
- How does an MCMC user estimate σ_∞^2?

Chapter 7 answers these questions. Here we focus on error assessment as measured by

$$\sigma_t^2 := \mathrm{var}\widehat{\lambda}_t, \tag{6.58}$$

where

$$\begin{aligned}\mathrm{var}\widehat{\lambda}_t &:= \mathrm{E}(\widehat{\lambda}_t - \lambda)^2 \\ &= \frac{1}{t^2}\sum_{j=1}^{t}\sum_{l=1}^{t} \mathrm{cov}[h(\mathbf{X}_j), h(\mathbf{X}_l)]\end{aligned}$$

and where

$$\text{cov}[h(\mathbf{X}_j), h(\mathbf{X}_l)] := \text{E}\{[h(\mathbf{X}_j) - \lambda]\,[h(\mathbf{X}_l) - \lambda]\}$$

denotes the covariance between $h(\mathbf{X}_j)$ and $h(\mathbf{X}_l)$.

If $\{\mathbf{X}_j,\ j \geq 0\}$ is an i.i.d. sequence, then

$$\text{cov}[h(\mathbf{X}_j), h(\mathbf{X}_l)] = \begin{cases} \text{var}h(\mathbf{X}_1) & \text{if } j = l \\ 0 & \text{if } j \neq l, \end{cases}$$

so that $\sigma_t^2 = t^{-1}\text{var}_\pi h(\mathbf{X}_1)$. By contrast, MCMC sampling relies on conditioning $\mathbf{X}_j$ on $\mathbf{X}_{j-1}$, which induces dependence, and therefore, $\text{cov}(\mathbf{X}_j, \mathbf{X}_l) \neq 0$ for some or all finite $|j - l| > 0$. It is easily seen that

$$\text{cov}[h(\mathbf{X}_j), h(\mathbf{X}_l)] = R_{j-l}, \tag{6.59}$$

where

$$R_{j-l} := \sum_{\mathbf{x}\in\mathcal{X}}\sum_{\mathbf{y}\in\mathcal{X}}\{[h(\mathbf{x}) - \lambda]\,[h(\mathbf{y}) - \lambda] \\ \times \pi(\mathbf{x})K^{(|j-l|)}(\mathbf{x}, \mathbf{y})\}.$$

Then (6.58) has the representation

$$\sigma_t^2 = \frac{1}{t^2}\sum_{j=1}^{t}\sum_{l=1}^{t} R_{j-l},$$

and, because $R_l = R_{-l}$,

$$\sigma_t^2 = \frac{1}{t}\left[R_0 + 2\sum_{l=1}^{t-1}\left(1 - \frac{l}{t}\right)R_l\right]. \tag{6.60}$$

The symmetric function $\{R_j\}$ is called the *autocovariance function* of $\{\mathbf{X}_j\}$. In addition to R_0/t because of variation in each $\mathbf{X}_j$, pairwise dependence between $h(\mathbf{X}_i)$ and $h(\mathbf{X}_j)$ for $j \neq l$ in (6.59) contributes $\frac{2}{t}\sum_{l=1}^{t-1}\left(1 - \frac{l}{t}\right)R_l$ to (6.60). If this summation is positive, then dependence inflates σ_t^2. If it is negative, σ_t^2 is less than would occur for i.i.d. observations. Moreover, unlike the independent case, $t\sigma_t^2$ is not a constant for finite t. In this case, a limiting result is of value.

Theorem 6.9.1

If a Markov chain is geometrically ergodic, then

$$t\sigma_t^2 \to \sigma_\infty^2 = \textit{constant} < \infty \qquad \text{as } t \to \infty. \tag{6.61}$$

■

The significance of (6.61) is that $\text{var}\widehat{\lambda}_t \approx t^{-1}\sigma_\infty^2$ for large t, so that variance estimation reduces to approximating a constant σ_∞^2. The existence of this finite limit depends crucially on the properties of $\{R_j\}$. Although MCMC sampling often induces positive correlation between $\mathbf{X}_j$ and $\mathbf{X}_l$, leading to $R_{j-l} > 0$ for finite $j - l$, this correlation tends to decrease to 0 as $|j - l|$ grows large. That is, $R_{j-l} \to 0$ as $|j - l| \to \infty$. Whereas positive correlation explains the inflation of the variance (6.60) over that for the i.i.d case, convergence to 0 is necessary but, by itself, not sufficient for Theorem 6.9.1 to hold. That limit requires that R_{j-l} converge to 0 sufficiently fast. Chapter 7 returns to this issue.

6.10 CHOOSING A NOMINATING KERNEL

To facilitate HM sampling, Section 6.1 introduced the concept of a nominating kernel. Subsequent sections and examples illustrated its use, emphasizing single-coordinate updating on each iteration. Sections 6.10.1 and 6.10.2 continue the discussion, focusing on updating more than one coordinate per iteration. Indeed, the structure of problems occasionally indicates that updating more than one coordinate on an iteration is the most beneficial strategy. Section 6.10.1 describes an approach that nominates a new state independent of the current state. Section 6.10.2 describes one way of nominating a new state based on the location of the current state in $\Re^m$. Whereas the former relies exclusively on *global neighborhood information,* the latter relies on *local neighborhood information.*

We again emphasize that ergodicity is a required property of any nominating kernel q. Moreover, one prefers kernels that induce geometrically or uniformly ergodic processes, are easy to implement, and have small unit generating costs. Although (6.8) reveals that the acceptance frequency increases as the variation in $\{\pi(\mathbf{y})/q(\mathbf{x}, \mathbf{y})\}$ decreases, choosing q to effect this outcome is difficult to achieve in practice. Expression (6.8) also reveals an additional property that must be satisfied to ensure ergodicity. For all $(\mathbf{x}, \mathbf{y}) \in \mathcal{X} \times \mathcal{X}$, the ratio $\pi(\mathbf{y})/q(\mathbf{x}, \mathbf{y})$ needs to be positive. This is especially important when $\mathcal{X} = \Re^m$ and $\pi(\mathbf{y}) \to 0$ as $|\mathbf{y}| := \left(y_1^2 + \cdots + y_m^2\right)^{1/2} \to \infty$. Otherwise, the acceptance probability can be 0, implying that a sample path can enter a state (on the previous step) in which it remains permanently as t increases without bound. This last property circumscribes the choice of q.

6.10.1 Independence Hastings-Metropolis Sampling

One simplifying choice takes

$$q(\mathbf{x}, \mathbf{y}) = q(\mathbf{y}) \qquad \forall (\mathbf{x}, \mathbf{y}) \in \mathcal{X} \times \mathcal{X}, \tag{6.62}$$

where $\{q(\mathbf{x})\}$ is an everywhere positive distribution on $\mathcal{X}$. Because a random sample from $\{q(\mathbf{x})\}$ on iteration j is independent of the sample path up to iteration $j - 1$, this technique is called *independence Hastings-Metropolis sampling* (IHMS). Focusing on a single rather than on multiple sampling distributions (one for each $\mathbf{x}$) allows whatever may be known about π to be brought to bear more easily on the selection of a nominating distribution.

Suppose that

$$\kappa(\pi, q) = \max_x \left[\frac{\pi(\mathbf{x})}{q(\mathbf{x})}\right]$$

is bounded from above. Then it is known (e.g., Tierney 1994, Corollary 4) that the induced Markov chain is uniformly ergodic, as in (6.57), with convergence factor $\rho = 1 - 1/\kappa(\pi, q)$. Also, it can be shown (e.g., Liu 1996a) that for both countably infinite- and finite-state spaces, the boundedness of $\kappa(\pi, q)$ leads to the variance bound

$$\sigma_\infty^2 \leq R_0[2\kappa(\pi, q) - 1], \tag{6.63}$$

where $R_0 := \text{var}_\pi h(\mathbf{X})$. The bounds on ρ and σ_∞^2 also hold for a continuous-state space if $\{\pi(\mathbf{x})\}$ and $\{q(\mathbf{x})\}$ are p.d.f.s on a closed and bounded region $\mathcal{X} \in \Re^m$ (based on results in Smith and Tierney 1996).

To nominate m coordinates, one option samples Y_i from a univariate distribution q_i for each $i = 1, \ldots, m$ so that the candidate $\mathbf{Y} := (Y_1, \ldots, Y_m)$ has the distribution

$$q(\mathbf{y}) = \prod_{i=1}^{m} q_i(y_i).$$

This formulation would allow global moves as described in Section 6.8.1. Although simple in concept, the resulting acceptance function $\{\alpha(\mathbf{x}, \mathbf{y})\}$ frequently fluctuates widely with an overall acceptance probability so low as to give the approach little appeal. The discussion in Example 6.4.1 confirms this observation.

A second option nominates an m-coordinate candidate from a multivariate distribution $\{q(\mathbf{x})\}$ that accounts at least partially for these dependencies and for which computationally efficient generating algorithms exist. For the continuous case, these include the multinormal and multivariate Student t distributions. For the discrete case, they include the multinomial distribution. Although an improvement over the first option, the acceptance function and, therefore, $\left\{\frac{\pi(\mathbf{x})}{q(\mathbf{x})}\right\}$ may again fluctuate widely unless $\{q(\mathbf{x})\}$ closely mimics $\{\pi(\mathbf{x})\}$. A third approach dampens these fluctuations, thereby increasing the acceptance rate. In some cases, it also induces independence among elements $\mathbf{X}_1, \ldots, \mathbf{X}_t$ on a sample path. The appendix to this chapter (Section 6.18) describes this alternative in considerable detail. Section 9.6 describes more details about independence sampling in the context of slice sampling.

6.10.2 Random-Walk Nominating Kernels

In contrast to IHMS, which uses no local information when nominating a state, *random-walk Hastings-Metropolis sampling* (RWHMS) relies heavily on the current location in selecting a candidate state. Consider the nominating kernel

$$q(\mathbf{x}, \mathbf{y}) := q\left(\frac{\mathbf{x} - \mathbf{y}}{\xi}\right) \qquad \xi > 0, \tag{6.64}$$

where $\{q(\mathbf{x})\}$ is a probability distribution with the property

$$q\left(\frac{\mathbf{x}-\mathbf{y}}{\xi}\right) = q\left(\frac{\mathbf{y}-\mathbf{x}}{\xi}\right). \tag{6.65}$$

This symmetry leads to the acceptance function

$$\alpha(\mathbf{x},\mathbf{y}) = \min\left[1, \frac{\pi(\mathbf{y})}{\pi(\mathbf{x})}\right].$$

On each step $j \geq 1$, one performs Algorithm RWHMS, which shows the *random-walk* character of the underlying stochastic process.

ALGORITHM

RWHMS

Purpose: To update a random walk using modified acceptance-rejection sampling as in Algorithm MARS (in Section 6.18) and HM sampling.

Method:

Using Algorithm MARS, randomly sample $\mathbf{Z}_j$ from $\{q(\mathbf{z})\}$ in (6.64).
Randomly sample U_j from $\mathcal{U}(0, 1)$.
If $U_j \leq \alpha(\mathbf{X}_{j-1}, \mathbf{X}_{j-1} + \xi\mathbf{Z}_j)$, $\mathbf{X}_j = \mathbf{X}_{j-1} + \xi\mathbf{Z}_j$.
Otherwise, $\mathbf{X}_j = \mathbf{X}_{j-1}$. ■

For small ξ, $\pi(\mathbf{X}_{j-1} + \xi\mathbf{Z}_j)$ varies little from $\pi(\mathbf{X}_{j-1})$, thereby inducing frequent acceptances. However, $\mathbf{X}_j$ also varies little from $\mathbf{X}_{j-1}$, leading to strong dependence among elements on a sample path. Large ξ causes $\mathbf{X}_{j-1} + \xi\mathbf{Z}_j$ to differ more substantially from $\mathbf{X}_{j-1}$, but $\pi(\mathbf{X}_{j-1} + \xi\mathbf{Z}_j)$ is also likely to differ noticeably from $\pi(\mathbf{X}_{j-1})$, reducing the acceptance rate and thereby inducing strong dependence among elements on a sample path. In essence, ξ is a *tuning parameter*, much like b in Algorithm MARS for the IHMS method in the appendix to this chapter.

As to ergodicity, HM sampling with $\{q(\mathbf{x},\mathbf{y})\}$ as in (6.64) and (6.65) on unbounded $\mathcal{X}$ is known to generate non-uniformly ergodic sample paths (Mengersen and Tweedie 1996). This property also holds when $\{q(\mathbf{x},\mathbf{y})\}$ in (6.64) is asymmetric in $\mathbf{x}$ and $\mathbf{y}$.

Theorem 6.10.1 (Roberts and Tweedie 1996, Proposition 3.1.)

Let π have the representation

$$\begin{aligned} \pi(\mathbf{x}) &= a(\mathbf{x})e^{b(\mathbf{x})} \\ &> 0 \qquad \mathbf{x} \in \Re^m, \end{aligned} \tag{6.66}$$

where a and b denote polynomials in $\mathbf{x}$ *with b of order* $l \geq 2$*. Let* b_l *denote the polynomial consisting of the terms of b of order l or higher, and*

$$b_l(\mathbf{x}) \to -\infty \qquad \text{as} \qquad \left(x_1^2 + \cdots + x_m^2\right)^{1/2} \to \infty.$$

Then repeated application of Algorithm RWHMS for $j = 1, 2, \ldots$ *with q bounded away from zero in some region around the origin* $(\mathbf{x} - \mathbf{y} = \mathbf{0})$ *induces a Markov chain that is geometrically ergodic.* ■

In practice, this symmetric random-walk approach offers considerable convenience (e.g., Example 8.3.2). Apart from ξ, it requires no tailoring to the properties of π. For $\mathcal{X} \in \Re^m$, the multinormal distribution where $X_1, \ldots, X_m$ are i.i.d., each from $\mathcal{N}(\mathbf{0}, \xi^2)$, provides one option in continuous space. For π more general than (6.66), Roberts and Tweedie (1996) also specify properties that ensure that repeated application of Algorithm RWHMS induces a geometrically ergodic chain.

6.11 CHAINS FAVORING SMALLER σ_∞^2

The choice of MCMC technique determines K, which determines σ_∞^2 and the convergence rate for (6.61). We next describe how a comparison of kernels may shed light on how each kernel performs relative to the other with regard to statistical accuracy.

Theorem 6.11.1 (Peskun 1973.)

If K_1 *and* K_2 *are two ergodic transition kernels with*

$$K_1(\mathbf{x}, \mathbf{y}) \geq K_2(\mathbf{x}, \mathbf{y}) \qquad \text{for all } \mathbf{y} \in \mathcal{X} \text{ and } \mathbf{x} \in \mathcal{X} \text{ with } \mathbf{x} \neq \mathbf{y},$$

then

$$\sigma_\infty^2(K_1) \leq \sigma_\infty^2(K_2).$$ ■

This result implies that larger off-diagonal transition probabilities induce a smaller asymptotic variance than smaller ones and, therefore, induce better statistical accuracy for any long sample-path length t.

For discrete-state spaces, an application of Theorem 6.11.1 arises in comparing random-coordinate Gibbs sampling as in Algorithm RGIBBS with random-coordinate HM sampling as in Algorithm RHM. On step $j \geq 1$, Algorithm RHM randomly selects coordinate I with probability $\frac{1}{m}$, randomly nominates a candidate from the full conditional distribution (6.16) for coordinate I restricted to $\mathcal{X} \backslash \{\mathbf{X}_{I,j-1}\}$, and then does HM acceptance-rejection.

PROOF Expression (6.40) gives K_{RGIBBS}. On iteration j, Algorithm RHM selects coordinate $I = i$ with nominating distribution

$$q_i(\mathbf{x}, \mathbf{x}_i(y)) = \begin{cases} \frac{\pi_i(y|\mathbf{x}_{-i})}{1-\pi_i(x_i|\mathbf{x}_{-i})} & \text{if } y \neq x_i \\ 0 & \text{elsewhere,} \end{cases}$$

and for all $\mathbf{x} \neq \mathbf{y}$ (see Exercise 4),

$$K_{\text{RHM}}(\mathbf{x}, \mathbf{y}) = \begin{cases} \frac{1}{m}\sum_{i=1}^{m} \frac{\pi_i(y_i|\mathbf{x}_{-i})}{1-\pi_i(x_i|\mathbf{x}_{-i})}\delta(\mathbf{y} - \mathbf{x}_i(y_i))I_{\mathcal{X}_i\backslash\{x_i\}} \min\left[1, \frac{1-\pi_i(x_i|\mathbf{x}_{-i})}{1-\pi_i(y_i|\mathbf{x}_{-i})}\right] \\ \qquad\qquad \text{if } \mathbf{y} \in \{\mathbf{x}_1(y_1), \ldots, \mathbf{x}_m(y_m)\} \\ 0 \qquad\qquad \text{elsewhere,} \end{cases} \tag{6.67}$$

where

$$I_{\mathcal{X}_i\backslash\{x_i\}} := \begin{cases} 1 & \text{if } y_i \neq x_i \in \mathcal{X}_i \\ 0 & \text{elsewhere.} \end{cases}$$

Because for all $\mathbf{x} \neq \mathbf{y}$, $K_{\text{RHM}}(\mathbf{x}, \mathbf{y}) - K_{\text{RGIBBS}}(\mathbf{x}, \mathbf{y}) \geq 0$, (6.68) holds by Theorem 6.11.1. ■

ALGORITHM **RHM**

Purpose: To generate a sample path $\mathbf{X}_1, \ldots, \mathbf{X}_t$ by HM sampling using the RCS rule.

Given: Initial state $\mathbf{X}_0$ and sample-path length t.

Definitions: $I :=$ selected coordinate, $j :=$ step, and $X_{ij} :=$ value of coordinate i on iteration j.

Method:

$j = 1$.
Until $j > t$:
  randomly sample I uniformly from $\left\{\frac{1}{m}, \ldots, \frac{1}{m}\right\}$
  randomly sample Y from

$$\left\{\frac{\pi_I(y \mid \mathbf{X}_{-I,j-1})}{1 - \pi_I(X_{I,j-1} \mid \mathbf{X}_{-I,j-1})}, \; y \in \mathcal{X}_I\backslash\{X_{I,j-1}\}\right\}$$

  randomly sample U from $\mathcal{U}(0, 1)$
  $A = \frac{\pi(\mathbf{X}_{I,j-1}(Y))}{\pi(\mathbf{X}_j)} \times \frac{1-\pi_I(X_{I,j-1}|\mathbf{X}_{-I,j-1})}{\pi_I(Y|\mathbf{X}_{-I,j-1})} \times \frac{\pi_I(X_{I,j-1}|\mathbf{X}_{-I,j-1})}{1-\pi_I(Y|\mathbf{X}_{-I,j-1})}$
  if $U \leq \min(1, A)$,
    $\mathbf{X}_j = \mathbf{X}_{I,j-1}(Y)$
  otherwise, $\mathbf{X}_j = \mathbf{X}_{j-1}$
  $j = j + 1$. ■

Theorem 6.11.2 (Liu 1996b.)

Let K_{RHM} denote the transition kernel induced by Algorithm RHM, *and K_{RGIBBS} the transition kernel induced by Algorithm* RGIBBS. *Then*

$$\sigma_\infty^2(K_{\text{RHM}}) \le \sigma_\infty^2(K_{\text{RGIBBS}}). \tag{6.68}$$

■

Theorem 6.11.2 reveals that for a discrete-state space, there always exists an HM sampling algorithm (RHM) that induces a large-sample variance of the sample mean no larger than the corresponding variance that obtains for random-coordinate Gibbs sampling (RGIBBS). This implies that the sample-path length required to achieve a specified high accuracy with Algorithm RHM is always less than or equal to that required for Algorithm RGIBBS. However, this property does not unequivocally make us prefer Algorithm RHM to Algorithm RGIBBS. Because it is easily seen that the unit generating cost is higher for RHM, the total cost of achieving the desired accuracy may not necessarily favor RHM. Also, the inequality (6.68) casts no light on which sampling procedure leads to faster convergence to equilibrium from an arbitrarily selected starting point $\mathbf{X}_0$. Section 6.2 addresses this issue.

EXAMPLE 6.11.1 STATISTICAL ACCURACY FOR CAPTURE-RECAPTURE

To illustrate how the choice of MCMC technique affects statistical accuracy and computing efficiency, we once again use the capture-recapture problem in Examples 6.4.1 and 6.6.1. This time we focus on estimating

$$\begin{aligned}\lambda &= \sum_{x_{r+1}=S_{r+1}}^{\infty} h(x_{r+1}) \int_{[0,1]^r} \pi(x_1, \ldots, x_r, x_{r+1}) \mathrm{d}x_1, \ldots, \mathrm{d}x_r \\ &= \text{mean population size for the posterior size distribution,}\end{aligned}$$

where

$$h(x_{r+1}) := x_{r+1}$$

and

$$\pi(x_1, \ldots, x_r, x_{r+1}) = f_{\text{posterior}}(x_1, \ldots, x_r, x_{r+1} \mid \mathbf{N}, \mathbf{M})$$

in (6.30). Table 6.2 displays population-size sample averages $\widehat{\lambda}_t$, sample variances $\widehat{R_0}$, and estimates, $\widehat{\sigma}_\infty^2$, of σ_∞^2 for three different MCMC sampling strategies.[5] Of particular interest,

$$\text{s.e.}(\widehat{\lambda}_t) := \sqrt{\widehat{\sigma}_\infty^2 / t},$$

[5]The batch-means method of Section 7.4 was used to estimate $\widehat{\sigma}_\infty^2$.

in the 99% confidence interval, estimates the asymptotic standard error of $\widehat{\lambda}_t$. As before, its relative error is estimated by

$$\text{r.e.}(\widehat{\lambda}_t) := \frac{\text{s.e.}(\widehat{\lambda}_t)}{\widehat{\lambda}_t}.$$

Although the three experiments give similar population size sample averages and similar sample variances, their $\widehat{\sigma}_\infty^2$ vary considerably. In particular, the two-coordinate HM experiment produced an estimate $5.932/1.546 = 3.837$ times as large as that for the two-coordinate Gibbs sampling experiment.

The computed acceptance rate of 0.3435 for HM sampling explains this difference. This modest rate means that $\mathbf{X}_j = \mathbf{X}_{j+1}$ frequently, thus inducing a high level of dependence among elements on its sample path. This dependence inflates the variance. Because the corresponding computing time ratio is 0.9962. It would take a sample path 3.837 times as long and $0.9962 \times 3.837 = 3.822$ as much cost to get the same accuracy for the sample average with the HM approach as with the Gibbs sampling approach updating two coordinates on each iteration.

A comparison of two-coordinate with all-coordinate updating for Gibbs sampling is equally revealing. It shows a ratio of estimated asymptotic variances of $2.164/15.46 = 0.1400$, favoring the all-coordinate approach. With a computing time ratio of 1.331, these results imply that it takes a sample-path length 0.1400 times, or approximately 1/7, as long to get the same accuracy for the sample average with the all-coordinates approach at $1.331 \times 0.1400 = 0.1863$ times the computing cost of the two-coordinate approach. The foregoing analysis unequivocally establishes an advantage for all-coordinate Gibbs sampling.

The entries in Table 6.2 also establish the need to take seriously the dependence among observations. For example, the ratio $\widehat{\sigma}_\infty^2/\widehat{R_0} = 2.164 \times 10^4/3433 = 6.304$ reveals that for all-coordinate Gibbs sampling, dependence inflates $\text{var}\widehat{\lambda}_t$ by more than a factor of 6. Example 7.7.1 returns to the estimation of the asymptotic variances for this problem. ■

TABLE 6.2 | *Sunfish MCMC statistical analysis of population size:* ($t = 10^6$)

Sampling technique	$\widehat{\lambda}_t \pm \Phi^{-1}(0.995)\text{s.e.}(\widehat{\lambda}_t)$	$\widehat{R}_0$	$\widehat{\sigma}_\infty^2$	Computing time ratio
Two coordinates				
Hastings-Metropolis	430.9 ± 2.000	3419	5.932×10^5	0.9962
Gibbs	431.1 ± 1.050	3442	1.546×10^5	1
All coordinates				
Gibbs	430.9 ± 0.4000	3433	2.164×10^4	1.331

TABLE 6.3 | *Eye color versus hair color for 592 subjects**

	Hair color				
Eye color	Black	Brunette	Red	Blond	Total
Brown	68	119	26	7	220
Blue	20	84	17	94	215
Hazel	15	54	14	10	93
Green	5	29	14	16	64
Total	108	286	71	127	592

*Source: Snee (1974).

EXAMPLE 6.11.2

EYE COLOR AND HAIR COLOR[6]

Table 6.3 displays data on 592 human subjects based on two attributes, eye color and hair color. Given these data, it is of interest to determine whether the two attributes are statistically independent. The table is a *two-way contingency table,* a frequently employed structure for statistically analyzing categorical data. More generally, consider a population of individuals characterized by k attributes where attribute i has I_i categories. For example, Table 6.3 has $k = 2$, eye color has $I_1 = 4$ categories, and hair color $I_2 = 4$. It is a special case of the more general k-way contingency table, sometimes called an $(I_1 \times \cdots \times I_k)$-contingency table.

For some subset of r attributes, where $r \in \{1, \ldots, k-1\}$, the structure provides a convenient framework for studying the extent to which the categories into which subjects fall for these r attributes are statistically independent of the categories into which they fall for the remaining $k-r$ attributes. In what follows, we describe how MCMC sampling provides an answer to this problem for $k = 2$. The concepts apply for $k > 2$ as well. For conciseness of notation, we take $I := I_1$ and $L := I_2$.

Let

$$N_{il} := \text{number of individuals in cell } (i, l)$$

$$N_{i.} := \textstyle\sum_{l=1}^{L} N_{il} \qquad \text{(row sum)}$$

$$N_{.l} := \textstyle\sum_{i=1}^{I} N_{il} \qquad \text{(column sum)}$$

$$N := \textstyle\sum_{l=1}^{L} N_{.l} = \sum_{i=1}^{I} N_{i.} \qquad \text{(total)}$$

$$\mathbf{N} := \begin{pmatrix} N_{11} & N_{12} & \cdots & N_{1L} \\ N_{21} & N_{22} & \cdots & N_{2L} \\ \vdots & \vdots & \ddots & \vdots \\ N_{I1} & N_{I2} & \cdots & N_{IL} \end{pmatrix} = \text{an } I \times L \text{ matrix with } N_{il} \text{ in row } i \text{ and column } l.$$

[6]This problem is taken from Snee (1974) and Diaconis and Sturmfels (1998).

The hypothesis of interest is H_0: Attributes 1 and 2 are statistically independent. Let

$$\mathbf{x} := \begin{pmatrix} x_{11} & x_{12} & \cdots & x_{1L} \\ x_{21} & x_{22} & \cdots & x_{2L} \\ \vdots & \vdots & \ddots & \vdots \\ x_{I1} & x_{I2} & \cdots & x_{IL} \end{pmatrix},$$

where the x_{il} are nonnegative integers for all $i = 1, \ldots, I$ and $l = 1, \ldots, L$. Let $\mathcal{X}$ denote the set of all nonnegative integer matrices $\mathbf{x}$ such that

$$\sum_{i=1}^{I} x_{il} = N_{.l}$$

and

$$\sum_{l=1}^{L} x_{il} = N_{i.}.$$

It is of interest to determine how "typical" the observed data matrix $\mathbf{N}$ is of the matrices in $\mathcal{X}$ under H_0. One approach based on the function

$$\begin{aligned} B(\mathbf{x}) &:= N \sum_{i=1}^{I} N_{i.}^{-1} \sum_{l=1}^{L} N_{.l}^{-1} \left(x_{il} - \frac{N_{i.} N_{.l}}{N} \right)^2 \\ &= N \sum_{i=1}^{I} \sum_{l=1}^{L} \left(\frac{x_{il}^2}{N_{i.} N_{.l}} - 1 \right) \end{aligned} \qquad \forall \mathbf{x} \in \mathcal{X} \qquad (6.69)$$

measures the dispersion of the entries in a matrix $\mathbf{x}$ relative to its fixed row and column sums. This measure facilitates a procedure for evaluating the credibility of H_0. Suppose H_0 is true and $\mathbf{N}$ is given. Then for $\mathbf{X}$ randomly drawn from $\mathcal{X}$,

$$\text{p-value} := \text{pr}[B(\mathbf{X}) > B(\mathbf{N}) \mid \mathbf{N}, H_0] \qquad (6.70)$$

gives the probability of observing $B(\mathbf{X})$ larger than $B(\mathbf{N})$ for the given data matrix in the sample. Small values of p-value raise doubts about the credibility of H_0.

Suppose that $\mathbf{X}$ has p.m.f. $\{\pi(\mathbf{x}),\ \mathbf{x} \in \mathcal{X}\}$ and let

$$h(\mathbf{x}) := \begin{cases} 1 & \text{if } B(\mathbf{x}) > B(\mathbf{N}) \\ 0 & \text{elsewhere} \end{cases} \qquad \forall \mathbf{x} \in \mathcal{X}.$$

Then (6.70) has the representation

$$\begin{aligned} \lambda &= \sum_{\mathbf{x} \in \mathcal{X}} h(\mathbf{x}) \pi(\mathbf{x}) \\ &= \text{proportion of matrices in } \mathcal{X} \\ &\quad \text{for which } B(\mathbf{x}) > B(\mathbf{N}) = 138.29. \end{aligned} \qquad (6.71)$$

Because $|\mathcal{X}| > 10^{15}$, exact evaluation of λ by total enumeration is not an option.

As an approximation, it is known that regardless of the distribution of $\mathbf{X}$ given $\mathbf{N}$ under H_0, $B(\mathbf{X})$ converges in distribution to a chi-squared random variable with $I(I-1)L(L-1)$ degrees of freedom as $N \to \infty$. Under H_0, the chi-squared p-value for $B(\mathbf{N}) = 138.29$ is less than 10^{-24}, suggesting that 138.29 is considerably larger than is likely to be encountered by random selection and thus giving little support to H_0. However, experience with this asymptotic approach indicates relatively slow convergence to the limiting distribution as N increases, thereby limiting one's ability to draw conclusions from results based on it. This slow convergence raises doubts about the chi-squared approximation.

MCMC offers an alternative approach to solution. It assumes that $\mathbf{X}$ has the discrete uniform p.m.f.

$$\pi(\mathbf{x}) := \begin{cases} |\mathcal{X}|^{-1} & \mathbf{x} \in \mathcal{X} \\ 0 & \text{elsewhere} \end{cases}$$

and generates a sample path of $m = I \times L$ matrices, $\mathbf{X}_1, \ldots, \mathbf{X}_t$, from $\{\pi(\mathbf{x})\}$; evaluates $B(\mathbf{X}_1), \ldots, B(\mathbf{X}_t)$; and uses these data to estimate the proportion λ. Note that this procedure is a form of resampling, as discussed in Section 2.8, in which row and column totals are held constant.

Because of the row and column constraints, no fewer than four coordinates can be randomly sampled on an iteration. In this regard, the problem provides a good example of *block-coordinate updating*. On each iteration $j \geq 1$, Algorithm CONTINGENCY randomly identifies a 2×2 submatrix of $\mathbf{X}_{j-1}$, randomly updates its four entries, and then assigns the updated matrix to $\mathbf{X}_j$. The identity

$$(x_{il} + \theta - N_{i.}N_{.l}/N)^2 = (x_{il} - N_{i.}N_{.l}/N)^2 + \theta\,[\theta + 2(x_{il} - N_{i.}N_{.l}/N)] \qquad (6.72)$$

allows $B(\mathbf{X}_j)$ to be computed recursively from $B(\mathbf{X}_{j-1})$ in time independent of I and L and, therefore, independent of m (see Exercise 5). ■

ALGORITHM

CONTINGENCY

Purpose: To generate a sample path for a two-way contingency table.

Given: Sample-path length t.

Method:

Set initial state $\mathbf{X}_0$.
$j = 1$.
Until $j > t$:
  randomly generate:
    row number I_1 from $\{1, \ldots, I\}$ with probability I^{-1}
    row number I_2 from $\{1, \ldots, I\} \backslash \{I_1\}$ with probability $(I-1)^{-1}$
    row number L_1 from $\{1, \ldots, L\}$ with probability L^{-1}
    row number L_2 from $\{1, \ldots, L\} \backslash \{L_1\}$ with probability $(L-1)^{-1}$

randomly generate $X_{I_1 j}, X_{I_2 j}, X_{L_1 j}, X_{L_2 j}$
$\mathbf{X}_j = \mathbf{X}_{j-1}$ with $X_{I_1 j}, X_{I_2 j}, X_{L_1 j}, X_{L_2 j}$,
replacing $X_{I_1 j-1}, X_{I_2 j-1}, X_{L_1 j-1}, X_{L_2 j-1}$ respectively
compute $B(\mathbf{X}_j)$ using recursion (6.72)
$j = j + 1$. ■

Algorithms PMO and ALL describe two procedures for updating the randomly selected 2×2 matrix. It is easily seen that Algorithm PMO allows four coordinate values to change by $+1$ or -1. By contrast, Algorithm ALL allows the values of the four coordinates to vary more widely on successive iterations.

ALGORITHM PMO

Purpose: To update the entries in a 2×2 matrix with fixed row and column sums.

Given: $X_{I_1 J_1, j-1}, X_{I_1 J_2, j-1}, X_{I_2 J_1, j-1}, X_{I_2 J_2, j-1}$.

Method:

Randomly generate U from $\mathcal{U}(0, 1)$.
If $U \le 0.50$, $A = 1$; otherwise, $A = -1$.
If $A \min(X_{I_1 J_2, j-1}, X_{I_2 J_1, j-1}) > 0$ or $A \min(X_{I_1 J_1, j-1}, X_{I_2 J_2, j-1}) < 0$:
$X_{I_1 J_1 j} = X_{I_1 J_1, j-1} + A$
$X_{I_1 J_2 j} = X_{I_1 J_2, j-1} - A$
$X_{I_2 J_1 j} = X_{I_2 J_1, j-1} - A$
$X_{I_2 J_2 j} = X_{I_2 J_2, j-1} + A$. ■

The two algorithms each imply a *random walk* process

$$\mathbf{X}_j = \mathbf{X}_{j-1} + \mathbf{B}_j \qquad j \ge 1,$$

where the $I \times L$ matrix $\mathbf{B}_j$ has zeros in all but four entries, and the chosen algorithm dictates how those four entries are to be sampled. In particular, Algorithm ALL induces the one-step transition kernel

$$K(\mathbf{x}, \mathbf{y}) = \frac{1}{I(I-1)L(L-1)} \sum_{i_1=1}^{I} \sum_{\substack{i_1=1 \\ i_2 \ne i_1}}^{I} \sum_{i_1=1}^{L} \sum_{\substack{l_2=1 \\ l_2 \ne l_1}}^{L} \left[\frac{1}{M_{i_1 i_2 l_1 l_2}(\mathbf{x})} \right] \qquad \forall \mathbf{x} \in \mathcal{X}, \tag{6.73}$$

where

$$M_{i_1 i_2 l_1 l_2}(\mathbf{x}) := 1 + \min(x_{i_1 l_1} + x_{i_1 l_2}, x_{i_1 l_1} + x_{i_2 l_1}, x_{i_2 l_1} + x_{i_2 l_2}, x_{i_1 l_2} + x_{i_2 l_2}).$$

ALGORITHM ALL

Purpose: To update the entries in a 2×2 matrix with fixed row and column sums.

Given: $X_{I_1 J_1, j-1}, X_{I_1 J_2, j-1}, X_{I_2 J_1, j-1}, X_{I_2 J_2, j-1}$.

Method:

$$W_{1.} = X_{I_1J_1,j-1} + X_{I_1J_2,j-1}.$$
$$W_{2.} = X_{I_2J_1,j-1} + X_{I_2J_2,j-1}.$$
$$W_{.1} = X_{I_1J_1,j-1} + X_{I_2J_1,j-1}.$$
$$W_{.2} = X_{I_1J_2,j-1} + X_{I_2J_2,j-1}.$$
$$W = W_{1.} + W_{2.}\ .$$
$$M = \min(W_{1.}, W_{2.}, W_{.1}, W_{.2}).$$
$$V_1 = \min(W_{1.}, W_{.1}).$$
$$V_2 = \min(W_{2.}, W_{.2}).$$
$$l^* = \begin{cases} 1 \text{ if } V_l \le V_2 \\ 2 \text{ if } V_l > V_2 \end{cases}$$
$$l' = l^*(\bmod 2) + 1.$$

Randomly generate Z from $\{0, 1, \ldots, M\}$ with probability $(M+1)^{-1}$.

$$X_{I_{l^*}L_{l^*}j} = Z.$$
$$X_{I_{l^*}L_{l'}j} = W_{l^*.} - Z.$$
$$X_{I_{l'}L_{l^*}j} = W_{.l^*} - Z.$$
$$X_{I_{l'}L_{l'}j} = W - X_{I_{l^*}L_{l^*},j} - X_{I_{l^*}L_{l'},j} - X_{I_{l'}L_{l^*},j}.$$ ■

Table 6.4 compares MCMC results for sample-path data based on Algorithm CONTINGENCY using Algorithms PMO and ALL. The sample proportions, 0.1636 with standard error $\sqrt{319.5/524288} = 0.02469$, and 0.1682 with standard error $\sqrt{1.721/524288} = 0.001812$, reveal more credibility for H_0 than the approximating chi-squared p-value of less than 10^{-24} indicates. The entries also show a variance ratio $319.5/1.721 = 185.6$, favoring Algorithm ALL. With comparable computing times, the results indicate that it would take approximately 185.6 as much computing time with Algorithm PMO to get the same standard error as with Algorithm ALL. The $\widehat{\sigma}_\infty^2$ are *batch-means variance* estimates (see Section 7.4) as described in Example 7.8.1. Algorithm ALL also induces a considerably shorter warm-up interval. Example 7.3.2 addresses the estimation of the warm-up interval.

Differences in correlation structures within sample paths are responsible for the better performance of Algorithm ALL. In particular, Algorithm PMO induces a variance inflation of $319.5/0.1368 = 2336$ in contrast to $1.721/0.1399 = 12.30$ for Algorithm ALL. Clearly, allowing the four coordinate values to vary more

TABLE 6.4 | ***Estimating eye color and hair color proportion: t = 524,288***

Algorithm	$\widehat{\lambda}_t \pm \Phi^{-1}(0.995)\text{s.e.}(\widehat{\lambda}_t)$	$\widehat{R}_0$	$\widehat{\sigma}_\infty^2$	Computing time ratio	Estimated warm-up interval
PMO	0.1636 ± 0.0876	0.1368	319.5	1	4000
ALL	0.1682 ± 0.0059	0.1399	1.721	1.057	50

widely on an update with the latter algorithm keeps the correlation lower between elements in a sample path.

6.12 GENERAL-STATE SPACES

MCMC applies to m-dimensional state spaces that are

1. Discrete.
2. Continuous.
3. For some $1 < l < m$, l dimensions are discrete and $m - l$ are continuous.

Collectively these encompass the *general-state-space formulation*. Sometimes the third configuration is called a *mixed-state space*. Earlier sections described the properties that a Markov chain on a discrete-state space must have in order for MCMC to be an effective estimating procedure. On a continuous-state space, a Markov chain must have these plus additional properties in order for MCMC to be equally effective. This section briefly describes the additional properties.

Unless explicitly stated to the contrary, we continue to assume that $\pi := \{\pi(\mathbf{x}) > 0,\ \mathbf{x} \in \mathcal{X}\}$, π being a p.m.f. when $\mathcal{X}$ is discrete and a p.d.f. when $\mathcal{X}$ is continuous. For each subset of states $\mathcal{A} \subseteq \mathcal{X}$, define the d.f.

$$F(\mathcal{A}) := \begin{cases} \int_{\mathcal{A}} \pi(\mathbf{x})\mathrm{d}\mathbf{x} & \text{if } \mathcal{X} \text{ is continuous} \\ \sum_{\mathbf{x}\in\mathcal{A}} \pi(\mathbf{x}) & \text{if } \mathcal{X} \text{ is discrete.} \end{cases}$$

As in the discrete case, the continuous case relies on ergodic properties to establish the efficacy of MCMC. Although the principal focus remains the transition from individual elements (states) $\mathbf{x} \in \mathcal{X}$ to individual elements $\mathbf{y} \in \mathcal{X}$, the properties of irreducibility, recurrence, and aperiodicity now must be couched in terms of transition from a single state, $\mathbf{x} \in \mathcal{X}$, to subsets of states, $\mathcal{A} \subseteq \mathcal{X}$, for which $\pi(\mathcal{A}) > 0$. Let $\mathcal{S}(\mathcal{X})$ denote the collection of all subsets of $\mathcal{X}$ with this property. Because $\pi(\mathbf{x}) > 0$ for all $\mathbf{x} \in \mathcal{X}$, this condition is easily satisfied in the discrete case. Because $F(\mathcal{A}) = 0$ for a single point, $\mathcal{A} = \{\mathbf{x}\}$, in the continuous case, we can no longer speak of point-to-point transition when discussing ergodicity.

In continuous one-dimensional space with $\mathcal{X} = [0, \infty)$, $\mathcal{S}(\mathcal{X})$ consists of all line segments and all collections of line segments for which $F(\mathcal{A}) > 0$. As examples,

- $\mathcal{A} = [a, b]$ for $0 \le a < b \le 1$ with

$$F(\mathcal{A}) = \int_a^b \pi(x)\mathrm{d}x > 0$$

- $\mathcal{A} = [a_1, b_1] \cup [a_2, b_2]$ for $0 \le a_1 < b_1 < a_2 < b_2 \le 1$ with

$$F(\mathcal{A}) = \int_{a_1}^{b_1} \pi(x)\mathrm{d}x + \int_{a_2}^{b_2} \pi(x)\mathrm{d}x > 0$$

- $\mathcal{A} = [a_1, b_1] \cup [a_2, b_2]$ for $0 \le a_1 < a_2 < b_1 < b_2 \le 1$ with
$$F(\mathcal{A}) = \int_{a_1}^{b_2} \pi(x)\mathrm{d}x > 0$$
- $\mathcal{A} = [a_1, b_1] \cap [a_2, b_2]$ for $0 \le a_1 < a_2 < b_1 < b_2 \le 1$ with
$$F(\mathcal{A}) = \int_{a_2}^{b_1} \pi(x)\mathrm{d}x > 0.$$

However, the subset $\mathcal{A} = [a_1, b_1] \cap [a_2, b_2]$ for $0 \le a_1 < b_1 < a_2 < b_2 \le 1$ with $F(\mathcal{A}) = 0$ does not have the desired properties. For the two-dimensional space $\mathcal{X} = [0, \infty)^2$, $\mathcal{S}(\mathcal{X})$ consists of all closed and bounded regions and all collections of closed and bounded regions $\mathcal{A}$ for which $F(\mathcal{A}) > 0$.

Mixed-State Space. Let $\mathcal{X}_{\text{dis}}$ denote a finite collection of points in $\Re^m$, let $\mathcal{X}_{\text{con}}$ denote a continuous region in $\Re^m$, and define $\mathcal{X} := \mathcal{X}_{\text{dis}} \cup \mathcal{X}_{\text{con}}$. Also define
$$\pi(\mathbf{x}) := \pi_{\text{dis}}(\mathbf{x}) + \pi_{\text{con}}(\mathbf{x}) \qquad \mathbf{x} \in \mathcal{X},$$
where
$$\pi_{\text{dis}}(\mathbf{x}) \begin{cases} > 0 & \text{if } \mathbf{x} \in \mathcal{X}_{\text{dis}} \\ = 0 & \text{elsewhere} \end{cases}$$
and
$$\pi_{\text{con}}(\mathbf{x}) > 0 \qquad \text{for all } \mathbf{x} \in \mathcal{X}_{\text{con}}.$$

Then $\mathcal{A} \subseteq \mathcal{S}(\mathcal{X})$ if and only if
$$F(\mathcal{A}) = \sum_{\mathbf{x} \in \mathcal{A} \cap \mathcal{X}_{\text{dis}}} \pi_{\text{dis}}(\mathbf{x}) + \int_{\mathcal{A} \cap \mathcal{X}_{\text{con}}} \pi_{\text{con}}(\mathbf{x})\mathrm{d}\mathbf{x} > 0.$$

Irreducibility. In the continuous case, irreducibility, recurrence, and aperiodicity are all based on the properties of the transition kernel
$$K(\mathbf{x}, \mathcal{A}) := \int_{\mathcal{A}} K(\mathbf{x}, \mathbf{y})\mathrm{d}\mathbf{y} \qquad \text{for all } \mathcal{A} \in \mathcal{S}(\mathcal{X}). \tag{6.74}$$

The j-step transition kernel takes the form
$$K^{(j)}(\mathbf{x}, \mathcal{A}) := \begin{cases} K(\mathbf{x}, \mathcal{A}) & j = 1 \\ \int_{\mathcal{X}} K^{(j-1)}(\mathbf{x}, \mathbf{y}) K(\mathbf{y}, \mathcal{A}) \mu(\mathbf{y})\mathrm{d}\mathbf{y} & j > 1 \end{cases}$$
with $K^{(j)}(\mathbf{x}, \mathcal{X}) = 1$ for all $j \ge 1$. Global balance takes the form
$$\pi(\mathcal{A}) = \int_{\mathcal{A}} \pi(\mathbf{x})\mathrm{d}\mathbf{x} = \int_{\mathcal{X}} \pi(\mathbf{y}) K(\mathbf{y}, \mathcal{A}) \mu(\mathbf{y})\mathrm{d}\mathbf{y} \quad \mathcal{A} \in \mathcal{S}(\mathcal{X}), \tag{6.75}$$
so that a π satisfying (6.75) is an invariant distribution with respect to K. For the chain to be irreducible, there must exist an integer $j \ge 1$ such that for all $\mathbf{x} \in \mathcal{X}$,
$$K^{(j)}(\mathbf{x}, \mathcal{A}) > 0 \qquad \text{for all } \mathcal{A} \in \mathcal{S}(\mathcal{X}) \text{ s.t. } \pi(\mathcal{A}) > 0.$$

We then say that K is π-irreducible. If $K(\mathbf{x}, \mathbf{y}) > 0$ for all $(\mathbf{x}, \mathbf{y}) \in \mathcal{X}^2$, then the chain clearly has this property.

Recurrence. If irreducibility holds, then, analogous to the result stated in Theorem 6.7.1, π is the unique invariant distribution of (6.75). We then say that the chain is *positive Harris recurrent.* In the discrete case, positive recurrence is equivalent to Harris recurrence. However, for general-state spaces, Harris recurrence is a stronger property, ruling out nuisance sets that can arise in a continuous-state space setting.

Aperiodicity. The significance of aperiodicity is best understood by first defining periodicity. A transition kernel K is periodic if for some integer $r \geq 2$ there exist nonempty disjoint subsets $\mathcal{A}_0, \mathcal{A}_1, \ldots, \mathcal{A}_r \in \mathcal{S}(\mathcal{X})$ such that for each $i = 0, 1 \ldots, r-1$ and all $\mathbf{x} \in \mathcal{A}_i$, $K(\mathbf{x}, \mathcal{A}_l) = 1$ for $l = (i+1)(\text{mod}\, r)$. For example, if $r = 2$, then for all $\mathbf{x} \in \mathcal{A}_0$ $K(\mathbf{x}, \mathcal{A}_1) = 1$ and for all $\mathbf{x} \in \mathcal{A}_1$, $K(\mathbf{x}, \mathcal{A}_0) = 1$, ensuring transitions from either $\mathcal{A}_0$ to $\mathcal{A}_1$ or $\mathcal{A}_1$ to $\mathcal{A}_0$ on each step. If no such r exists, K is aperiodic.

Ergodicity. In the general-state-space case, irreducibility, positive Harris recurrence, and aperiodicity, together with global balance, define ergodicity. Then, analogous to Theorem 6.7.2 for the discrete case, the limiting behavior in (6.52) and (6.53) holds for all $\mathcal{A} \in \mathcal{S}(\mathcal{X})$ with $K^{(j)}(\mathbf{x}, \mathcal{A}) - \pi(\mathcal{A})$ replacing $K^{(j)}(\mathbf{x}, \mathbf{y}) - \pi(\mathbf{y})$. Moreover, we say that the chain is geometrically ergodic if there exist a positive function $\{M(\mathbf{x})\}$ and $\rho \in [0, 1)$ such that for each $\mathbf{x} \in \mathcal{X}$,

$$\left|K^{(j)}(\mathbf{x}, \mathcal{A}) - \pi(\mathcal{A})\right| \leq M(\mathbf{x})\rho^j \quad \text{for all } \mathcal{A} \in \mathcal{S}(\mathcal{X}) \quad j \geq 1. \tag{6.76}$$

If there exist a finite positive constant M^* and $\rho \in [0, 1)$ such that for each $\mathbf{x} \in \mathcal{X}$,

$$\left|K^{(j)}(\mathbf{x}, \mathcal{A}) - \pi(\mathcal{A})\right| \leq M^*\rho^j \quad \text{for all } \mathcal{A} \in \mathcal{S}(\mathcal{X}) \quad j \geq 1, \tag{6.77}$$

then we say that the chain is uniformly ergodic.

6.13 POLYNOMIAL CONVERGENCE

Recall that the j-step transition kernel, $K^{(j)}$, of a finite-state ergodic Markov chain converges to the equilibrium distribution, π, at a geometric rate. This property suffices for $\sigma_t^2 \approx \sigma_\infty^2/t$ for large t. However, for continuous-state and countably infinite-state chains, there exist ergodic transition kernels that converge at a polynomial, rather than a geometric, rate and for which σ_t^2 converges more slowly than t^{-1}.

As illustration, suppose π is symmetric around $\mathbf{x} = \mathbf{0}$ and has tails that decrease as $|\mathbf{x}| := \left(x_1^2 + \cdots + x_m^2\right)^{-(\alpha+1)/2} \to \infty$ for some $\alpha > 0$. We say that π with this property is a *heavy-tailed distribution*. For bounded h, λ and $\text{var}_\pi h(\mathbf{X}) = R_0$ exist for all positive α. Although they may not exist for some unbounded h, the current discussion assumes that they do.

Although for π with heavy tails there exist transition kernels, K, that induce ergodicity, they converge at a polynomial, rather than geometric, rate; that is,

$$|K^{(j)}(\mathbf{x},\mathbf{y}) - \pi(\mathbf{y})| \le c(\mathbf{x}) j^{-\alpha} \qquad j = 1, \ldots, m,$$

for some positive $c(\mathbf{x})$ for all initial states $\mathbf{x} \in \mathcal{X}$. This bound implies a decidedly slower mode of convergence than (6.55). Nevertheless, if $\alpha > 2$, it can be shown that the limit (6.61) holds so that $\sigma_t^2 \approx \sigma_\infty^2/t$ for large t. However, for $\alpha \in (1, 2)$,

$$t^{\alpha-1}\sigma_t^2 \to \nu = \text{constant} < \infty \qquad \text{as } t \to \infty. \tag{6.78}$$

This limit has important implications:

- It holds for both bounded and unbounded h.
- For long sample paths, $\sigma_t^2 \approx \nu/t^{\alpha-1}$, revealing that statistical error in $\widehat{\lambda}_t$ decreases at a rate slower than t^{-1}.
- Because (6.61) does not hold, the Central Limit Theorem (6.56) does not apply, precluding the use of normal theory as a basis for assessing error.

A Markov chain with the limit (6.78) is said to have the property of *long-range dependence*. More often than not, a problem exhibiting this property requires an exceptionally long sample path to produce a solution that has an acceptable level of statistical accuracy. For example, this limitation has plagued MC-based analyses of Internet traffic flow where file size frequently has a heavy-tailed distribution and sample paths exhibit long-range dependence. Moreover, methods such as those discussed in Section 7.4 do not apply for estimating σ_∞^2 when long-range dependence holds.

6.14 DISCRETE-EVENT SYSTEMS

To estimate λ in (2.1) with g as in (6.1), HM sampling requires that the distribution π be known up to a constant. Gibbs sampling requires availability of the full conditional distributions associated with π. However, problems exist in which it is of interest to estimate λ, but analytical forms are unknown for π and its full conditional distributions. This is especially true in *discrete-event systems* where changes in state occur at random points in time and are intrinsic features of model behavior. Nevertheless, it is possible to sample randomly from other relevant distributions that collectively induce sample paths whose marginal distribution is the unknown π. Models with these limitations occur most notably in *queueing systems* where customers compete for limited resources and the purpose of inquiry is to characterize waiting time, queue length, and resource utilization. Although their distributions are unknown, the givens of the problem usually include distributions for customer interarrival times and resource service times. Together with the remaining problem specifications, sampling from the latter distributions induces sample paths for waiting times, queue length, and resource use, and these data serve as the basis for statistical inference about their equilibrium distributions and moments. The following example illustrates the principal concept.

EXAMPLE 6.14.1

SINGLE-SERVER SANDWICH SHOP

Consider a sandwich shop with a single sandwich maker. Interarrival times between successive customers are i.i.d. random variables with known distribution $\{f_a(z),\ z \geq 0\}$, and sandwich-preparation (service) times are i.i.d. random variables with known distribution $\{f_s(z),\ z \geq 0\}$. The sequences of customer interarrival and sandwich-preparation times are independent, and the sandwich maker serves customers in the order of arrival. It is of interest to characterize customer waiting times by their distribution $\{\pi_W(w), w \geq 0\}$ and mean waiting time λ_W. Although analytical results exist for this model (e.g., Gross and Harris 1974), it provides an instructive starting point for our relatively concise description of how MCMC applies to queueing models.

This distribution of waiting time illustrates the general-state space concept of Section 6.12. In particular,

$$\pi_W(x) := \pi_{W,\text{dis}}(x) + \pi_{W,\text{con}}(x) \qquad x \geq 0,$$

where

$$\pi_{W,\text{dis}}(x) \begin{cases} > 0 & \text{if } x = 0 \\ = 0 & \text{if } x > 0 \end{cases}$$

and

$$\pi_{W,\text{con}}(x) \begin{cases} > 0 & \text{if } x > 0 \\ = 0 & \text{elsewhere.} \end{cases}$$

Here $\pi_{W,\text{dis}}(0)$ denotes the probability that an arriving customer finds the sandwich maker idle and does not have to wait. As a consequence,

$$\lambda_W = \int_0^\infty x\pi_{W,\text{con}}(x)\text{d}x.$$

Let $Y_1, Y_2, \ldots$ denote a sequence of interarrival times from f_a, and $Z_1, Z_2, \ldots$, a sequence of sandwich preparation times from f_s. Let W_j denote the waiting time of the jth arrival for $j \geq 1$. Then the recursion of Lindley (1952),

$$W_{j+1} = \begin{cases} W_j - Y_j + Z_j & \text{if } Z_j > Y_j - W_j \\ 0 & \text{if } Z_j \leq Y_j - W_j \end{cases} \qquad j = 1, 2 \ldots, \tag{6.79}$$

provides a convenient basis for randomly generating waiting times. The assignments $W_0 = 0$ and $Y_1 = 0$ prime the sampling experiment and correspond to an arrival to an empty and idle sandwich shop at the beginning of MC sampling. The conditioning of W_j on W_{j-1} makes $\{W_j,\ j \geq 1\}$ a Markov process.

In order for the sample path $W_1, \ldots, W_t$ to be useful for statistical inference, f_a and f_s must satisfy specific conditions. Let

$$\mu_a := \int_0^\infty z f_a(z)\text{d}z$$
$$= \text{mean interarrival time}$$

and

$$\mu_s := \int_0^\infty z f_s(z) dz$$
$$= \text{mean sandwich preparation time.}$$

Then customer waiting time has a unique equilibrium distribution π_W if and only if $\mu_s < \mu_a < \infty$. Moreover, for all possible starting states (W_0, Y_1),

$$W_t \xrightarrow{d} \pi_W \text{ w.p.1} \qquad \text{as } t \to \infty,$$

where the rate of convergence decreases as μ_s/μ_a increases toward 1 and increases as μ_s/μ_a decreases.

For λ_W to be finite, an additional condition must be met. Given that $\mu_s < \mu_a$, a necessary and sufficient condition for $\lambda_W < \infty$ is that $\int_0^\infty z^3 f_s(z)dz < \infty$ (e.g., Daley 1967). This condition also ensures that $\{W_j, j \geq 1\}$ is uniformly ergodic. In the absence of this condition, $\lambda_W = \infty$, implying that the mean is not a meaningful descriptor of waiting time. Moreover, the waiting-time process $\{W_j, j \geq 1\}$ is not even geometrically ergodic, implying a convergence rate that is slower than geometric in t.

Analogous conditions are necessary and sufficient to ensure the existence of equilibrium distributions and convergence for many queueing models. These include models with multiple servers and queueing disciplines other than first-come-first-served. Because these problems rarely have analytical solutions, the ability to induce sample paths with marginal equilibrium distribution π_W, even though π_W is unknown, gives MCMC its appeal. More complex queueing systems are not as concisely represented as in (6.79). Nor does an analogous recursive expression exist for generating sample paths for queue length, the number of customers waiting for the sandwich maker. Nevertheless, more detailed algorithms offer ways of generating these sample paths. We describe the concept for the single-server model merely to illustrate how these algorithms work.

As illustration, four *elementary events* characterize the single-server sandwich shop problem:

A. A customer arrives

B. A sandwich is started

C. A sandwich is completed

D. A customer departs with a sandwich

Because events C and D always occur at the same time, we can combine them into a single *compound event* CD. However, event B is a *conditional event* that occurs at the same time as event A if there are no waiting customers (event AB) or at the same time as event CD if there are waiting customers (event CDB). In this system, the occurrence of an event A, AB, CD, or CDB denotes a change in state.

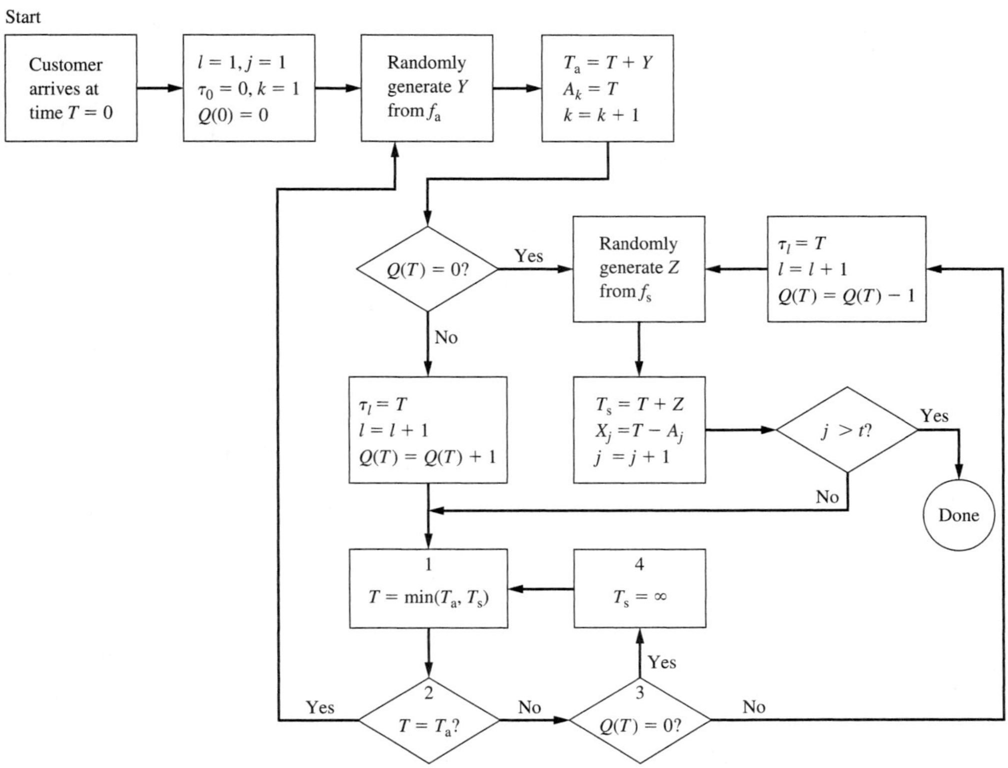

FIGURE 6.6 | *Single-server sandwich shop flowchart*

Let

$$T := \text{simulated time.}$$

$$Q(T) := \text{number of customers waiting for the sandwich maker at time } T \text{ (queue length).}$$

$$\tau_l := \text{time at which queue length changes for the } l\text{th time.}$$

$$t := \text{number of waiting times to be collected (sample-path length).}$$

Figure 6.6 shows the corresponding logic for randomly generating sample paths for queue length and waiting time. Execution begins with an arrival to an empty shop and an idle sandwich maker and continues until t customers have received sandwiches and departed. Note that these initial conditions induce $W_1 = 0$ and $Q(0) = 0$, convenient but not necessarily the only initial conditions that can be imposed. Block 1 determines the time of the next event. If $T = T_a$ is true in block 2, then an arrival (event A) occurs. Otherwise, a customer leaves the shop with a sandwich (event CD).

In the setting of Figure 6.6, the triplet $(Q(T), T_a, T_s)$ defines the state of the system, where $Q(T) \in \{0, 1, \ldots\}$, $0 \leq T_a < \infty$, and $0 \leq T_s < \infty$. At simulated time T, an event causes a transition from state $(Q(T), T_a, T_s)$ according to:

Event	Condition	To state
AB	$T = T_a$ and $Q(T) = 0$	$(Q(T), T_a + Y, T_s + Z)$
A	$T = T_a$ and $Q(T) > 0$	$(Q(T) + 1, T_a + Y, T_s)$
CD	$T = T_s$ and $Q(T) = 0$	$(Q(T), T_a, \infty)$
CDB	$T = T_s$ and $Q(T) > 0$	$(Q(T) - 1, T_a, T_s)$

Uniform ergodicity ensures that consistent estimates of π_W and λ_W based on $W_1, \ldots, W_t$ are available in a manner analogous to previous statistical analyses in this chapter. For example,

$$\widehat{\lambda}_{Wt} = \frac{1}{t - k_W} \sum_{j=k_W+1}^{t} W_j$$

estimates λ_W after deleting the first $k_W < t$ waiting times to reduce the bias that results from the initial conditions.

Let $\{\pi_Q(q);\ q = 0, 1, \ldots\}$ denote the equilibrium distribution of queue length. That is, $\pi_Q(q)$ is the probability that q customers are waiting at a randomly selected point in time. As an estimator of $\pi_Q(q)$, we have

$$\widehat{\pi}_{Qt}(q) := \frac{1}{\tau_{I(t)} - \tau_{k_Q}} \sum_{\substack{l=k_Q+1 \\ \text{s.t. } Q(\tau_{l-1})=q}}^{I(t)} (\tau_l - \tau_{l-1}) \qquad q = 0, 1, \ldots, \tag{6.80}$$

which, after deleting a warm-up interval τ_{k_Q}, is the ratio of the sum of the total simulated time with queue length q and total simulated time $\tau_{I(t)}$. As an estimate of mean queue length $\lambda_Q := \sum_{q=0}^{\infty} q\pi_Q(q)$, we have

$$\begin{aligned}\widehat{\lambda}_{Qt} &:= \frac{1}{\tau_{I(t)} - \tau_{k_Q}} \int_{\tau_{I(t)} - \tau_{k_Q}}^{\tau_{I(t)}} Q(z)dz \\ &= \frac{1}{\tau_{I(t)} - \tau_{k_Q}} \sum_{l=k_Q+1}^{I(t)} \int_{\tau_{l-1}}^{\tau_l} Q(z)dz \\ &= \sum_{q=0}^{\infty} q\widehat{\pi}_{Qt}(q).\end{aligned}$$

Note that $\widehat{\lambda}_{Qt}$ has alternative representations. It is a time average over the random time interval $\left[\tau_{k_Q}, \tau_{I(t)}\right)$. Because $\{Q(\tau),\ \tau \geq 0\}$ is constant over each time interval $\tau_l - \tau_{l-1}$ for $l = 1, \ldots, I(t)$, $\widehat{\lambda}_{Qt}$ also can be represented as an expectation with respect to the sample queue-length p.m.f. $\{\widehat{\pi}_{Qt}(q);\ q = 0, 1, \ldots\}$. The condition $\mu_s < \mu_a$ ensures the existence of π_Q, and the condition $\int_0^\infty z^3 f_s(z)dz < \infty$ ensures that for each $q = 0, 1, \ldots,$

$$\widehat{\pi}_{Qt}(q) \xrightarrow{d} \pi_q \ \text{ w.p.1} \qquad \text{as } t \to \infty$$

and

$$\widehat{\lambda}_{Qt} \xrightarrow{d} \lambda_Q \ \text{ w.p.1} \qquad \text{as } t \to \infty.$$

The single-server queueing model is the conceptually simplest of delay systems. More complex models require more detail than Figure 6.6 provides and, therefore, more effort devoted to model development. Indeed, the level of detail can quickly overwhelm a modeler's ability to articulate its logical flow explicitly. Two modeling concepts, the *event-scheduling approach,* and the *process-interaction approach,* significantly reduce this burden (e.g., Fishman 2001). Special-purpose *discrete-event simulation* applications software incorporates either one or both of these into their logical constructs. This software includes Arena, AutoMod, AweSim, GPSS/H, ProModel, Taylor, and SLX, descriptions of which are accessible on the Internet.

This software also improves computational efficiency, especially in determining the time and type of the next event. In Figure 6.6, simulated time advance occurs at block 1. It requires a comparison of T_a, the time of the next arrival, and T_s, the time at which the next sandwich is completed. If block 2 determines that the next event is sandwich completion (event CD) and no additional customers are waiting (block 3), then block 4 assigns $T_s = \infty$, ensuring an arrival (event A) as the next event.

An extension of the model reveals how computing time, to identify the next event, increases with problem size (m). Suppose that $m \geq 1$ sandwich makers work in the shop and T_{si} denotes the time at which server i completes her sandwich. Then $T = \min(T_a, T_{s1}, \ldots, T_{sm})$ determines the time of the next event and identifies when it is to be executed. If each determination relies on a direct comparison of $T_a, T_{s1}, \ldots, T_{sm}$, then computing time to make this determination is proportional to $m + 1$. The relatively high frequency with which this determination occurs makes this approach relatively costly in queueing systems that have many servers (large m). Therefore, most discrete-event simulation applications software incorporates *Henriksen's Algorithm,* which maintains the list of event times in a way that reduces computing time to be proportional to $\log m$ (Henriksen 1983). ■

6.15 FIRST-PASSAGE TIMES

The ability to generate a sample path whose marginal distribution is an unknown distribution π significantly increases the range of application of the MC method. However, an important distinction exists between this and earlier sampling experiments. When π is known up to a constant, HM sampling can employ many different nominating kernels to effect sampling. Each induces a transition kernel, which, provided that it satisfies global balance, leads to the desired steady-state properties. If steady-state behavior is the sole interest, then the form of this transition kernel is of incidental importance. When *conditional behavior* is the principal issue, a potential computing problem does exist. Then the choice of sampling mechanisms must necessarily be restricted to those that induce a transition kernel that has the requisite conditional properties. This is certainly the case in the discrete-event environment of Section 6.14. First-passage times illustrate the role that these properties play.

Let $\{\mathbf{X}_j,\ j \geq 0\}$ denote an ergodic Markov chain with state space $\mathcal{X}$. For some $\mathbf{x} \in \mathcal{X}$ and subset $\mathcal{Y} \subset \mathcal{X}$, it is occasionally of interest to describe the behavior of the *first-passage time*

$$\begin{aligned} J &:= \min(j \geq 1 : \ \mathbf{X}_j \in \mathcal{Y} \mid \mathbf{X}_0 = \mathbf{x}) \\ &= \text{first time the chain enters region } \mathcal{Y} \text{ given that it is} \\ &\quad\ \text{initially in state } \mathbf{x}. \end{aligned} \tag{6.81}$$

If $\mathcal{X}$ is discrete, then $\mathcal{Y}$ can be a single state $\mathbf{y} \in \mathcal{X}$. In the special case of $\mathbf{x} = \mathbf{y}$, J is called the *recurrence time* for state $\mathbf{x}$. If $\mathcal{X}$ is continuous, then $\mathcal{Y}$ can be any subset of $\mathcal{X}$ for which $\int_{\mathcal{Y}} \pi(\mathbf{y})d\mathbf{y} > 0$.

As illustration, suppose that the objective in Example 6.14.1 is to estimate the distribution of the number of customers that enter the shop until the sandwich maker becomes idle for the first time. Clearly this is a recurrence time representable as

$$J := \min(j \geq 1 : \ W_j = 0 \mid W_0 = 0)$$

because there is positive probability that an arriving customer does not have to wait. Alternatively, the objective may be to estimate the characteristics of

$$\begin{aligned} J &:= \min(j \geq 1 : \ W_j > w \mid W_0 = 0) \\ &= \text{first arrival that waits longer that } w \text{ for service.} \end{aligned}$$

If in a particular problem the transition kernel $\{K(\mathbf{x}, \mathbf{y}),\ (\mathbf{x}, \mathbf{y}) \in \mathcal{X}^2\}$ is known and $\mathcal{X}$ is discrete, then the first-passage time distribution can be computed by recursion (e.g., Kulkarni 1995, Chapter 4). When the kernel is unknown, MCMC offers a relatively straightforward approach to estimation that applies more generally to general-state spaces. It randomly generates n independent sample paths, each beginning in the same initial state $\mathbf{x}$, and terminates path i in the specified region $\mathcal{Y}$ on step $J^{(i)}$. Then elementary statistical methodology, as in Section 2.1, applies to the data $J^{(1)}, \ldots, J^{(n)}$.

Although the warm-up interval plays no role here, it may do so in some problems where starting the experiment in a particular state $\mathbf{x}$ is difficult. One strategy for discrete-state spaces starts replication i in an easily implementable state $\mathbf{z} \in \mathcal{X}$, samples along the path until $\mathbf{x}$ is entered, samples until $\mathcal{Y}$ is entered, and then takes

$$J^{(i)}(\mathbf{x}, \mathcal{Y}) = J^{(i)}(\mathbf{z}, \mathcal{Y}) - J^{(i)}(\mathbf{z}, \mathbf{x}),$$

where

$$\begin{aligned} J^{(i)}(\mathbf{w}, \mathcal{W}) := &\ \text{first-passage time from state } \mathbf{w} \in \mathcal{X} \text{ to} \\ &\ \text{region } \mathcal{W} \subset \mathcal{X} \text{ on sample-path } i, \end{aligned}$$

as the first-passage time from $\mathbf{x}$ to $\mathcal{Y}$. Different $\mathbf{z}$ can be used on different replications. Because any sampling method is acceptable for moving from $\mathbf{z}$ to $\mathbf{x}$, choosing an MC method and a $\mathbf{z}$ that saves steps is desirable on this portion of the sample path. If the relevant transition kernel $\{K(\mathbf{x}, \mathbf{y}),\ (\mathbf{x}, \mathbf{y}) \in \mathcal{X}^2\}$ is known explicitly, then other improvements in efficiency are available (e.g., Ross and Schechner 1985).

First-passage times need not themselves be the topic of principal interest. As illustration, it may be of interest in Example 6.14.1 to characterize

$$\begin{aligned} J &:= \min\,[l \geq 1 :\ Q(\tau_l) = 0] \\ &= \text{first state change in queue length that results in an empty queue.} \end{aligned}$$

Then on each sample path, τ_J is the datum of interest. Note that this quantity is essentially a continuous first-passage time.

6.16 ABSORBING STATES

Some modeling formulations call for a Markov chain $\{\mathbf{X}_j,\ j \geq 0\}$ with an *absorbing region* $\mathcal{Z} \subset \mathcal{X}$ such that

$$\text{pr}(\mathbf{X}_j \in \mathcal{Z} \mid \mathbf{X}_0 = \mathbf{x}_0 \in \mathcal{X} \backslash \mathcal{Z}) \to 1 \qquad \text{as } j \to \infty. \tag{6.82}$$

That is, once the chain enters $\mathcal{Z}$, it remains there forever. In the discrete case, $\mathcal{Z}$ can be a single *absorbing state* **z**. In contrast to irreducible Markov chains where the process $\{\mathbf{X}_j\}$ visits each state infinitely often with probability 1 as $j \to \infty$, a Markov chain satisfying (6.82) has a one-step transition kernel $\{K(\mathbf{x}, \mathbf{y}),\ (\mathbf{x}, \mathbf{y}) \in \mathcal{X}^2\}$ such that for each $\mathbf{x} \in \mathcal{X} \backslash \mathcal{Z}$,

$$\left.\begin{aligned} &K^{(j)}(\mathbf{x}, \mathcal{Z}) \to 1 \\ &K^{(j)}(\mathbf{x}, \mathbf{y}) \to 0 \qquad \text{for all } \mathbf{y} \text{ in } \mathcal{X} \backslash \mathcal{Z} \end{aligned}\right\} \quad \text{as } j \to \infty.$$

It is frequently of interest to characterize the chain's first-passage time

$$J := \min(j \geq 1 :\ \mathbf{X}_j \in \mathcal{Z} \mid \mathbf{X}_0 = \mathbf{x}_0 \in \mathcal{X} \backslash \mathcal{Z}).$$

For discrete-state spaces, the distribution of J can be evaluated recursively, provided $\{K(\mathbf{x}, \mathbf{y})\}$ is known explicitly (e.g., Kulkarni 1995, p. 167). When it is not known but other relevant distributions are given, then, as in Section 6.15, MCMC becomes a viable method for randomly generating $J^{(1)}, \ldots, J^{(n)}$ on n independent replications. Each starts in state $\mathbf{x}_0 \in \mathcal{X} \backslash \mathcal{Z}$. Statistical inference as in Section 2.1 allows one to estimate the properties of first-passage time. As illustration, we describe a discrete-event simulation evolving in *discrete time*. It reveals an interesting nuance not present in the continuous-time version of the problem.

EXAMPLE 6.16.1 THINKING AND DINING PHILOSOPHERS

Five philosophers are sitting at a round table. Each has a bowl of rice in front of him. There is a single chopstick between neighboring philosophers, each of whom spends his time either thinking or eating. When a philosopher ends a thinking period (gets hungry), he first picks up the chopstick to his left, next the chopstick to the right. Then he begins eating for a period of time. When his hunger is satisfied, he puts down the chopsticks and begins thinking again. A philosopher can begin eating only if the chopsticks on each side of him are idle. Otherwise, he must wait until they are idle.

Although a whimsical setting, this scenario's principal feature of sharing limited resources characterizes many practical problems. Most notably, this concept is illustrated in parallel-executing computer processors that share some memory and *critical regions* of code. A critical region is a section of code (chopstick) that distributed tasks (philosophers) need to execute without interruption. It is of interest to characterize the time that a task waits before being allowed to execute the code and to characterize the elapsed time to *deadlock*, the time until all processors are simultaneously prohibited from executing the code. Large waiting times are undesirable, and a deadlock can be catastrophic.

If time were continuous, then deadlock would not occur. If time is discrete, then the priority order for using resources may induce a deadlock. With this in mind, assume that

i. Thinking times are discrete i.i.d. random variables from the uniform distribution on $\{1, \ldots, 10\}$.

ii. Eating times are discrete i.i.d. random variables from the uniform distribution on $\{1, \ldots, 10\}$.

iii. If two neighboring philosophers become hungry at the same time, the one to the right of the idle chopstick (between them) has priority.

iv. At time $T = 0$, all philosophers begin a thinking period.

Assumptions **i** through **iii** ensure that a deadlock eventually occurs.

To characterize this system, it is of interest to estimate

- Mean, variance, and distribution function of waiting time.
- Mean, variance, and distribution function of time to deadlock.

For philosopher i on step $j \geq 0$, let

$$S_{ij} = \begin{cases} \text{thinking} & \text{if thinking} \\ \text{eating} & \text{if eating} \\ \text{lwait} & \text{if waiting for the left chopstick} \\ \text{rwait} & \text{if waiting for the right chopstick} \end{cases}$$

$$T_{ij} = \begin{cases} \text{time at which thinking stops} & \text{if } S_{ij} = \text{thinking} \\ \text{time at which eating stops} & \text{if } S_{ij} = \text{eating} \\ \infty & \text{if } S_{ij} = \text{lwait or rwait}. \end{cases}$$

Figure 6.7 describes an event-scheduling approach to simulating the philosophers problem, where

$$\text{etime} := T + L$$
$$\text{ttime} := T + M,$$

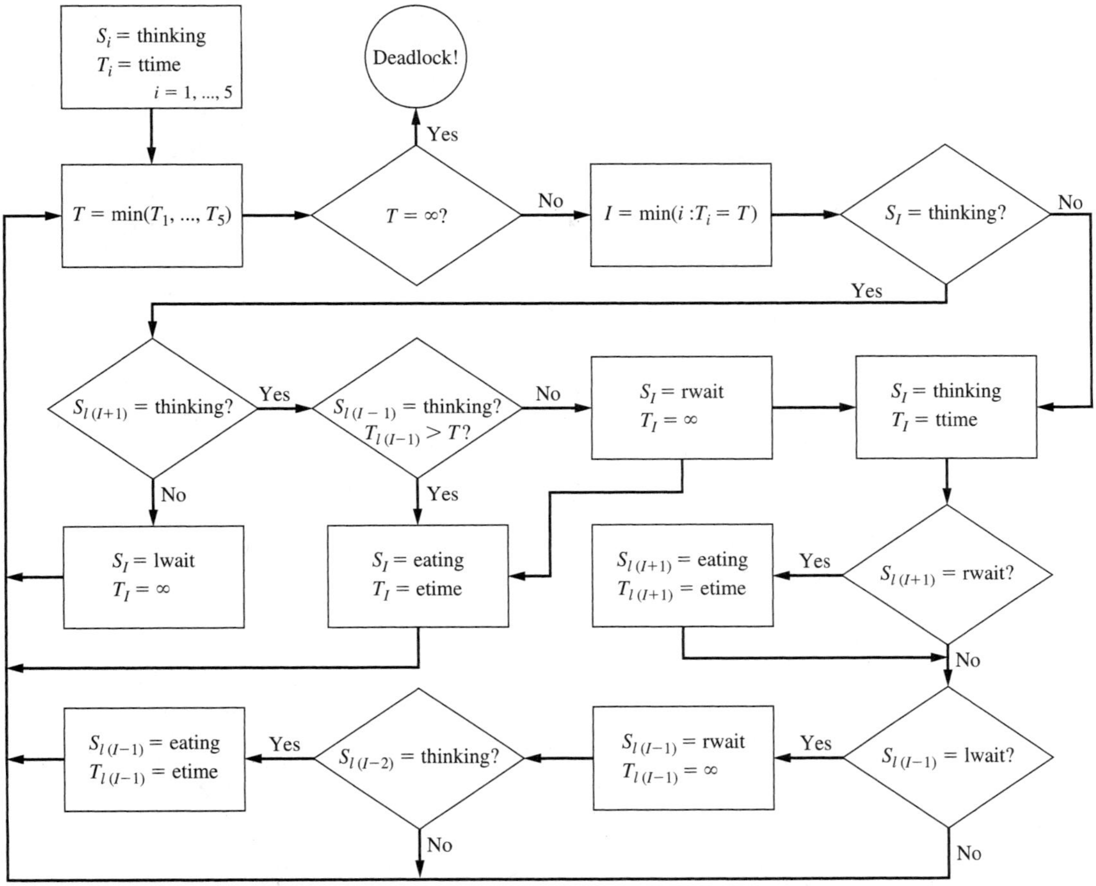

FIGURE 6.7 | *Thinking and dining philosophers flowchart*

T is simulated time, and L and M are independent samples from the uniform distribution on $\{1, \ldots, 10\}$. State change $j + 1$ occurs at time

$$T := \min(T_{1j}, \ldots, T_{5j}).$$

It changes the 10-dimensional state space representation from $(S_{1j}, \ldots, S_{5j}, T_{1j}, \ldots, T_{5j})$ to $(S_{1,j+1}, \ldots, S_{5,j+1}, T_{1,j+1}, \ldots, T_{5,j+1})$. The subscript j is implicit in the figure. The constants, thinking, eating, lwait, and rwait can be assigned any four distinct numerical values. The sampling experiment ends when the system enters state (rwait, rwait, rwait, rwait, rwait, $\infty, \infty, \infty, \infty, \infty$), which is deadlock.

Table 6.5 shows the summary statistics for deadlock time for $n = 1{,}000$ independent replications, each starting with the beginning of a thinking period for each philosopher. Figure 6.8 shows the empirical histogram. As one accuracy

TABLE 6.5 | *Deadlock time sample statistics*

Average $(\bar{\lambda}_n)$	12,206,434.11
Standard deviation $\left(\sqrt{s_n^2}\right)$	12,160,076.60
Coefficient of variation	0.996202
Minimum	9
Maximum	108,995,040
Number of independent observations (n)	1,000

Sample complementary d.f.	
pr$(T > t)$	t
0.99	100231
0.95	672730
0.90	1217204
0.80	2643560
0.70	4663827
0.60	6424099
0.50	8660615
0.40	11817483
0.30	15215449
0.20	19432718
0.10	27361117
0.05	34948573
0.01	51746034

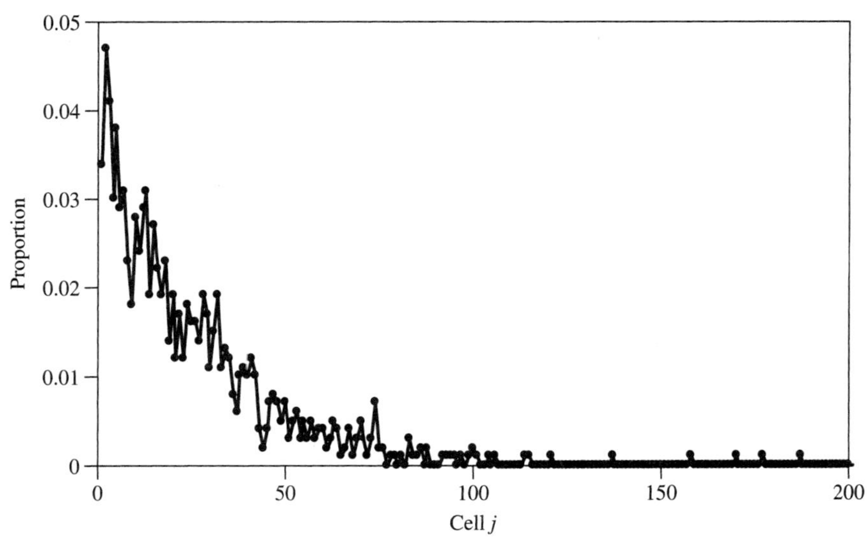

Cell j contains deadlock times in the interval $[500{,}000(j-1), 500{,}000j]$

FIGURE 6.8 | *Deadlock time histogram*

assessment, we have the .99 confidence interval for the mean deadlock time as

$$\bar{\lambda}_n \pm \tau_{n-1}(0.995)\sqrt{s_n^2/n} = 12,206,434 \pm 990,563.$$

A comparison of the entries in Table 6.5 with $\{e^{-t/\bar{\lambda}_n},\ t \geq 0\}$ suggests that the exponential distribution well approximates the empirical distribution. ■

6.17 LESSONS LEARNED

- When generating a sample directly from π is difficult, Markov chain Monte Carlo (MCMC) provides techniques for generating a sample path $\mathbf{X}_0, \mathbf{X}_1, \ldots, \mathbf{X}_{k-1}, \mathbf{X}_k, \ldots, \mathbf{X}_{t+k-1}$, where for k sufficiently large, $\mathbf{X}_k, \ldots, \mathbf{X}_{t+k-1}$ each have the desired distribution π.
- Generation usually begins with an $\mathbf{X}_0$ that is not from π. Then $\mathbf{X}_j$ is sampled from a conditional distribution, given $\mathbf{X}_{j-1}$ for $j \geq 1$. The one-step probability transition probability distribution must be endowed with certain properties so that the induced Markov process is ergodic, the condition required for the distribution of $\mathbf{X}_j$ to converge to π.
- Hastings-Metropolis (HM) sampling and Gibbs sampling are commonly employed methods for MCMC generation.
- On each generating step or iteration, just one coordinate is frequently selected and updated. The Deterministic Coordinate Selection (DCS) Rule selects coordinates in a fixed deterministic order, whereas the Random Coordinate Selection (RCS) Rule randomly selects the coordinate. By sampling from one-dimensional conditional distributions in this way, MCMC is able to produce sample paths without exceptional cost.
- Discrete-event simulation is a form of MCMC that generates events at random points in time. It is commonly used in the analysis of queueing systems.
- Another form of MCMC generates sample paths that begin and end in specified states. The ending state is frequently called the absorbing state. Ergodicity is not an issue here.

6.18 APPENDIX: MODIFIED ACCEPTANCE-REJECTION SAMPLING

Recall the earlier account of the acceptance-rejection method in Section 3.6 for randomly generating a sample from a univariate distribution. In principle, this method's extension to sampling from a multivariate distribution is relatively straightforward. In practice, it has a limitation: implementation requires knowledge of the maximum, $\kappa(\pi, q)$, or at least an upper bound on it. This information is rarely available for multivariate problems.

IHMS creates an opportunity for overcoming this limitation but, as we shall see, at the cost of frequently inducing dependence between elements on the sample path. Let $\{\eta(\mathbf{x}),\ \mathbf{x} \in \mathcal{X}\}$ denote an everywhere positive m-dimensional distribution, let b denote a positive constant, and suppose Algorithm MARS randomly nominates a state.

ALGORITHM **MARS**

Purpose: To sample $\mathbf{Y}$ randomly from q.

Given: Initial value b_0.

Method:

$b = b_0$.
(∗) Until $U \le \pi(\mathbf{Y})/b\eta(\mathbf{Y})$:
randomly sample $\mathbf{Y}$ from η and U from $\mathcal{U}(0,1)$.
Return $\mathbf{Y}$.

■

It is easily seen that the induced nominating distribution is

$$q(\mathbf{y}) := \begin{cases} \frac{\pi(\mathbf{y})}{bL(b)} & \text{if } \mathbf{y} \text{ satisfies } \pi(\mathbf{y}) \le b\eta(\mathbf{y}) \\ \frac{\eta(\mathbf{y})}{L(b)} & \text{otherwise} \end{cases}$$
$$= \frac{\min\left[\pi(\mathbf{y})/b,\ \eta(\mathbf{y})\right]}{L(b)} \qquad \mathbf{y} \in \mathcal{X}, \tag{6.83}$$

where

$$L(b) := b^{-1}\int_{\mathcal{C}} \pi(\mathbf{z})d\mathbf{z} + \int_{\overline{\mathcal{C}}} \eta(\mathbf{z})d\mathbf{z} \tag{6.84}$$
$$\mathcal{C} := \{\mathbf{y} \in \mathcal{X} :\ \pi(\mathbf{y}) \le b\eta(\mathbf{y})\}$$
$$\overline{\mathcal{C}} := \mathcal{X}\backslash\mathcal{C}.$$

Observe that $\left\{\frac{\pi(\mathbf{x})}{q(\mathbf{x})}\right\}$ exhibits less variation than $\left\{\frac{\pi(\mathbf{x})}{\eta(\mathbf{x})}\right\}$. In particular, $\frac{\pi(\mathbf{x})}{q(\mathbf{x})} = bL(b)$ for all $\mathbf{x} \in \mathcal{C}$.

Algorithm MARS extends the acceptance-rejection method by replacing the unknown $\kappa(\pi, \eta)$ with a user-specified b. We call this extension *modified acceptance-rejection sampling* (MARS). Some writers call it *rejection sampling* (e.g., Tierney 1994).

This development leads to the following theorem.

Theorem 6.18.1 (Tierney 1994.)

If q in (6.83) is the nominating kernel for HM sampling, then the corresponding acceptance function is

$$\alpha(\mathbf{x},\mathbf{y}) = \begin{cases} 1 & \text{if } \mathbf{x} \in \mathcal{C} \text{ and } \mathbf{y} \in \mathcal{X} \\ \frac{b\eta(\mathbf{x})}{\pi(\mathbf{x})} & \text{if } \mathbf{x} \in \overline{\mathcal{C}} \text{ and } \mathbf{y} \in \mathcal{C} \\ \min\left[1,\ \frac{\pi(\mathbf{y})\eta(\mathbf{x})}{\pi(\mathbf{x})\eta(\mathbf{y})}\right] & \text{if } \mathbf{x} \in \overline{\mathcal{C}} \text{ and } \mathbf{y} \in \overline{\mathcal{C}}. \end{cases} \tag{6.85}$$

■

PROOF Direct substitution of (6.83) into (6.8) leads to (6.85). ■

For all $\mathbf{x} \neq \mathbf{y}$, the acceptance function (6.85) induces the corresponding one-step transition kernel

$$K(\mathbf{x}, \mathbf{y}) = \begin{cases} \frac{\min[\pi(\mathbf{y})/b, \eta(\mathbf{y})]}{L(b)} & \text{if } \mathbf{x} \in \mathcal{C} \text{ and } \mathbf{y} \in \mathcal{X} \backslash \{\mathbf{x}\} \\ \frac{\pi(\mathbf{y})}{bL(b)} & \text{if } \mathbf{x} \in \overline{\mathcal{C}} \text{ and } \mathbf{y} \in \mathcal{C} \\ \min\left[1, \frac{\pi(\mathbf{y})\eta(\mathbf{x})}{\pi(\mathbf{x})\eta(\mathbf{y})}\right] \frac{\eta(\mathbf{y})}{L(b)} & \text{if } \mathbf{x} \in \overline{\mathcal{C}} \text{ and } \mathbf{y} \in \overline{\mathcal{C}} \backslash \{\mathbf{x}\}. \end{cases} \tag{6.86}$$

For Algorithm MARS,

$$\text{pr}\,[U \leq \pi(\mathbf{Y})/b\eta(\mathbf{Y}) \mid \mathbf{Y} = \mathbf{y}] = \min[1, \pi(\mathbf{y})/b\eta(\mathbf{y})],$$

so the mean number of trials until success on the MARS step is

$$\begin{aligned} \gamma(b, \pi, \eta) &:= \frac{1}{\text{pr}\,[U \leq \pi(\mathbf{Y})/b\eta(\mathbf{Y})]} \\ &= \frac{1}{L(b)}. \end{aligned} \tag{6.87}$$

As to the upper bound (6.63) on σ_∞^2,

$$\begin{aligned} \kappa(\pi, q) &:= \max_{\mathbf{y} \in \mathcal{X}} \left[\frac{\pi(\mathbf{y})}{q(\mathbf{y})}\right] \\ &= L(b) \max\,[b, \kappa(\pi, \eta)]\,. \end{aligned}$$

The standard acceptance-rejection method takes $b = \kappa(\pi, \eta)$. This implies

- $\mathcal{C} = \mathcal{X}$.
- $\gamma(b, \pi, \eta) = \kappa(\pi, \eta)$.
- Acceptance w.p.1 in the HM step.
- For all $\mathbf{y} \in \mathcal{X}$, $q(\mathbf{y}) = \pi(\mathbf{y})$ and $K(\mathbf{x}, \mathbf{y}) = \pi(\mathbf{y})$, so $\mathbf{X}_1, \mathbf{X}_2, \ldots$ are i.i.d. from π and, therefore, $\sigma_\infty^2 = R_0$ in (6.63).

Because $\kappa(\pi, \eta)$ is rarely known, assigning $b = \kappa(\pi, \eta)$ is usually not possible. Therefore, it is of interest to see how IHMS using MARS behaves as a function of b. For $b > \kappa(\pi, \eta)$, it is easily seen that

$$\begin{aligned} \mathcal{C} &= \mathcal{X} \\ \overline{\mathcal{C}} &= \emptyset \\ L(b) &= b^{-1} \\ \gamma(b, \pi, \eta) &= b \\ \kappa(\pi, q) &= 1 \\ \sigma_\infty^2 &= R_0. \end{aligned}$$

Therefore, acceptance on the HM step occurs w.p.1, and $\mathbf{X}_1, \mathbf{X}_2, \ldots$ form an i.i.d. sequence from π. However, the mean number of trials until success on the MARS step increases with b.

If $b < \kappa(\pi, \eta)$, then

$$\begin{aligned} \overline{\mathcal{C}} &\neq \emptyset \\ L(b) &> b^{-1} \\ \gamma(b, \pi, \eta) &< b \\ \kappa(\pi, q) &= L(b)\kappa(\pi, \eta) \\ &> b^{-1}\kappa(\pi, \eta), \end{aligned}$$

so

$$\begin{aligned} \sigma_\infty^2 &\leq R_0[2\kappa(\pi, q) - 1] \\ &\leq R_0[2b^{-1}\kappa(\pi, \eta) - 1]. \end{aligned}$$

In this case, the mean number of trials until success on the MARS step is less than for standard acceptance-rejection, but the bound on σ_∞^2 exceeds R_0 and increases with decreasing b.

For some $\zeta > 0$,

$$t^* := \text{largest integer less than or equal to } \sigma_\infty^2/\zeta$$

denotes the minimal number of IHMS iterations required in a sample path to achieve $\mathrm{var}\widehat{\lambda}_t \leq \zeta$ for small ζ. Then $\gamma(b, \pi, \eta)t^*$, the total minimal mean number of trials required on the MARS steps to satisfy this criterion, has the upper bound

$$\gamma(b, \pi, \eta)t^* \leq \left\{2 \max\left[b, \kappa(\pi, \eta)\right] - \frac{1}{L(b)}\right\} \times \frac{R_0}{\zeta}. \tag{6.88}$$

Because $b = \kappa(\pi, \eta)$ gives $\gamma(b, \pi, \eta)t^* = \kappa(\pi, \eta)R_0/\zeta$, a comparison with upper bound (6.88) reveals that standard acceptance-rejection allows one to compute an estimate with specified accuracy in total generating time less than or equal to that offered by IHMS using MARS. If, as frequently is the case, $\kappa(\pi, \eta)$ is unknown and high accuracy is required (i.e., ζ is small), then a long sample path is often required, and it pays to conduct a search to estimate $\kappa(\pi, \eta)$ and to assign the estimate to b.

One approach merely adds the statement

$$b_0 = \max\left[b, \frac{\pi(\mathbf{y})}{\eta(\mathbf{Y})}\right] \tag{6.89}$$

after step (*) in Algorithm MARS. If $b_0 < \kappa(\pi, \eta)$, then this addition has the effect of increasing b toward $\kappa(\pi, \eta)$ as sampling progresses. If $b_0 > \kappa(\pi, \eta)$, then $b = b_0$ on all sample generations. Because this assignment results in a higher-than-desired rejection rate on the MARS step, a judicious choice is to make b_0 small; for example, $b_0 = 1$.

Let us call the amended Algorithm AMARS. When π and η are discrete, Algorithm AMARS with $b_0 = 1$ and Algorithm MARS with $b_0 = \kappa(\pi, \eta)$ generate sequences of accepted values that differ at finitely many points w.p.1 as $t \to \infty$ (Caffo, et al. 2002). That is, the two sequences are essentially the same. This result does not quite hold for continuous-state spaces. However, if Algorithm AMARS generates $\mathbf{X}_1, \ldots, \mathbf{X}_t$ and the objective is to estimate λ, then without the HM step (Caffo, et al. 2002),

$$\frac{t^{-1}\sum_{j=1}^{t} h(\mathbf{X}_j) - \lambda}{\sqrt{R_0/t}} \xrightarrow{\text{d}} \mathcal{N}(0, 1) \qquad \text{as } t \to \infty.$$

When using Algorithm AMARS, a tendency for the empirical acceptance rate to approach 1 on successive HM steps indicates that b is close to $\kappa(\pi, \eta)$ and that the dependence among $\mathbf{X}_1, \ldots, \mathbf{X}_t$ is diminishing. If $b_0 > 1$ is initially assigned, presumably to speed up convergence, then a failure of b to change value from the very first generation suggests that $b_0 > \kappa(\pi, \eta)$. If $\varphi \neq 1$ but is finite in (6.2), the developments in this appendix continue to apply, with π/φ replacing π everywhere.

Because $\kappa(\pi, \eta)$ frequently is an increasing function of dimension m, IHMS with MARS may, in practice, turn out to induce a high unit generating cost. Then it pays to consider other methods, including single-coordinate updating. The latter likely will induce greater dependence than IHMS with MARS, but with a lower unit generating cost.

6.19 EXERCISES

1. Show that Algorithm RGIBBS induces a transition kernel that satisfies detailed balance.
2. Show that the conditional distribution of θ in (6.43) is negative binomial.
3. Given the Beta prior (6.32), show that the full conditional distribution of each independent p_i in (6.42) is $\mathcal{B}e(N_i + a, \theta - N_i + b)$.
4. Derive the one-step transition kernel in (6.67).
5. Derive the identity (6.72) and show how it allows $B(\mathbf{X}_j)$ to be computed recursively from $B(\mathbf{X}_{j-1})$ in time independent of I and L and, therefore, independent of m.

6.20 HANDS-ON EXERCISES

The hands-on exercises in this chapter familiarize students with the mechanics of MCMC. Exercises H.1 through H.12 focus on HM and Gibbs sampling, whereas Exercises H.13 and H.14 illustrate how MCMC applies to discrete-event systems. The problems challenge students to generate sample paths whose elements have the desired distribution and, in the case of discrete-event systems, also have the desired conditional distributions. Later, Chapter 7 describes methods for evaluating the accuracy of estimates based on these and related data.

Hastings-Metropolis Sampling

H.1. Apply HM sampling to estimate λ in (2.1) with

$$g(x, y) = h(x, y)\pi(x, y) \qquad 0 \le x \le \frac{\pi}{4}, \frac{\pi}{2} \le y \le \frac{3\pi}{4},$$

where

$$\left.\begin{aligned} h(x, y) &:= Cx^2y^2 \ln(x + y) \\ \pi(x, y) &:= C^{-1} \sin(x + y) \\ C &:= \int_0^{\frac{\pi}{4}} \int_{\frac{\pi}{2}}^{\frac{3\pi}{4}} \sin(x + y)\mathrm{d}x\mathrm{d}y. \end{aligned}\right\} \tag{6.90}$$

Note that g is the integrand in Exercise 2.H.1.

a. Specify nominating distributions for sampling variable x and variable y separately.

b. Show that the one-step transition kernels satisfy detailed balance.

c. Using the DCS rule, describe and execute an algorithm that randomly generates at least three sample paths, each with a different starting point. Use these data to determine a suitable warm-up interval. Justify your choice.

d. After discarding the warm-up interval, randomly generate a sample path of $t = 10^6$ steps and use it to estimate λ and the corresponding acceptance ratio.

e. Use the *correlogram,*

$$\widehat{\rho}_j := \frac{\widehat{R}_j}{\widehat{R}_0} \qquad j = 0, 1 \ldots, 100, \tag{6.91}$$

where

$$\widehat{R}_j := \frac{1}{t-j} \sum_{l=1}^{t-j} \left[h(X^{(l)}, Y^{(l)}) - \widehat{\lambda}_t\right] \left[h(X^{(l+j)}, Y^{(l+j)}) - \widehat{\lambda}_t\right],$$

to characterize the extent of dependence among points on the sample path. Is your choice of warm-up interval consistent with the correlogram? Explain your answer.

H.2. Using the RCS rule, repeat Exercises H.1c through H.1e. Comment on the similarities and differences in results here with those in Exercise H.1.

H.3. Apply HM sampling to estimate λ in (2.1) with

$$g(x, y) = h(x, y)\pi(x, y) \qquad 0 \le x \le \frac{\pi}{2}, \quad 0 \le y \le 1,$$

where

$$\left.\begin{aligned} h(x, y) &:= C \sin^2\left(x + \tfrac{\pi}{2} y\right) \\ \pi(x, y) &:= C^{-1} x e^{xy} \\ C &:= \int_0^{\frac{\pi}{2}} \int_0^1 x e^{xy} dx dy. \end{aligned}\right\} \tag{6.92}$$

Note that g is the integrand in Exercise 2.H.3.

a. Specify nominating distributions for sampling variable x and variable y separately.

b. Show that the resulting one-step transition kernels each satisfy detailed balance.

c. Using the DCS rule, describe an algorithm and use it to generate at least three random sample paths, each with a different starting point. Use these data to determine a suitable warm-up interval. Justify your choice.

d. After discarding the warm-up interval, randomly generate a sample path of $t = 10^6$ steps and use it to estimate λ.

e. Use the correlogram (6.91) to characterize the extent of dependence among points on the sample path.

H.4. Using the RCS rule, repeat Exercises H.3c through H.3e. Comment on the similarities and differences in results here with those in Exercise H.3.

H.5. Describe an HM sampling plan that uses the DCS rule to estimate

$$\lambda := \frac{1}{|\mathcal{X}|}\sum_{\mathbf{x}\in\mathcal{X}} g(\mathbf{x}) \quad \mathbf{x} := (x_1, x_2, x_3, x_4), \tag{6.93}$$

where $\mathcal{X}$ denotes all points in the polytope (2.94) and

$$g(\mathbf{x}) := \begin{cases} \frac{1}{4}\sum_{i=1}^{4}\sum_{j=1}^{4}(x_i - x_j)^2 & \mathbf{x} \in \mathcal{X} \\ 0 & \text{elsewhere.} \end{cases}$$

a. Evaluate the acceptance function explicitly.

b. Randomly generate at least three sample paths, each with a different starting point, to determine a suitable warm-up interval. Justify your choice.

c. After discarding the warm-up interval, randomly generate a sample path of $t = 10^6$ steps and use it to estimate λ and the acceptance rate.

d. Use the correlogram (6.91) to characterize the extent of dependence among points on the sample path.

H.6. Repeat Exercises H.5a through H.5c using the RCS rule.

Gibbs Sampling

H.7. Apply Gibbs sampling to estimate λ in (2.1) for h in (6.90).

a. Derive the conditional p.d.f.s from the joint p.d.f. in (6.90).

b. Describe algorithms for randomly sampling from $\{\pi(y \mid x)\}$ and $\{\pi(x \mid y)\}$.

c. Show that the one-step transition kernels each satisfy detailed balance.

d. Using the DCS rule, randomly generate at least three sample paths, each with a different starting point, to determine a suitable warm-up interval. Justify your choice.

e. After discarding the warm-up interval, randomly generate a sample path of $t = 10^6$ steps and use it to estimate λ.

f. Use the correlogram (6.91) to characterize the extent of dependence among points on the sample path, and compare it with the correlogram in Exercise H.1.

H.8. Using the RCS rule, repeat Exercises H.7d through H.7f. Comment on the similarities and differences in results here with those in Exercises H.1, H.2, and H.7.

H.9. Apply Gibbs sampling to estimate λ in (2.1) for h in (6.92).

a. Derive the conditional p.d.f.s corresponding to the joint p.d.f. in (6.92).

b. Describe algorithms for randomly sampling from $\{\pi(y \mid x)\}$ and $\{\pi(x \mid y)\}$.

c. Show that the resulting transition kernels each satisfy detailed balance.

d. Using the DCS rule, randomly generate at least three sample paths, each with a different starting point, to determine a suitable warm-up interval. Justify your choice.

e. After discarding the warm-up interval, randomly generate a sample path of $t = 10^6$ steps and use it to estimate λ.

f. Use the correlogram (6.91) to characterize the extent of dependence among points on the sample path, and compare it with the correlogram in Exercise H.3.

H.10. Using the RCS rule, repeat Exercises H.9d through H.9f. Comment on the similarities and differences in results here with those in Exercises H.3, H.4, and H.7.

H.11 Repeat Exercise H.5, using the DCS rule with Gibbs sampling.

a. Describe an algorithm based on the following representation. Let

$$S_{lj} = \sum_{i=1} a_{ij} X_{ij} \qquad\qquad l = 1, \ldots, r, \;\; j = 1, 2, \ldots,$$

and suppose $S_{l,j-1} < 1$ for all $l = 1, \ldots, r$. On step j, let $X_{2j} = X_{2,j-1}$, $X_{3j} = X_{3,j-1}$, $X_{4j} = X_{4,j-1}$. It can be shown that all constraints are satisfied if and only if $X_{1j} \in \left[A_{1j}, B_{1j}\right]$, where

$$A_{1j} := \min_{1 \le l \le r,\; a_{1l} < 0} \left(\frac{1 - S_{l,j-1} + a_{1l} X_{1,j-1}}{a_{1l}} \right)$$

$$B_{1j} := \max_{1 \le l \le r,\; a_{1l} > 0} \left(\frac{1 - S_{l,j-1} + a_{1l} X_{1,j-1}}{a_{1l}} \right).$$

An analogous property holds for coordinates 2, 3, and 4. It is easily seen that this procedure is Gibbs sampling.
Let

$$i(j) := (j - 1) \pmod 4 + 1.$$

Given an initial state $X_{10}, \ldots, X_{40}$, satisfying $S_{l0} < 1$ for each $l = 1, \ldots, r$, it can be shown that if the DCS rule is employed to update $X_{i(j)j}$ by sampling from $\mathcal{U}(A_{i(j)j}, B_{i(j)j})$, then $\left(X_{1j}, X_{2j}, X_{3j}, X_{4j}\right)$ has a distribution that converges to the uniform distribution on $\mathcal{X}$ as $j \to \infty$.

b. Randomly generate at least three sample paths, each with a different starting point, to determine a suitable warm-up interval. Justify your choice.

c. After discarding the warm-up interval, randomly generate a sample path of $t = 10^6$ steps and use it to estimate λ and the acceptance rate.

d. Use the correlogram (6.91) to characterize the extent of dependence among points on the sample path.

H.12. Repeat Exercise H.11, using the RCS rule.

Discrete-Event Systems

Exercises in this section allow considerable discretion about how to perform a study that employs MCMC. In addition to warm-up analysis and sample-path length, you are to choose the number of experiments, their input, and the order in which they are to be executed to arrive at an answer.

H.13. Interarrival times to the sandwich shop in Example 6.14.1 are i.i.d. from the exponential distribution with mean 2.7 minutes. The shop offers a roast beef sandwich with honey mustard, lettuce, and tomato, and its preparation times are i.i.d. from the distribution

$$f_s(z) = \begin{cases} \beta^{-1} e^{-(z-z_0)/\beta} & z \geq z_0 \\ 0 & \text{elsewhere,} \end{cases} \tag{6.94}$$

where $z_0 = 30$ seconds is a setup time for every customer and $\beta = 2$ minutes. The shop owner wants to reduce customer waiting time by speeding up sandwich preparation (reducing β).

Describe and perform a sequence of MCMC experiments to estimate how much mean service time would have to be reduced, if at all, in order for $\text{pr}(W_j \geq 2.50) = 0.05$, where W_j denotes the waiting time of customer j in equilibrium. As a first step, show that an equilibrium distribution exists. Give the interarrival time distribution and the current total service-time distribution.

H.14. Three monasteries have the same basic setup as in Example 6.16.1, except for one difference: the "Well-Balanced Monastery" has equal mean thinking times and mean eating times. At the "Gluttonous Monastery," philosophers have a mean eating time of 12 and a mean thinking time of 8. Conversely, at the "Cerebral Monastery," philosophers have a mean eating time of 8 and a mean thinking time of 12. Although mean eating times differ and mean thinking times differ, each philosopher experiences 20-minute cycles of eating and thinking on average.

Using Figure 6.7 as a guide, perform $n = 1{,}000$ independent replications each for the Gluttonous and Cerebral Monasteries. Describe how the difference of habits affects deadlock times when compared to the profile in Table 6.5. What moral do you draw from this example?

6.21 REFERENCES

Berger, J. O. (1985). *Statistical Decision Theory and Bayesian Analysis*, Springer-Verlag, New York.

Besag, J. (1974). Spatial interaction and the statistical analysis of lattice systems, *Journal of the Royal Statistical Society B*, **36**, 192–225.

Caffo, B. S., J. G. Booth, and A. C. Davison (2002). Empirical supremum rejection sampling, *Biometrika,* **89,** 745–754.

Castledine, B. (1981). A Bayesian analysis of multiple-capture sampling for a closed population, *Biometrika,* **67,** 197–210.

Daley, D. J. (1967). The serial correlation coefficients of waiting times in a stationary single server queue, *Journal of the Australian Mathematical Society,* **8,** 683–699.

Diaconis, P., and B. Sturmfels (1998). Algebraic algorithms for sampling from conditional distributions, *Annals of Statistics,* **26,** 363–397.

Fishman, G. S. (2001). *Discrete-Event Simulation: Modeling, Programming, and Analysis,* Springer-Verlag, New York.

George, E. I., and C. P. Robert (1992). Capture-recapture estimation via Gibbs sampling, *Biometrika,* **79,** 677–683.

Gross, D., and C. M. Harris (1974). *Fundamentals of Queueing Theory,* Wiley, New York.

Hastings, W. K. (1970). Monte Carlo sampling methods using Markov chains and their applications, *Biometrika,* **57,** 92–109.

Henriksen, J. O. (1983). Even management: A tutorial, *Proceedings of the 1983 Winter Simulation Conference,* IEEE and SCS, Piscataway, N.J., 543–551.

Kulkarni, V. G. (1995). *Modeling and Analysis of Stochastic Systems,* Chapman and Hall, London.

Lindley, D. V. (1952). The theory of queues with a single server, *Proceedings of the Cambridge Philosophical Society,* **48,** 277–289.

Liu, J. S. (1996a). Metropolized independent sampling with comparisons to rejection sampling and importance sampling, *Statistics and Computing,* **6,** 113–119.

Liu, J. S. (1996b). Peskun's theorem and a modified discrete-time Gibbs sampler, *Biometrika,* **83,** 681–682.

Mengersen, K. L., and R. L. Tweedie (1996). Rates of convergence of the Hastings-Metropolis algorithms, *Annals of Statistics,* **24,** 101–121.

Metropolis, N., A. W. Rosenbluth, M. N. Rosenbluth, A. H. Teller, and E. Teller (1953). Equations of state calculations by fast computing machines, *Journal of Chemical Physics,* **21,** 1087–1092.

Peskun, P. H. (1973). Optimum Monte-Carlo sampling using Markov chains, *Biometrika,* **60,** 607–612.

Roberts, G. O., and J. S. Rosenthal (1997). On convergence rates of Gibbs samplers for uniform distributions, *Annals of Applied Probability,* **8,** 1291–1302.

Roberts, G. O., and R. L. Tweedie (1996). Geometric convergence and the central limit theorems for multidimensional Hastings and Metropolis algorithms, *Biometrika,* **83,** 95–110.

Ross, S. M., and Z. Schechner (1985). Using simulation to estimate first passage distribution, *Management Science,* **31,** 224–234.

Smith, R. L., and L. Tierney (1996). Exact transition probabilities for the independence Metropolis sampler, Department of Statistics, University of North Carolina at Chapel Hill, to appear in *Bernoulli.*

Snee, R. D. (1974). Graphical display of two-way contingency tables, *American Statistician,* **38,** 9–12.

Tierney, L. (1994). Markov chains for exploring posterior distributions, *Annals of Statistics,* **22,** 1701–1762.

CHAPTER 7

MCMC Sample-Path Analysis

MCMC generates sample paths from which a numerical estimate of an unknown quantity, λ, can be computed. The paths also contain data from which an assessment of how well the estimate approximates λ can be obtained. This assessment is crucial for deciding how much confidence to place in the estimate in subsequent decision making. Although this need for assessment is no different than that for IMC-based estimates, the methodology for achieving it differs. This chapter describes the principal steps for designing and executing a sampling experiment that leads to this MCMC error assessment.

Four quantities play central roles:

$\pi_0 :=$ distribution for sampling the initial state $\mathbf{X}_0$

$k :=$ iteration number on the same sample path on which to begin data collection for estimation

$n :=$ number of independent replications

$t :=$ total number of observations to collect for estimating λ on all replications

Two alternative methods of assessment for MCMC exist. One is based on $n > 1$ independently generated sample paths of length t', the other on a single sample path ($n = 1$) sample path of length $t = nt'$. Each has benefits and limitations. We describe both methods.

7.1 MULTIPLE INDEPENDENT REPLICATIONS

Algorithm INDREPS describes the principal steps for estimating λ from data collected on $n > 1$ independent replications, each with sample-path length t'. Using standard statistical methodology for i.i.d. observations, it computes n independent sample averages $\bar{\lambda}_{t'}^{(1)}, \ldots, \bar{\lambda}_{t'}^{(n)}$, one for each independent replication, and uses them to compute a grand sample average $\bar{\lambda}_{nt'}$. It next computes the sample variance $s^2_{nt'}$ of a $\bar{\lambda}^{(l)}_{t'}$, the standard error and relative error of $\bar{\lambda}_{nt'}$, and a confidence interval (CI) for λ, where $\tau_{n-1}(1-\delta/2)$ denotes the $1-\delta/2$ critical value of the Student t distribution with $n-1$ degrees of freedom.

ALGORITHM **INDREPS**

Purpose: To estimate λ by $\bar{\lambda}_{nt'}$ from data on iterations $k, k+1, \ldots, k+t'-1$ on each of n independent sample paths, each of length t', and to estimate the error of approximation.

Given: $\pi_0, k, n, t', \delta := 1 -$ confidence level ($\in (0, 1)$).

Method:

Until $l > n$:
 choose an initial state $\mathbf{X}^{(l)}_0$ from π_0
 randomly generate $\mathbf{X}^{(l)}_1, \ldots, \mathbf{X}^{(l)}_{k+t'-1}$
 $\bar{\lambda}^{(l)}_{t'} = \frac{1}{t'}\sum_{j=k}^{k+t'-1} h(\mathbf{X}^{(l)}_j)$
 $l = l+1.$

Compute:

$$\bar{\lambda}_{nt'} = \frac{1}{n}\sum_{l=1}^{n} \bar{\lambda}^{(l)}_{t'}.$$

$$s^2_{nt'} = \frac{1}{n-1}\sum_{l=1}^{n}\left[\left(\bar{\lambda}^{(l)}_{t'}\right)^2 - n\widehat{\lambda}^2_{t'}\right].$$

$$\text{s.e.}(\bar{\lambda}_{nt'}) = \sqrt{\frac{s^2_{nt'}}{n}}.$$

$$\text{r.e.}(\bar{\lambda}_{nt'}) = \frac{\text{s.e.}(\bar{\lambda}_{nt'})}{\bar{\lambda}_{nt'}},$$

approximating

$$100 \times (1-\delta)\%\ \text{CI} = [\bar{\lambda}_{nt'} - \tau_{n-1}(1-\delta/2) \times \text{s.e.}(\bar{\lambda}_{nt'}),$$
$$\bar{\lambda}_{nt'} + \tau_{n-1}(1-\delta/2) \times \text{s.e.}(\bar{\lambda}_{nt'})].$$ ■

Under relatively general conditions, the estimator $\bar{\lambda}_{nt'}$ has the properties: As $t' \to \infty$,

P27. $\bar{\lambda}_{nt'} \to \lambda$ w.p.1

P28. $\sqrt{t'}\left(\bar{\lambda}_{nt'} - \lambda\right) \xrightarrow{d} \mathcal{N}(0, \sigma^2_\infty)$

P29. $(n-1)t's^2_{nt'} \to \sigma^2_\infty \chi^2_{n-1}$ w.p.1, where χ^2_{n-1} is a chi-squared random variable with $n-1$ degrees of freedom.

P30. $\dfrac{\bar{\lambda}_{nt'}-\lambda}{\sqrt{s^2_{nt'}/n}} \xrightarrow{d} \tau_{n-1}$, where τ_{n-1} denotes the Student t distribution with $n-1$ degrees of freedom.

Limit P30 establishes the asymptotic validity of the confidence interval in Algorithm INDREPS. Based on independent replications, this approach has the virtue of estimating error in $\bar{\lambda}_{nt'}$ by relatively elementary statistical methods, analogous to those described in Section 2.1 for IMC. However, it has a limitation. For $\bar{\lambda}_{nt'}$ to be relatively free of the influence of the initial state $\mathbf{X}_0$ generated from distribution $\pi_0 \neq \pi$ on each replication, the algorithm discards $k-1$ data points on each replication for a total of $n(k-1)$ observations.

To keep the cost of truncation relatively low when k is large, there is a tendency in practice to make n small and t' large. However, small n implies few degrees of freedom for $s^2_{nt'}$ regardless of the size of t'. Because the length of the confidence interval is proportional to the critical value, $\tau_{n-1}(1-\delta/2)$, which is a strictly decreasing function of n, small n implies greater length. Because $\tau_{n-1}(1-\delta/2) \to \Phi^{-1}(1-\delta/2)$ as $n \to \infty$, the ratios

n	$1-\delta$	$\frac{\tau_{n-1}(1-\delta/2)}{\Phi^{-1}(1-\delta/2)}$
11	0.01	1.23
	0.05	1.14
21	0.01	1.10
	0.05	1.06
31	0.01	1.07
	0.05	1.04

indicate that $n = 31$ induces a relatively small increase in critical value over that for the standard normal distribution.

Although the CI in Algorithm INDREPS remains valid for fixed n as $t = nt'$ increases, this error assessment, based on a fixed n, creates another limitation: because property P29 reveals that $s^2_{nt'}$ remains a random variable as t' increases for fixed n, $t's^2_n$ is not a consistent estimator of σ^2_∞. That is,

$$t's^2_n \nrightarrow \sigma^2_\infty \text{ w.p.1} \qquad \text{as } t' \to \infty.$$

Clearly, a consistent estimator for σ^2_∞ would be regarded as preferable. This can be accomplished by allowing n, as well as t', to increase as $t = nt'$ becomes large. But again, one needs to weigh the benefit of a better estimate of $\text{var}\bar{\lambda}_{nt'}$ versus the additional cost of discarding $n(k-1)$ observations for n warm-up intervals.

7.2 SINGLE SAMPLE PATH

The number of discarded data points can be reduced significantly by estimating λ from data collected on a single ($n = 1$) MCMC-generated sample path of length $t = nt'$. Algorithm SINREP describes the principal steps.

ALGORITHM **SINREP**

Purpose: To estimate λ by $\widehat{\lambda}_t$ from data on iterations $k, k+1, \ldots, k+t-1$ on a single sample path ($n = 1$) and to assess the error of approximation.

Given: $\pi_0, k, t, \delta :=$ 1− confidence level ($\in (0, 1)$).

Method:

Choose an initial state $\mathbf{X}_0$ from π_0.
Randomly generate $\mathbf{X}_1, \ldots, \mathbf{X}_{k+t-1}$.
Compute $\widehat{\lambda}_t = \frac{1}{t}\sum_{j=k}^{k+t-1} h(\mathbf{X}_j)$.
Use $h(\mathbf{X}_k), \ldots, h(\mathbf{X}_{k+t-1})$ to compute $\widehat{\sigma}_\infty^2$ as an estimate of the asymptotic variance σ_∞^2 (6.61) and its degrees of freedom $L-1$.
Compute:

$$\text{s.e.}(\widehat{\lambda}_t) = \sqrt{\frac{\widehat{\sigma}_\infty^2}{t}}.$$

$$\text{r.e.}(\widehat{\lambda}_t) = \frac{\text{s.e.}(\widehat{\lambda}_t)}{\widehat{\lambda}_t},$$

approximating

$$100 \times (1-\delta)\% \text{ CI} = [\widehat{\lambda}_t - \tau_{L-1}(1-\delta/2) \times \text{s.e.}(\widehat{\lambda}_t), \\ \widehat{\lambda}_t + \tau_{L-1}(1-\delta/2) \times \text{s.e.}(\widehat{\lambda}_t)].$$ ■

Under relatively general conditions, the estimator $\widehat{\lambda}_t$ has the following properties: As $t \to \infty$,

P31. $\widehat{\lambda}_t \to \lambda$ w.p.1.

P32. $\sqrt{t}(\widehat{\lambda}_t - \lambda) \xrightarrow{\text{d}} \mathcal{N}(0, \sigma_\infty^2)$.

While discarding only $k-1$ data points, Algorithm SINREP calls for a more elaborate procedure than Algorithm INDREPS for estimating σ_∞^2. Section 7.4 describes how the *batch-means method* estimates σ_∞^2 and its degrees of freedom $L-1$. Together with subsequent sections, it presents a justification for the asymptotic validity of the confidence interval for λ.

Simplicity of technique clearly favors Algorithm INDREPS, especially if the warm-up interval is not too large. When substantial dependence exists within a sample path and, therefore, a long warm-up interval is required, the need in Algorithm SINREP to account just once for the influence of initial conditions and the availability of software for estimating σ_∞^2 (see Section 7.9) offers an equally compelling argument for this single sample-path approach.

Recall that k is an estimate of what it takes for initial conditions to have negligible influence on the data used to estimate λ. If k is too small, then initial conditions influence $\bar{\lambda}_{t'}^{(l)}$ on every replication $l = 1, \ldots, n$ in Algorithm INDREPS. That is,

$$\mathrm{E}\left(\bar{\lambda}_{t'}^{(l)} \mid \mathbf{X}_0 \text{ from } \pi_0 \neq \pi\right) \neq \lambda.$$

Moreover, this influence remains in $\bar{\lambda}_{nt'}$, regardless of how large n is for a fixed t'. By contrast, Algorithm SINREP with a single run of length $t = nt'$ after truncation produces $\widehat{\lambda}_t$ with (often substantially) less bias caused by the influence of initial conditions. That is,

$$\left|\mathrm{E}(\widehat{\lambda}_t | \mathbf{X}_0) - \lambda\right| \leq \left|\mathrm{E}\left(\bar{\lambda}_{nt'} | \mathbf{X}_0^{(1)}, \ldots, \mathbf{X}_0^{(n)}\right) - \lambda\right|.$$

In this regard, an underestimate of the warm-up interval in Algorithm INDREPS is more serious than in Algorithm SINREP.

This introduction makes clear that two important statistical challenges need to be overcome before a valid error assessment can be made for $\bar{\lambda}_{nt'}$ in Algorithm INDREPS and for $\bar{\lambda}_t$ in Algorithm SINREP. One is the estimation of a warm-up interval $k - 1$ and the other is the estimation of σ_∞^2 in the single sample-path approach in Algorithm INDREPS. The remainder of this chapter addresses these problems. Section 9.7 in the next chapter describes an alternative technique that eliminates the influence of initial conditions on a sample path. Although that technique is, at present, more conceptual than operational, further research may broaden the class of problems to which it can be effectively applied.

7.3 ESTIMATING THE WARM-UP INTERVAL

When $\mathbf{X}_0$ is drawn from an initializing distribution $\pi_0 \neq \pi$, $\mathbf{X}_1, \mathbf{X}_2, \ldots$ do not have the equilibrium distribution π. However, ergodicity ensures that as $j \to \infty$, $\mathbf{X}_j$ converges in distribution to π. By truncating $\mathbf{X}_1, \ldots, \mathbf{X}_j$ from the sample path for some $j = k - 1$ such that for $j \geq k$ the distribution of $\mathbf{X}_j$ differs negligibly from π, the remaining sample path $\mathbf{X}_k, \ldots, \mathbf{X}_{k+t-1}$ can be regarded as being in equilibrium and thus suitable for estimating λ with little or no influence from initial conditions. This section describes graphical techniques for estimating k. They are easily implemented and apply to a wide range of stochastic processes. Later, Section 9.7 describes *coupling from the past,* a technique that, when it applies, eliminates error caused by initial conditions. Although the latter technique currently is justified for a considerably more limited class of stochastic processes than the graphical methods presented here, it may yet develop into a more widely applicable methodology.

A preliminary analysis provides data for estimating the warm-up interval $k-1$. It consists of n_0 independent replications, each of length t_0, where replication l begins in state $\mathbf{X}_0^{(l)}$, generated from the initializing distribution π_0. Usually, t_0 is chosen much smaller than the anticipated t required to estimate λ to within a desired accuracy.

Let

$$\lambda_{j\text{cond}} := \frac{1}{n_0}\sum_{l=1}^{n_0} \pi_0\left(\mathbf{x}_0^{(l)}\right) \mathrm{E}_{K^{(j)}\left(\mathbf{x}_0^{(l)},\mathbf{x}_j\right)} h(\mathbf{X}_j)$$

$$= \text{average across } n_0 \text{ replications of the conditional means of } h(\mathbf{X}) \text{ for initial states } \mathbf{X}_0^{(l)} = \mathbf{x}_0^{(l)} \text{ for } l = 1, \ldots, n_0.$$

Given that ergodicity holds and that h obeys certain regularity conditions, then

$$\mathrm{E}_{K^{(j)}(\mathbf{x}_0,\mathbf{x}_j)} h(\mathbf{X}_j) \to \lambda \qquad \text{as } j \to \infty,$$

so under relatively general conditions,

$$\lambda_{j\text{cond}} \to \lambda \text{ w.p.1} \qquad \text{as } j \to \infty. \tag{7.1}$$

A plot of the sequence $\lambda_{1\text{cond}}, \lambda_{2\text{cond}}, \ldots, \lambda_{j\text{cond}}$ would provide a picture of this convergence. Because the sequence is unknown, we plot the conditional sample averages

$$\bar{\lambda}_{j\text{cond}} = \frac{1}{n_0}\sum_{l=1}^{n_0} h\left(\mathbf{X}_j^{(l)}\right) \qquad j = 1, \ldots, t_0.$$

Then

$$\bar{\lambda}_{j\text{cond}} - \lambda = \underbrace{\bar{\lambda}_{j\text{cond}} - \lambda_{j\text{cond}}}_{\text{sampling error}} + \underbrace{\lambda_{j\text{cond}} - \lambda}_{\text{systematic error}}$$

reveals two sources of error when estimating λ by $\bar{\lambda}_{j\text{cond}}$. In the present context, *systematic error* arises from the dependence of each sample path on its initial state. *Sampling error* arises from variation induced by random sampling on each path. Because sampling error in $\bar{\lambda}_{j\text{cond}}$ decreases as $n_0^{-1/2}$, large n_0 tends to reduce the magnitude of sampling fluctuations and, therefore, sampling error, but it leaves systematic error unaffected for each fixed j; hence, the motivation to choose a sufficiently long warm-up interval $k - 1$.

Of most importance for fixed n_0, as $j \to t_0$ and $t_0 \to \infty$,

$$\bar{\lambda}_{j\text{cond}} - \lambda \to Z \qquad \text{w.p.1}, \tag{7.2}$$

where Z denotes a random variable, independent of initial conditions on the n_0 replications, with mean zero and variance R_0/n_0. Recall from Section 6.9 that R_0 denotes the equilibrium variance $R_0 := \mathrm{var}_\pi h(\mathbf{X})$.

These properties suggest a strategy for estimating the warm-up interval. Because $\mathbf{X}_1^{(l)}, \mathbf{X}_2^{(l)}, \ldots$ are not in equilibrium, we anticipate that $\lambda_{j\text{cond}} \neq \lambda$ for $j = 1, 2 \ldots$. However, as j increases, (7.2) implies that we should observe successive $\bar{\lambda}_{j\text{cond}}$ fluctuating around a constant but unknown mean λ. A graph reveals this stationarity for some iterations $j \geq k$ for $k \ll t'$, suggesting negligible dependence is present. The notation $k \ll t_0$ denotes a k much smaller than t_0 so that stationarity is discernible over more than a few iterations in the neighborhood of t_0.

In principle, $k-1$, becomes the warm-up interval so that on subsequent runs, data collection for estimating λ begins on iteration k. If $\{\bar{\lambda}_{j\text{cond}};\ j = 1, \ldots, t_0\}$ reveals no such k, then common practice increases the sample-path length to some $t_1 > t_0$ on each of the n_0 replications and looks for stationarity in $\{\bar{\lambda}_{j\text{cond}};\ j = 1, \ldots, t_1\}$. Because in equilibrium $\bar{\lambda}_{j\text{cond}}$ has standard error $\sqrt{R_0/n_0}$, stationarity, when it exists, is more easily discernible when n_0 is large. Nevertheless, a small n_0 often suffices to detect stationarity, as Examples 7.3.1 and 7.3.2 show.

Algorithm WARM-UP describes a procedure for estimating k. Perusal of the superimposed n_0 sample paths in a single plot may reveal convergence to a local region in which random variation occurs independent of j. This is especially helpful when the initial values $h(\mathbf{X}_0^{(l)}), \ldots, h(\mathbf{X}_0^{(n_0)})$ are widely disparate. Because $\bar{\lambda}_{j\text{cond}}$ smooths away the variation across replications, a graph of $\{\bar{\lambda}_{j\text{cond}};\ j = 1, \ldots, t_0\}$ makes stationarity more discernible, when it is present.

ALGORITHM WARM-UP

Purpose: To generate graphs for a warm-up analysis.

Given: n_0 := number of independent replications, t_0 := sample-path length.

Method:

$l = 1$.
Until $l > n_0$:
  randomly generate an initial state $\mathbf{X}_0$ from $\pi_0^{(l)}$
  randomly generate $h(\mathbf{X}_1^{(l)}), \ldots, h(\mathbf{X}_{t_0}^{(l)})$
  $l = l + 1$.
$j = 1$.
Until $j > t_0$:
  compute:

$$\bar{\lambda}_{j\text{cond}} := \frac{1}{n_0}\sum_{l=1}^{n_0} h(\mathbf{X}_j^{(l)})$$

$$s^2_{j\text{cond}} := \frac{1}{n_0 - 1}\left[\sum_{l=1}^{n_0} h^2(\mathbf{X}_j^{(l)}) - n_0\bar{\lambda}^2_{j\text{cond}}\right]$$

  $j = j + 1$.

Plot options:

- All sample paths $\{h(\mathbf{X}_j^{(l)});\ j = 1, \ldots, t_0\}$ for $l = 1, \ldots, n_0$ on a single plot
- Graph of $\{\bar{\lambda}_{j\text{cond}};\ j = 1, \ldots, t_0\}$
- Graph of $\left\{\sqrt{s^2_{j\text{cond}}};\ j = 1, \ldots, t_0\right\}$

■

On iteration j, the statistic $s^2_{j\text{cond}}$ estimates the conditional variance of $h(\mathbf{X})$ across n_0 replications for initial states $\mathbf{X}_0^{(l)} = \mathbf{x}_0^{(l)}$ for $l = 1, \ldots, n_0$. Under relatively general conditions,

$$s^2_{j\text{cond}} \to R_0 \text{ w.p.1} \qquad \text{as } j \to \infty.$$

A graph of $\left\{\sqrt{s^2_{j\text{cond}}};\ j = 1, \ldots, t_0\right\}$ provides an additional basis for detecting stationarity in at least two important circumstances:

- If the n_0 independent replications begin in states $\mathbf{X}_0^{(1)}, \ldots, \mathbf{X}_0^{(n_0)}$ such that $h\left(\mathbf{X}_0^{(1)}\right), \ldots, h\left(\mathbf{X}_0^{(n_0)}\right)$ span virtually the entire range of $\{h(\mathbf{x})\}$, then $s^2_{j\text{cond}}$ tends to be large for small j and to decrease as j increases, again exhibiting stationarity for sufficiently large j.
- If all replications begin in the same state $\mathbf{X}_0$, then $s^2_{j\text{cond}}$ will be close to zero for small j, increase as j tends to increase, and eventually exhibit stationarity for large j.

We emphasize that trends toward stationarity in

$$\{\bar{\lambda}_{j\text{cond}}\} \quad \text{and} \quad \left\{\sqrt{s^2_{j\text{cond}}}\right\}$$

are rarely monotone. First, sampling fluctuations of order $n_0^{-1/2}$ characterize both sample functions so that their fluctuations may obscure trends. Second, the actual conditional functions $\{\lambda_{j\text{cond}};\ j = 1, \ldots, t_0\}$ and

$$\left\{\mathrm{E}\left(\sqrt{s^2_{j\text{cond}}} \,\middle|\, \mathbf{X}_0^{(1)}, \ldots, \mathbf{X}_0^{(n)}\right);\ j = 1, \ldots, t_0\right\}$$

may themselves not be monotone. Although the first property encourages a large n_0 to smooth away sampling fluctuations, at each j, a relatively small n_0 often can provide sufficient information to detect where trend ends and stationarity prevails. Examples 7.3.1 and 7.3.2 support this assertion.

Unlike an algebraic expression that, for example, produces a single value for $\bar{\lambda}_{j\text{cond}}$ as an estimate of the true $\lambda_{j\text{cond}}$, using graphs leaves considerable discretion to the analyst to choose a value for k. When confronted with the same graphs, different analysts may pick different values. This often reflects differing attitudes toward underestimating a suitable warm-up interval and need not have a major effect on statistical error in subsequent MCMC experiments. However, because the ensuing estimation of λ via multiple independent replications requires that data for the first $k-1$ iterations be discarded on each replication, different choices for k inevitably affect the corresponding computing times.

EXAMPLE 7.3.1 CAPTURE-RECAPTURE: WARM-UP ANALYSIS

The objective in capture-recapture Examples 6.4.1 and 6.11.1 is to estimate the mean of the posterior distribution of θ, the size of the sunfish population in Gordy Lake in Michigan. Figure 6.3, based on Hastings-Metropolis sampling, and Figure 6.4, based on Gibbs sampling, show estimated posterior distributions of θ.

Two coordinates are updated on each iteration. Moreover, Table 6.2 displays results for the posterior mean population size based on sample paths generated by these procedures and an additional Gibbs sampling experiment that updated all coordinates on each iteration. Lastly, Figure 6.5 shows sample paths for three widely spaced initial states: $\mathbf{x}_0^{(1)} = 138$, the minimal possible population size; $\mathbf{x}_0^{(2)} = 457$, the maximum likelihood estimate of θ; and $\mathbf{x}_0^{(3)} = 600$, a value believed to be be much larger than the mean of θ. In theory, wide dispersion should produce sample paths that straddle the equilibrium mean, thereby reducing the possibility of starting all n_0 replications so far from the mean that none exhibit a tendency toward stationary fluctuations for any iteration $j \in \{1, \ldots, t_0\}$. This implies three distinct degenerate distributions

$$\left\{\pi_0(\mathbf{x}) := \delta\left(\mathbf{x}_0^{(l)}\right),\ \mathbf{x} \in \mathcal{X}\right\}$$

for $l = 1, \ldots, 3$, where $\delta(\cdot)$ denotes the Dirac delta function in (6.21).

In practice, a warm-up analysis is usually performed for only the MCMC method chosen for sampling. For illustration, however, we describe analyses for all three methods. By way of review, the plots in Figure 6.5, restricted to iterations 1 through 100, are designed to reveal the differences in convergence rates to equilibrium. Those in Figure 7.1 for $\{\bar{\lambda}_{j\text{cond}};\ j = 1, \ldots, t_0 = 300\}$ average the fluctuations over the $n_0 = 3$ independent replications for each iteration j, thereby giving a smoother plot from which to discern if stationarity sets in, and on which iteration this occurs. For two-coordinate Hastings-Metropolis updating, Figure 7.1a

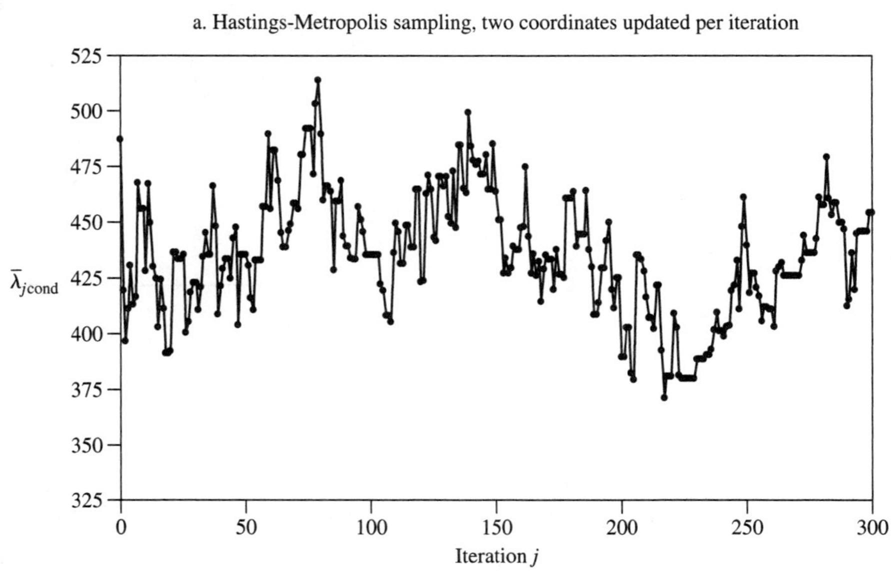

FIGURE 7.1 | *Warm-up analysis for sunfish population size*

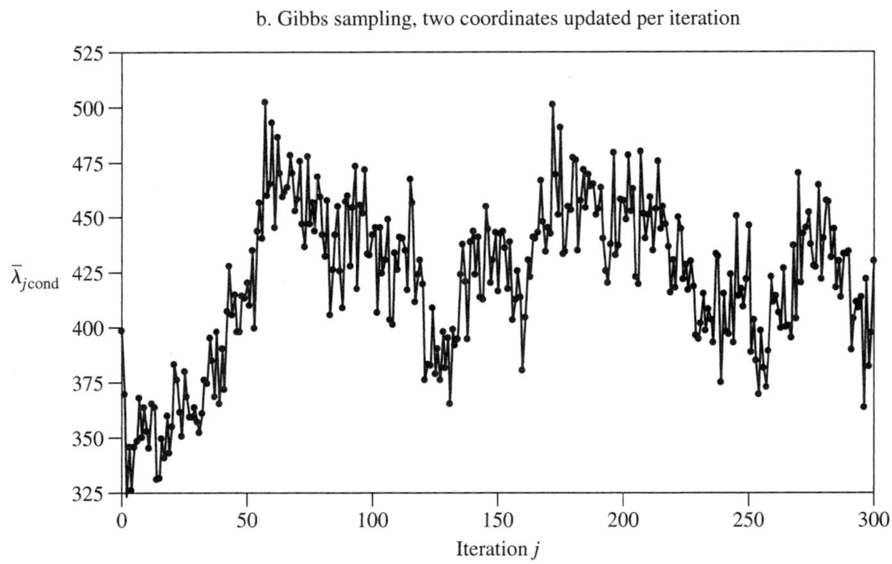

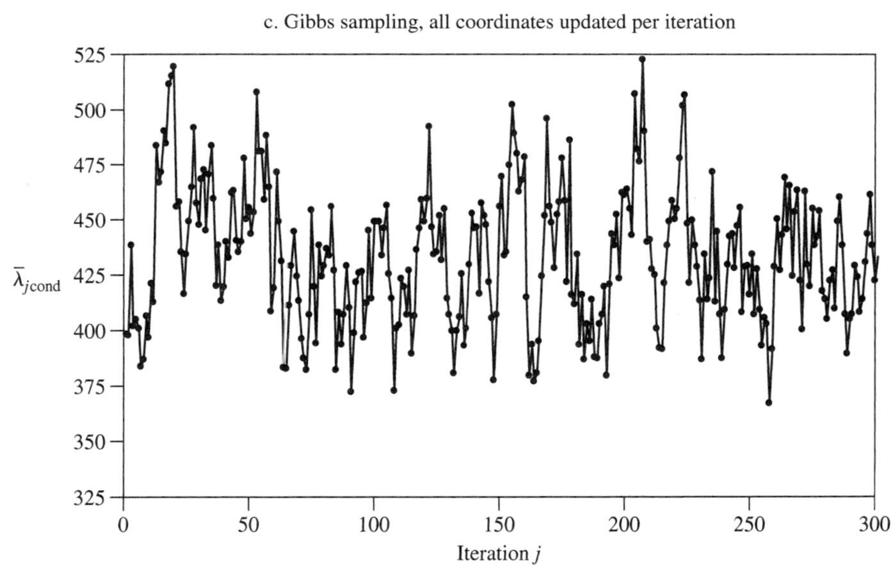

FIGURE 7.1 | *Continued*

suggests stationarity for $j \geq k = 26$. For Gibbs sampling with two-coordinate updating, Figure 7.1b suggests stationarity for $j \geq k = 76$. For Gibbs sampling with all-coordinate updating, Figure 7.1c suggests this behavior for $j \geq k = 11$. These "eyeball" estimates possibly overstate k, but by negligible amounts.

Although the results tend to favor Gibbs all-coordinate updating as having the smallest warm-up interval, recall that this problem has $m = 15$ coordinates. As a consequence, HM sampling with two-coordinate updating implies that $2(k-1) =$

50 sample generations need to be discarded, whereas Gibbs sampling with two-coordinate updating with $k = 76$ implies that $2(k-1) = 150$ sample generations need to be discarded. Likewise, all-coordinate updating with $k = 11$ suggests that $15(k-1) = 150$ be discarded. Because all these methods have approximately the same unit generating times, the three estimated warm-up intervals k in this example suggest that the total computing time is least for HM sampling and is roughly the same for each Gibbs sampling option. ■

EXAMPLE 7.3.2

EYE-HAIR COLOR: WARM-UP ANALYSIS

Recall the contingency table in Example 6.11.2, where it is of interest to estimate the proportion λ in (6.71) using Algorithms CONTINGENCY and ALL with

$$h(\mathbf{x}) := \begin{cases} 1 & \text{if } B(\mathbf{x}) > B(\mathbf{N}) \\ 0 & \text{elsewhere} \end{cases}$$

$$\pi(\mathbf{x}) := |\mathcal{X}|^{-1} \qquad \mathbf{x} \in \mathcal{X}.$$

As a basis for choosing k, $n_0 = 5$ independent replications, each of $t_0 = 1{,}000$ iterations, were executed. In lieu of randomly sampling the initial states from a common initializing distribution, the deterministic assignments $\mathbf{X}_0^{(1)} = \mathbf{x}_0^{(1)}, \ldots, \mathbf{X}_0^{(5)} = \mathbf{x}_0^{(5)}$ were chosen such that

$$B\left(\mathbf{x}_0^{(1)}\right) = 0.096342$$

$$B\left(\mathbf{x}_0^{(2)}\right) = 138.29$$

$$B\left(\mathbf{x}_0^{(3)}\right) = 352.82$$

$$B\left(\mathbf{x}_0^{(4)}\right) = 587.90$$

$$B\left(\mathbf{x}_0^{(5)}\right) = 800.78,$$

revealing wide variation below and above $B(\mathbf{N}) = 138.29$. Because $h(\mathbf{x}) \in \{0, 1\}$, this is an instance in which it is more revealing to focus the warm-up analysis on the n_0 independent sample paths $\left\{B\left(\mathbf{X}_j^{(l)}\right);\ j = 1, \ldots, t_0\right\}$, their sample conditional-mean sequence

$$B_{j\text{cond}} := \frac{1}{n_0} \sum_{l=1}^{n_0} B\left(\mathbf{X}_j^{(l)}\right) \qquad j = 1, \ldots, t_0,$$

and their corresponding sample conditional-variance sequence

$$s_{j\text{cond}}^2 := \frac{1}{n_0 - 1} \left[\sum_{l=1}^{n_0} B^2\left(\mathbf{X}_j^{(l)}\right) - n_0 B_{j\text{cond}}^2 \right] \qquad j = 1, \ldots, t_0.$$

Figure 7.2a shows that starting from widely disparate initial states, all five paths converge to a local neighborhood as j increases. In particular, for $j \geq k = 300$, $B_{j\text{cond}}$ in Figure 7.2b appears stationary. In Figure 7.2c, $s^2_{j\text{cond}}$ shows fluctuations decreasing in magnitude as j increases, again converging to a stationary appearance for $j \geq k = 300$. ■

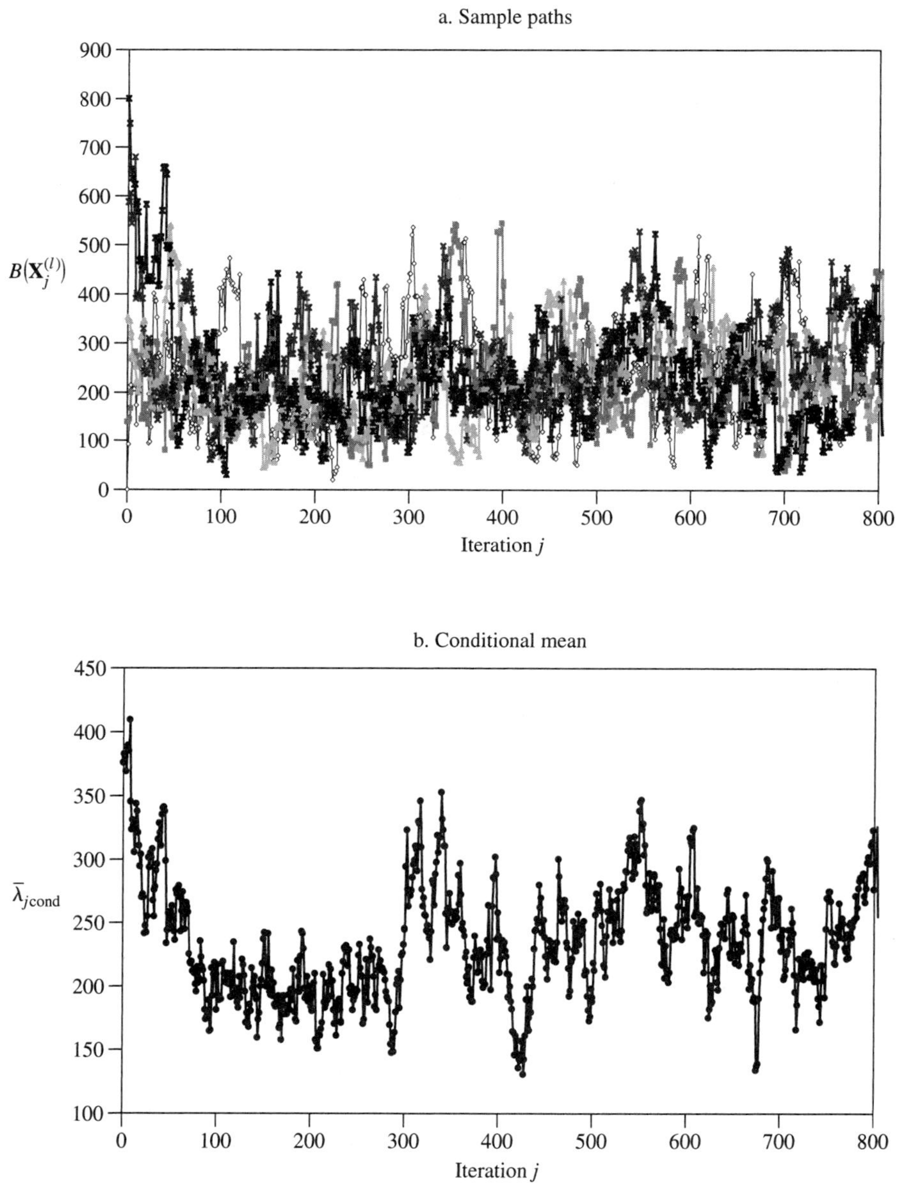

FIGURE 7.2 | *Warm-up analysis for eye color and hair color contingency table*

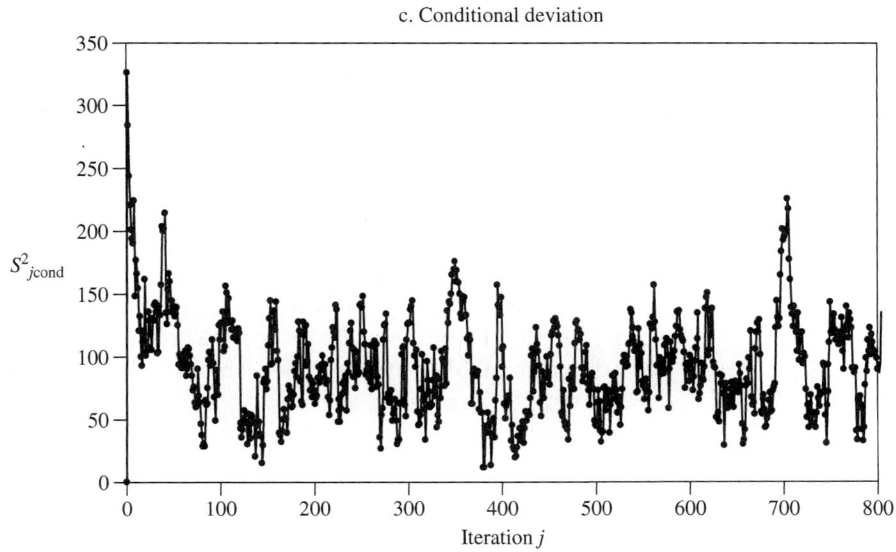

FIGURE 7.2 | *Continued*

EXAMPLE 7.3.3

CAPTURE-RECAPTURE: MULTIPLE REPLICATIONS

Once a warm-up interval is chosen, Algorithm INDREPS provides a blueprint for estimating λ, based on multiple independent replications. As illustration, consider the estimation of the mean of the posterior distribution of population size in the capture-recapture problem (Examples 6.4.1 and 7.3.2) using Hastings-Metropolis sampling and updating two coordinates per iteration. With $k = 26$, $n = 31, t' = 45,056$, and $t = nt' = 1,396,736$, Algorithm INDREPS produced

$$\begin{aligned}
\bar{\lambda}_{nt'} \pm \tau_{30}(0.995)\sqrt{\bar{\sigma}^2_\infty/t} &= 430.1 \pm 1.8 \\
s^2_{nt'} &= 15.54 \\
\bar{\sigma}^2_\infty := t' s^2_{nt'} &= 7.002 \times 10^5 \\
\text{s.e.}\left(\bar{\lambda}_{nt'}\right) = \sqrt{\frac{s^2_{nt'}}{n}} &= 0.7080 \\
\text{r.e.}\left(\bar{\lambda}_{nt'}\right) = \frac{\text{s.e.}\left(\bar{\lambda}_{nt'}\right)}{\bar{\lambda}_{nt'}} &= 0.001646.
\end{aligned}$$

The statistic $\bar{\sigma}^2_\infty$ estimates the asymptotic variance σ^2_∞ in the present experiment.

For Algorithm SINREP with $t = 10^6$, Table 6.2 reveals a single sample-path estimate $\widehat{\lambda}_t = 430.9$. As an estimate of σ^2_∞, it has $\widehat{\sigma}^2_\infty = 5.932 \times 10^5$, so that the standard error is $\sqrt{\frac{\widehat{\sigma}^2_\infty}{t}} = 0.7701$, the relative standard error is 0.001797, and the approximating 99% confidence interval is 430.9 ± 2.0. These results are comparable to those for Algorithms INDREPS with regard to mean estimation and error assessment. However, INDREPS discarded $26 \times 31 = 811$ iterations for warm-up whereas SINREP discarded only 26.

Although discarding this number of observations is a relatively small price to pay for the convenience of using relatively simple statistical methods in Algorithm INDREPS, other problems with considerably larger estimated warm-up intervals may make one reluctant to incur the corresponding penalty. In this case, single sample-path analysis as in Algorithm SINREP significantly reduces the amount of truncation. However, it relies on a more advanced method for estimating σ_∞^2. The next several sections describe one such technique for which easily implementable software is available. Unless otherwise stated, we assume that the sample paths being analyzed are relatively free of the influence of initial conditions. ■

7.4 BATCH-MEANS METHOD

This technique applies to geometrically and uniformly ergodic Markov chains and, therefore, would be inappropriate for chains exhibiting long-range dependence, as described in Section 6.13. Let

$$b(t) := \text{a positive integer} < t$$

$$l(t) := \text{largest integer} \leq t/b(t)$$

and

$$a(t) := l(t)b(t). \tag{7.3}$$

The *batch-means method* partitions a sample path $\{Z_j := h(\mathbf{X}_j);\ j = 1, \ldots, t\}$ into $l(t)$ non-overlapping batches

$$\underbrace{Z_1, \ldots, Z_{b(t)}}_{\text{batch 1}},\ \underbrace{Z_{b(t)+1}, \ldots, Z_{2b(t)}}_{\text{batch 2}},\ \ldots\ \underbrace{Z_{[l(t)-1]b(t)+1}, \ldots, Z_{a(t)}}_{\text{batch } l(t)}$$

and computes the batch averages

$$Y_{jb(t)} := \frac{1}{b(t)} \sum_{i=1}^{b(t)} Z_{(j-1)b(t)+i} \qquad j = 1, \ldots, l(t). \tag{7.4}$$

It also computes

$$W_{l(t)b(t)} := \frac{1}{l(t)-1} \sum_{i=1}^{l(t)} \left(Y_{jb(t)} - \widehat{\lambda}_{a(t)}\right)^2 \tag{7.5}$$

as an estimate of $\sigma_{b(t)}^2 := \operatorname{var} Y_{jb(t)}$, the variance of a batch average. Lastly, it takes $\sqrt{V_{a(t)}/t}$, where

$$V_{a(t)} := b(t) W_{l(t)b(t)},$$

as an estimate of the asymptotic standard error $\sqrt{\sigma_\infty^2/t}$ of $\widehat{\lambda}_t$, and

$$\left[\widehat{\lambda}_t - \tau_{l(t)-1}(1-\delta/2)\sqrt{V_{a(t)}/t},\ \widehat{\lambda}_t + \tau_{l(t)-1}(1-\delta/2)\sqrt{V_{a(t)}/t}\right]$$

as an approximating $100 \times (1-\delta)$ percent confidence interval for λ.

The motivation for this method of error assessment arises from the relationship between the sample average $\widehat{\lambda}_t$ and the batch averages $Y_{1b(t)}, \ldots, Y_{l(t)b(t)}$. Because

$$\widehat{\lambda}_{a(t)} = \frac{1}{l(t)} \sum_{j=1}^{l(t)} Y_{jb(t)},$$

it is easily seen that

$$\mathrm{var}\widehat{\lambda}_{a(t)} = \frac{1}{l(t)} \sigma_{b(t)}^2 \left(1 + \psi_{l(t)b(t)}\right),$$

where

$$\psi_{l(t)b(t)} := \frac{2}{l(t)} \sum_{i=1}^{l(t)} \sum_{\substack{j=1 \\ j \neq i}}^{l(t)} \mathrm{cov}\left(Y_{ib(t)}, Y_{jb(t)}\right).$$

As batch size $b(t)$ increases, dependence among batch averages tends to decrease so that for all batches $i \neq j$, $\mathrm{cov}(Y_{ib(t)}, Y_{jb(t)}) \to 0$. If convergence occurs with sufficient speed and the number of batches $l(t)$ does not increase too rapidly, then $\psi_{l(t)b(t)} \to 0$ as $b(t) \to \infty$. This implies

$$a(t)\mathrm{var}\widehat{\lambda}_{a(t)} \to b(t)\sigma_{b(t)}^2 \to \sigma_\infty^2 \qquad \text{as } b(t) \to \infty \text{ and } t \to \infty,$$

where $\sigma_{b(t)}^2 = \mathrm{var}\widehat{\lambda}_{b(t)}$, so for large t and $b(t)$, $\sigma_{b(t)}^2 \approx \sigma_\infty^2/b(t)$. Therefore, an estimate of $\sigma_{b(t)}^2$ also provides a way of estimating σ_∞^2.

These limiting properties give the batch-means method considerable appeal in practice. Indeed, it is not unusual to encounter a recommendation to partition a sample path into 30 non-overlapping batches, for example, and take batch size to be $b(t) = \lfloor t/30 \rfloor$. However, this advice is less than complete. It provides no way to assess the extent to which the resulting batch averages are independent of each other. Without this assessment, there is no way to measure how well $V_{a(t)}$ approximates σ_∞^2 and, therefore, how accurate $\widehat{\lambda}_n$ is as an estimator of λ. The remainder of this chapter describes a batch-means procedure that allows this assessment for $V_{a(t)}$ and ultimately for $\widehat{\lambda}_n$ on a single sample path.

The error of approximation in $V_{a(t)}$ materializes from

$$V_{a(t)} - \sigma_\infty^2 = \underbrace{V_{a(t)} - \mathrm{E}V_{a(t)}}_{\text{sampling error}} + \underbrace{\mathrm{E}V_{a(t)} - \sigma_\infty^2}_{\text{systematic error}}. \tag{7.6}$$

In the present context, systematic error concerns how well $b(t)\sigma_{b(t)}^2$ approximates σ_∞^2 for batch size $b(t)$. Sampling error concerns the sampling variation in $V_{a(t)}$ around σ_∞^2 that random sampling inevitably induces for finite t. Whereas systematic error is strictly an issue of mathematical approximation, sampling error is entirely the result of randomness. If systematic error is negligible, then the earlier described distribution theory provides the basis for computing an asymptotically valid confidence interval for λ.

It is of interest to see how systematic error dissipates with increasing batch size. Rewriting (7.5) as

$$W_{l(t)b(t)} := \frac{1}{l(t)-1}\sum_{i=1}^{l(t)} \left[\left(Y_{jb(t)} - \lambda\right) + \left(\lambda - \widehat{\lambda}_{a(t)}\right)\right]^2$$

leads to

$$\begin{aligned} \mathrm{E}\,V_{a(t)} &= \frac{a(t)}{l(t)-1}\left(\sigma^2_{b(t)} - \sigma^2_{a(t)}\right) \\ &= b(t)\sigma^2_{b(t)} + \frac{b(t)\sigma^2_{b(t)}}{l(t)-1} - \frac{a(t)\sigma^2_{a(t)}}{l(t)-1} \\ &\to \sigma^2_\infty \qquad \text{as } b(t) \to \infty, \end{aligned}$$

so that systematic error decreases with increasing batch size.

The magnitude of sampling error depends critically on the number of batches $l(t)$. For fixed $l(t)$ and under relatively general conditions,

$$\sqrt{l(t)}\left[V_{a(t)} - \mathrm{E}\,V_{a(t)}\right] \to Z \text{ w.p.1} \qquad \text{as } b(t) \to \infty,$$

where Z denotes a random variable with mean zero and whose variance is a constant independent of t. Therefore, sampling error in $V_{a(t)}$ exhibits fluctuations of order $[l(t)]^{-1/2}$ for large t.

Every implementation of the batch-means method requires a procedure that dictates how batch size $b(t)$ increases with sample-path length t, subject to the constraint that the number of batches $l(t)$ is nondecreasing in t. Because the validity of the method depends on making systematic error negligible in (7.6), an implementation should also include a way to make this assessment. For increasing path lengths $t_1 < t_2 < \cdots \le t$, one approach computes $V_{a(t_1)}, V_{a(t_2)}, \ldots, V_{a(t)}$ and analyzes this sequence for trend, either increasing or decreasing. For conciseness, we hereafter write $V_j := V_{a(t_j)}$. The absence of trend for some $j > j^*$ implies that systematic error is negligible. However, sampling fluctuations remain present in $V_{a(j^*)}, V_{a(j^*+1)}, \ldots$.

Sections 7.5, 7.6, and 7.7 describe and illustrate rules for increasing $b(t)$ with t. In what follows, assume that *interim reviews* are performed at path lengths $t_{j+1} = 2t_j$ for $j \ge 1$, given t_1. This makes it possible to create an implementation in which $V_{a(t_1)}, V_{a(t_2)}, \ldots, V_{a(t)}$ are computable in time proportional to t and space proportional to $\log_2 t$. The single sample-path examples to follow all use an implementation called LABATCH.2 (see Section 7.9), which exploits this feature.

7.5 FNB RULE

The FNB Rule provides one way to benefit from the asymptotic independence among batches. For some positive integer l_1, the rule fixes the number of batches (hence, FNB Rule) at $l(t) = l_1$ and then allows batch size to grow with t as

$$b(t) = \text{largest integer} \le t/l_1.$$

Under relatively weak conditions, the following is known:
As $t \to \infty$ (Brillinger 1973),

P33. $\frac{Y_{1b(t)}-\lambda}{\sqrt{\sigma_\infty^2/b(t)}}, \ldots, \frac{Y_{l_1 b(t)}-\lambda}{\sqrt{\sigma_\infty^2/b(t)}}$
converges to a sequence of l_1 i.i.d. random variables from $\mathcal{N}(0, 1)$.

P34. $\frac{(l_1-1)V_{a(t)}}{\sigma_\infty^2} \xrightarrow{d} \chi^2_{l_1-1}$.

P35. $\frac{\widehat{\lambda}_t-\lambda}{\sqrt{V_{a(t)}/t}} \xrightarrow{d} \tau_{l_1-1}$.

Uniformly ergodic chains satisfy these conditions, and so do geometrically ergodic chains for bounded h. Chains exhibiting polynomial convergence do not necessarily satisfy the conditions (see Section 6.13). Property P35 provides a basis for computing an asymptotically valid confidence interval for λ.

The simplicity of the FNB Rule accounts for its appeal. However, as before for independent replications, a balance needs to be struck between l_1 and $b(t)$ so that sufficient degrees of freedom $l_1 - 1$ result for a useful confidence interval and a sufficiently large batch size $b(t)$ makes the batch averages roughly independent. As mentioned earlier, in lieu of additional information about the problem, l_1 is frequently chosen in the neighborhood of 30. But that tells us little about how well $V_{a(t)}$ estimates σ_∞^2.

Property P34 reveals a limitation of the FNB Rule:

$$V_{a(t)} \nrightarrow \sigma_\infty^2 \quad \text{w.p.1} \quad \text{as } t \to \infty,$$

so $V_{a(t)}$ lacks the desirable statistical property of consistency. This is analogous to the issue raised in Section 7.1 in the discussion on independent replications. Although the distribution theory on which the confidence interval for λ is based takes this limiting behavior into account, it will always be desirable to employ a consistent estimator of σ_∞^2.

EXAMPLE 7.5.1

CAPTURE-RECAPTURE ACCURACY ASSESSMENT: FNB RULE

Again consider the estimation of the mean of the posterior distribution of sunfish population size. In Example 6.4.1, the Hastings-Metropolis method updated two coordinates on each iteration to generate a sample path of $t = 22 \times 31 \times 2^{11} =$ 1,396,736 observations. Application of the FNB Rule to these data for $l_1 = 31$ batches and initial batch size $b_1 = 22$ produced a sequence of estimates $V_1, \ldots, V_{12}$ with $a(t_j) = t_j$, so that all t_j observations were used to compute each V_j.

Figure 7.3 displays $\sqrt{V_1}, \ldots, \sqrt{V_{12}}$. It reveals an increasing trend up to review 6 at iteration $t_6 = 31{,}824$ and batch size $b(t_6) = 704$ and an absence of trend thereafter. This implies that 704 or more observations per batch suffice to make systematic error negligible in this example.

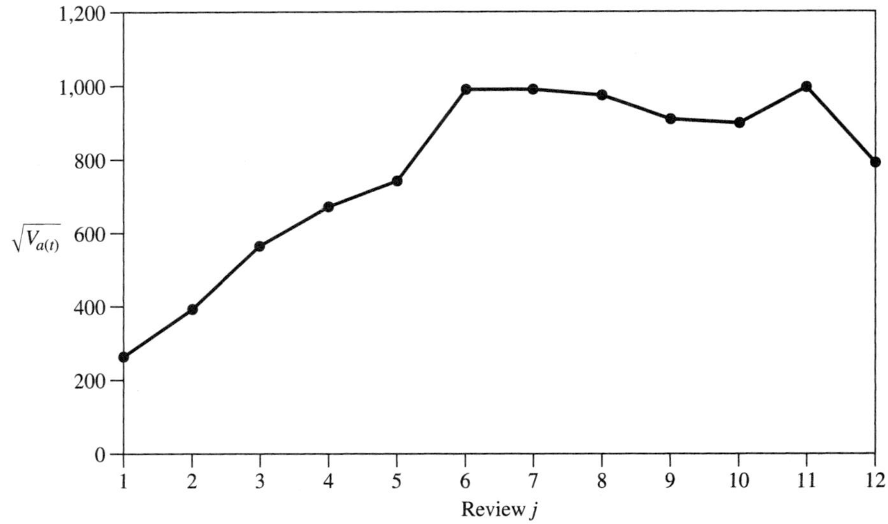

FIGURE 7.3 | *Estimating σ_∞^2 for sunfish sample posterior mean size: FNB Rule*

As an error assessment for $\widehat{\lambda}_t$, we have

$$\begin{aligned}\widehat{\lambda}_t \pm \tau_{30}(0.995)\text{s.e.}\left(\widehat{\lambda}_t\right) &= 430.4 \pm 1.8\\ \widehat{\sigma_\infty^2} := V_{12} &= 6.241 \times 10^5\\ \text{s.e.}\left(\widehat{\lambda}_t\right) &= 0.6685\\ \text{r.e.}\left(\widehat{\lambda}_t\right) &= 0.001553.\end{aligned}$$

These results closely resemble those for multiple replications in Example 7.3.3.

The most important lesson here is that a comparative analysis of $V_1, \ldots, V_{12}$ for some $t_1 < t_2 < \cdots < t_{12}$ is essential to check for the presence of systematic error. ■

7.6 SQRT RULE

The SQRT Rule provides one way of overcoming the limitation of the FNB Rule. It takes $b(t)$ and $l(t)$ each approximately proportional to $t^{1/2}$ so that both the number of batches and the batch size grow with increasing sample-path length. Under relatively general conditions, the following is known:
As $t \to \infty$,

P36. $V_{a(t)}$ is a consistent estimator of σ_∞^2. That is,

$$V_{a(t)} \to \sigma_\infty^2 \text{ w.p.1.} \tag{7.7}$$

P37. $\frac{\widehat{\lambda}_t - \lambda}{\sqrt{V_{a(t)}/t}} \xrightarrow{d} \mathcal{N}(0, 1)$.

Most notable is the consistent estimator of σ_∞^2.

The SQRT Rule has at least one additional appealing property: a random variable W from $\mathcal{N}(0, 1)$ has moments

$$\begin{aligned} \mathrm{E}W &= 0 \\ \mathrm{E}(W - \mathrm{E}W)^2 &= 1 \\ \mathrm{E}(W - \mathrm{E}W)^3 &= 0 \\ \mathrm{E}(W - \mathrm{E}W)^4 &= 3. \end{aligned}$$

By choosing $b(t)$ and $l(t)$ each proportional to $t^{1/2}$, the SQRT Rule ensures the fastest convergence of these first four moments to those for a standardized normal variate (Chien et al. 1997). Therefore, for any fixed t, this rule reduces the error of approximation of using the normal distribution as the basis for a confidence interval for λ (Chien 1989).

Although the SQRT Rule overcomes the limitation of the FNB Rule with respect to consistency in (7.7), it has a limitation of its own. Let the subscripts $_{\mathrm{FNB}}$ and $_{\mathrm{SQRT}}$ denote the corresponding rules. Because $b_{\mathrm{SQRT}}(t)$ increases more slowly than $b_{\mathrm{FNB}}(t)$, its dependence among batch averages decreases more slowly so that for two batches $i \neq j$ based on sample-path length t,

$$|\,\mathrm{corr}(Y_{ib_{\mathrm{SQRT}}(t)}, Y_{jb_{\mathrm{SQRT}}(t)})\,| \geq |\,\mathrm{corr}(Y_{ib_{\mathrm{FNB}}(t)}, Y_{jb_{\mathrm{FNB}}(t)})\,|\,.$$

This implies more dependence between batches i and j when using the SQRT Rule, a critical issue when $\{h(\mathbf{X}_k), \ldots, h(\mathbf{X}_{k+t-1})\}$ in the sample path are highly dependent.

Estimation of σ_∞^2 using the SQRT Rule for the same sample path analyzed in Example 7.5.1 illustrates this limitation. For the same initial assignments $l_1 = 31$ $b_1 = 22$, and sample-path length $t = 31 \times 22 \times 2^{11} = 1{,}396{,}736$, the SQRT Rule produced an increasing sequence $\sqrt{V_1} < \cdots < \sqrt{V_{12}}$. This implies the presence of systematic error in all V_j, making it clear that the SQRT Rule itself does not dominate the FNB Rule in performance. The next section describes one way to combine the best of both rules.

7.7 ABATCH RULE

The ABATCH Rule offers a compromise that reduces importance of the limitations inherent in the FNB and SQRT Rules. Let $0 < t_1 < t_2 < \cdots \leq t$ denote a sequence of increasing sample-path lengths, and let H denote the hypothesis that for some t_j, the batch averages $Y_{1b_{\mathrm{ABA}}(t_j)}, \ldots, Y_{l_{\mathrm{ABA}}(t_j)b_{\mathrm{ABA}}(t_j)}$ are statistically independent. The subscript $_{\mathrm{ABA}}$ denotes the ABATCH Rule (Fishman and Yarberry 1997, Fishman 1997 and 2000). The rule tests H on successive interim reviews $j = 1, 2, \ldots$ based on sample-path lengths $t_1, t_2, \ldots$. The principal steps are:

1. On review 1, form batch averages using the FNB Rule.

2. On review $j \geq 1$:

If H is rejected, use the FNB Rule on review $j + 1$.

Otherwise use the SQRT Rule on review $j + 1$.

If H is rejected on review j, then switching to the FNB Rule enables batch size to grow as fast as possible on review $j + 1$, thus allowing the remaining dependence to dissipate as fast as possible. If H is accepted on review j, then a switch to the SQRT Rule allows both batch size and degrees of freedom to increase at equal rates. By testing H on successsive reviews, the ABATCH Rule takes into consideration the possibility of Type II error (acceptance of H when it is false). In contrast to the SQRT Rule, which ignores this potential error, the ABATCH Rule allows the procedure for increasing $l(t_j)$ and $b(t_j)$ to correct itself.

One effect of the ABATCH Rule is to make batch size and number of batches random variables on interim reviews $j > 1$. To acknowledge the randomness, we denote the former by B_j and the latter by L_j on review j. As illustration of how this randomness comes about, suppose that the first review occurs at $t_1 = l_1 b_1$, $B_1 = b_1$, and $L_1 = l_1$, where b_1 and l_1 are user specified, and subsequent reviews occur at $t_{j+1} = 2t_j$ for $j \geq 1$. Then

$$B_{j+1} \doteq \begin{cases} t_{j+1}/L_j & \text{if H is rejected} \\ \sqrt{2}B_j & \text{if H is accepted} \end{cases}$$

and

$$L_{j+1} \doteq \begin{cases} L_j & \text{if H is rejected} \\ \sqrt{2}L_j & \text{if H is accepted,} \end{cases}$$

where the $\doteq$ notation indicates approximation induced here by truncating to integers.

These assignments turn out to be convenient. In particular (7.3), the total number of observations used to estimate σ_∞^2 on review $j + 1$, is

$$\begin{aligned} a(t_{j+1}) &\doteq L_{j+1}B_{j+1} \\ &\doteq \begin{cases} \frac{2^j l_1 b_1}{L_j} \times L_j & \text{if H is rejected} \\ 2L_j B_j & \text{if H is accepted} \end{cases} \\ &\doteq \begin{cases} 2^j l_1 b_1 & \text{if H is rejected} \\ 2a(t_j) & \text{if H is accepted.} \end{cases} \end{aligned}$$

By iteration on $j - 1, j - 2, \ldots$, it follows that

$$a(t_j) \doteq L_j B_j = 2^{j-1} l_1 b_1.$$

Some choices for t, l_1, and b_1 are better than others. For example, a software implementation of the ABATCH Rule, called LABATCH.2 in the appendix to this chapter (Section 7.9) does the following:

- Selects from Table 7.1 an initial batch size b_1 and an initial number of batches l_1 that maximize

  $$\frac{a(t)}{t} = \text{proportion of } t \text{ observations used to estimate } \sigma_\infty^2$$

  and

  $$\begin{aligned} J(t) &:= 1 + \text{largest integer} \leq \frac{\log\left(t/l_1 b_1\right)}{\log 2} \\ &= \text{total number of interim reviews.} \end{aligned}$$

TABLE 7.1 | *(l_1, b_1) for LABATCH.2**

b_1	l_1	b_1	l_1
1	3	38	54
2	3, 6	39	55
3	4	41	58, 87
4	6	43	61
5	7, 14, 21, 28	44	62, 93
7	10, 15, 20, 25, 30, 35	46	65
8	11	48	51, 68, 85
9	13	49	69
10	14, 21, 28	50	71
12	17, 34, 51, 68, 85	51	60, 72, 84, 96
14	15, 20, 25, 30, 35	53	75
15	21, 28	54	57
17	24, 36, 48, 60, 72, 84, 96	55	78
19	27, 54	56	79
20	21, 28	58	82
21	25, 30, 35	60	68, 85
22	31, 62, 93	61	86
24	34, 51, 68, 85	62	66
25	28	63	89
26	37	65	92
27	38, 57	66	93
28	30, 35	67	95
29	41, 82	68	72, 84, 96
31	44, 66, 88	70	99
32	45	72	85
33	47	82	87
34	36, 48, 60, 72, 84, 96	84	85
36	51, 68, 85	85	96
37	52	88	93

*Source: Fishman (1997).

- Performs interim reviews at $t_1 = l_1 b_1$ and $t_j = 2t_{j-1}$ for $j = 2, \ldots, J(t)$.
- For $a(t) = t_{J(t)}$, generates $V_1, \ldots, V_{J(t)}$, a graph of which can be used to assess the extent to which systematic error remains as the number of reviews j increases.
- Under relatively general conditions, produces an asymptotically valid $100 \times (1 - \delta)\%$ confidence interval

$$\left[\hat{\lambda}_t - \tau_{L_{J(t)}-1}(1 - \delta/2)\sqrt{V_{J(t)}/t},\ \hat{\lambda}_t + \tau_{L_{J(t)}-1}(1 - \delta/2)\sqrt{V_{J(t)}/t}\right]$$

for λ (Yarberry 1993, Fishman and Yarberry 1997).

LABATCH.2 has two additional features:

- If a user chooses $t = 2^{\alpha} l_1 b_1$ as the sample-path length, for any integer $\alpha > 1$ and underlined (l_1, b_1) in Table 7.1, then $a(t) = t_{J(t)} = t$. That is, LABATCH.2 uses all the data to compute the final $V_{J(t)}$ on review $J(t)$.
- For $t \geq 500$, choosing any (l_1, b_1) from the table induces the lower bound $\frac{a(t)}{t} \geq .898$, ensuring that at least 90 percent of the data are used to estimate σ_∞^2.

EXAMPLE 7.7.1 CAPTURE-RECAPTURE ACCURACY ASSESSMENT: ABATCH RULE

The results in Table 6.2 for the capture-recapture problem were computed by LABATCH.2 using the ABATCH rule and $t = 10^6$, which resulted in $a(t)/t = 0.9851$ of the sample path being used to compute $V_{a(t)}$ and $J(t) = 11$ interim reviews. A second MCMC experiment, with results presented in Table 7.2 used $t = 2^{19} \times 3 \times 1 = 1{,}572{,}864$, which produced $a(t)/t = 1$ and $J(t) = 20$, thus using the total sample path to compute $V_{J(t)}$ and increasing the number of interim reviews. They occurred at $t_j = l_1 b_1 2^{j-1}$ for $j = 1, \ldots, 20$, where $l_1 = 3$ and $b_1 = 1$, giving more opportunities for detecting trend caused by systematic error.

In addition to results for the posterior mean population size, Table 7.2 displays results for the posterior probabilities $\text{pr}(\theta \leq 332)$ and $\text{pr}(\theta \leq 561)$. Recall that 332 and 561 were the estimated 0.025 and 0.975 quantiles, respectively, for the posterior distribution in Figures 6.3 and 6.4. Collectively, these graphs demonstrate LABATCH.2's capacity to analyze sample paths on more than one stochastic process during a single sampling experiment. For the mean population size, estimates of λ and σ_∞^2 are comparable to those in Example 7.5.1.[1]

As mentioned earlier, the validity of the confidence intervals in Table 7.2 depends on the extent to which systematic error in $V_{J(t)}$ is negligible. Perusal of a graph of $\sqrt{V_1}, \ldots, \sqrt{V_{20}}$ in Figure 7.4a allows us to make this judgment for the mean. It reveals that for Hastings-Metropolis with two-coordinate updating per iteration, $\sqrt{V_j}$ tends to increase for $j \leq 11$ and thereafter fluctuates in a stable manner. The pattern suggests an absence of systematic error in $V_{20} = 7.767 \times 10^5$

[1] For an analysis of the differences in standard errors for each sampling method, see the discussion following Table 6.2.

TABLE 7.2 | ***Sunfish population size sample-path analysis: $t = 1{,}572{,}864$; $a(t) = t$; $J(t) = 20$; ABATCH Rule***

λ	Sampling technique	No. of batches L_{20}	Batch size B_{20}	$\widehat{\lambda}_t \pm \tau_{L_{20}-1}(0.995) \times$ s.e.$(\widehat{\lambda}_t)$	r.e.$(\widehat{\lambda}_t)$	$\widehat{R}_0$	$\widehat{\sigma}^2_\infty = V_t$
Mean							
	Two coordinates						
	Hastings-Metropolis	48	32,768	430.2 ± 1.90	1.633×10^{-3}	3402	7.767×10^5
	Gibbs	256	6,144	431.2 ± 0.800	7.092×10^{-4}	3430	1.471×10^5
	All coordinates						
	Gibbs	384	4,096	430.7 ± 0.300	2.795×10^{-4}	3423	2.280×10^4
$\text{pr}(\theta \le 332)$							
	Two coordinates						
	Hastings-Metropolis	96	16,384	0.02577 ± 0.00244	3.599×10^{-2}	0.02512	1.353
	Gibbs	384	4,096	0.02478 ± 0.00112	1.755×10^{-2}	0.02415	0.2957
	All coordinates						
	Gibbs	192	8,192	0.02550 ± 0.000490	7.345×10^{-4}	0.02483	0.05523
$\text{pr}(\theta \le 561)$							
	Two coordinates						
	Hastings-Metropolis	96	16,384	0.9760 ± 0.00275	1.072×10^{-3}	0.02307	1.721
	Gibbs	128	12,288	0.9750 ± 0.00130	5.047×10^{-4}	0.02437	0.3808
	All coordinates						
	Gibbs	384	4,096	0.9754 ± 0.000550	2.287×10^{-4}	0.02396	0.07829

as an estimate of σ_∞^2. For $j = 11$, LABATCH.2 computed batch size as $B_{11} = 384$ and number of batches as $L_{11} = 8$, based on $t_{11} = 3{,}072$.

Figure 7.4a also shows different convergence rates based on Gibbs sampling. For two-coordinate Gibbs sampling, it suggests negligible systematic error for $j \geq 9$ with $B_9 = 64$ and $L_9 = 12$. For all-coordinate Gibbs sampling, it also suggests $j \geq 9$, but with $B_9 = 48$ and $L_9 = 16$. Figures 7.4b and 7.4c allow similar analysis for the two quantiles. For each of the MCMC sampling plans, the scans suggest that V_{20} is an estimate of σ_∞^2 free of systematic error. ■

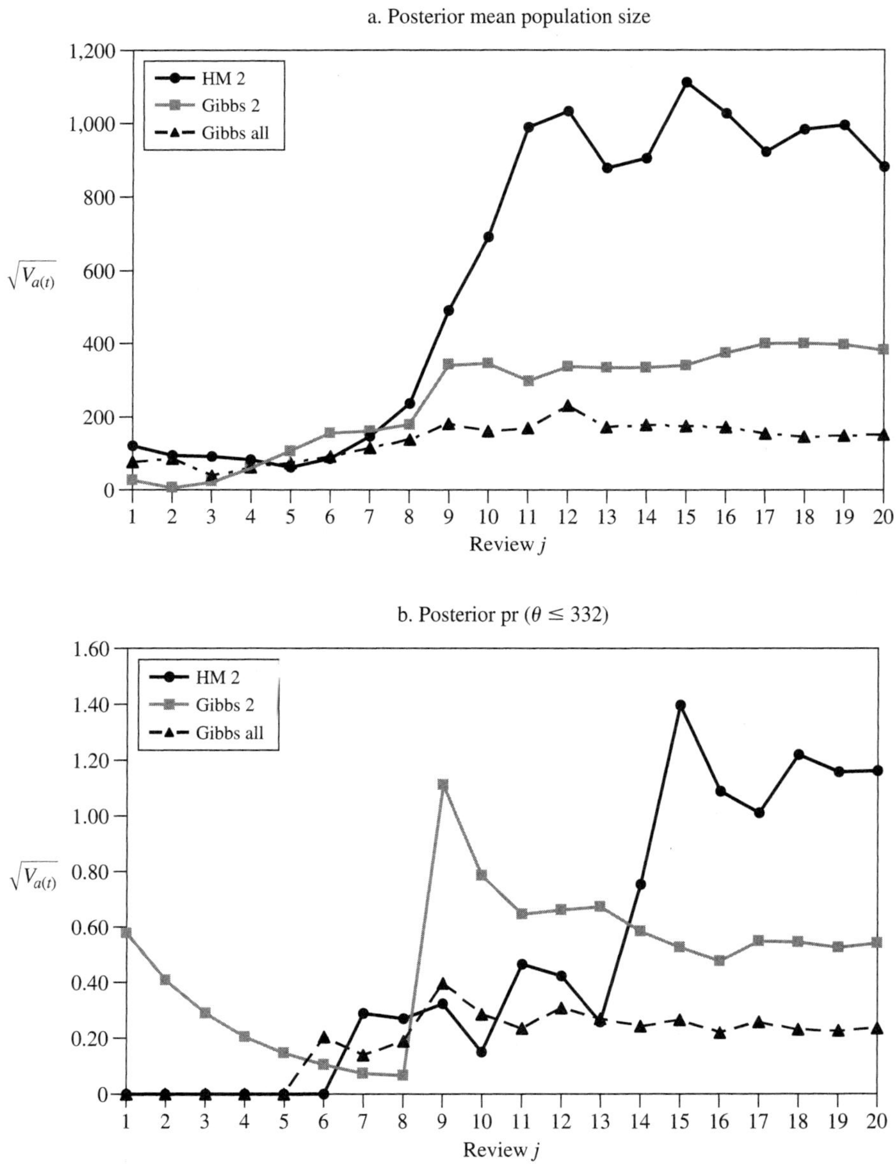

FIGURE 7.4 | *Estimating σ_∞^2 for sunfish sample posterior mean size: ABATCH Rule*

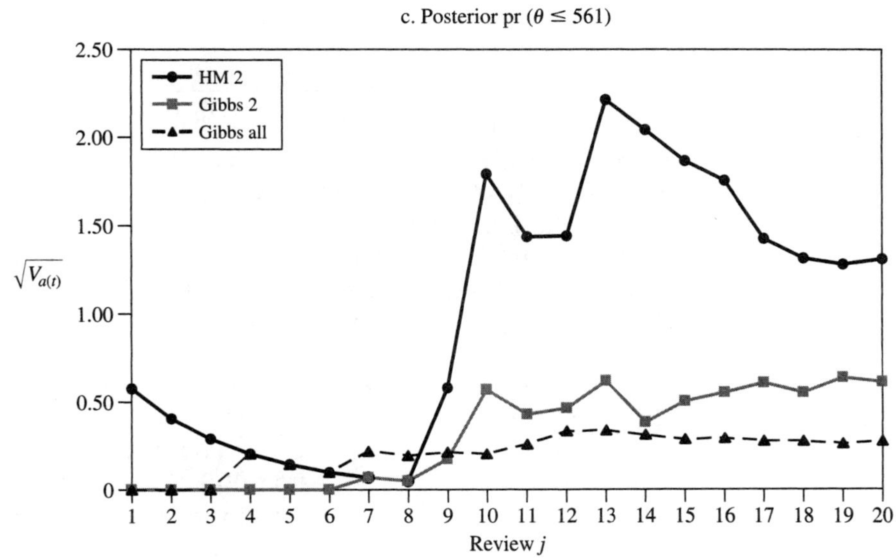

FIGURE 7.4 | *Continued*

7.8 TESTING FOR INDEPENDENCE

We now turn to testing H. Let $b := b(t)$ and $l := l(t)$. Because of stationarity, $Y_{1b}, \ldots, Y_{lb}$ in (7.4) have the *autocorrelation function*

$$\begin{aligned} Q_{|i-j|} &:= \operatorname{corr}(Y_{ib}, Y_{jb}) \\ &= \frac{\operatorname{cov}(Y_{ib}, Y_{jb})}{\sqrt{\operatorname{var} Y_{ib} \operatorname{var} Y_{jb}}} \quad i, j = 1, \ldots, l. \end{aligned}$$

Recall from Section 6.9 that $\{Z_j = h(\mathbf{X}_j),\ j \geq 0\}$ has autocovariance function

$$R_{i-j} = \operatorname{cov}(Z_i, Z_j) \quad i, j = 1, 2, \ldots.$$

Its corresponding autocorrelation function is $\{R_j/R_0;\ j = \pm 1, \pm 2, \ldots\}$. Then

$$Q_{|i-j|} = \frac{1}{R_0 b^2} \sum_{b_1=1}^{b} \sum_{b_2=1}^{b} R_{b(i-j)+b_1-b_2},$$

establishing the intimate connection that exists between the correlation properties of the batches $\{Y_{ib}\}$ and those of the original stochastic process $\{Z_j,\ j \geq 1\}$.

The *von Neumann ratio* provides one way of testing H (von Neumann 1941, Young 1941). Let

$$\begin{aligned} C_l &:= 1 - \frac{\sum_{i=1}^{l-1} (Y_{ib} - Y_{i+1,b})^2}{2 \sum_{i=1}^{l} (Y_{ib} - \widehat{\lambda}_t)^2} \\ &= \widehat{Q}_1 + \frac{(Y_{1b} - \widehat{\lambda}_t)^2 + (Y_{lb} - \widehat{\lambda}_t)^2}{2 \sum_{i=1}^{l} (Y_{ib} - \widehat{\lambda}_t)^2}. \end{aligned} \tag{7.8}$$

Because of stationarity,

$$\widehat{Q}_1 := \frac{\sum_{i=1}^{l-1}(Y_{ib} - \widehat{\lambda}_t)(Y_{i+1,b} - \widehat{\lambda}_t)}{\sum_{i=1}^{l}(Y_{ib} - \widehat{\lambda}_t)^2}$$

estimates the first lagged autocorrelation Q_1. Inspection reveals that C_l is itself an estimator of Q_1 adjusted for end effects that diminish in importance as the number of batches l increases. If $\{Z_i,\ i \geq 0\}$ has a monotone decreasing autocorrelation function, then in the present setting Q_1 is positive and decreases monotonically to zero as batch size b increases. If for some l and b, H is true for $Y_{1b}, \ldots, Y_{lb}$, then $Q_1 = 0$, and more generally, the monotone property implies $\text{corr}(Y_{ib}, Y_{jb}) = 0$ for all $i \neq j$.

As batch size b increases with sample-path length t, the batch averages become jointly normally distributed. Then $Q_j = 0$, for all $j > 0$, implies that the batch averages are independent. The von Neumann ratio test for H exploits this property. For i.i.d. $\frac{Y_{1b}-\mu}{\sigma_b}, \ldots, \frac{Y_{lb}-\mu}{\sigma_b}$ from $\mathcal{N}(0, 1)$ for l as small as eight, C_l has mean zero, variance $(l-2)/(l^2-1)$, and a distribution close to the standard normal. Although the batch averages may be non-normal for the initial batch size b_1, they become increasingly normal as the sample-path length t_j increases, making the normality assumption more tenable on successive interim reviews. The speed of convergence is a consequence of the symmetry that characterizes the distribution of C_l, a property absent from the corresponding distribution of Q_1.

Under the assumption of a monotone decreasing autocorrelation function for $\{Z_i\}$, a *one-sided test* of size β is in order. In particular, if

$$C_l \leq \Phi^{-1}(1-\beta)\sqrt{(l-2)/(l^2-1)},$$

then H is accepted. Otherwise, it is rejected. The one-sided test protects against accepting H when the sample autocorrelation, C_l, between Y_{ib} and Y_{jb} for all $|i - j| = 1$ is "too large" to be attributed to sampling variation. LABATCH.2 uses this test.

On interim review j, LABATCH.2 also computes the corresponding p-value as

$$P_j := 1 - \Phi\left(\sqrt{\frac{L_j^2 - 1}{L_j - 2}}\, C_{L_j}\right) \qquad j = 1, \ldots, J(t).$$

It serves as a credibility index by estimating the probability of observing a C_{L_j}, under H, larger than the one computed from $Y_{1B_j}, \ldots, Y_{L_jB_j}$. A small P_j implies low credibility for H.

LABATCH.2 makes its decisions as follows:

1. If $P_j < \beta$ on review j for a user-specified β, then it uses the SQRT Rule on review $j+1$.
2. Otherwise, it uses the FNB Rule.

All the examples in the present account use $\beta = .10$.

Suppose that $\{Z_i\}$ has an autocorrelation function that exhibits damped harmonic behavior around the zero axis or that $\{Z_i\}$ is an i.i.d. sequence. In this case, the one-sided test accepts H when negative correlation is present. LABATCH.2 provides information that enables a user to infer the presence of either of these cases and to take advantage of this knowledge.

Recall that $Z_1, \ldots, Z_t$ are the MCMC generated data. Upon completing its last review $J(t)$, it computes the p-value

$$P_t' := 1 - \Phi\left(\sqrt{\frac{t^2-1}{t-1}}C_t'\right),$$

where

$$C_t' := 1 - \frac{\sum_{i=1}^{t}(Z_i - Z_{i+1})^2}{2(t-1)R_0}$$

$$\widehat{R}_0 := \frac{1}{t-1}\sum_{i=1}^{t}\left(Z_i - \widehat{\lambda}_t\right)^2,$$

for the hypothesis H. It also computes the sample standard deviation $\sqrt{\widehat{R}_0}$ and a $100 \times (1-\delta)$ percent interval

$$\left[\widehat{\lambda}_t \pm \tau_{t-1}(1-\delta/2)\sqrt{\widehat{R}_0/t}\right]. \tag{7.9}$$

Whenever $P_t' > \beta$, the specified significance level for testing H, three possibilities arise:

- $\{Z_i\}$ is an i.i.d. sequence.
- $\{Z_i\}$ has a damped harmonic autocorrelation function.
- This p-value merely exhibits a random aberration.

To investigate how likely any of these are, we proceed as follows:

- If as j increases $\sqrt{V_j}$ fluctuates around $\sqrt{\widehat{R}_0}$, displaying no discernible trend, then it is not unreasonable to assume that $\{Z_i\}$ is an i.i.d. sequence and to use the approximating confidence interval (7.9) for λ. This approach offers a statistically better estimate of σ_∞^2 with greater degrees of freedom.
- If $P_t' > \beta$ but $\sqrt{V_j}$ tends to decrease with increasing j and, in particular, $V_{J(t)} \ll \widehat{R}_0$, then there is reason to believe that $\{Z_i\}$ has an autocorrelation function that fluctuates around zero, implying the presence of negative as well as positive correlation. In this case, using the confidence interval for review $J(t)$ reaps the potential benefits of the negative correlation, usually inducing a shorter interval length for the confidence interval for λ.
- If $P_t' > \beta$ but $\sqrt{V_j}$ tends to increase with increasing j and, in particular, $V_{J(t)} \gg \widehat{R}_0$, then there is reason to believe that this p-value reflects a random aberration.

EXAMPLE 7.8.1

EYE/HAIR COLOR ACCURACY ASSESSMENT

Table 6.4 displays the results of MCMC sampling based on Algorithms CONTINGENCY and ALL for a sample-path length $t = 2^{19} = 524{,}288$. The graph in Figure 7.5a suggests that $\sqrt{V_j}$ exhibits stationary behavior for all reviews $j \geq 11$, implying that $\widehat{\sigma}^2_\infty = 1.721$ in Table 6.4 is relatively free of systematic error. When the graph is compared to $\widehat{R}_0 = 0.3740$, it also shows the considerable variance inflation that dependence among observations induces.

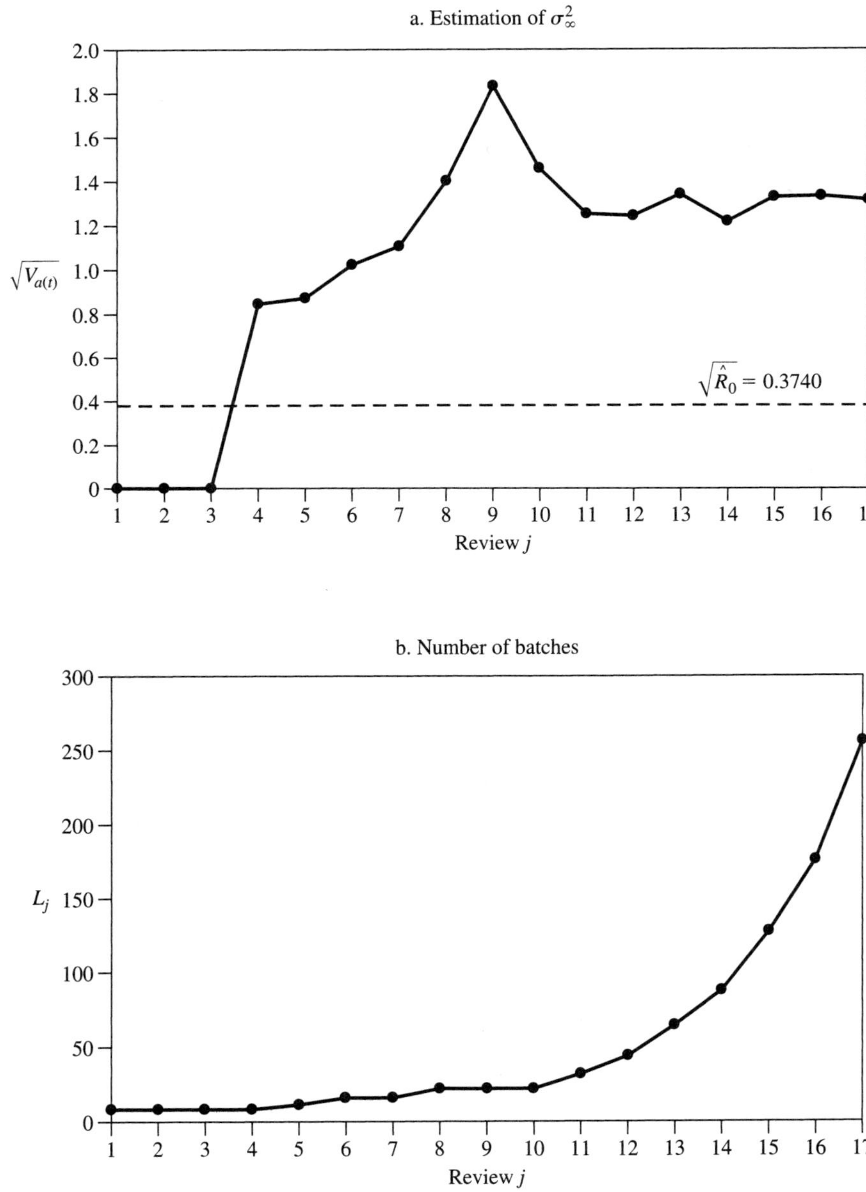

FIGURE 7.5 | *Estimating σ^2_∞ for the eye/hair color proportion*

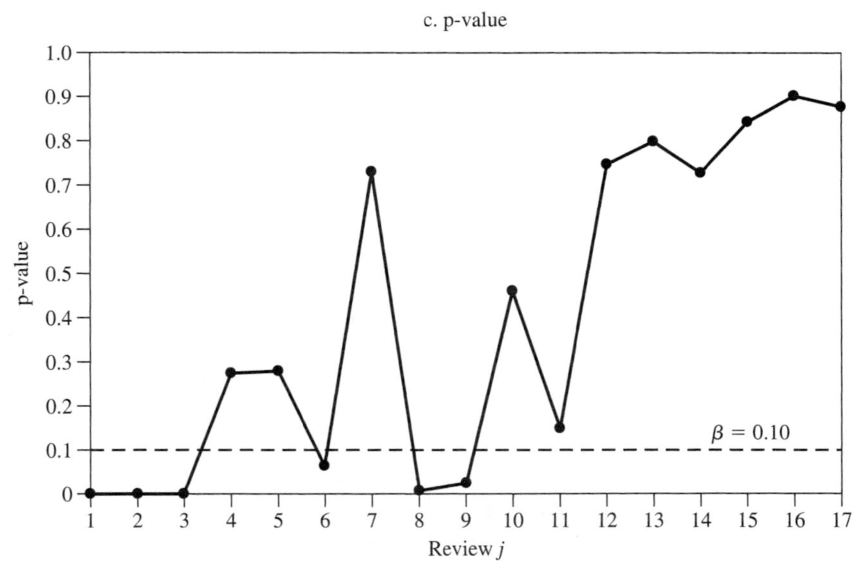

FIGURE 7.5 | *Continued*

Figures 7.5b and 7.5c illustrate how testing H on each review affects growth in the number of batches L_j. The horizontal line segments in Figure 7.5b correspond to p-values in Figure 7.5c that lie below the $\beta = 0.10$ significance level. For reviews $j \geq 11$, no p-value lies below 0.10, a sample result consistent with the expected number of rejections $(17 - 11 + 1) \times 0.10 = 0.70$ among reviews 11 through 17. ■

7.9 APPENDIX: LABATCH.2

LABATCH.2 is a collection of computer programs that can perform statistical analyses simultaneously on up to 60 sample sequences (Fishman 1997). Available in C, FORTRAN, and SIMSCRIPT II.5, it can be downloaded from the Internet site:

http://www.or.unc.edu/~gfish/labatch.2.html

LABATCH.2 offers two modes of implementation. One integrates the program into an executing data-generating program to analyze the evolving data on repeated subroutine calls. The other takes data from an existing file as input. In addition to minimizing user effort, space requirements for the first option are proportional to $\log_2 t$ rather than to t. The second option permits statistical analysis of stored data, regardless of source and date of generation, thereby making LABATCH.2 applicable in a considerably wider range of settings. It also allows user interaction with the executing program via screen displays of interim estimates.

A call from a C / FORTRAN / SIMSCRIPT II.5 main program to

BATCH_MEANS(IN_UNIT,OUT_UNIT,T,S_NUM,PSI_VECTOR, DELTA,RULE,BETA,L_UPPER,SCREEN) (7.10) implements LABATCH.2, where Table 7.3 defines its arguments. As an example, suppose these arguments assume the values:

$$\begin{aligned}
\text{IN_UNIT} &= 0\\
\text{OUT_UNIT} &= 15\\
\text{T} &= 1000\\
\text{S_NUM} &= 2\\
\text{PSI_VECTOR} &= \text{pointer to data vector}\\
\text{DELTA} &= 0.01\\
\text{RULE} &= 1\\
\text{BETA} &= 0.10\\
\text{L_UPPER} &= 30\\
\text{SCREEN} &= 0
\end{aligned} \tag{7.11}$$

Then LABATCH.2 processes S_NUM = 2 sequences in T = 1000 iterative calls from the user's main program, computes 100 × (1 − DELTA) = 99 percent confidence intervals for the sample averages, and writes the output to a file called c.15 fort.15 SIMU15 (IN_UNIT = 0, OUT_UNIT = 15). RULE = 1 causes LABATCH.2 to employ the ABATCH rule to determine batch size on each review, and BETA = 0.10 directs it to test for independence of batch averages at the 0.10 significance level. LABATCH.2 begins its first iteration with the number of batches no greater than L_UPPER = 30. SCREEN = 0 suppresses the interactive screen feature. In practice, all but the values of T and S_NUM can be set once, and the subroutine can be used repeatedly for different settings of these two parameters.

Setting BETA = 1 causes LABATCH.2 to employ the FNB Rule, whereas setting BETA = −1 causes it to employ the SQRT Rule. By setting RULE = 0, a fourth rule called LBATCH can also be implemented. It employs the FNB Rule until H is accepted for the first time, and thereafter it employs the SQRT Rule (see Fishman 1997).

Assigning a positive number to IN_UNIT directs LABATCH.2 to take data from that file number for input. For example, IN_UNIT = 30 implies files titled fort.30 in the FORTRAN implementation and c.30 in the C implementation. This option requires a single call to BATCH_MEANS to begin processing the entire file. L_UPPER provides a user with an option to control the upper bound on possible values of l_1 that LABATCH.2 chooses from Table 7.1.

SCREEN = 1 activates LABATCH.2's interactive feature that, after each interim review j, displays partial results on the PC monitor's screen for min(6, S_NUM) sample paths being processed. These include $\widehat{\lambda}_{t_j}$, $\sqrt{V_j}$, and r.e. $(\widehat{\lambda}_{t_j}) := \sqrt{V_j/t_j}/\widehat{\lambda}_{t_j}$. It then asks whether it should continue or terminate the run, and it pauses to await the user's response. This procedure enables a user to curtail execution early if he or she concludes that the V_j have become stationary and the last r.e.s are sufficiently small.[2]

[2]For a more complete description, see Fishman (1997) or Fishman (2000).

TABLE 7.3 | ***LABATCH.2 arguments***

User-specified input	Definition	Notation in chapter
IN_UNIT	:= 0 if data are repeatedly to be transferred from a main program	...
	:= 30 if data are to be read one vector at a time from input file c.15 / fort.15 / SIMU15 (30 is merely an example)	
OUT_UNIT	:= designated unit for writing output	...
T	:= total number of observations	t
S_NUM	:= number of sample series	...
PSI_VECTOR	:= pointer to vector of S_ NUM sample values	...
DELTA	:= desired significance level for confidence intervals (0.01 or 0.05 suggested)	δ
RULE	:= 1 if the ABATCH Rule is used	
	:= 0 if the LBATCH Rule is used	
BETA	:= significance level for testing for independence (0.10 suggested) If BETA = 1, each review uses the FNB Rule. = −1, each review uses the SQRT Rule.	β
L_UPPER	:= upper bound on initial number of batches (3 <= L_UPPER <= 100)	...
SCREEN	:= 1 if interim review estimates are to be displayed on the screen	
	:= 0 otherwise	

7.10 LESSONS LEARNED

- MCMC experiments usually begin in arbitrarily selected states.
- The elements in MCMC-generated sample paths are usually dependent.
- MCMC-generated data present two principal challenges: estimating the warm-up interval, k, that must elapse before the data are relatively free (because of dependence) of initial conditions and estimating the statistical accuracies of sample averages.

- Graphs based on several independent sample paths generated in preliminary experiments provide a basis for estimating the warm-up interval.
- There are two approaches to estimating the statistical accuracy of the sample mean: one based on n independent replications each with sample-path length t', and the other based on a single sample path of length nt'.
- The multiple-path approach calls for truncating k observations at the beginning of each sample path for a total of nk observations, whereas the single-path approach requires a total of k observations to be truncated.
- The multiple-path approach relies on relatively elementary statistical theory to estimate accuracy, whereas the single-path approach requires a more sophisticated statistical methodology. LABATCH.2 is an implementation of the latter approach.

7.11 HANDS-ON EXERCISES

As described in Chapter 6, the Deterministic Coordinate Selection (DCS) and Random Coordinate Selection (RCS) Rules provide two alternative ways of selecting coordinates for MCMC updating, and Hastings-Metropolis (HM) and Gibbs sampling provide two alternatives for randomly generating each point on a sample path. Moreover, an experiment can be designed to generate multiple independent sample paths on each of which a warm-up interval needs to be deleted but relatively elementary statistical analysis can be performed. Or it can be designed to generate a single sample path on which a warm-up interval needs to be deleted once but where error analysis requires more sophisticated methods. Collectively these alternatives present the user with a choice of eight different options for designing an MCMC experiment. Exercises H.1 through H.12 and H.22 through H.33 provide a basis for comparing the efficiency of the options. By efficiency here, we mean an option that achieves a specified standard error in less computing time than the alternative option. For purposes of comparison, record the computing times for each exercise.

Of course, MCMC sampling methods other than HM and Gibbs are possible. Exercise H.43 for discrete-event systems provides an illustration.

Multiple Independent Replications

Perform an MCMC experiment that generates $n = 100$ independent sample paths, each with $k+t'$ iterations, where k is the warm-up interval computed in the corresponding hands-on exercises in Chapter 6, and $t' = 10{,}000$. Compute an estimate of λ and standard and relative errors of the estimate. Perform these steps for the problems described in:

H.1. Exercise 6.H.1

H.2. Exercise 6.H.2

H.3. Exercise 6.H.3

H.4. Exercise 6.H.4

H.5. Exercise 6.H.5

H.6. Exercise 6.H.6

H.7. Exercise 6.H.7

H.8. Exercise 6.H.8

H.9. Exercise 6.H.9

H.10. Exercise 6.H.10

H.11. Exercise 6.H.11

H.12. Exercise 6.H.12

Comparing Coordinate-Selection Rules

Based on standard errors and computing times here and in Chapter 6 and on other information gathered in Chapter 6, identify and justify your choice as to whether DCS/HM sampling or RCS/HM sampling is more desirable in:

H.13. Exercises H.1 and H.2

H.14. Exercises H.3 and H.4

H.15. Exercises H.5 and H.6

Based on standard errors and computing times, identify and justify your choice as to whether DCS/Gibbs sampling or RCS/Gibbs sampling is more desirable in:

H.16. Exercises H.7 and H.8

H.17. Exercises H.9 and H.10

H.18. Exercises H.11 and H.12

Comparing Coordinate-Selection Rules and MCMC Sampling Methods

H.19. For Exercises H.1, H.2, H.7, and H.8, order the sampling plans, DCS/HM sampling, RCS/HM sampling, DCS/Gibbs sampling, and RCS/Gibbs sampling, based on overall efficiency. Justify your ordering.

H.20. For Exercises H.3, H.4, H.9, and H.10, order the sampling plans, DCS/HM sampling, RCS/HM sampling, DCS/Gibbs sampling, and RCS/Gibbs sampling, based on overall efficiency. Justify your ordering.

H.21. For Exercises H.5, H.6, H.11, and H.12, order the sampling plans, DCS/HM sampling, RCS/HM sampling, DCS/Gibbs sampling, and RCS/Gibbs sampling, based on overall efficiency. Justify your ordering.

Single Sample Path

These exercises call for the estimation of λ and corresponding standard and relative errors from a single sample path. Statistical analysis may be performed using the C or FORTRAN version of LABATCH.2, which is obtainable at the Internet site listed in Section 7.9. It may also be performed using any other applications software designed for this purpose.

Perform an MCMC sampling experiment that generates a single sample path ($n = 1$) of $k + t$ iterations, where $t = 10^6$ and k is the warm-up interval computed in the corresponding problem in Chapter 6. Compute an estimate of λ and evaluate its standard and relative errors. Perform these steps for the problem specifications in:

H.22. Exercise 6.H.1

H.23. Exercise 6.H.2

H.24. Exercise 6.H.3

H.25. Exercise 6.H.4

H.26. Exercise 6.H.5

H.27. Exercise 6.H.6

H.28. Exercise 6.H.7

H.29. Exercise 6.H.8

H.30. Exercise 6.H.9

H.31. Exercise 6.H.10

H.32. Exercise 6.H.11

H.33. Exercise 6.H.12

Comparing Coordinate-Selection Rules

Based on standard errors and computing times, identify and justify your choice as to whether DCS/HM sampling or RCS/HM sampling is more efficient in:

H.34. Exercises H.22 and H.23

H.35. Exercises H.24 and H.25

H.36. Exercises H.26 and H.27

Based on standard errors and computing times, identify and justify your choice as to whether DCS/Gibbs sampling or RCS/Gibbs sampling is more desirable in:

H.37. Exercises H.28 and H.29

H.38. Exercises H.30 and H.31

H.39. Exercises H.32 and H.33

Comparing Coordinate-Selection Rules and MCMC Sampling Methods

H.40. For Exercises H.22, H.23, H.28, and H.29, order the sampling plans, DCS/HM sampling, RCS/HM sampling, DCS/Gibbs sampling, and RCS/Gibbs sampling, based on efficiency. Justify your ordering.

H.41. For Exercises H.24, H.25, H.30, and H.31, order the sampling plans, DCS/HM sampling, RCS/HM sampling, DCS/Gibbs sampling, and RCS/Gibbs sampling, based on efficiency. Justify your ordering.

H.42. For Exercises H.26, H.27, H.32, and H.33, order the sampling plans, DCS/HM sampling, RCS/HM sampling, DCS/Gibbs sampling, and RCS/Gibbs sampling, based on efficiency. Justify your ordering.

Discrete-Event Systems

H.43. Estimate the standard and relative errors for each estimate in Exercise 6.H.13. Use these errors to justify your choice of the β that meets the requirement $\text{pr}(W_j \geq 2.50) = 0.05$.

7.12 REFERENCES

Brillinger, D. (1973). Estimation of the mean of a stationary time series, *Journal of Applied Probability,* **10,** 157–174.

Chien, C. -H., D. Goldsman, and B. Melamed (1997). Large sample results for batch means, *Management Science,* **43,** 1288–1295.

Chien, C. -H. (1989). Small sample theory for steady state confidence intervals, Technical Report 37, Department of Operations Research, Stanford University.

Fishman, G. S. (1997). LABATCH.2 for analyzing sample path data, TR-97/04, Operations Research Department, University of North Carolina, Chapel Hill, available at http://www.or.unc.edu/~gfish/labatch.2.html.

Fishman, G. S. (2000). *Discrete-Event Simulation: Modeling, Programming, and Analysis,* Springer-Verlag, New York.

Fishman, G. S., and L. S. Yarberry (1997). An implementation of the batch means method, *INFORMS Journal on Computing,* **9,** 296–310.

von Neumann, J. (1941). The mean square difference, *Annals of Mathematical Statistics,* **12,** 153–162.

Yarberry, L. S. (1993). Incorporating a dynamic batch size selection mechanism in a fixed-sample-size batch means procedure, unpublished Ph.D. dissertaion, Department of Operations Research, University of North Carolina at Chapel Hill.

Young, L. C. (1941). On randomness of order sequences, *Annals of Mathematical Statistics,* **12,** 293–300.

CHAPTER 8

Optimization via MCMC

This chapter describes how MCMC sampling can be applied to the problem of minimizing a function of m variables. Although some of the methodology is similar to that in Chapter 6, other parts call for a considerably different perspective to solve the problem.

For a function H on a finite or countably infinite state space $\mathcal{X}$ with unknown finite global minimum

$$H_{\min} := \min_{\mathbf{x} \in \mathcal{X}} H(\mathbf{x}), \tag{8.1}$$

MCMC provides techniques, most notably, *simulated annealing,* for estimating $H_{\min}$ and for estimating elements of the unknown *optimal solution set*

$$\mathcal{X}_{\min} := \{\mathbf{x} \in \mathcal{X}\colon H(\mathbf{x}) = H_{\min}\}. \tag{8.2}$$

Presumably, $\mathcal{X}$ is so large that exact evaluation is not an option. Sometimes H is called the *objective function.*

MC techniques for estimating minima come under the heading of *stochastic optimization.* As mentioned in Chapter 1, the techniques are of two types:

- **Random Deterministic (RD).** On each iteration, it:
  - RD1. Randomly generates a location in the feasible space.
  - RD2. Applies a *deterministic rule* to decide on the next location on the path to solution.
- **Random Random (RR).** On each iteration, it:
  - RR1. Randomly generates a location in the feasible space.
  - RR2. Applies a *randomized rule* to decide whether to move to the new location on the path to solution.

The RR procedures include:[1]

- Genetic algorithms: Procedures that emulate the biological evolutionary processes of reproduction and mutation
- Neural networks: Procedures that incorporate what has been learned from past search, into a strategy for future search
- Simulated annealing: A procedure based on Hastings-Metropolis (HM) sampling with an acceptance function (6.8) that varies with *temperature*

This chapter begins by describing the conceptually simplest of these, simulated annealing. It next describes three of its enhancements, simulated tempering (Section 8.5), swapping (Section 8.6), and stochastic tunneling (Section 8.7). Lastly, Section 8.8 describes genetic algorithms and their relationship to the previously introduced concepts.

Simulated annealing is a sequential search technique that relies on HM sampling (see Section 6.1) to nominate states $\mathbf{x}$ in $\mathcal{X}$ for comparison with previously visited states with respect to $H(\mathbf{x})$. By modifying its acceptance function at the beginning of successive stages, simulated annealing induces a sample path that converges to an element in $\mathcal{X}_{\min}$ with probability converging to one as the sample path's length increases without bound.

For more than half a century, variants of this methodology have provided solutions to problems in the physical sciences, most notably in condensed-matter physics. More recently, the methodology has been extended to solving problems in *combinatorial optimization* (Kirkpatrick et al. 1983, Černy 1985). For a large class of these basic problems, including graph coloring, linear arrangement matching, quadratic assignment, and scheduling, simulated annealing often finds solutions with errors comparable to those observed when more specialized deterministic algorithms are employed.

Simulated annealing's appeal derives from its relative simplicity of concept and its ease of implementation. It relies minimally on special structure in H. This contrasts with the detailed exploitation of the structure of H, on which many deterministic methods rely for their efficacy. Its limitation is the relatively large amount of computing time required for some problems to provide a solution close to one in the $\mathcal{X}_{\min}$. By "close to," we mean a solution $\mathbf{x}^*$ such that $H(\mathbf{x}^*)$ is within a specified neighborhood of $H_{\min}$. Nevertheless, simulated annealing's demonstrated successes in producing solutions in areas as diverse as condensed-matter physics, protein folding, and microchip layout attest to its value in practice.

Annealing means to heat and then to cool. Sections 8.1 and 8.3 describe how simulated annealing and, in particular, its search technique incorporate procedures, analogous to heating and cooling, to coerce a sample path to converge to a solution in the optimal solution set. Examples 8.3.1 and 8.3.2 testify to the technique's breadth of application.

[1]For more detailed descriptions of genetic algorithms, neural networks, and other techniques, see Spall (2003).

Two examples illustrate how simulated annealing achieves its solution. Example 8.3.1 concerns the placement of microchips on circuit boards. Example 8.3.2 addresses the problem of *protein folding:* identifying the three-dimensional structure of proteins in the human genome. Both problems have explicit constraints that any meaningful solution must satisfy.

Relatively straightforward in concept, the constraints in the microchip problem are easily understood and applied. However, those for the protein folding problem arise from chemical and biological interactions among atoms and molecules. Some knowledge of these interactions is essential before their influence on MCMC sampling can be appreciated. The example includes a not inconsequential description of these issues. Although this makes the example longer than all others in this text, it also demonstrates to the reader the sophisticated way in which MCMC sampling can contribute to a significant scientific problem.

8.1 SIMULATED ANNEALING

To conduct its search for an optimal solution, simulated annealing employs a specialized form of HM sampling, first described in Metropolis et al. (1953) and principally used in computational chemistry, physics, and more recently, computational biology to identify three-dimensional configurations of atoms and molecules that minimize energy.[2] Kirkpatrick, et al. (1983) first described its applicability for solving a considerably wider range of problems encountered in other areas of combinatorial optimization. These include integrated circuit layout, facilities layout, image processing, and code design, as well as the more conceptual problems of graph partitioning, graph coloring, matching, quadratic assignment, and scheduling.

Most notable is simulated annealing's introduction of randomness into what more often than not is a strictly deterministic problem. For a p.m.f. $\pi := \{\pi(\mathbf{x}), \mathbf{x} \in \mathcal{X}\}$ and reversible and ergodic (irreducible and aperiodic) nominating kernel $q := \{q(\mathbf{x}, \mathbf{y}), (\mathbf{x}, \mathbf{y}) \in \mathcal{X}^2\}$ as in Section 6.1, recall that

- If $\mathbf{X}_0$ is has distribution π, then $\mathbf{X}_1, \mathbf{X}_2, \ldots$, generated by HM sampling, each have the equilibrium distribution π.
- If $\mathbf{X}_0$ is arbitrarily assigned a value in $\mathcal{X}$ and $\mathbf{X}_1, \mathbf{X}_2, \ldots$ are generated by HM sampling, then the distribution of $\mathbf{X}_j$ converges to π as $j \to \infty$.

In contrast to the problem of estimating λ where $\pi(\mathbf{x})$ arises as a component of the factorization (2.23), the present setting allows a choice for π that caters to the search for an optimal solution. In particular, for some $T > 0$, let[3]

$$\pi(\mathbf{x}) := \frac{e^{-H(\mathbf{x})/T}}{\sum_{\mathbf{z}\in\mathcal{X}} e^{-H(\mathbf{z})/T}} \qquad \mathbf{x} \in \mathcal{X}, \tag{8.3}$$

[2]Because $H(\mathbf{x})$ is the potential energy at site $\mathbf{x}$ in physics, H is sometimes called the *potential energy surface* (PES).

[3]Provided that H has the form in Theorem 9.3.1, then π is a Gibbs distribution (9.9) with an explicit notation $T := \tau/a$, where τ denotes absolute temperature and a the Boltzmann constant.

and let

$$\pi(\mathcal{X}_{\min}) := \sum_{\mathbf{x} \in \mathcal{X}_{\min}} \pi(\mathbf{x}). \tag{8.4}$$

Large $\pi(\mathcal{X}_{\min})$ implies that in equilibrium there is a high probability of finding $\mathbf{X}_j$ in $\mathcal{X}_{\min}$ and, hence, a solution to the problem.

In addition to having maxima at each $\mathbf{x} \in \mathcal{X}_{\min}$, definition (8.3) endows π with properties that facilitate the search for an optimal solution. For example,

$$\pi(\mathbf{x}) \to \begin{cases} |\mathcal{X}|^{-1} & \text{if } \mathbf{x} \in \mathcal{X} \\ 0 & \text{elsewhere} \end{cases} \quad \text{as } T \to \infty,$$

revealing an approximately uniform distribution for large T. Conversely,

$$\pi(\mathbf{x}) \to \begin{cases} |\mathcal{X}_{\min}|^{-1} & \text{if } \mathbf{x} \in \mathcal{X}_{\min} \\ 0 & \text{elsewhere} \end{cases} \quad \text{as } T \to 0,$$

revealing a concentration of probability mass on the optimal solution set for small T. Most importantly,

$$\begin{aligned} \pi(\mathcal{X}_{\min}) &= \frac{1}{1 + |\mathcal{X}_{\min}|^{-1} \sum_{\mathbf{x} \in \mathcal{X} \setminus \mathcal{X}_{\min}} e^{-[H(\mathbf{x}) - H_{\min}]/T}} \\ &\to \begin{cases} \frac{|\mathcal{X}_{\min}|}{|\mathcal{X}|} & \text{as } T \to \infty \\ 1 & \text{as } T \to 0 \end{cases} \end{aligned} \tag{8.5}$$

and

$$\frac{d\pi(\mathcal{X}_{\min})}{dT} < 0 \qquad 0 \le T < \infty.$$

For $|\mathcal{X}_{\min}| \ll |\mathcal{X}|$, large T implies a low probability of being in a globally or locally optimal state. Small T allows a high probability of being in such a state. Simulated annealing exploits both properties. However, it must also contend with the possibility that the probability of being in a locally optimal state may be comparable to, although less than, the probability of being in a globally optimal state. This situation makes the search more difficult.

The most commonly encountered form of simulated annealing specifies π as in (8.3) and a symmetric nominating kernel $\{q(\mathbf{x}, \mathbf{y}) = q(\mathbf{y}, \mathbf{x}),\ (\mathbf{x}, \mathbf{y}) \in \mathcal{X}^2\}$. Given $\mathbf{X}_{j-1}$, $\mathbf{Y}$ from $\{q(\mathbf{X}_{j-1}, \mathbf{y})\}$, and U from $\mathcal{U}(0, 1)$ on iteration j, the HM method assigns

$$\mathbf{X}_j = \begin{cases} \mathbf{Y} & \text{if } U \le e^{-[H(\mathbf{Y}) - H(\mathbf{X}_{j-1})]/T} \\ \mathbf{X}_{j-1} & \text{otherwise.} \end{cases} \tag{8.6}$$

Note that if $H(\mathbf{Y}) < H(\mathbf{X}_{j-1})$, then $\mathbf{X}_j = \mathbf{Y}$ with probability 1. However, if $H(\mathbf{Y}) \ge H(\mathbf{X}_{j-1})$, then $\mathbf{X}_j = \mathbf{Y}$ with probability $e^{-[H(\mathbf{Y}) - H(\mathbf{X}_{j-1})]/T}$ and $\mathbf{X}_j = \mathbf{X}_{j-1}$ with probability $1 - e^{-[H(\mathbf{Y}) - H(\mathbf{X}_{j-1})]/T}$. Therefore, the procedure does not always assign to $\mathbf{X}_j$ the value that makes $H(\mathbf{X}_j)$ decrease.

This last property is especially important when H has local minima. If the assignment rule were

$$\mathbf{X}_j = \begin{cases} \mathbf{Y} & \text{if } H(\mathbf{Y}) < H(\mathbf{X}_{j-1}) \\ \mathbf{X}_{j-1} & \text{if } H(\mathbf{Y}) \geq H(\mathbf{X}_{j-1}), \end{cases} \tag{8.7}$$

then a sample path would have a substantial probability of becoming lodged permanently in a local minimum, thus frustrating any attempt to reach an optimal solution.

Because a maximum in π corresponds to a minimum in H, the discussion in Section 6.8.1 regarding multimodality also applies here. Tall, steep peaks in π imply deep, steep wells in H. Most notably, they tend to slow convergence to equilibrium. In the present context, this means a tendency to remain at an $\mathbf{x}$ corresponding to a local minimum for a substantial number of iterations. As before, the properties of K affect these sojourn times. Because we are using HM sampling for a given π, the choice of the nominating kernel, q, determines these properties.

For given $\mathbf{x} \in \mathcal{X}$, common practice specifies a $\{q(\mathbf{x}, \mathbf{y}), \mathbf{y} \in \mathcal{X}\}$ that randomly generates only local moves in the neighborhood of $\mathbf{x}$. This tends to raise the acceptance rate, but displacements between successive moves are small. More rarely, global moves are also permitted. These allow larger displacements from $\mathbf{x}$ but tend to lower the acceptance rate. Global moves also tend to raise the cost per iteration of generating a candidate, $\mathbf{Y}$, that satisfies all constraints; that is, to generate a $\mathbf{Y} \in \mathcal{X}$.

8.2 NOMINATING KERNELS

Sections 6.10 through 6.10.2 and 6.11 describe the intricacies of nominating kernels. Here we describe the role that they play in simulated annealing. In addition to the symmetric case, $q(\mathbf{x}, \mathbf{y}) = q(\mathbf{y}, \mathbf{x})$, simulated annealing also applies to the asymmetric case, $q(\mathbf{x}, \mathbf{y}) \neq q(\mathbf{y}, \mathbf{x})$ for some $\mathbf{x} \neq \mathbf{y}$, provided we replace the acceptance criterion in (8.6) by

$$\alpha(\mathbf{x}, \mathbf{y}) := \min\left\{1, \mathrm{e}^{-[H(\mathbf{y})-H(\mathbf{x})]/T} \times \frac{q(\mathbf{y}, \mathbf{x})}{q(\mathbf{x}, \mathbf{y})}\right\}.$$

This form is less frequently seen in practice, and unless otherwise stated, we hereafter assume that q is a symmetric nominating kernel.

Simulated annealing frequently nominates a new state, $\mathbf{Y} = \mathbf{y}$, in the vicinity of the old state, $\mathbf{X}_{j-1} = \mathbf{x}$. Random-walk nominating kernels are common (e.g., Section 6.10.2). At any fixed temperature, T, small moves are more likely to be accepted than large ones if $H(\mathbf{x})$ is close to a minimum. Section 8.4 describes this relationship between local minima and acceptance rates. A common practice randomly selects a coordinate I with probability $\frac{1}{m}$ and then generates $Y_I = X_{I,j-1} + \Delta$, where Δ is random perturbation sampled from a distribution symmetric about $X_{I,j-1}$ and which has its probability mass concentrated close to

$X_{I,j-1}$. Then the nominated location is $\mathbf{Y} = (Y_1, \ldots, Y_r)$, where $Y_i = X_{i,j-1}$ for all $i \neq I$. Examples 8.3.1 and 8.3.2 illustrate the concept.

More rarely, simulated annealing generates global moves, sometimes using the independence kernel in Section 6.10.1. Although this approach allows for a more far-ranging search in $\mathcal{X}$ on each iteration, global moves tend to have lower acceptance rates and higher unit generating costs than local ones. This is especially true for acceptance rates when H varies widely in magnitude over $\mathcal{X}$. The notably higher unit generating costs arise when $\mathcal{X}$ is not a product space. This is a consequence of the greater computing time required to generate a nominee that satisfies all the constraints that define membership in $\mathcal{X}$.

8.3 COOLING

On any iteration $j \geq 1$, a large value of T increases the probability in (8.3) that $\mathbf{X}_j = \mathbf{Y}$ in (8.6), whereas a small value decreases that probability. These properties become important when sampling begins in an arbitrarily assigned state $\mathbf{X}_0 = \mathbf{x}_0 \in \mathcal{X}$, which is common practice. Whereas for fixed T uniform ergodicity holds for $\{\mathbf{X}_j,\ j \geq 1;\ \mathbf{X}_0 = \mathbf{x}_0 \in \mathcal{X}\}$, the value of T also influences the speed of convergence to equilibrium, a critical issue in any stochastic search procedure. In particular, the speed of convergence tends to increase as T increases, but the corresponding equilibrium distribution is then more uniform on $\mathcal{X}$. Conversely, small T induces slow convergence to an equilibrium distribution concentrated on a small region in $\mathcal{X}$ that includes $\mathcal{X}_{\min}$.

Most importantly, for any fixed T and for $\mathbf{x}_0 \in \mathcal{X}$,

$$\mathrm{pr}(\mathbf{X}_j \in \mathcal{X}_{\min} \mid \mathbf{X}_0 = \mathbf{x}_0) \not\to 1 \qquad \text{as } j \to \infty.$$

This implies that there is no guarantee of finding a globally optimal solution, regardless of the length of the sample path, unless we find some way of progressively reducing T. That is precisely what simulated annealing does. This section describes issues relevant to the specification of T. Section 8.4 returns to the topic of balance between convergence to equilibrium and an equilibrium distribution whose probability mass is substantially concentrated on $\mathcal{X}_{\min}$.

The procedure for reducing T is called *cooling*. To coax the sample path $\mathbf{X}_0 = \mathbf{x}_0, \mathbf{X}_1, \mathbf{X}_2, \ldots$ into converging to an element of $\mathcal{X}_{\min}$, we need to choose T as a nondecreasing function of iteration number j, which for all $\mathbf{x}_0 \in \mathcal{X}$ ensures

$$\mathrm{pr}(\mathbf{X}_j \in \mathcal{X}_{\min} \mid \mathbf{X}_0 = \mathbf{x}_0) \to 1 \qquad \text{as } T \to 0 \text{ and } j \to \infty. \tag{8.8}$$

In the present case, this would also imply

$$H(\mathbf{X}_j) \to H_{\min} \ \text{w.p.1} \qquad \text{as } T \to 0 \text{ and } j \to \infty.$$

Let T_j denote the value of T on iteration $j \geq 1$, and suppose that $T_1 > T_2 > \ldots$, where $T_j \to 0$ as $j \to \infty$. Given that q is itself ergodic and reversible, it is known that for all $\tau \ge d$, where

$d :=$ the maximum of the depths of all states (as measured from the maximal $H(\mathbf{x})$), that are local but not global minima,

(8.8) holds if and only if (Geman and Geman 1984, Hajek 1988)

$$T_j \ge \frac{\tau}{\ln(j+1)} \qquad j = 1, 2, \ldots. \tag{8.9}$$

This implies that a sequence of temperatures, $\{T_j\}$, that decreases too rapidly precludes the guarantee in (8.8), regardless of how long a sample path is generated. That is, cooling occurs too rapidly. Conversely, a $\{T_j\}$ that lies substantially above the bound in (8.9) for $\tau = d$ ensures the desired convergence, but at a slow rate.

A commonly employed method of sequentially lowering T relies on an r-stage *cooling schedule* $(\tau_1, l_1), (\tau_2, l_2), \ldots, (\tau_r, l_r)$, where $\tau_1 > \tau_2 > \cdots > \tau_r$ and where $\tau_r \to 0$ as $r \to \infty$, and $l_0 := 0 < l_1 < l_2 < \cdots < l_r$ are integers. Then the assignment

$$T_j = \tau_k \qquad l_{k-1} < j \le l_k \tag{8.10}$$

keeps T_j constant within a stage but reduces it as j moves from one stage to the next. Presumably, the cooling schedule is chosen so that given $\mathbf{X}_{l_{k-1}}$ on iteration l_{k-1}, $\mathbf{X}_j, \mathbf{X}_{j+1}, \ldots, \mathbf{X}_{l_k}$ for some j such that $l_{k-1} < j \le l_k$ approximately have the *local equilibrium distribution*

$$\pi_k(\mathbf{x}) = \begin{cases} \dfrac{e^{-H(\mathbf{x})/\tau_k}}{\sum_{\mathbf{z}\in\mathcal{X}} e^{-H(\mathbf{z})/\tau_k}} & \mathbf{x} \in \mathcal{X} \\ 0 & \text{elsewhere} \end{cases} \tag{8.11}$$

for the assigned values of τ_k. That is, convergence to an equilibrium distribution occurs in each stage k for given τ_k. If the cooling schedule satisfies (8.9), then the limit (8.8) holds as the number of stages r increases without bound. Also, for fixed k

$$\pi_k(\mathbf{x}) \to \begin{cases} \frac{1}{|\mathcal{X}_{\min}|} & \text{if } \mathbf{x} \in \mathcal{X}_{\min} \\ 0 & \text{elsewhere} \end{cases} \quad \text{as } \tau_k \to 0. \tag{8.12}$$

As a methodology for mathematical problem solving, the concept of annealing originated in the physical sciences and, in particular, in statistical mechanics as applied to interacting particle systems. The manufacture of a steel ingot provides a good example. An ingot is a solid block whose atoms have arranged themselves in a crystalline structure $\mathbf{x}$ with corresponding energy $H(\mathbf{x})$. Heating the ingot sufficiently converts it to a molten state, energizing its atoms (increasing $H(\mathbf{x})$) so that all possible crystalline configurations are equally likely. As the molten ingot cools, its atoms lose energy. If cooling proceeds at a sufficiently slow rate, the atoms coalesce into a configuration that gives the solidified ingot its greatest structural strength (global minimal energy $H_{\min}$).

If cooled too rapidly, the atoms coalesce to a crystalline configuration $\mathbf{x}$ corresponding to one of the local minima in $\{H(\mathbf{x})\}$, endowing the solidified ingot with less structural strength. That is, the sample path generated thereafter by simulated annealing is highly unlikely to climb out of this local trough. This occurs because of the correspondingly very low acceptance rate in HM sampling at each stage. The depths of these local minima determine τ in (8.9). If T drops too slowly, then it takes considerable time to reach the minimal energy state. Because cost increases with iterations, this is unappealing. Hereafter, we occasionally refer to T as temperature.

8.3.1 Initial Temperature

Expression (8.9) indicates that T initially must be no less than $\tau/\ln 2$. In practice, τ is unknown. Although it can be calculated from knowledge of H, indications are that for problems of combinatorial optimization this calculation is unlikely to occur in time polynomial in problem size (e.g., Kern 1993). Therefore, we need an alternative approach that increases T to a sufficiently high temperature (heating) so that (8.9) is not initially violated.

To indicate when T is sufficiently large, one approach relies on the empirical acceptance rate of HM sampling, computed from sample-path data. As T increases, π in (8.3) becomes more uniform, and the acceptance rate in (6.8) increases. Most importantly, an empirical acceptance rate close to 1 implies that the unobservable π is close to uniform. Algorithm SAINITIAL exploits these properties.

ALGORITHM **SAINITIAL**

Purpose: To identify an appropriate starting temperature τ_1.

Given: $\gamma > 1$ and $\tau'_1 > 0$.

Method:

1. $k = 1$.
2. Repeat:
   a. choose an initial state $\mathbf{X}_0$
   b. run a preliminary simulated annealing sampling experiment with temperature $T = \tau'_k$
   c. compute the empirical acceptance rate
   d. $k = k + 1$
   e. $\tau'_k = \gamma \tau'_{k-1}$

   until the acceptance rate is close to 1.
3. Assign τ'_k to τ_1. ■

8.3.2 Temperature Gradient

Given $\tau_1 > 0$, one practice for choosing $\tau_2, \tau_3, \ldots$ relies on the assignment

$$\tau_{k+1} = \alpha \tau_k \qquad 0 < \alpha < 1 \qquad k = 1, 2, \ldots, \tag{8.13}$$

where typically $[0.80 \leq \alpha \leq 0.99]$. For example, Kirkpatrick et al. (1983) use $\alpha = 0.90$.

8.3.3 Stage Length

Because each successive temperature lowering in simulated annealing slows convergence to the new equilibrium distribution, $l_1 < l_2 < \ldots$ should be taken so that

$$l_{k+1} - l_k < l_{k+2} - l_{k+1} \qquad k = 1, 2, \ldots. \tag{8.14}$$

After the kth stage, a perusal of the sample path $H(\mathbf{X}_{l_k+1}), \ldots, H(\mathbf{X}_{l_k})$ provides a basis for assessing the adequacy of $l_k - l_{k-1}$ as a warm-up interval. Multiple independent replications, as in Section 7.3, all starting with the same initial value $\mathbf{X}_{l_{k-1}}$, provide a more comprehensive basis for making this assessment.

8.3.4 Stopping Rule

No statistical theory comparable to that for MCMC estimation of λ in Chapters 2, 6, and 7 exists for assessing the accuracy of an estimate of $H_{\min}$ generated by simulated annealing. Although a notable limitation, the availability of sample-path data $H(\mathbf{X}_1), \ldots, H(\mathbf{X}_{l_k})$ allows one to devise heuristic *stopping rules* to terminate sampling. For example:

- If the numbers of successes on sample paths on three successive stages fall below a specified minimal number of acceptances, the system is considered frozen, and sampling terminates (e.g., Kirkpatrick et al. 1983).
- If for some specified $k - s > 0$, $H(\mathbf{X}_{l_{k-s+1}}) = H(\mathbf{X}_{l_{k-s+2}}) = \cdots = H(\mathbf{X}_{l_k})$, sampling terminates at the end of stage k (e.g., Aarts and Korst 1989, p. 59). That is, no change in the value of the objective function has occurred during the last s successive stages.
- Let $\mathbf{Z}_1, \ldots, \mathbf{Z}_r$ denote the final states on r independent replications based on the same cooling schedule. If $H(\mathbf{Z}_1), \ldots, H(\mathbf{Z}_r)$ show a propensity to concentrate in a relatively small neighborhood, then sampling terminates (e.g., Kolinski and Skolnick 1994b; also see Section 8.3.2).

Algorithm ANNEALING describes the principal steps for simulated annealing to terminate sampling with confidence that $\mathbf{X}_j \in \mathcal{X}_{\min}$ for a single exercise of the cooling schedule.

ALGORITHM ANNEALING

Purpose: To estimate $H_{\min}$ and an element in $\mathcal{X}_{\min}$.

Given: $\alpha, r, \{l_1 < l_2 < \cdots\}$.

Method:

1. Use Algorithm SAINITIAL to determine the maximal heating temperature τ_1.
2. $j = 1$ and $k = 1$.
3. Assign an initial state $\mathbf{X}_0$, $\widehat{H}_{\min} = H(\mathbf{X}_0)$, and $\mathbf{Z} = \mathbf{X}_0$.

4. Repeat:
   a. while $k \le r$:
      b. while $j \le l_k$:
         c. randomly sample $\mathbf{Y}$ from a symmetric $\{q(\mathbf{X}_{j-1}, \mathbf{y})\}$
         d. randomly sample U from $\mathcal{U}(0, 1)$
         e. if $U \le \min\left\{1, e^{-[H(\mathbf{Y})-H(\mathbf{X}_{j-1})]/\tau_k}\right\}$, $\mathbf{X}_j = \mathbf{Y}$
         f. otherwise, $\mathbf{X}_j = \mathbf{X}_{j-1}$
         g. if $H(\mathbf{X}_j) < \widehat{H}_{\min}$, $\widehat{H}_{\min} = H(\mathbf{X}_j)$ and $\mathbf{Z} = \mathbf{X}_j$
         h. $j = j + 1$
      i. $k = k + 1$
      j. $\tau_k = \alpha \tau_{k-1}$

   until the stopping rule is satisfied or $k > r$.
5. $\left(\widehat{H}_{\min}, \mathbf{Z}\right)$ is the estimated solution. ■

EXAMPLE 8.3.1

CIRCUITS AND MICROCHIPS[4]

The design of a digital computer consists of a collection of connected *logic circuits* located on *microchips* connected to each other by *wires*. The placement of circuits within and among chips is a principal determinant of the computer's speed and reliability. Although a chip can accommodate a substantial number of circuits, there are constraints on what constitutes a desirable layout. As the number of circuits increases on a chip, the circuits become more densely packed. Then electrical intereference may result among wires in close proximity within a circuit and among circuits, thereby jeopardizing the quality of signal transmission within and between circuits.

Sparsely packed circuits on a chip require signals to travel longer distances between them, thus inducing longer transmission times between circuits or, equivalently, reducing overall computing speed. Moreover, each connection between a circuit on one chip and a circuit on another chip requires inclusion of a relatively slow, power-consuming circuitry.

The assignment of circuits to chips must be done in a way that limits interference, keeps signal transmission times small between circuits on the same chip, and keeps the number of required signals (connections) low between chips. As illustration, suppose there is a total of m circuits on two microchips. For $i = 1, \dots, m$, let

$$x_i := \begin{cases} 1 & \text{if circuit } i \text{ is on chip 1} \\ -1 & \text{if circuit } i \text{ is on chip 2,} \end{cases} \tag{8.15}$$

so that $\mathcal{X} = \{-1, 1\}^m$ and $|\mathcal{X}| = 2^m$. Also, for $i \ne j = 1, \dots, m$, let

$$a_{ij} := \begin{cases} 1 & \text{if circuits } i \text{ and } j \text{ are connected} \\ 0 & \text{if circuits } i \text{ and } j \text{ are not connected} \end{cases}$$

[4]This example is taken from Kirkpatrick et al. (1983).

and assume that there is, at most, one connection between any two circuits. Because

$$a_{ij}\left|\frac{x_i - x_j}{2}\right| = \begin{cases} 1 & \text{if circuits } i \text{ and } j \text{ are on different chips} \\ 0 & \text{otherwise,} \end{cases}$$

we have

$$\begin{aligned} N &:= \sum_{i=1}^{m-1}\sum_{j=i+1}^{m} a_{ij}\left|\frac{x_i - x_j}{2}\right| \\ &= \text{number of signals that cross from one chip to the other} \end{aligned} \tag{8.16}$$

and

$$\sum_{i=1}^{m} x_i = \text{difference in numbers of circuits on chips 1 and 2.} \tag{8.17}$$

Two types of cost arise, one for each signal connecting the two microchips (b_1) and the other (b_2) reflecting the imbalance between the numbers of circuits on each chip. This latter cost is quadratic in $\sum_{i=1}^{m} x_i$. Then the objective is to assign circuits to chips in a way that minimizes the cost function

$$\begin{aligned} H(\mathbf{x}) &:= b_1 \sum_{i=1}^{m-1}\sum_{j=i+1}^{m} a_{ij}\left|\frac{x_i - x_j}{2}\right| + b_2\left(\sum_{i=1}^{m} x_i\right)^2 \\ &= \frac{b_1}{2}\sum_{i=1}^{m-1}\sum_{j=i+1}^{m} a_{ij} + \left(2b_2 - \frac{b_1}{2}\right)\sum_{i=1}^{m-1}\sum_{j=i+1}^{m} a_{ij}x_i x_j. \end{aligned} \tag{8.18}$$

This problem is a special case of the *graph partitioning problem* in graph theory. A related problem that minimizes the total "wire length" of connections is a special case of the *Steiner tree problem* (e.g., Hwang et al. 1992).

As illustration of how the cooling concept works in this problem, suppose there is a total of $m = 5{,}000$ circuits on the two chips so that the state space has size $|\mathcal{X}| = 2^{5000} \approx 10^{1505}$, clearly beyond the capacity of total enumeration. Each signal connecting the two chips requires two pins so that the total number of pins is $2N$. Figure 8.1 displays distributions of $2N$ for selected T. With the exception of $T = 1.05$, simulated annealing generated a sample path for each T. These data were used to compute an approximation to the equilibrium for that T. The temperature $T = \infty$ corresponds to sampling $x_i = 1$ and $x_i = -1$, each with probability $\frac{1}{2}$ for each i. Its distribution is concentrated roughly in the neighborhood of 5,800 pins. As T decreases, the mode of the distribution moves to the left and the distribution becomes more concentrated. In particular, $T = 0.10$ corresponds to solutions that require about 500 pins, or about 8.62% of the modal value for $T = \infty$. The results for $T = 1.05$ were obtained by following rule (8.7) on multiple replications. Its modal value is about three times that for $T = 0.10$.

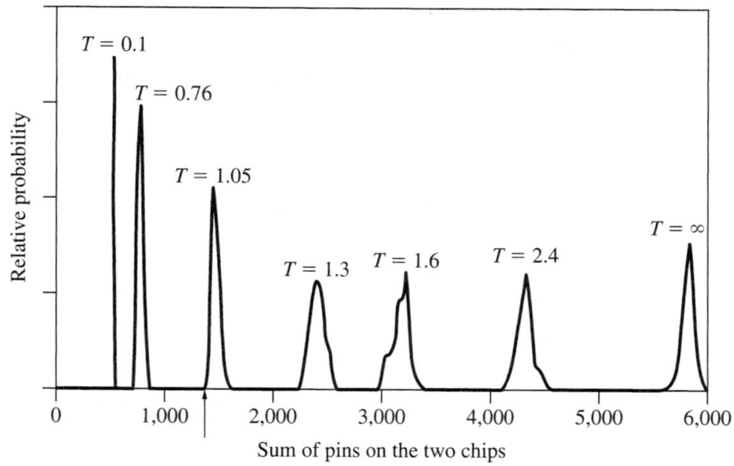

FIGURE 8.1 | *Distributions of pins for circuit placement problem*

Nominating Distributions. Given $\mathbf{X}_{j-1} = \mathbf{x}$, the nominating distribution took the form

$$q(\mathbf{x}, \mathbf{y}) := \frac{1}{m}\sum_{i=1}^{m} \delta(\mathbf{y} - \mathbf{x}_i(-x_i)) \qquad \mathbf{y} \in \mathcal{X},$$

where (6.21) defines $\delta(\cdot)$ and (6.14) defines $\mathbf{x}_i(\cdot)$. That is, on iteration j circuit I was randomly selected with probability $\frac{1}{m}$, and the nominated state was

$$\mathbf{Y} = \begin{cases} \mathbf{x}_I(-1) & \text{if } x_I = 1 \\ \mathbf{x}_I(1) & \text{if } x_I = -1. \end{cases}$$

From the random-walk perspective, we have $Y_I = (X_{I,j-1} + 3) \pmod 4 - 1$.

Temperature Schedule. At the beginning of each stage, temperature decreased according to (8.21) with $\alpha = 0.90$. At stage k, the number of iterations, $l_k - l_{k-1}$, was the lesser of $100 \times$ number of circuits $= 500{,}000$ or the number of trials until each circuit experienced at least 10 switches (i.e., $\mathbf{X}_{j-1} = 1, \mathbf{X}_j = -1$ or $\mathbf{X}_{j-1} = -1, \mathbf{X}_j = 1$).

There are multiple solutions to this problem. Each corresponds to a distinct layout of circuits on the two chips. Some layouts are infeasible because they violate density and other constraints.[5] ■

EXAMPLE 8.3.2

PROTEIN FOLDING

Although simulated annealing, in one of several versions, has been a standard tool of analysis in condensed matter physics and physical chemistry for many

[5]For an illustration of circuit placement by simulated annealing, see Kirkpatrick et al. (1983).

years, its introduction to problems in the biological sciences, especially *bioinformatics,* has been relatively recent. Nevertheless, it shows considerable promise for providing useful information about protein structure, especially protein folding. The problem is to identify how the building blocks of a protein array themselves in three-dimensional space. With an energy function defined over all three-dimensional structures that are consistent with the laws of chemistry and biology, simulated annealing searches for the configuration with the minimal energy.

This concise statement of the problem belies the complex environment in which the simple principle of simulated annealing is to be applied. To make the example as complete as possible within the context of a textbook on MCMC, we have included considerable detail on chemical and biological structure. A reader more focused on simulated annealing may elect to skip ahead to the "Modeling" heading in the example and then refer back to the detailed description as needed.

Every living organism has a *genome* of genetic material consisting of pairs of chromosomes, each composed of molecules of deoxyribonucleic acid (DNA). For example, the human genome has 23 pairs. First drafts of the complete human genome sequence were published in 2001. Although a major advance in scientific knowledge, they provide merely a first step in understanding and preventing disease in humans. Every cell in a particular living organism contains the same DNA. A cell encodes the information that its machinery uses to make *proteins,* the working molecules of an organism. These proteins perform most of the organism's daily operations, including metabolism, cellular repair, cell-to-cell signaling, and cellular reproduction. Although identical in every cell of a particular organism, different cells use the DNA information to make different proteins. The process is called *gene expression.*

A *protein* is a linear chain (*polymer*) composed of *amino acids.*[6] Globular proteins, which are compact, approximately spherical molecules, are the principal agents of biological action in a cell. Our discussion applies to globular proteins. They include:

- *Enzymes* that accelerate the rate of chemical reaction in a cell
- *Immunoglobulins* that serve as the first line of defense against bacterial and viral infections
- *Transport proteins* that carry materials across cell membranes
- *Hormones* (e.g., insulin) that control cell functions such as metabolism, cell repair, and cell reproduction

A *polypeptide chain* is a sequence of amino acids. Each amino acid has a central carbon atom (C_α) linked to four groups, a basic amino group ($-NH_2$), an acidic carboxyl group ($-COOH$), a hydrogen atom (H), and a group called a *side chain* that uniquely identifies the amino acid. Links between amino acids are called *peptide bonds.*

[6]For topics related to protein structure, see Branden and Tooze (1999) and Rawn (1989).

a. Primary structure

Alanine-Glutamic acid -Valine-Threonine-Aspartic acid-Proline-Glycine

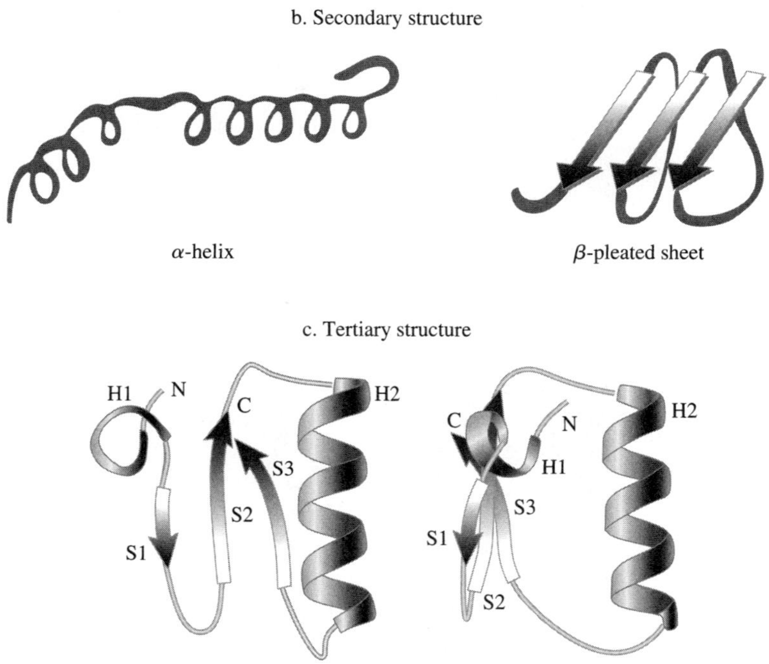

FIGURE 8.2 | *Examples of protein primary, secondary, and tertiary structures*

The *primary structure* of a protein consists of its sequence of amino acids sometimes called *residues*. This sequence dictates the protein's *conformation* (three-dimensional shape). Figure 8.2a provides an example. A primary structure customarily has from 100 to many thousand amino acid residues. The sequence of C_α atoms and peptide bonds forms the backbone of the protein, called the *main chain*. Links between groups are called *covalent bonds*.

The *secondary structure* of a protein consists of its regularly repeating conformations of polypeptide chains. The composition of secondary structures can include *α-helices, β-pleated sheets,* and *turns* or *bends,* as in Figure 8.2b. The *tertiary structure* of a protein consists of its three-dimensional biologically active *native conformation*. It includes all helices, sheets, and *folds,* which reveal how the sequence's component parts (*α-helices* and *β-sheets*) are packed. Every tertiary structure has a C terminal and an N terminal, which identify its beginning and end respectively.

Figure 8.2c shows *orthogonal views* of the tertiary structures of a protein in ribbon representation. H1 denotes an α-helix of one revolution and H2, an α-helix of three revolutions. S1, S2, and S3 represent three β-strands forming a β-sheet with loops between them. The diagram in Figure 8.2c suppresses the atomic detail of the protein structure but provides information about the spatial orientation of the amino acids and demonstrates peptide sequence directionality and the overall protein folding.

Native Conformations. Although knowledge of the human genome provides the basis for identifying primary structure in proteins in human cells, it provides little information about secondary and tertiary structures. Efforts to identify these and, ultimately, the protein's native conformation constitute the *protein folding problem*, and a principal motivation for solving it comes from medical science. To enhance the ability to design drugs to cure human disease, a knowledge of these three-dimensional shapes is essential (e.g., Swope 2001). For example, several drugs may have the potential for inhibiting the effect of proteins essential for the transmission of a virus. The proteins of interest are those on the *disease pathway,* the biochemical processes and molecules involved in the causes and symptoms of the disease. The objective is to find a protein on the critical part of the pathway that can *bind* with the drug molecule. This binding provides a means of interrupting the progress of the disease. Once identified, a screening of drug molecules is conducted to identify the one that best interacts with the protein to block the disease.

Simulated Annealing. To perform this kind of search, it is essential to know the three-dimensional shapes of proteins on the critical pathway. X-ray crystallography and nuclear magnetic resonance (NMR) spectroscopy are two traditional techniques for deriving this information. Both have limitations. As its name indicates, X-ray crystallography requires a crystalline structure to reveal information. Because proteins are not normally crystalline, highly concentrated solutions are needed to create crystals. These are frequently difficult and expensive to obtain. The ability of NMR to discern conformational structure extends to proteins with no more than about 120 residues. Although these as well as other methods have identified primary structures (sequencing) for many human proteins, relatively few of these proteins' secondary and tertiary structures are known.

Alternative approaches that rely on computer-based modeling are now active areas of research for providing this knowledge. One approach, based on *molecular dynamics,* incorporates a high level of detail for molecular and atomic structures and their dynamics into a model, which then simulates the biochemical processes by which protein folding occurs (e.g., Karplus and Pitsko 1990). Because of its detailed dynamics, this technique can produce useful insights into secondary and tertiary structures, but at a cost of considerable computing time. An alternative approach foregoes this level of detail in favor of uncovering more gross but, nevertheless, useful information in considerably less computing time. Simulated annealing has become a useful tool for identifying these gross structures.

The application of simulated annealing to the protein-folding problem begins with a model, usually a graph. Figure 8.3 displays a fragment of a simplified two-dimensional main chain and side chain which, by virtue of its linkages,

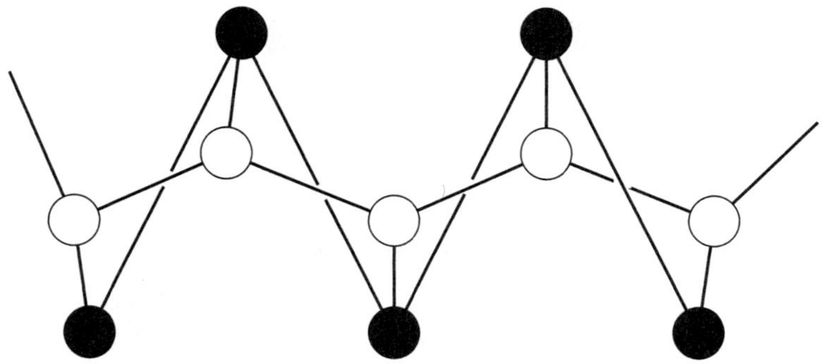

FIGURE 8.3 | *Model of short expanded protein fragment*
(• := side chain centers, ∘ := C_α positions)

forms a connected graph. Because the large number of atoms and the myriad of interactions among them make it computationally intractable to work toward a solution with all these relationships accounted for explicitly, a first step groups atoms on the main chain and on the side chains, being careful to account for critical spatial properties of residues and more generally for the principal interactions among atoms that influence protein shape. These include:

- Spacing of residues
- Restrictions on the backbone's (main chain's) rotational degrees of freedom
- Rigid structure of peptide bonds
- Fixed *chirality* of C_α atoms

Chirality. Chirality characterizes a particular type of asymmetry, a three-dimensional shape that cannot be superimposed on its mirror image.

Rigidity and Spacing. Every carboxyl oxygen in an α-helix in the polypeptide backbone is bonded to a hydrogen atom, and these weak bonds are approximately parallel to the long axis of the helix. Amino acid residues are spaced at intervals of 0.15 nanometers along the helix path, and each complete turn of the helix contains 3.6 residues. One nanometer or, equivalently, one Ångstrom is 10^{-9} meters.

Rigidity and Rotational Freedom. Figure 8.4 provides a basis for characterizing the rigidity and rotational degrees of freedom in a main-chain polypeptide. The gray panels demarcate two peptide units, each going from one C_α atom to the next, except for the first. All atoms in a unit are spatially fixed in three dimensions with bond lengths and angles approximately the same for corresponding pairs of atoms in different units; hence, the rigidity within units. Covalent bonds at the C_α atom link the units together, and each unit as a whole can rotate around these bonds. By convention, ϕ describes angular rotation around the

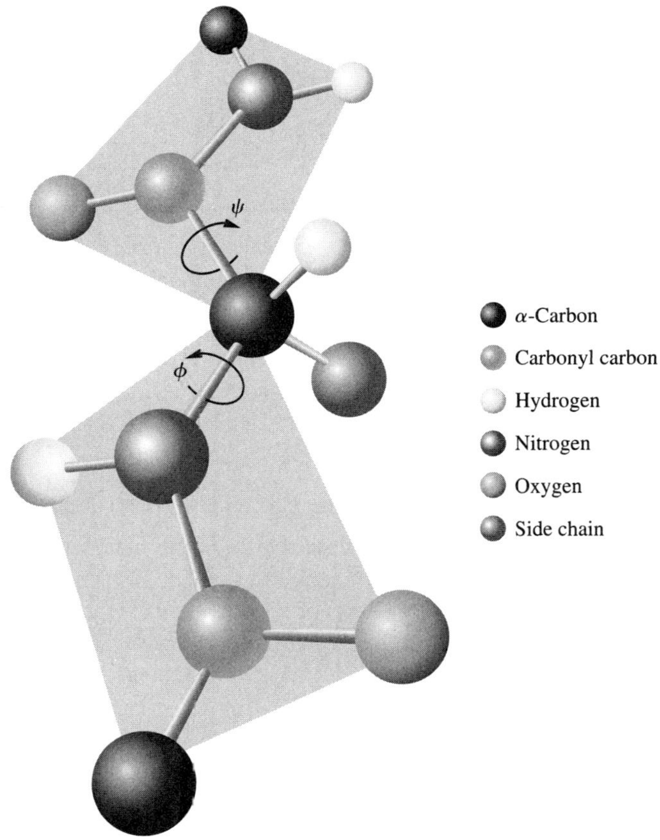

FIGURE 8.4 | *Backbone rigidity and rotational freedom*

N-C_α bond. Also, ψ describes angular rotation around the C_α-C' bond for the same C_α, where C' denotes the carbon atom bonded to the oxygen atom.

A library of statistical information accumulated from previous crystallographic and NMR studies can substantially aid in simulated annealing's search for a protein's three-dimensional shape. This often includes distributions and lower and upper bounds on distances and angles of rotation. For example:

- *Ramachandran plots* identify regions in which the ϕ and ψ angles can exist for α-helices and for β-sheets (e.g., Branden and Tooze 1999, Rawn 1989).
- It is known that a hydrogen bond exists between the ith and $i + 3$rd residue in a sequence and that this bond varies little in length and in angle between particles i and $i + 3$ for different i.
- Because some proteins can manifest two different secondary structures, length distributions are sometimes bimodal.
- Unless the backbone is relatively rigid (small variation in distances), it cannot maintain an α-helix shape.

Studies also show that certain amino acid sequences have a propensity to form particular local conformations called *rotamers*. Although this tendency does not guarantee that the particular rotamers are preserved when several amino acid sequences are chained together in a polypeptide chain, they nevertheless provide a valuable starting state for sampling. For example, certain rotamers are known to have low energy. Because simulated annealing favors a conformation with minimal energy, starting with these low-energy rotamers is a judicious strategy.

Modeling. A protein fragment, a segment of a total protein chain, can be modeled as a graph $\mathcal{G}$ with each site (vertex, node) representing an atom or group of atoms, and each link (edge, arc), a bond between sites. Formally, $\mathcal{G} = (\mathcal{V}, \mathcal{E})$, where $\mathcal{V}$ denotes the set of sites on the main and side chains and $\mathcal{E}$ denotes the set of links between sites.

In the absence of *constraints,* the fragment can assume many different three-dimensional shapes or conformations. The constraints among atoms and the distributions of distances and angles available from past studies dictate the region in $\mathcal{X}$ in which the *native conformation,* the shape that the fragment actually assumes, is most likely to be found. The objective is to identify the native conformation, or at least its principal features.

Let $\mathcal{L}$ denote a three-dimensional cubic lattice in $\Re^3$ with span or mesh size d, usually of the order of 1 to 2 Ångstroms. Discretizing the problem facilitates search, but a successful search requires d sufficiently small so that discrete results well approximate the presumed continuous reality. Each site $i = 1, \ldots, m$ on the graph has a distinct location $\mathbf{x}_i := (x_{1i}, x_{2i}, x_{3i}) \in \mathcal{L}$. One site, usually at the beginning or the end of the fragment, is fixed in location, and the challenge is to find a conformation in $\mathcal{L}$ that satisfies a collection of constraints $\mathcal{K}_1, \ldots, \mathcal{K}_r$ that accounts for known spatial restrictions among site locations (e.g., Kolinski and Skolnick 1994a).

Constraints. We use mathematical examples to illustrate the role of constraints. Each constraint, $\mathcal{K}_i$, identifies the set of points in $\Re^3$ that meet a specified condition. For example, suppose that

$$\mathcal{K}_1 := \left\{(x_{1i}, x_{2i}, x_{3i}) \in \Re^3 : \ x_{1i}^2 + x_{2i}^2 + x_{3i}^2 \le a^2\right\}$$

denotes all points in $\Re^3$ that lie inside or on the circle centered at $(0, 0, 0)$ with radius a (> 0), and

$$\mathcal{K}_2 := \{(x_{1i}, x_{3i}) \in \Re^2 : \ bx_{1i} + cx_{3i} \ge d\}$$

denotes all points in $\Re^2$ that satisfy $x_{3i} \ge \frac{d-bx_{1i}}{c}$ if $c > 0$ and $x_{3i} \le \frac{d-bx_{ii}}{c}$ if $c < 0$. Then $\mathcal{K}_1 \cap \mathcal{K}_2$ denotes all points in $\Re^3$ that satisfy both constraints, and $\mathcal{L} \cap \mathcal{K}_1 \cap \mathcal{K}_2$ denotes all points on the lattice $\mathcal{L}$ that satisfy both constraints. In solving a problem with these constraints, the solution is restricted to points in the *feasible set* $\mathcal{L} \cap \mathcal{K}_1 \cap \mathcal{K}_2$.

In the context of protein structure, the constraints include restrictions on distances and angles between particles. Let $\mathbf{z}_{ij} := \mathbf{x}_i - \mathbf{x}_j$ for some $i \ne j \in \mathcal{V}$. For pairs of sites both close and far apart, the constraints consist of limitations on distance $|\mathbf{z}_{ij}| < a_{ij}$, for specified a_{ij}. They also include bounds on angles between the $\mathbf{z}_{ij}$ vectors in terms of *dot products* $\mathbf{z}_{ij} \cdot \mathbf{z}_{ik}$ and *cross products* $\mathbf{z}_{ij} \times \mathbf{z}_{ik}$ for $j \neq k$. For example, in the case of an α-helix, these restrictions ensure that the amino acid chain conforms to that shape. In actuality, constraints on dot products and cross products correspond to bounds on the lengths of vectors induced by $\mathbf{z}_{ij}$ and $\mathbf{z}_{ik}$. Sometimes a dot product is called a scalar or inner product, and a cross product is called a vector or outer product.

Let H denote a score function defined for all $\mathbf{x} \in \mathcal{L}$. Then for all $\mathbf{x}$ in the feasible set, simulated annealing seeks a solution that minimizes H. Recall that the feasible set $\mathcal{L} \cap \mathcal{K}_1 \cap \mathcal{K}_2$ denotes the collection or subset of all $\mathbf{x} \in \mathcal{L}$ that satisfy the r constraints $\mathcal{K}_1, \ldots, \mathcal{K}_r$. When viewed as a problem of interacting particles (to be described in Section 9.3.1), H becomes an energy function, and energy minimization is the objective. For example, the energy function (Kolinski and Skolnick 1998)

$$H(\mathbf{x}) = H_{\mathbf{C}_\alpha}(\mathbf{x}) + H_{\text{H-bond}}(\mathbf{x}) + H_{\text{SR}}(\mathbf{x}) + H_{\text{LR}}(\mathbf{x}) \tag{8.19}$$

contains contributions from the protein's C_α backbone, the alignment of its hydrogen bonds (CH-bond), short-range (SR) interactions among particles, and long-range (LR) interactions among particles. Whereas $H_{\mathbf{C}_\alpha}(\mathbf{x})$ and $H_{\text{H-bond}}(\mathbf{x})$ are sequence independent, $H_{\text{SR}}(\mathbf{x})$ and $H_{\text{LR}}(\mathbf{x})$ depend on the particular residue sequence. For example, $H_{\text{SR}}(\mathbf{x})$ includes the energy associated with rotamers, which reflect local conformal propensities of subsequences of amino acid residues.

The values that the elements in (8.19) assume are directly related to the constraints $\mathcal{K}_1, \ldots, \mathcal{K}_r$. For example, $H_{\mathbf{C}_\alpha}$ increases (decreases) as the number of violated constraints that explicitly include C_α atoms increases (decreases). Because generating a sample path with all constraints satisfied would be a formidable task, one alternative generates a sample path where on each iteration an HM-nominated state is rejected if it violates a constraint and otherwise is subjected to HM acceptance-rejection (e.g., Kolinski and Skolnick 1998).

In practice, different investigators rely on different score functions, some of which may admit more than one optimal solution. Moreover, it is known that the length of the fragment taken for study and the range of the interaction explicitly accounted for the constraints $\mathcal{K}_1, \ldots, \mathcal{K}_r$ can lead to different optimal configurations for different energy functions. On the bright side, simulated annealing, when applied to protein fragments whose three-dimensional structures are known, produces solutions in good agreement with these known native conformations.[7]

Folding a Protein. Once a model is formulated, good practice calls for testing its capacity to predict known protein structures. As illustration, consider the prediction of the secondary and tertiary structures of a 60-residue fragment of the

[7]For examples, see Bryant and Lawrence (1993), Kolinski and Skolnick (1994b, 1998), Kolinski et al. (1998), and Shimada et al. (2001).

folded B domain of staphylococal protein A (Kolinski and Skolnick 1994a, 1994b). These references are the basis for the description that follows. *Motifs* are formed by packing side chains from adjacent $\alpha-$helices and $\beta-$strands close to each other. A *domain* consists of several motifs for which folding takes place independently of foldings for other domains. Hence, tertiary structure may be decomposed into problems of independently estimating foldings in the individual domains; hence, the focus on a single domain.[8]

For the present problem, NMR studies indicate that the secondary structure contains three helices: helix I for residues 10 through 19, helix II for residues 25 through 37, and helix III for residues 42 through 55. The studies also indicate a hairpin turn crossing the C terminal at about 30 degrees. For a model based entirely on sequence and interaction information and using a coarse lattice ($d =$ 1.70 Ångstroms), 45 independent runs were executed, each starting in a randomly generated configuration. An energy function somewhat more elaborate than that conceptualized in (8.19) was employed.

Nominating Distributions. Proposed moves (transitions) were more versatile than those in Example 8.3.1. They included:

a. A random modification of a rotamer representation of a randomly selected amino acid (Figure 8.5a).

b. A two-bond move, subject to angle constraints (Figure 8.5b).

c. A four-bond kink move, similar to the two-bond moves. A library of all possible four-bond moves was used (Figure 8.5c).

d. Eight- and ten-bond moves. A fragment of the chain moves a distance that is small enough to avoid crossing any other portion of the chain (Figure 8.5d).

e. A small random displacement of a larger part of the chain, starting from a randomly selected segment extending to the end of the chain. In contrast to moves a through d, most of the side-chain rotamers remain the same because of the rigid body-like translation of the subchain.

For an m-site chain, each MCMC iteration j consisted of a random mixture of m nominations of types a and b; $m-4$ of type c; $m-8$ of one kind and $m-10$ of another kind of type d; and one of type e. Motivation for the alternative types and random mixture arose from a desire to include local moves (types a, b, and to some extent c) and to include global transitions (types d and e) that move about already assembled secondary structure (Skolnick 2001).

Once a move type was selected on iteration j, a small random perturbation of the conformation on iteration $j-1$ was made using "Brownian dynamics."[9]

Temperature Schedule. Every protein has a temperature above which it unfolds into an expanded form and below which it begins to fold. We call this the

[8] For a study focusing on motif, see Bryant and Lawrence (1993).

[9] For details, see Kolinski and Skolnick (1994a) and references therein.

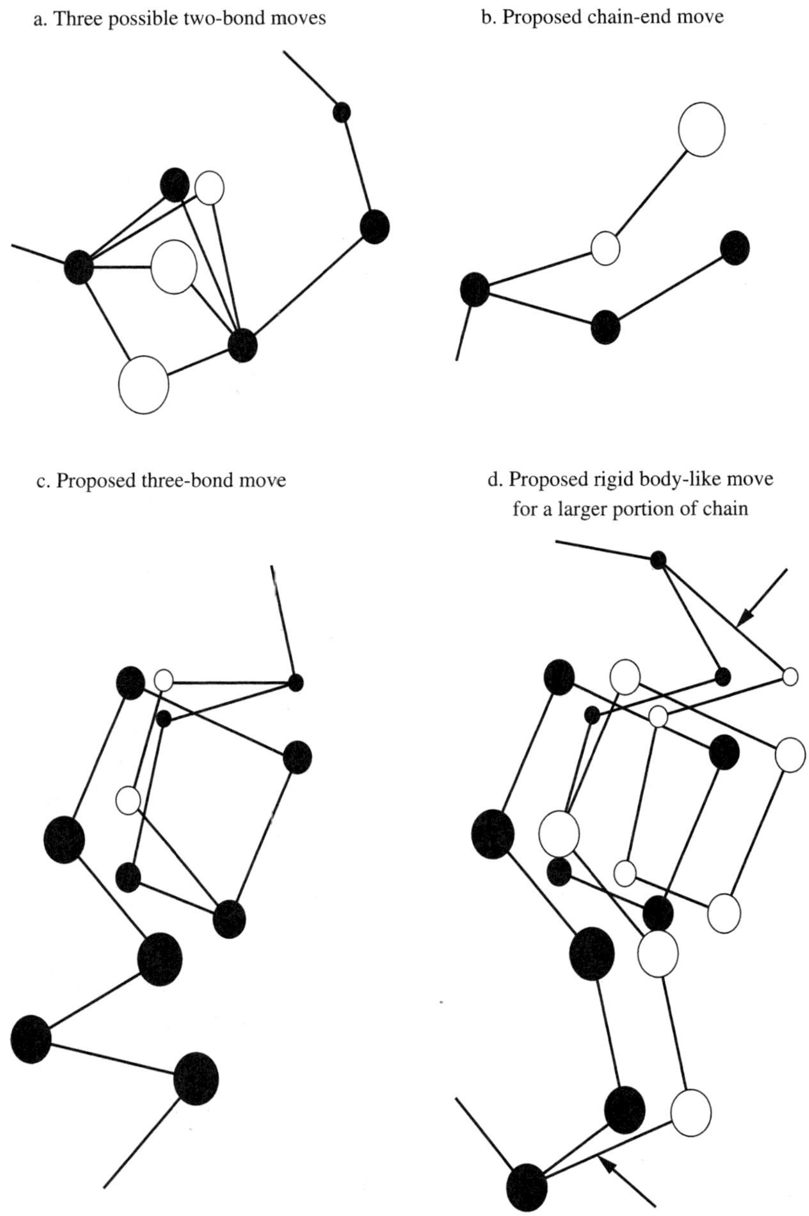

FIGURE 8.5 | *Nominated moves for protein folding example*
Source: Kolinski and Skolnick (1998).

renatured temperature. Preliminary HM sampling was performed to obtain a rough estimate of this temperature (Section 8.3.1). Then simulated annealing generated multiple independent sample paths, each beginning from a different starting conformation. For those paths that appeared trapped in local minima, the temperature was raised above the estimated renaturated temperature, and simulated

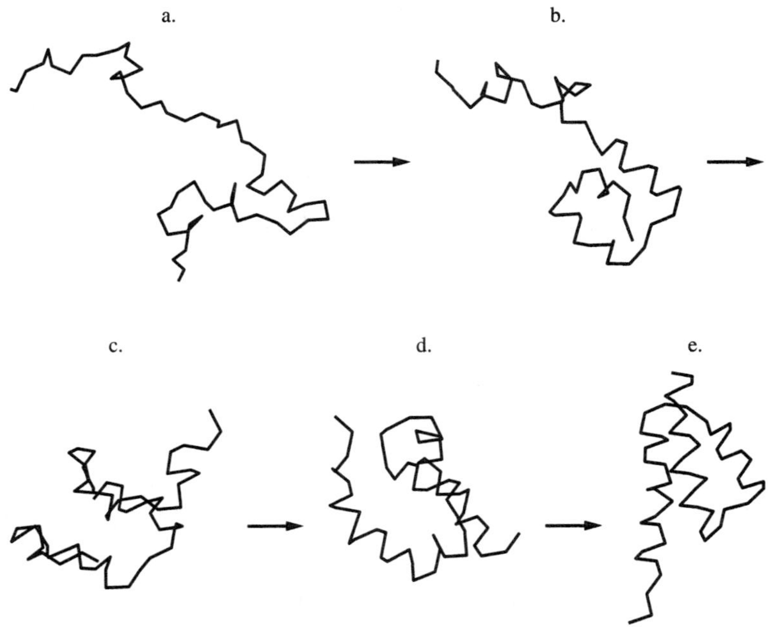

FIGURE 8.6 | *Sequential snapshots of simulated protein folding*

annealing continued. The newly raised temperature increased the acceptance rate in HM sampling, thus allowing more frequent escapes from local minima.

Results for the Coarse Lattice. Thirty of the runs successfully folded to three helices. Once folded, they retained that shape on their sample paths until run completion. The remaining 15 either collapsed into irregular shapes or formed helices that did not endure until run completion. Presumably, longer sample paths at the lower end of the temperature schedule would have reduced the number of nonconverging runs.

Let $\widehat{H}_{\min}$ denote the smallest energy observed on a sample path. Among the 30 sample paths that ended with three helices, 19 corresponded to the correct topology (found) in the NMR studies. These had an average $\widehat{H}_{\min}$ of -181 and smallest $\widehat{H}_{\min}$ of -225. The remaining 11 with average $\widehat{H}_{\min}$ of -198 and smallest $\widehat{H}_{\min}$ of -153 did not converge to the correct topology. Figure 8.6 shows a sequence of snapshots of the C_α trace from a single sample path.

Let $\mathcal{V}(C_\alpha)$ denote the set of C_α atoms on the main chain. For each $i \in \mathcal{V}(C_\alpha)$, let L_i denote the NMR-provided location of the ith C_α atom, and let $\widehat{L}_i$ denote its simulated annealing–generated location. Then

$$\widehat{\text{rms}} := \left[|\mathcal{V}(C_\alpha)|^{-1} \sum_{i \in \mathcal{V}(C_\alpha)} \left(\widehat{L}_i - L_i \right)^2 \right]^{1/2}$$

provides an assessment of how well the model with simulated annealing has performed relative to the exact C_α atom locations identified by NMR. For the coarse lattice, $\widehat{\text{rms}} = 4.3$ Ångstroms.

Results for the Finer Lattice. In general, the paths on the coarse lattice yielded much of the correct secondary structure but not of the tertiary structure. Continued simulated annealing with a finer lattice ($d = 1.21$ Ångstroms) produced more accurate predictions of tertiary structure. It relied on a modified but not dissimilar nominating regime and defined an iteration in much the same way as for the coarse lattice. Twenty-seven independent runs were made with initial states corresponding to the minimal energy configurations on the five sample paths with smallest $\widehat{H}_{\text{min}}$ in the coarse lattice simulations. At least three runs were made with each of the five initial configurations. The 27 sample paths all produced helices I, II, and III for their minimal energy configurations with average $\widehat{H}_{\text{min}}$ of -209 and standard deviation 8. Nineteen configurations have $\widehat{H}_{\text{min}}$ below the average. For these 19, $\widehat{\text{rms}}$ ranged from 2.55 to 3.42 Ångstroms with average $\widehat{\text{rms}}$ of 3.3. Because the individual $\widehat{\text{rms}}$ has a standard deviation of 2.5 Ångstroms, this 3.3 Ångstroms result represents a significant improvement over the $\widehat{\text{rms}}$ for the coarse lattice. ■

8.4 ACCELERATING CONVERGENCE

As past sections and the examples make clear, the effective use of simulated annealing depends on the ability to balance two countervailing phenomena. Recall the convergence bound in (6.57) and let ρ_k denote the convergence factor corresponding to temperature τ_k. As temperature drops from τ_k to τ_{k+1}, the convergence factor increases from ρ_k to ρ_{k+1}. That is, $\rho_k < \rho_{k+1}$. Therefore, convergence to the equilibrium distribution is slower at stage $k+1$ than at stage k. However, the equilibrium distribution, π_{k+1}, concentrates more of its probability mass on $\mathcal{X}_{\text{min}}$ than π_k does. As a consequence, the challenge facing the user is to conceive of a temperature schedule, $(\tau_1, l_1), \ldots, (\tau_r, l_r)$, that allows for this growth in concentration but does not slow down convergence to a self-defeating rate. The need for this balance pervades all stochastic optimization based on simulated annealing and similar methods.

Local minima in H can strongly influence the speed of convergence. An analysis of the acceptance rate in the HM step in Algorithm ANNEALING at successive temperatures reveals the role these temperatures play in retarding movement. Recall that $\alpha_k(\mathbf{x}, \mathbf{y})$ denotes the acceptance rate on iteration j, given $\mathbf{X}_{j-1} = \mathbf{x}$ and $\mathbf{Y} = \mathbf{y}$ has been randomly selected from the symmetric nominating distribution, $\{q(\mathbf{x}, \mathbf{y}),\ \mathbf{y} \in \mathcal{X}\}$. Let

$$\mathcal{X}(\mathbf{w}) := \{\mathbf{y} \in \mathcal{X} : H(\mathbf{y}) > H(\mathbf{w})\}.$$

Then on the HM step,

$$\begin{aligned}\text{pr}(\text{accept} \mid \mathbf{X}_{j-1} = \mathbf{x}) &= \sum_{\mathbf{y}\in\mathcal{X}} \alpha_k(\mathbf{x},\mathbf{y})q(\mathbf{x},\mathbf{y}) \\ &= \sum_{\mathbf{y}\in\mathcal{X}(\mathbf{x})} \frac{\pi_k(\mathbf{y})}{\pi_k(\mathbf{x})} q(\mathbf{x},\mathbf{y}) + \sum_{\mathbf{y}\in\mathcal{X}\setminus\mathcal{X}(\mathbf{x})} q(\mathbf{x},\mathbf{y}).\end{aligned}$$

Let $\mathcal{X}_{\text{local}}$ denote the subset of $\mathcal{X}$ that contains all sites at which H has local minima. Because $\{q(\mathbf{x},\mathbf{y}), \mathbf{y} \in \mathcal{X}\}$ often assigns positive probability only to local moves in the neighborhood of $\mathbf{x}$, it frequently occurs that

$$q(\mathbf{x},\mathbf{y}) = 0 \qquad \text{for all } \mathbf{y} \in \mathcal{X}\setminus\mathcal{X}(\mathbf{x}) \text{ and } \mathbf{x} \in \mathcal{X}_{\text{local}},$$

so that

$$\text{pr}(\text{accept} \mid \mathbf{X}_{j-1} = \mathbf{x}) = \sum_{\mathbf{y}\in\mathcal{X}(\mathbf{x})} e^{-[H(\mathbf{y})-H(\mathbf{x})]/\tau_k} q(\mathbf{x},\mathbf{y}) \qquad \mathbf{x} \in \mathcal{X}_{\text{local}}. \tag{8.20}$$

Difficulties leaving $\mathbf{X}_{j-1} = \mathbf{x} \in \mathcal{X}_{\text{local}}$ can arise in two related ways:

- Suppose H has deep wells with steep sides. For fixed k, $e^{-[H(\mathbf{y})-H(\mathbf{x})]/\tau_k}$ and, therefore, (8.20) are small for sites $\mathbf{y} \in \mathcal{X}(\mathbf{x})$ and $\mathbf{x} \in \mathcal{X}_{\text{local}}$.
- For particular sites $\mathbf{y} \in \mathcal{X}(\mathbf{x})$ and $\mathbf{x} \in \mathcal{X}_{\text{local}}$, $e^{-[H(\mathbf{y})-H(\mathbf{x})]/\tau_k}$ and, therefore, (8.20) decrease as k increases (as temperature decreases).

This analysis makes clear the critical way in which the temperature schedule influences convergence and sets the stage for describing other methods that accelerate convergence.

Although many optimization problems exist for which simulated annealing takes a relatively short computing time to generate solutions with acceptable accuracy, there are many others for which the computing time requirements are so excessive as to make solution impractical. However, all is not lost. Recall that every simulated annealing application consists of an alternating sequence of nominated displacements (new locations) and random decision rules, based on HM sampling to determine whether to move to this new location. Several methodologies exist that augment simulated annealing in ways that significantly reduce the computing time to generate a solution with acceptable accuracy. They include:

- Simulated tempering
- Swapping
- Stochastic tunneling
- Genetic algorithms

Simulated tempering is described in Section 8.5, swapping in Section 8.6, stochastic tunneling in Section 8.7, and genetic algorithms in Section 8.8. All four aim to reduce the probability of becoming trapped for a prolonged number of iterations at a local minimum of H. Each affects this reduction by modifying the way in which temperature, or some transformed temperature, changes as a sample path evolves.

Applications of simulated tempering, swapping, stochastic tunneling, and genetic algorithms to problems with known solutions have shown that these techniques converge to the correct solutions faster than simulated annealing. However, little rigorous theory exists at present in support of this behavior.[10] In what follows, we describe the concepts that have led to these empirical successes.

8.5 SIMULATED TEMPERING

In contrast to simulated annealing, where temperature decreases slowly according to a fixed schedule, *simulated tempering* allows temperature to vary on each iteration. Recall that an iteration of simulated annealing consists of two steps. The first randomly generates a candidate location in $\mathcal{X}$, and the second randomly decides whether to move to the nominated location. Moreover, it performs iteration j for a fixed temperature T_j that decreases slowly at iterations $j = l_1 < \cdots < l_r$ according to a prespecified temperature schedule, $\tau_1 > \cdots > \tau_r$. Simulated tempering adds two additional steps to each iteration.[11] Given the temperature T_j at the end of step 2 on iteration $j+1$, the third step selects an integer L from $\{1, \ldots, r\}$, and the fourth decides whether to assign $T_{j+1} = \tau_L$ or $T_{j+1} = T_j$. That is, the temperature on the next iteration may change from τ_I to τ_L. Some implementations make these selections according to a deterministic schedule. Others use random selection on either or both additional steps.

As before, let

$$\mathbf{X}_j := \text{location at the beginning of iteration } j+1 \qquad j \geq 0.$$

Whereas a sample path generated by simulated annealing can easily become trapped in a local minimum of H, well-designed simulated tempering increases the probability of an escape. That is, by allowing temperature to vary, it creates a sample path, $\mathbf{X}_1, \mathbf{X}_2, \ldots$, with considerably more variation than would occur in conventional simulated annealing alone.

Not all temperature-generating methods induce the desired variation. For example, if $T_j = T_I$ and the difference $\tau_L - \tau_I > 0$ is large, then progress made toward the global minimum of H would be lost on the next iteration when $T_{j+1} = \tau_L$ would induce a more volatile environment. A schedule that restricts changes in (reciprocal) temperature to nearest-neighbor pairs tends to control this volatility, provided that for each $l = 1, \ldots, r-1$, $\tau_l - \tau_{l+1}$ is not too large.

Most accounts of simulated tempering are based on *reciprocal temperature*,[12]

$$\beta_i := \frac{1}{\tau_i} \qquad i = 1, \ldots, r.$$

[10] An exception for a special case of simulated tempering is Madras and Zheng (2003).

[11] See Geyer and Thompson (1995), Marinari and Parisi (1992), and Lyubartsov et al. (1992).

[12] Reciprocal temperature fits the definition of an auxiliary variable. See Section 9.6.

Let

$$\nu_i > 0$$
$$\nu_1 + \cdots + \nu_r = 1$$

denote the chosen equilibrium p.m.f. for reciprocal temperature. Then as $j \to \infty$, $\mathbf{X}_j$ has a distribution that converges to

$$\pi_{\mathrm{ST}}(\mathbf{x}) = \nu_1\pi_1(\mathbf{x}) + \cdots + \nu_r\pi_r(\mathbf{x}) \qquad \forall \mathbf{x} \in \mathcal{X}, \tag{8.21}$$

where (8.11) defines $\{\pi_k(\mathbf{x})\}$ for each k.

On step 3 of each iteration, one procedure for simulated tempering randomly selects an integer from a nominating p.m.f.

$$s_{il} \geq 0 \qquad l = 1, \ldots, r$$
$$s_{i1} + \cdots + s_{ir} = 1$$

and then applies HM acceptance-rejection on step 4. For

$$A_{il}(\mathbf{x}) := \frac{\pi_l(\mathbf{x})\nu_l}{\pi_i(\mathbf{x})\nu_i}, \tag{8.22}$$

Algorithm TEMPERING generates an estimate of $H_{\min}$ using a randomized selection rule and a randomized decision rule for temperature updating on each iteration.

ALGORITHM TEMPERING[13]

Purpose: To estimate $H_{\min}$ and an element in $\mathcal{X}_{\min}$ by simulated tempering.

Given: Reciprocal temperatures $\beta_1 < \cdots < \beta_r$ and functions $\{A_{il}(\mathbf{x})\}$, $\{\nu_i\}$, and $\{s_{il}\}$.

Definitions:

I := reciprocal temperature index at beginning of an iteration.
L := nominated reciprocal temperature index.

Method:

1. Assign an initial state $\mathbf{X}_0 \in \mathcal{X}$, an integer $I \in \{1, \ldots, r\}$, $\widehat{H}_{\min} = H(\mathbf{X}_0)$, and $\mathbf{Z} = \mathbf{X}_0$.
2. $\beta = \beta_I$.
3. $j = 1$.
4. Repeat:

   displacement move

   a. randomly select a candidate $\mathbf{Y}$ from a symmetric $\{q(\mathbf{X}_{j-1}, \mathbf{y})\}$
   b. randomly sample U from $\mathcal{U}(0, 1)$
   c. $\mathbf{X}_j = \mathbf{X}_{j-1}$
   d. if $U \leq e^{-\beta[H(\mathbf{Y})-H(\mathbf{X}_{j-1})]}$, $\mathbf{X}_j = \mathbf{Y}$
   e. if $H(\mathbf{X}_j) < \widehat{H}_{\min}$, $\widehat{H}_{\min} = H(\mathbf{X}_j)$ and $\mathbf{Z} = \mathbf{X}_j$

[13]Based on Geyer and Thompson (1995).

temperature move

f. randomly sample L from $\{s_{Il};\ l = 1, \ldots, r\}$
g. randomly sample V from $\mathcal{U}(0, 1)$
h. evaluate $A_{IL}(\mathbf{X}_j)$
i. if $V \leq \min\left[1, \frac{s_{LI}}{s_{IL}} A_{IL}(\mathbf{X}_j)\right]$, $\beta = \beta_L$
j. $j = j + 1$

until terminated by stopping rule.

5. $\left(\widehat{H}_{\min}, \mathbf{Z}\right)$ is the estimated solution. ■

Because Algorithm TEMPERING alternates between displacement and temperature moves, the induced Markov chain is not reversible. However, amending the algorithm to incorporate any of the schemes

- displacement, temperature, displacement, displacement, temperature, displacement, ...
- temperature, displacement, temperature, temperature, displacement, temperature, ...
- displacement with probability $\frac{1}{2}$ or temperature with probability $\frac{1}{2}$

would ensure reversibility. As mentioned, earlier reversibility facilitates the analysis of convergence (Section 9.1). However, schemes like that in Algorithm TEMPERING are common practice.

Whereas simulated annealing alone aims to achieve the local equilibrium distribution $\{\pi_k(\mathbf{x})\}$ in (8.11) at each temperature τ_k of the schedule, simulated tempering aims to achieve equilibrium just once. However, it is important to note again that the choice of $\{\nu_i\}$ strongly influences the form of the equilibrium distribution π and the speed of convergence. For a given (reciprocal) temperature schedule, $\{\nu_i\}$ should be chosen to ensure temperature mixing on successive iterations and to ensure that the equilibrium distribution, π_{ST}, concentrates a substantial amount of its probability mass on points in $\mathcal{X}_{\min}$. Mixing temperatures speeds up convergence on a sample path from an initial state to the steady state, whereas concentrating probability mass increases the probability of obtaining an estimate, $H(\mathbf{X}_j)$, close to the optimum, $H_{\min}$. Because these two objectives are antithetic to each other, the choice of $\{\nu_i\}$ has to strike a balance between them.

Algorithm TEMPERING makes clear that the setup for simulated tempering is greater than for simulated annealing, requiring a user also to specify a nominating p.m.f., $\{s_{il}\}$, and and a mixing p.m.f., $\{\nu_i\}$. One proposal for the temperature nominating p.m.f.,

$$\begin{aligned} s_{12} &= 1 \\ s_{ij} &= \begin{cases} \frac{1}{2} & j = i-1, i+1 \\ 0 & \text{elsewhere} \end{cases} \qquad i = 2, \ldots, r-1 \\ s_{r,r-1} &= 1, \end{aligned}$$

induces a nearest-neighbor random walk on the integers $1, \ldots, r$, with reflecting barriers at both ends. A slightly amended proposal, $s_{11} = s_{12} = s_{r,r-1} = s_{rr} = \frac{1}{2}$, leads to $\frac{s_{li}}{s_{il}} = 1$ for all i and l so that the acceptance probability for the proposed reciprocal temperature reduces to $\min[1, A_{IL}(\mathbf{X}_j)]$ on iteration j.

The choice of a mixing distribution, $\{\nu_i\}$, is more complex. For $i = 1, \ldots, r$, let

$$c_i(\mathbf{x}) := \mathrm{e}^{-\beta_i H(\mathbf{x})}$$

and

$$c_i := \sum_{\mathbf{x}\in\mathbf{X}} \mathrm{e}^{-\beta_i H(\mathbf{x})}.$$

If one were to choose

$$\nu_i := \frac{c_i}{c_1 + \cdots + c_r} \qquad i = 1, \ldots, r, \tag{8.23}$$

then

$$A_{il}(\mathbf{x}) = \mathrm{e}^{-(\beta_l - \beta_i)H(\mathbf{x})} \qquad \forall i, l = 1, \ldots, r,$$

eliminating the need to know $c_1, \ldots, c_r$ explicitly. The corresponding equilibrium distribution would be

$$\pi_{\mathrm{ST}}(\mathbf{x}) = \frac{c_1(\mathbf{x}) + \cdots + c_r(\mathbf{x})}{c_1 + \cdots + c_r}. \tag{8.24}$$

Unfortunately, this assignment reveals that the hottest temperature, τ_1, is most likely to be assigned, and the coolest, τ_r, is least likely to be assigned. Moreover for some $1 \le k < r$, the limit

$$\pi_{\mathrm{ST}}(\mathbf{x}) \to \frac{c_1(\mathbf{x}) + \cdots + c_k(\mathbf{x})}{c_1 + \cdots + c_k} \qquad \text{as } \tau_{k+1} \to 0$$

shows how this property is exacerbated as some temperatures in the schedule become very low. Although a convenience for computation, (8.23) would not induce the desired mixing of temperatures and, therefore, the desired mobility for the sample path.

An alternative choice, $\nu_i = \frac{1}{r}$ for each $i = 1, \ldots, r$, makes each temperature equally likely and thus ensures the desired mixing. It also induces the equilibrium distribution

$$\pi_{\mathrm{ST}}(\mathbf{x}) = \frac{1}{r}\left[\frac{c_1(\mathbf{x})}{c_1} + \cdots + \frac{c_r(\mathbf{x})}{c_r}\right],$$

which for some $1 \le k < r$ has the limit

$$\pi_{\mathrm{ST}}(\mathbf{x}) \to \frac{1}{r}\left[\frac{c_1(\mathbf{x})}{c_1} + \cdots + \frac{c_k(\mathbf{x})}{c_r}\right] + \begin{cases} \frac{r-k}{r}\frac{1}{|\mathcal{X}_{\min}|} & \text{if } \mathbf{x} \in \mathcal{X}_{\min} \\ 0 & \text{if } \mathbf{x} \in \mathcal{X}\backslash\mathcal{X}_{\min} \end{cases} \qquad \text{as } \tau_{k+1} \to 0.$$

This limit reveals a concentration of probability mass at points in $\mathcal{X}_{\min}$ when temperatures $\tau_{k+1}, \ldots, \tau_r$ are very low. Because $A_{il}(\mathbf{x})$ is now a function of c_i and c_l, a method is required to estimate these quantities.

Let

$$\nu_i := \frac{a_i c_i}{a_1 c_1 + \cdots + a_r c_r} \qquad i = 1, \ldots, r, \tag{8.25}$$

where $a_1, \ldots, a_r$ are positive constants. Then (8.22) reduces to

$$A_{il}(\mathbf{x}) = \frac{a_l}{a_i} e^{-(\beta_l - \beta_i)H(\mathbf{x})}. \tag{8.26}$$

Suppose we fix $a_1 = 1$ and choose $a_2, \ldots, a_r$ to satisfy the recursion

$$a_i = a_{i-1}\frac{c_{i-1}}{c_i} \qquad i = 2, \ldots, r. \tag{8.27}$$

In addition to inducing the desired mixing p.m.f., $\{\nu_i = \frac{1}{r};\ i = 1, \ldots, r\}$, the recursion provides a convenient way to estimate $a_2, \ldots, a_r$ recursively. Algorithm LEVEL describes one way to do this. Because $c_1 > \cdots > c_r$, an ascending initial assignment, $b_1 < \cdots < b_r$, is preferable.

ALGORITHM LEVEL

Purpose: To estimate $a_2, \ldots, a_r$ recursively.

Method:

1. Initialize $b_1, \ldots, b_r$ and set $a_1 = 1$.
2. Repeat:
   a. $f_1 = \cdots = f_r = 0$
   b. Execute Algorithm TEMPERING for a fixed number of iterations with $a_i = b_i$ for $i = 1, \ldots, r$ as input and increment f_L by 1 if the condition at step 4h is true
   c. $b_1 = 0$
   d. $b_{i+1} = \frac{\max(1, f_i)}{\max(1, f_{i+1})} \times \frac{a_{i+1}}{a_i}$ for $i = 1, \ldots, r-1$

   until $f_1, \ldots, f_r$ are approximately equal. ■

The ratio of maxima in step 2d facilitates updating. Executing this algorithm for several preliminary short runs provides the basis for subsequently executing Algorithm TEMPERING for a considerably longer run with the estimated $\left\{\nu_i \approx \frac{1}{r}; i = 1, \ldots, r\right\}$. It is also noteworthy from (8.26) that choosing a nearest-neighbor random walk for assignments increases the acceptance rate, as does keeping the increments $\beta_{i+1} - \beta_i$ sufficiently small.

8.6 SWAPPING

Swapping extends the simulated tempering concept to r sample paths, where β_i is assigned to path i throughout the sampling experiment.[14] The objective is to generate $\mathbf{X}_j^{[1]}, \ldots, \mathbf{X}_j^{[r]}$ jointly from the equilibrium distribution

$$\pi_{\text{SWAP}}\left(\mathbf{x}^{[1]}, \ldots, \mathbf{x}^{[r]}\right) = \prod_{i=1}^{r} \pi_i\left(\mathbf{x}^{[i]}\right), \tag{8.28}$$

[14] For example Bhatnagar and Randall (2004), Falcioni and Deem (1999), and Hansmann (1997).

where (9.31) defines π_i, and the superscript $^{[i]}$ denotes path i. Swapping algorithms execute two actions:

- **Displacement Move.** Choose a path I from $\{1, \ldots, r\}$ and generate $\mathbf{X}_j^{[I]}$ by HM sampling.
- **Swap Move.** Choose a path I from $\{1, \ldots, r-1\}$ and decide whether to swap $\mathbf{X}_j^{[I]}$ and $\mathbf{X}_j^{[I+1]}$.

One form of the overall procedure alternates between displacement and swap moves. Another randomly decides to perform a displacement with user-specified probability p or a swap move with probability $1-p$. A displacement move may select path $I = (j-1) \pmod r + 1$ on iteration j or may randomly select I from $\{1, \ldots, r\}$ with user-specified (possibly unequal) probabilities. Likewise, swap moves may use either a deterministic or a randomized decision rule.

The effect is similar to the randomization of the reciprocal temperature in simulated tempering. However, there are now r sample paths being generated, each exposed to all r reciprocal temperatures $\beta_1, \ldots, \beta_r$. Let

$$\begin{aligned} B_j &:= \min \left[1, \frac{\pi\left(\mathbf{X}_j^{[1]}, \ldots, \mathbf{X}_j^{[I-1]}, \mathbf{X}_j^{[I+1]}, \mathbf{X}_j^{[I]}, \ldots, \mathbf{X}_j^{[r]}\right)}{\pi\left(\mathbf{X}_j^{[1]}, \ldots, \mathbf{X}_j^{[I-1]}, \mathbf{X}_j^{[I]}, \mathbf{X}_j^{[I+1]}, \ldots, \mathbf{X}_j^{[r]}\right)} \right] \\ &= \min \left[1, \frac{\pi_I\left(\mathbf{X}_j^{[I+1]}\right)\pi_{I+1}\left(\mathbf{X}_j^{[I]}\right)}{\pi_I\left(\mathbf{X}_j^{[I]}\right)\pi_{I+1}\left(\mathbf{X}_j^{[I+1]}\right)} \right] \\ &= \min \left[1, \frac{c_I\left(\mathbf{X}_j^{[I+1]}\right)c_{I+1}\left(\mathbf{X}_j^{[I]}\right)}{c_I\left(\mathbf{X}_j^{[I]}\right)c_{I+1}\left(\mathbf{X}_j^{[I+1]}\right)} \right]. \end{aligned}$$

Algorithm SWAP describes the main features of the procedure. It alternates between displacement and swap moves and randomly selects path I on a swap move with probability $\frac{1}{r-1}$.

ALGORITHM **SWAP**[15]

Purpose: To estimate $H_{\min}$ and an element in $\mathcal{X}_{\min}$ by swapping.

Given: Reciprocal temperatures $\beta_1 < \cdots < \beta_r$.

Method:

1. For $i = 1, \ldots, r$, assign an initial state $\mathbf{X}_0^{[1]}$ and β_i to path i.
2. Randomly select path I from $\{1, \ldots, r\}$ with probability $\frac{1}{r}$.
3. $j = 1$, $\widehat{H}_{\min} = H(\mathbf{Z})$, and $\mathbf{Z} = \mathbf{X}_0^{[1]}$.
4. Repeat:
    a. for all $l \neq I$, $\mathbf{X}_j^{[l]} = \mathbf{X}_{j-1}^{[l]}$

[15]Sometimes $\beta_1 = 0$, for which $\pi_1(\mathbf{x}) = |\mathcal{X}|^{-1}$ for all $\mathbf{x} \in \mathcal{X}$. This guarantees acceptance on path 1.

displacement move

b. generate $\mathbf{X}_j^{[I]}$ by HM sampling with reciprocal temperature β_I

c. if $H(\mathbf{X}_j^{[I]}) < \widehat{H}_{\min}$, then $\widehat{H}_{\min} = H(\mathbf{X}_j^{[I]})$ and $\mathbf{Z} = \mathbf{X}_j^{[I]}$

swap move

d. randomly select path I from $\{1, \ldots, r-1\}$ with probability $\frac{1}{r-1}$

e. $\mathbf{Y} = \mathbf{X}_j^{[I]}$

f. randomly sample U from $\mathcal{U}(0, 1)$

g. if $U \le B_j$, then $\mathbf{X}_j^{[I]} = \mathbf{X}_j^{[I+1]}$ and $\mathbf{X}_j^{[I+1]} = \mathbf{Y}$

h. $j = j + 1$

select next path to update

i. randomly select path I from $\{1, \ldots, r\}$ with probability $\frac{1}{r}$

until terminated by stopping rule.

5. $(\widehat{H}_{\min}, \mathbf{Z})$ is the estimated solution. ■

Although characteristic of swapping procedures used in practice, Algorithm SWAP does not induce a reversible Markov chain. However, amending the algorithm to incorporate any of the following plans would ensure a reversible chain.

- displacement, swap, displacement, displacement, swap, displacement, ...
- swap, displacement, swap, swap, displacement, swap, ...
- displacement with $p = \frac{1}{2}$ and swap with $1 - p = \frac{1}{2}$

Reversibility facilitates the analysis of convergence (see Section 9.1 and Madras and Zheng 2003).

Because

$$B_j = \min\left[1, e^{(\beta_{I+1}-\beta_I)\left[H\left(\mathbf{x}_j^{[I]}\right)-H\left(\mathbf{x}_j^{[I+1]}\right)\right]}\right],$$

a swap occurs with probability 1 if $H(\mathbf{X}_j^{[I]}) \ge H(\mathbf{X}_j^{[I+1]})$. After this swap, the lower temperature path $I+1$ has the site with the lower energy. If $H(\mathbf{X}_j^{[I]}) < H(\mathbf{X}_j^{[I+1]})$, the the probability of a swap is a decreasing function of $\beta_{I+1} - \beta_I > 0$, again arguing for small increments in the reciprocal temperature schedule.

Although swapping has many features in common with simulated tempering, it also has several distinct properties. They include:

- There is no need to estimate the unknown denominators $c_1, \ldots, c_r$.
- It is easily seen that for $1 \le k < r$:

$$\pi_{\text{SWAP}}\left(\mathbf{x}^{[1]}, \ldots, \mathbf{x}^{[r]}\right) \to \begin{cases} \frac{1}{|\mathcal{X}_{\min}|^{r-k}} \prod_{i=k+1}^{r} \pi_i\left(\mathbf{x}^{[i]}\right) & \text{if } \mathbf{x}^{[l]} \in \mathcal{X}_{\min},\ l = 1, \ldots, r-k \\ & \text{and } \mathbf{x}^{[l]} \in \mathcal{X},\ l = k+1, \ldots, r \qquad \text{as } \tau_k \to 0, \\ 0 & \text{elsewhere} \end{cases}$$

which notably differs from $\{\pi_{\text{ST}}(\mathbf{x})\}$.

- Because of swapping, all paths are less likely to become trapped in local minima.
- Convergence to equilibrium for all r paths in swapping takes more computing time than for one path in simulated tempering.
- Swapping procedures similar to Algorithm SWAP that use ergodic, but slowly mixing, nominating kernels may, nevertheless, be rapidly mixing (see Sections 6.8.2 and 9.2).[16]
- If convergence to equilibrium is rapidly mixing for a swapping algorithm, then the corresponding simulated tempering algorithm is also rapidly mixing (Zheng 2003).
- Because[17]

$$\frac{1}{r}\sum_{i=1}^{r}[1-\pi_i(\mathcal{X}_{\min})] \ge \left\{\prod_{i=1}^{r}[1-\pi_i(\mathcal{X}_{\min})]\right\}^{1/r} \ge \prod_{i=1}^{r}[1-\pi_i(\mathcal{X}_{\min})],$$

  it is easily seen that in equilibrium,

$$\begin{aligned}\mathrm{pr}\big(\widehat{H}_{\min} = H_{\min}|\mathrm{swap}\big) &= 1 - \prod_{i=1}^{r}[1-\pi_i(\mathcal{X}_{\min})] \\ &\ge 1 - \frac{1}{r}\sum_{i=1}^{r}\left[1-\pi_i\left(\mathcal{X}_{\min}\right)\right] \\ &= \mathrm{pr}(\widehat{H}_{\min} = H_{\min}|\text{simulated tempering}).\end{aligned}$$

  Because it is more probable to be in $\mathcal{X}_{\min}$ on iteration j for swapping than for simulated tempering, it should take less computing time for at least one path to enter $\mathcal{X}_{\min}$.

Some implementations of the swapping concept are called *parallel tempering,* a recognition that updating r paths can be done concurrently in a parallel computing environment (e.g., Earl and Deem 2004). It is easily seen that parallel tempering is equivalent to generating r sample paths, each of which randomizes the reciprocal temperature.

8.7 STOCHASTIC TUNNELING

Let $\mathbf{a}$ denote an element of $\mathcal{X}$, and let $\{G(\mathbf{x}|\mathbf{a}),\ \mathbf{x} \in \mathcal{X}\}$ denote a function for which the locations of all its local minima and all its global minima occur at the same sites in $\mathcal{X}$ as those of the original objective function H.[18] For a fixed temperature T, *stochastic tunneling* employs HM sampling with G in place of H to search for the global minimum, where G is chosen to be a positive-valued function such

[16]For example, see Madras and Zheng (2003).

[17]The first inequality follows from the property that for nonnegative numbers, the arithmetic mean is greater than or equal to the geometric mean.

[18]For example, Barhen et al. (1997) and Wenzel and Hamacher (1999).

that if $H(\mathbf{y}) \ge H(\mathbf{x})$ for some $\mathbf{x}$ and $\mathbf{y}$ in $\mathcal{X}$,

$$G(\mathbf{y}|\mathbf{a}) - G(\mathbf{x}|\mathbf{a}) \le H(\mathbf{y}) - H(\mathbf{x}). \tag{8.29}$$

Because this implies

$$e^{-[G(\mathbf{y}|\mathbf{a})-G(\mathbf{x}|\mathbf{a})]/T} \ge e^{-[H(\mathbf{y})-H(\mathbf{x})]/T},$$

stochastic tunneling based on G has a higher acceptance rate than simulated annealing. In effect, G reduces the heights of the peaks in H and their steepness relative to $\mathbf{a}$. The function G is sometimes called a *deformation* of H.

As illustration, suppose that at the beginning of iteration j,

$$\begin{aligned}\mathbf{Z}_j := {} & \text{site in } \mathcal{X} \text{ at which } H \text{ has its smallest minimum} \\ & \text{on the sample path } \mathbf{X}_0, \mathbf{X}_1, \ldots, \mathbf{X}_{j-1},\end{aligned}$$

and let

$$G(\mathbf{x}|\mathbf{Z}_j) = 1 - e^{-\gamma[H(\mathbf{x})-H(\mathbf{Z}_j)]}$$

for some fixed *tunneling parameter* $\gamma \in (0, 1]$. Then for fixed $H(\mathbf{y}) \ge H(\mathbf{x}) \ge H(\mathbf{Z}_j)$, the acceptance rate reduces to

$$e^{-[G(\mathbf{y}|\mathbf{Z}_j)-G(\mathbf{x}|\mathbf{Z}_j)/T]} = \exp\left\{\left[e^{-\gamma H(\mathbf{x})} - e^{-\gamma H(\mathbf{y})}\right]/T'_j\right\},$$

where the *effective temperature,*

$$T'_j := Te^{-\gamma H(\mathbf{Z}_j)} \qquad j \ge 0, \tag{8.30}$$

is monotone nondecreasing in j as $j \to \infty$.

We see that the effective temperature, T'_j, increases at random points during the MC experiment, namely, each time that $H(\mathbf{Z}_{j-1}) > H(\mathbf{Z}_j)$, in contrast to the decreasing temperature schedule for simulated annealing alone.

If we think of $\mathcal{X}_{\min}$ as a collection of absorbing states (see Section 6.16), then a carefully chosen G should induce

$$\mathbf{Z}_j \to \mathbf{z} \in \mathcal{X}_{\min} \ \text{w.p.1} \qquad \text{as } j \to \infty.$$

It then follows that

$$T'_j \to Te^{-\gamma H_{\min}} \ \text{w.p.1} \qquad \text{as } j \to \infty.$$

As a consequence, $\mathbf{X}_j$ has the limiting equilibrium distribution

$$\pi_{\text{TUN}}(\mathbf{x}) = \frac{e^{-\left\{1-e^{-\gamma[H(\mathbf{x})-H_{\min}]}\right\}/T}}{\sum_{\mathbf{y}\in\mathcal{X}} e^{-\left\{1-e^{-\gamma[H(\mathbf{y})-H_{\min}]}\right\}/T}}.$$

In particular,

$$\pi_{\text{TUN}}(\mathbf{x}) = \frac{1}{|\mathcal{X}_{\min}| + \sum_{\mathbf{y}\in\mathcal{X}\setminus\mathcal{X}_{\min}} e^{-\left\{1-e^{-\gamma[H(\mathbf{y})-H_{\min}]}\right\}/T}} \qquad \mathbf{x} \in \mathcal{X}_{\min}.$$

As is easily seen, stochastic tunneling relies on information accumulated up to the beginning of iteration j to attenuate the heights of peaks in the original objective function H. Moreover, it relies on γ to modify the steepness of ascent and descent.

For G in (8.29), benchmark comparisons between simulated annealing and stochastic tunneling for the traveling salesman problem,[19] allowing only local moves for updating, revealed greater computational efficiency for stochastic tunneling (Wenzel and Hamacher 1999). In the same study, a comparison that allowed more global moves showed little difference between the techniques, presumably because of the considerable computing time required to check that the symmetric nominating distribution, $\{q(\mathbf{x}, \mathbf{y}),\ \mathbf{y} \in \mathcal{X}\}$, generated $\mathbf{Y}$s that satisfied all constraints that defined membership in $\mathcal{X}$.

The deformation (8.29) does not improve computational efficiency for all problems. For example, when applied to a problem of protein folding, it induced a flat G with many relatively shallow minima, thus impeding convergence to an element of $\mathcal{X}_{\min}$ (Schug et al. 2003). An alternative deformation,

$$G(\mathbf{x}, \mathbf{Z}_j) := \ln\left\{\gamma[H(\mathbf{x}) - H(\mathbf{Z}_j)] + \sqrt{\gamma^2\,[H(\mathbf{x}) - H(\mathbf{Z}_j)]^2 + 1}\right\}, \tag{8.31}$$

led to considerably computationally more efficient solutions. Whereas (8.29) induces a random sequence of the successively lower effective temperatures (8.30), (8.31) attained its success when employed with a prespecified temperature schedule. If $\mathbf{Y}$ is randomly drawn from a symmetric nominating distribution, on iteration j, then the acceptance probability on the HM step given $\mathbf{X}_{j-1}$ and $\mathbf{Z}_j$ is

$$\min\left\{1, e^{-[G(\mathbf{Y},\mathbf{Z}_j) - G(\mathbf{X}_{j-1},\mathbf{Z}_j)]/T_j}\right\},$$

again revealing the importance of an appropriate temperature schedule. The corresponding local equilibrium distribution at temperature τ_k is as in (8.11), with G replacing H.

8.8 GENETIC ALGORITHMS

As their name implies, genetic algorithms denote a class of procedures that mimic natural evolution as they search for an element in $\mathcal{X}_{\min}$ or, equivalently, for the minimum $H_{\min}$. Each algorithm applies the genetic operators *mutation, crossover,* or *inversion* to a randomly selected population of *chromosomes* (points in $\mathcal{X}$) to effect movement toward the *fittest* chromosomal compositions (points in $\mathcal{X}_{\min}$). As originally conceived in the context of genetic algorithms, a chromosome denotes a string of (0-1) bits. A mutation flips one or more bits on a chromosome, from 0 to 1 or 1 to 0. A crossover denotes the operation whereby two chromosomes,

[19]Exercises H.3 and H.4 describe the traveling salesman problem.

paired for reproduction, exchange corresponding subsequences of bits. An inversion denotes the reversal of a substring of bits on a chromosome.[20]

In practice, genetic algorithms are also applied to more general data structures where each element, $\mathbf{x} := (x_1, \ldots, x_m) \in \mathcal{X}$, denotes a chromosome with m possibly different kinds of attributes, each of which can assume many different values. In this context, it suffices to note the following:

- A mutation operation replaces $\mathbf{x}$ by a new element $\mathbf{y}$, some or all of whose attributes differ from $\mathbf{x}$. For example, $\mathbf{y} = (x_1, \ldots, x_{k-1}, y_k, x_{k+1}, \ldots, x_m)$ merely alters the value of attribute k.
- A crossover operation exchanges values of a subset of corresponding attributes of $\mathbf{x}$ and $\mathbf{y}$. For example,

$$\begin{array}{ccc} \text{old} & & \text{new} \\ (x_1, \ldots, x_m) & \rightarrow & (x_1, \ldots, x_k, y_{k+1}, y_{k+2}, \ldots, y_m) \\ (y_1, \ldots, y_m) & \rightarrow & (y_1, \ldots, y_k, x_{k+1}, x_{k+2}, \ldots, x_m). \end{array}$$

- In the case of $x_i \in \{0, 1\}$ for each $i = 1, \ldots, m$, an inversion operation selects a bit string, $x_{k+1}, \ldots, x_l$, in $\mathbf{x} = (x_1, \ldots, x_m)$ for $1 \le k \le l \le m$, and replaces $\mathbf{x}$ by $\mathbf{y} = (x_1, \ldots, x_k, x_l, x_{l-1}, \ldots, x_{k+1}, x_{l+1}, \ldots, x_m)$. That is, it reverses the bit string. The inversion operation serves a similar role in more general data structures where the attributes are all of the same kind. For example, Exercises H.3 and H.4 illustrate the inversion operation for the traveling salesman problem wherein the attributes of each chromosome form an ordered list of locations.

Although all genetic algorithms embody these three concepts, the decision rules for employing them in practice vary with different implementations. The present account describes one MCMC approach to implementing mutation and crossover operations where HM sampling provides the randomized decision rules for change or no change in the values of chromosomal attributes. Within this setup there also exist alternative selection rules and nominating kernels that may fit a particular problem better than the ones we present here.[21]

An MCMC genetic algorithm starts by randomly sampling r points, $\mathbf{X}_0^{[1]}, \ldots, \mathbf{X}_0^{[r]}$, independently from a distribution π_0 on $\mathcal{X}$ and assigning them as the initial states on each of r paths. In biological terms, this is equivalent to assigning initial values to the attributes of each of the r chromosomes. The algorithm also initially assigns reciprocal temperature β_i to chromosome or path i for $i = 1, \ldots, r$, where as before, $\beta_1 < \cdots < \beta_r$. On each successive iteration, it iteratively attempts to perform either a mutation or a crossover on one or more chromosomes or, equivalently, to update the sample paths.

At each iteration $j > 0$,

- The algorithm randomly nominates a candidate for mutation with probability p_{mut}, or it nominates a crossover with probability $1 - p_{\text{mut}}$.

[20] The formalisms for applying these conceptualizations are described in Holland (1975), Goldberg (1989), and Davis (1991).

[21] See Liang and Wong (2000, 2001a, 2001b).

- For a mutation:

M1. Path I is randomly selected from $\{1, \ldots, r\}$ with probability $\frac{1}{r}$. That is, each chromosome is equally likely to be selected.

M2. A proposed new location, $\mathbf{Y} = \mathbf{y}$, is randomly sampled from a nominating distribution $\{q_{\text{mut}}(\mathbf{X}_{j-1}^{[I]}, \mathbf{y}),\ \mathbf{y} \in \mathcal{X}\}$.

M3. $\mathbf{X}_j^{[l]} = \mathbf{X}_{j-1}^{[l]}$ for all $l \neq I$.

M4. A sample value U is randomly drawn from $\mathcal{U}(0, 1)$, and

$$\mathbf{X}_j^{[I]} = \begin{cases} \mathbf{Y} & \text{if } U \leq \alpha_{\text{mut}}(\mathbf{X}_{j-1}^{[I]}, \mathbf{Y}) \\ 0 & \text{elsewhere,} \end{cases}$$

where

$$\alpha_{\text{mut}}(\mathbf{X}_{j-1}^{[I]}, \mathbf{Y}) := \min\left[1, \frac{\pi_I(\mathbf{Y})}{\pi_I(\mathbf{X}_{j-1}^{[I]})} \cdot \frac{\frac{1}{r}}{\frac{1}{r}} \cdot \frac{q_{\text{mut}}(\mathbf{Y}, \mathbf{X}_{j-1}^{[I]})}{q_{\text{mut}}(\mathbf{X}_{j-1}^{[I]}, \mathbf{Y})}\right] \tag{8.32}$$

and (8.11) defines π_I. Usually, q_{mut} is designed to nominate local moves for individual coordinates and to be symmetric; for example,

$$q_{\text{mut}}(\mathbf{x}, \mathbf{y}) := q_{\text{mut}}(\mathbf{x} - \mathbf{y}) = q_{\text{mut}}(\mathbf{y} - \mathbf{x}).$$

Then

$$\alpha_{\text{mut}}(\mathbf{X}_{j-1}^{[I]}, \mathbf{Y}) = \min\left\{1, \mathrm{e}^{-\beta_I\left[H(\mathbf{Y}) - H(\mathbf{X}_{j-1}^{[I]})\right]}\right\}.$$

- For a crossover:

C1. Path I is randomly selected from $\{1, \ldots, r\}$ with probability $\{\eta_i(\mathbf{Z}_{j-1}),\ i = 1, \ldots, r\}$, where

$$\mathbf{Z}_{j-1} := (\mathbf{X}_{j-1}^{[1]}, \ldots, \mathbf{X}_{j-1}^{[r]})$$

$$\eta_i(\mathbf{z}) := \begin{cases} \dfrac{\mathrm{e}^{-\beta^* H(\mathbf{x}_i)}}{\sum_{k=1}^{r} \mathrm{e}^{-\beta^* H(\mathbf{x}_k)}} & i = 1, \ldots, r \\ 0 & \text{elsewhere} \end{cases}$$

$$\mathbf{z} := (\mathbf{x}_1, \ldots, \mathbf{x}_r) \qquad \mathbf{x}_i \in \mathcal{X},\ \ i = 1, \ldots, r$$
$$\beta^* \in \{\beta_1 < \cdots < \beta_r\},$$

and path L is randomly selected from $\{1, \ldots, I-1, I+1, \ldots, r\}$ with probability $\frac{1}{r-1}$. Then for given $\mathbf{Z}_{j-1} = \mathbf{z} \in \mathcal{X}^r$, paths I and L have the joint conditional distribution

$$\eta_{il}(\mathbf{z}) := \begin{cases} \dfrac{\mathrm{e}^{-\beta^* H(\mathbf{x}_i)} + \mathrm{e}^{-\beta^* H(\mathbf{x}_l)}}{(r-1)\sum_{k=1}^{r} \mathrm{e}^{-\beta^* H(\mathbf{x}_k)}} & \forall i \neq l \\ 0 & \text{elsewhere.} \end{cases}$$

In contrast to mutation, a crossover assigns the highest probability of selection to the (i, l) 2-tuple for which either $H(\mathbf{x}_i)$ or $H(\mathbf{x}_l)$ is a minimum.

That is, the selection for crossover tends to favor the chromosome that is the fittest. Moreover, this probability increases with β^*. Liang and Wong (2001a) suggest a value of β^* close to β_1.

C2. $\mathbf{X}_j^{[k]} = \mathbf{X}_{j-1}^{[k]}$ for all $k \neq I$ or L.

C3. An attribute M is randomly selected from $\{1, \ldots, m\}$, with probability $\frac{1}{m}$.

C4. The nominated state is:

$$\mathbf{Z}_{ILMj} := \mathbf{Z}_{j-1} \text{ with } \mathbf{Y}_{ILM} \text{ and } \mathbf{Y}_{LIM} \text{ replacing } \mathbf{X}_{j-1}^{[I]} \text{ and } \mathbf{X}_{j-1}^{[L]},$$

where

$$\mathbf{Y}_{ILM} := \left(X_{1,j-1}^{[I]}, \ldots, X_{M,j-1}^{[I]}, X_{M+1,j-1}^{[L]}, \ldots, X_{m,j-1}^{[L]}\right)$$
$$\mathbf{Y}_{LIM} := \left(X_{1,j-1}^{[L]} \ldots, X_{M,j-1}^{[L]}, X_{M+1,j-1}^{[I]}, \ldots, X_{m,j-1}^{[I]}\right).$$

C5. A sample value U is randomly drawn from $\mathcal{U}(0, 1)$, and

$$\left(\mathbf{X}_j^{[I]}, \mathbf{X}_j^{[L]}\right) = \begin{cases} (\mathbf{Y}_{ILM}, \mathbf{Y}_{LIM}) & \text{if } U \leq \alpha_{\text{cross}}(\mathbf{Z}_{j-1}, \mathbf{Z}_{ILMj}) \\ 0 & \text{elsewhere,} \end{cases}$$

where

$$\alpha_{\text{cross}}(\mathbf{Z}_{j-1}, \mathbf{Z}_{ILMj}) := \min \left[1, \frac{\pi_I(\mathbf{Y}_{ILM})\pi_L(\mathbf{Y}_{LIM})}{\pi_I\left(\mathbf{X}_{j-1}^{[I]}\right)\pi_L\left(\mathbf{X}_{j-1}^{[L]}\right)} \times \frac{\frac{1}{m} q_{\text{cross}}\left(\mathbf{Z}_{ILMj}, \mathbf{Z}_{j-1}\right)}{\frac{1}{m} q_{\text{cross}}\left(\mathbf{Z}_{j-1}, \mathbf{Z}_{ILMj}\right)}\right]$$

and

$$q_{\text{cross}}(\mathbf{Z}_{j-1}, \mathbf{Z}_{ILMj}) := \frac{1}{m}\eta_{IL}(\mathbf{Z}_{j-1}).$$

Some genetic algorithms complete each iteration by attempting a swap between two current values, $\mathbf{X}_j^{[1]}, \ldots, \mathbf{X}_j^{[r]}$, on the r paths (e.g., Liang and Wong 2001b). The swap procedure is essentially the same as that beginning with step 4d in Algorithm SWAP.

If $p_{\text{mut}} = 1$ and $r = 1$, then no crossovers occur and the genetic algorithm just described reduces to a single stage of simulated annealing at a constant temperature. If $p_{\text{mut}} = 1$ and $r > 1$, it reduces to swapping as in Section 8.6. However, it is important to note that crossovers are the most distinguishing feature of genetic algorithms. Among the r chromosomes, the corresponding selection procedure tends to favor the one with the smallest $H(\mathbf{x})$ among the r paths.

If $p_{\text{mut}} = 0$, then only crossovers would occur. In this case, one can easily show that for arbitrarily selected initial states on the r paths, not all states are reachable as the paths evolve so that the induced Markov chain on $\mathcal{X}$ is not irreducible and, therefore, not ergodic. A not unusual practice is to assign p_{mut} a value in [0.20, 0.40] (e.g., Liang and Wong 2000, 2001a, 2001b).

An alternative crossover procedure randomly samples M from $\{1, \ldots, m\}$ with probability $\frac{1}{m}$, randomly samples N from $\{1, \ldots, m\}\backslash\{M\}$ with probability $\frac{1}{m-1}$, and then attempts to exchange the corresponding values of coordinates $1 + \min(M, N)$ through $\max(M, N)$ on chromosomes I and L.

8.9 SEARCHING FOR MORE THAN THE MINIMUM

In addition to facilitating the search for $H_{\min}$ and elements of $\mathcal{X}_{\min}$, HM sampling can reveal other relationships among the elements of $\{H(\mathbf{x})\}$ that facilitate problem solution. In particular, problems exist for which it is of interest to find those $\mathbf{x} \in \mathcal{X}$ for which $\gamma \le \frac{H(\mathbf{x})}{H_{\min}} < 1$ for some specified $\gamma \in (0, 1)$. Sometimes these solutions are more convenient to work with in subsequent analyses than those $\mathbf{x} \in \mathcal{X}_{\min}$. Other times, the objective function H is itself random and, therefore, there may exist $H(\mathbf{x})$ ($< H_{\min}$) that statistical testing would show not to be significantly different from $H_{\min}$. The following example illustrates how any one of the techniques described in this chapter apply to solving this formulation.

EXAMPLE 8.9.1 VARIABLE SUBSET SELECTION IN REGRESSION ANALYSIS

Given a data set consisting of N independent observations on $m+1$ ($< N$) variables labeled $0, 1, \ldots, m$, one wants to fit a function of the (independent) variables $1, \ldots, m$ to predict the values of variable 0, sometimes called the *response*. Usually more than one candidate function exists, and it is of interest to estimate how well each alternative predicts these responses. The functions may be linear or nonlinear, and they may be based on different subsets of variables. Here we consider linear functions and focus on identifying subsets of independent variables whose linear regressions predict the responses better than the vast majority of possible subsets.

Although one can regress the m variables on variable 0 and come up with an all-inclusive linear regression function, at least two issues motivate a more comprehensive study:

- If m is very large, then the resulting regression function may become cumbersome to employ in practice. Parsimony is always a virtue whenever it can be established without major degradation in predictive power.
- *Collinearity* may exist among the m variables. Because its presence tends to inflate the variances of the estimated regression coefficients, it is always desirable to work with variables that exhibit little, if any, collinearity.

Two variables, Z_1 and Z_2, are *exactly collinear* if $a_1 Z_1 + a_2 Z_2 = a_0$ holds for some constants $a_0,\ a_1,\ a_2$ for all observations. The presence of this property prevents application of linear regression to a problem containing both Z_1 and Z_2. These variables are said to be *approximately collinear* if the equality holds for some, but not all, observations. In this case, the variances of the estimated coefficients tend to increase as the correlation between Z_1 and Z_2 increases.

Let

$$\begin{aligned}
Y_i &:= \text{response of observation } i \\
x_i &:= \begin{cases} 1 & \text{if variable } i \text{ is included in the fitted regression} \\ 0 & \text{otherwise} \end{cases} \\
\mathbf{x} &:= (x_1, \ldots, x_m) \\
\mathbf{x}_{\text{all}} &:= \mathbf{x} \text{ with } x_1 = \cdots = x_m = 1 \quad \text{(all variables are included)} \\
\mathcal{J}(\mathbf{x}) &:= \{i : \ x_i = 1;\ i = 1, \ldots, m\} \\
\widehat{Y}_i(\mathbf{x}) &:= \text{fitted value for variable subset model } \mathcal{J}(\mathbf{x}) \text{ for observation } i.
\end{aligned}$$

One criterion measures how well the fitted regression for a given subset of variables predicts the values of the responses in the original data set. Then one indicator of how well subset $\mathcal{J}(\mathbf{x})$ performs is[22]

$$B(\mathbf{x}) := \frac{\text{MSE}\left[\widehat{Y}_i(\mathbf{x})\right]}{\sigma^2},$$

where

$$\begin{aligned}
\text{MSE}\left[\widehat{Y}_i(\mathbf{x})\right] &:= \sum_{i=1}^{N} \left[\widehat{Y}_i(\mathbf{x}) - \text{E}Y_i\right]^2 \\
&= \sum_{i=1}^{N} \left[\widehat{Y}_i(\mathbf{x}) - \text{E}\widehat{Y}_i(\mathbf{x}_{\text{all}})\right]^2 \\
\sigma^2 &:= \tfrac{1}{N} \sum_{i=1}^{N} (Y_i - \text{E}Y_i)^2 \\
&= \sum_{i=1}^{N} \left[Y_i - \text{E}\widehat{Y}_i(\mathbf{x}_{\text{all}})\right]^2,
\end{aligned}$$

under the assumption that the full model yields an unbiased estimator, $\widehat{Y}_i\left(\mathbf{x}_{\text{all}}\right)$, of Y_i. That is, $\text{E}\widehat{Y}_i\left(\mathbf{x}_{\text{all}}\right) = \text{E}Y_i$. Good subsets have small $B(\mathbf{x})$.

Because $B(\mathbf{x})$ is a function of unobservable parameters, it is usually estimated by

$$H(\mathbf{x}) := \frac{\text{RSS}(\mathbf{x})}{\widehat{\sigma}^2} + 2\sum_{i=1}^{N} x_i - N,$$

[22]For example, Mallows (1973) and Weisberg (1985, p. 216).

where[23]

$$\begin{aligned} \mathrm{RSS}(\mathbf{x}) &:= \sum_{i=1}^{N} \left[\widehat{Y}_i(\mathbf{x}) - \widehat{Y}_i(\mathbf{x}_{\text{all}})\right]^2 \\ &= \text{residual sum of squares when comparing subset } \mathcal{J}(\mathbf{x}) \\ &\quad \text{with all variable inclusion} \\ \widehat{\sigma}^2 &:= \frac{1}{N-m} \sum_{i=1}^{N} \left[Y_i - \widehat{Y}_i(\mathbf{x}_{\text{all}})\right]^2. \end{aligned}$$

Because linear regression assumes that the responses $Y_1, \ldots, Y_N$ in the original data are random, and because each $H(\mathbf{x})$ is a function of these responses, each $H(\mathbf{x})$ is also random. Therefore, it is of interest to test if the differences between $H_{\min}$ and the $H(\mathbf{x})$ for other subsets are statistically significant. If not, then one of these alternative subsets may be preferable for subsequent work, especially if it contains fewer variables. Mallows (1973) describes the statistical theory for testing.

For the full model, $\mathbf{x}_{\text{all}}$ with $x_1 + \cdots + x_m = 1$, $H(\mathbf{x}_{\text{all}}) = m$. If for some $\sum_{i=1}^{m} x_i = k < m$, the coefficients of variables in the subset $\{1, \ldots, m\} \backslash \mathcal{J}(\mathbf{x})$ are 0, then $H(\mathbf{x}) \approx k$, the approximating symbol indicating that $H(\mathbf{x})$ is a random quantity. Statistical theory provides procedures for testing this hypothesis and for choosing among alternative subsets.

The variable-subset selection process in regression analysis has two major components: a criterion by which the benefits of alternative subsets can be compared and a procedure for finding the subsets that this criterion indicates are best. The search for the best fit takes one of three forms:

Forward selection	Start with a model of $p < m$ variables and attempt to add a variable. Begin with $p = 1$.
Backward elimination	Start with $p = m$ variables and attempt to delete one variable.
Stepwise	Start with $p = 1$ variable and attempt to add, delete, or exchange variables.

Each approach is iterated until its prespecified stopping rule is satisfied. In the forward-selection and stepwise procedures, a variable whose simple correlation with the responses is large in absolute value is a likely candidate for inclusion.

These well-established procedures are valuable tools for finding good subsets for small and moderate m. In total, 2^m subsets exist, including the null case $x_1 = \cdots = x_m = 0$, and at worst, total enumeration across all 2^m possible subsets is possible. However, this becomes less of an option as m increases. For example, $m = 20$ would have more than a million subsets, whereas $m = 30$ would have more than a billion. Although the availability of additional information may allow an analyst to identify potentially good subsets for statistical testing, it may not be

[23]The usual notation for this criterion is C_p, where $p = x_1 + \cdots + x_m$. The notation used here keeps the exposition consistent with that in other sections in this chapter.

sufficient to cull the subsets down to a manageable size for subsequent uses. When it is not, the techniques of this chapter can be helpful in identifying potentially good subsets. The following case study shows one way this can be done.

Liang and Wong (2000) apply an MCMC genetic algorithm to subset selection for a linear regression analysis relating 1973 Minnesota accidents per million vehicle miles (response) to the following 11 variables:

i	Z_i
1	length of segment (miles)
2	estimated average daily traffic count (thousands) estimated
3	truck volume as a percent of total volume
4	speed limit (in 1973, prior to imposition of 55 mph limit)
5	lane width (feet)
6	width (feet) of outer shoulder on the roadway
7	number of freeway-type interchanges per mile in the segment
8	number of interchanges with signals per mile in the segment
9	number of access points per mile in the segment
10	total number of traffic lanes in both directions
11	type of highway

The data set was from Weisberg (1985).

Although a complete m-variable regression in the reference gave multiple correlation 0.87, the t values corresponding to estimated coefficients were all less than 2. Values this small suggest that these coefficients do not differ from 0 at significance levels 0.01 or less, a result consistent with our earlier comment that collinearity induces large variance for the estimated coefficient. This motivated a search for subsets that would eliminate collinearity.

Variable 1 was constrained to be in all regressions so that $2^{10} = 1{,}024$ subsets are of interest for the $m = 10$ remaining variables. Although MCMC genetic algorithms and similar procedures aim at solving problems for much larger m, the present example allowed total enumeration and graphing in Figure 8.7a of all $\{H(\mathbf{x}) : \mathbf{x} \in \{0, 1\}^{10}\}$ as a check on the results of the application.

Liang and Wong (2000) performed an MCMC experiment using a genetic algorithm for $r = 5$ paths, $t = 15{,}000$ iterations, $p_{\text{mut}} = 0.25$, single-coordinate mutation, uniformly sampled M for cross, and temperature schedule

$$T_1 = 5, \quad T_i = T_{i-1} - \frac{5}{r}, \qquad i = 2, \ldots, 5.$$

In terms of reciprocal temperatures, this gave

$$\beta_1 = 0.20, \ \beta_2 = 0.25, \ \beta_3 = 0.33, \ \beta_4 = 0.50, \ \beta_5 = 1.0.$$

The sample path corresponding to $\beta_1 = 1$ contained 457 distinct values of $H(\mathbf{x})$. The sample acceptance rates were 0.59, 0.60, and 0.76 for mutations, crossovers, and swaps respectively.

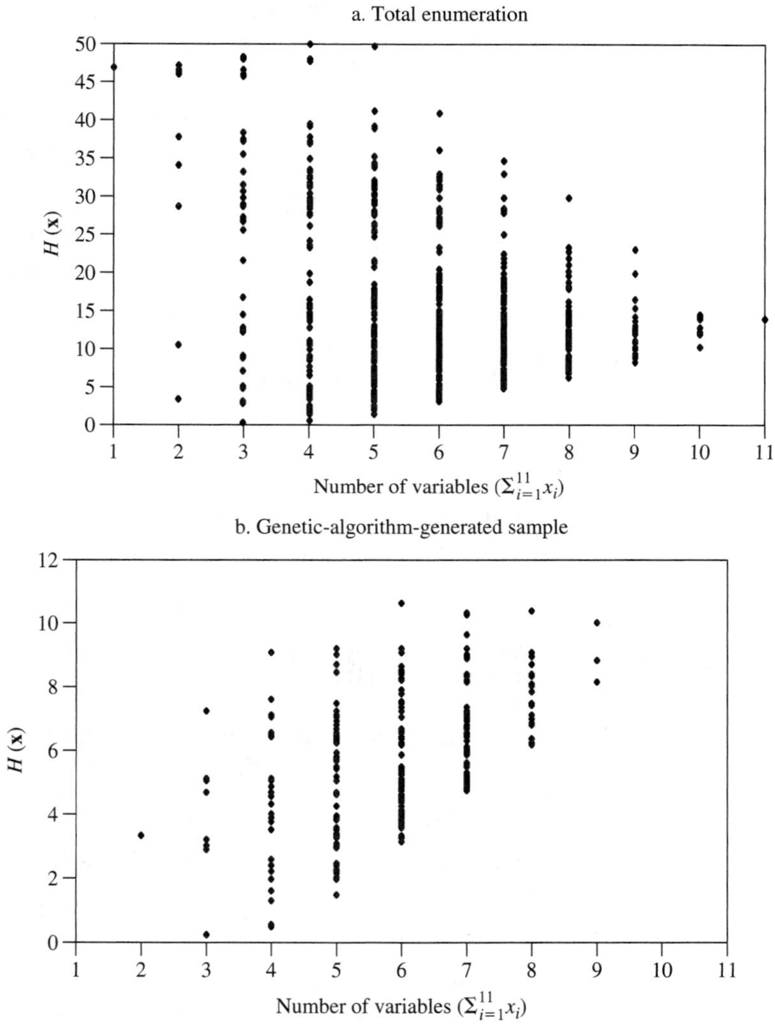

FIGURE 8.7 | *$H(\mathbf{x})$ for subsets*
Source: Liang (2005).

Figure 8.7b shows the sample $H(\mathbf{x})$ as a function of the number of variables, $x_1 + \cdots + x_m$. It contains all $H(\mathbf{x}) \leq 2$ in the complete enumeration in Figure 8.7a but none for $H(\mathbf{x}) \geq 10$. Because $\sum_{\mathbf{x}\in\mathcal{X}} \pi(\mathbf{x}) = 0.999$, where $\{\pi(\mathbf{x})\}$ uses reciprocal temperature $\beta_1 = 1$, it is clear that the sampling was highly representative. Table 8.1 lists all sample $H(\mathbf{x}) \leq 2$. Interestingly, the smallest subset corresponds to $H_{\min}$. However, we emphasize that this ordering may not hold in other subset-selection problems for which $H(\mathbf{x})$ is random.

An MCMC experiment based exclusively on swapping, analogous to that in Algorithm SWAP, was also run for the same r, t, and the given temperature schedule. It gave essentially the same results, but at a higher unit computing

TABLE 8.1 | *Best subset selection*

$H(\mathbf{x})$	ρ^2	$\mathcal{J}(\mathbf{x})$
0.23567	0.701	$\{1, 4, 9\}$
0.48521	0.718	$\{1, 4, 8, 9\}$
0.55953	0.717	$\{1, 3, 4, 9\}$
1.32905	0.710	$\{1, 4, 9, 10\}$
1.50657	0.727	$\{1, 3, 4, 8, 9\}$
1.58577	0.707	$\{1, 2, 4, 9\}$
1.62894	0.707	$\{1, 4, 7, 9\}$
1.96553	0.703	$\{1, 4, 5, 9\}$

Source: Weisberg (1985).
$\rho :=$ multiple correlation coefficient

cost per iteration. The greater cost arose from the updating of all r paths on each iteration. ■

8.10 LESSONS LEARNED

- Simulated annealing relies on MCMC with Hastings-Metropolis (HM) sampling to estimate the minimum of a function. It is relatively simple to apply, requiring little knowledge of the properties of the function being minimized. However, it can take substantial time to compute an estimate within an acceptable neighborhood of the true minimum.
- Simulated annealing requires an initial starting state, a temperature schedule, and a stopping rule as input.
- The temperature schedule specifies how long the sample paths should be at each of a sequence of successively lower temperatures.
- The stopping rule gives a criterion by which to terminate sampling and choose the estimate of the minimum.
- Starting simulated annealing at a relatively high temperature and lowering the temperature gradually tends to keep the sample path from becoming trapped in a local minimum.
- Simulated tempering, swapping, stochastic tunneling, and genetic algorithms are designed to speed up convergence to equilibrium.

8.11 HANDS-ON EXERCISES

The following exercises provide hands-on experience in solving optimization problems by simulated annealing. A solution should include the estimated optimal value, its corresponding configuration, and the number of iterations. It should also include descriptions of the initial configuration, temperature schedule, and stopping rule. When multiple runs are employed to solve a particular problem, comparisons should be made among the results. Similarities and differences should be explained.

Generating a Random Permutation. In what follows, it will sometimes be convenient to generate a random ordering of the integers $1, \ldots, L$. This may occur when choosing an initial configuration. Algorithm PERM describes how to do this.

ALGORITHM **PERM**

Purpose: To generate a random permutation of the integers $1, \ldots, N$.

Given: $I_j = j$ for $j = 1, \ldots, N$.

Method:

$l = 1$.
Until $l > N$:
randomly sample U from $\mathcal{U}(0, 1)$
$J = l + \lfloor (N - l + 1)U \rfloor$
$a = I_l, \; I_l = I_J, \; I_J = a$, and $l = l + 1$.
Return $(I_1, \ldots, I_N)$. ■

H.1. Resource Allocation.[24] An open lignite mine in Galicia, Spain, will close in the near future. The 25 × 25–square kilometer area disturbed by the mine's construction and operation originally consisted of 57% pine forest, 29% shrubs, and 14 % water. European environmental law requires the mine owners to restore the area to its original pre-mining situation, as closely as possible. Although restoration must conform to the original 57, 29, and 14% distribution, there is no requirement, for example, that an area cleared of pine forest must be restored as pine forest.

For present purposes, assume the disturbed area is a square partitioned into 100 cells of equal area. Figure 8.8 shows the square, where shading indicates the three different types of terrain. It also displays the cost per cell for each type of restoration. Past restoration experience has led to the higher unit cost for pine forest than for shrubs, because the former needs fertilizer and maintenance annually, whereas the latter requires maintenance every three years.

For $i = 1, \ldots, 10, \; j = 1, \ldots, 10$, and $k = 1, 2, 3$, let

$$x_{ijk} := \begin{cases} 1 & \text{if cell } (i, j) \text{ is restored to type } k \\ 0 & \text{elsewhere,} \end{cases}$$

[24]Based on Aerts and Heuvelink (2002).

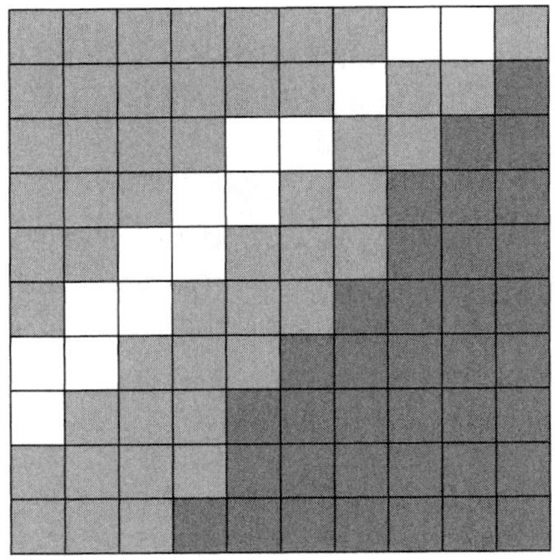

Terrain	Restoration cost: Pine forest	Shrubs	Water
	750	400	1,500
	2,700	1,450	6,000
	5,550	3,000	13,500

FIGURE 8.8 | *Terrain and restoration cost per cell*

subject to

$$x_{ij1} + x_{ij2} + x_{ij3} = 1,$$

where $k = 1$ denotes pine forest, $k = 2$ shrubs, and $k = 3$ water.

a. Formulate the cost minimization for restoring the area, explicitly listing the objective function and all constraints.

b. Formulate a simulated annealing procedure for solving the problem. One approach: on each iteration randomly select two cells and attempt to switch their types.

c. Determine an appropriate initial temperature T and justify your choice.

d. Solve the problem using a temperature schedule with $\alpha = 0.80$, $l_0 := 0$, and $l_k = l_{k-1} + 1{,}000$ for $k = 1, 2, \ldots$.

e. Repeat the exercise using a temperature schedule in (8.21) with $\alpha = 0.20$, $l_0 := 0$, and $l_k = l_{k-1} + 10{,}000$ for $k = 1, 2, \ldots$. This speeds up the rate at which temperature decreases, but at the risk of ending up in a local minimum.

f. Repeat the exercise using a temperature schedule with $\alpha = 0.80$, $l_0 := 0$, and $l_k = l_{k-1} + 10{,}000$ for $k = 1, 2, \ldots$. This reduces the risk of ending up in a local minimum but takes much longer to reach the desired solution.

g. Describe your overall conclusions. Include graphs depicting the optimal configurations for each part and a graphs depicting the sample paths of cost versus iteration number for all three parts. Also include a discussion of when convergence occurred for each part.

H.2. Exercise H.1 focuses on restoration cost. However, spatial patterns also affect the desirability of a particular configuration. For example, large compact areas of one type are more attractive than a patchy configuration of all three types. This compactness can be encouraged by rewarding configurations whose neighboring cells have the same land use. For $i = 1, \ldots, 10$; $j = 1, \ldots, 10$; and $k = 1, 2, 3$; let

$$b_{ijk} := x_{i-1,jk} + x_{i+1,jk} + x_{i,j-1,k} + x_{i,j+1,k},$$

where

$$\begin{aligned} x_{i-1,jk} &:= 0 && \text{if } i = 1 \\ x_{i+1,jk} &:= 0 && \text{if } i = 10 \\ x_{i,j-1,k} &:= 0 && \text{if } j = 1 \\ x_{i,j+1,k} &:= 0 && \text{if } j = 10. \end{aligned}$$

If $\{H_1(\mathbf{x})\}$ denotes the cost function in Exercise H.1, then for some $\beta > 0$ the revised objective function for simulated annealing is

$$H_2(\mathbf{x}) = H_1(\mathbf{x}) - \beta \sum_{i=1}^{10} \sum_{j=1}^{10} \sum_{k=1}^{3} b_{ijk} x_{ijk} \qquad \mathbf{x} \in \{0, 1\}^{100}.$$

a. Repeat parts a through g of Exercise H.1 for the revised land restoration problem for each $\beta \in \{1, 3, 5, 10\}$.

b. Comment on the relationship between reward and increased restoration cost. Explain your answer in detail, relating your results to those in Exercise H.2 for zero reward.

H.3. Traveling Salesman Problem.[25] Let $(y_1, z_1), \ldots, (y_N, z_N)$ denote N points in $\Re^2$, and let $\mathcal{P}$ denote the collection of all permutations of the integers $1, \ldots, N$. Then for some permutation $(i_1, \ldots, i_N) \in \mathcal{P}$,

$$((y_{i_1}, z_{i_1}), (y_{i_2}, z_{i_2})), \ldots, ((y_{i_j}, z_{i_j}), (y_{i_{j+1}}, z_{i_{j+1}})), \ldots, ((y_{i_{N-1}}, z_{i_{N-1}}), (y_{i_N}, z_{i_N}))$$

defines a path connecting all N points, and

$$\begin{aligned} D(i_1, \ldots, i_N) := {} & \sqrt{(y_{i_1} - y_{i_N})^2 + (z_{i_1} - z_{i_N})^2} \\ & + \sum_{j=1}^{N-1} \sqrt{(y_{i_j} - y_{i_{j+1}})^2 + (z_{i_j} - z_{i_{j+1}})^2} \end{aligned} \qquad (8.33)$$

[25] Based on Černy (1985).

defines the length of the path. It is of interest to find a permutation in $\mathcal{P}$ that minimizes $D(i_1, \ldots, i_N)$; that is, to find the shortest route for visiting all N nodes. Known as the *traveling salesman problem,* it arises in many areas under many guises. For present purposes, it is a convenient formulation for demonstrating how simulated annealing works in practice. However, more computationally efficient deterministic algorithms exist for solving this problem in practice.

a. How many permutations are there for this problem?

Suppose there are $N = 100$ independently and randomly generated sites uniformly distributed on the circle with radius 1 and centered at $(0, 0)$. The following property provides a way to generate these points.

Points Uniformly Distributed on a Circle. If W_1 and W_2 are i.i.d. from $\mathcal{N}(0, 1)$, then (Z_1, Z_2), where

$$Z_1 = \frac{W_1}{\sqrt{W_1^2 + W_2^2}} \quad \text{and} \quad Z_2 = \frac{W_2}{\sqrt{W_1^2 + W_2^2}},$$

is uniformly distributed on a circle of radius 1 and centered at $(0, 0)$.

b. Given a permutation $I_1, \ldots, I_N$, it is of interest to generate a new permutation using the following algorithm.

ALGORITHM TSP1

Purpose: To generate a new permutation in the traveling salesman problem by simulated annealing.

Given: $I_1, \ldots, I_N$.

Method:

Randomly select an integer K_1 from $\{1, \ldots, N\}$.
Randomly select an integer K_2 from $\{1, \ldots, N\}\backslash\{K_1\}$.
$a = \min(K_1, K_2)$ and $b = \max(K_1, K_2)$.

$$J_l := \begin{cases} I_l & l = 1, \ldots, a-1 \\ I_{a+b-l} & l = a, a+1, \ldots, b \\ I_l & l = b+1, \ldots, N. \end{cases}$$

$J_1, \ldots, J_N$ is the new permutation.

Describe in words how this algorithm changes the path. Calculate the probability of selecting the same two nodes as on the previous selection. ■

c. For the randomly generated points and employing the algorithm in part b, solve the traveling salesman problem using simulated annealing.

d. Explain how the constraint to be on the circle helps you determine if you have found the optimal solution. For example, what would happen if you repeated the exercise with a different set of randomly generated sites?

H.4. It is known that the inversion operation in Exercise H.3 may limit the speed of convergence (e.g., Chatterjee et al. 1996). An alternative coordinate selection rule to TSP1 is intended to ameliorate this problem by ensuring that every point gets selected in N successive iterations. At the start of sampling, the assignment $K_1 = 1$ is made. Then one executes:

ALGORITHM TSP2

Purpose: To generate a new permutation in the traveling salesman problem by simulated annealing.

Given: $I_1, \ldots, I_N$ and $K_1 = 1$.

Method: If $K_1 > N$, $K_1 = 1$.
Randomly select an integer K_2 from $\{1, \ldots, N\} \backslash \{K_1\}$.
$a = \min(K_1, K_2)$ and $b = \max(K_1, K_2)$.

$$J_l := \begin{cases} I_l & l = 1, \ldots, a-1 \\ I_{a+b-l} & l = a, \ldots, b \\ I_l & l = b+1, \ldots, N. \end{cases}$$

$J_1, \ldots, J_N$ is the new permutation.
$K_1 = K_1 + 1$. ■

a. Solve the problem posed in Exercise H.3 using Algorithm TSP2.

b. Compare the solution with that of Exercise H.3 with regard to estimated optimal solutions, final configurations, and number of iterations.

H.5. Non-Uniform Point Density. Let $(Y_1, Z_1), \ldots, (Y_{50}, Z_{50})$ denote independent random points uniformly distributed in the square

$$\mathcal{S}_1 := \{(y, z) : \; 0 \le y \le 1, \; 1.5 \le z \le 2.5\},$$

and let $(Y_{51}, Z_{51}), \ldots, (Y_{100}, Z_{100})$ denote independent random points uniformly distributed in the square $\mathcal{S}_2 := \left\{(y, z) : \frac{3}{4} \le y \le 2\frac{3}{4}, 0 \le z \le 2\right\}$. Note that the points in $\mathcal{S}_1 \cup \mathcal{S}_2$ are not uniformly distributed. In particular, they tend to be more dense in the intersection $\mathcal{S}_1 \cap \mathcal{S}_2$. It is of interest to study how this non-uniformity affects simulated annealing.

Randomly generate these points and, using Algorithm TSP2, apply simulated annealing to solve the traveling salesman problem.

Exercises H.6 through H.10 provide opportunities to experiment with enhancements to simulated annealing. They allow considerable latitude in design so as to encourage students to explore the consequences of assigning different values to parameters and, in some cases, different algorithmic logic. Comparing outcomes for these different settings for a class of students offers one way of gaining perspective about performance of the many alternative designs.

Simulated Tempering

H.6. Describe and execute an MCMC experiment based on simulated tempering for solving the resource allocation problem in Exercise H.1. Compare the solution and required computing time with those of Exercise H.1.

H.7. Describe and execute an MCMC experiment based on simulated tempering for solving the traveling salesman problem in Exercise H.3. Compare the solution and required computing time with those of Exercise H.3.

Stochastic Tunneling

H.8. Describe and execute an MCMC experiment based on stochastic tunneling for solving the resource allocation problem in Exercise H.1. Compare the solution and required computing time with those of Exercise H.1.

H.9. Describe and execute an MCMC experiment based on stochastic tunneling for solving the traveling salesman problem in Exercise H.3. Compare the solution and required computing time with those of Exercise H.3.

Genetic Algorithms

H.10. Every genetic algorithm has a population size r, a reciprocal temperature schedule $\beta_1 < \cdots < \beta_r$, and a mutation probability p_{mut}. For the resource allocation problem in Exercise H.1, select several sets of values for these parameters and apply a genetic algorithm to solve the exercise for each set using mutation and crossover operations. Compare the solutions and computing times for each set with those for Exercise H.1. Let $r_{\min}$ and $r_{\max}$ denote the smallest and largest values for r respectively. Choose the range of values so that $\frac{r_{\max}}{r_{\min}} \geq 100$.

8.12 REFERENCES

Aarts, E., and J. Korst (1989). *Simulated Annealing and Boltzmann Machines*, Wiley, New York.

Aerts, J. C. J., and G. B. M. Heuvelink (2002). Using simulated annealing for resource allocation, *Int. J. Geographical Information Science,* **16,** 571–587.

Barhen, J., V. Protopopescu, and D. Reister (1997). TRUST: A deterministic algorithm for global optimization, *Science,* **276,** 1094–1097.

Bhatnagar, N., and D. Randall (2004). Torpid mixing of simulated tempering on the Potts model, College of Computing, Georgia Institute of Technology, Atlanta.

Branden, C., and J. Tooze (1999). *Introduction to Protein Structures,* second edition, Garland, New York.

Bryant, S. H., and C. E. Lawrence (1993). An empirical energy function for threading protein sequences through the folding motif, *Proteins: Structure, Function, and Genetics,* **16,** 92–112.

Chatterjee, S., C. Carrera, and L. A. Lynch (1996). Genetic algorithms and traveling salesman problems, *European Journal of Operational Research,* **93,** 490–510.

Černy, V. (1985). Thermodynamical approach to the traveling salesman problem, *Journal of Optimization Theory and Applications,* **45,** 41–51.

Davis, L. (1991). *Handbook of Genetic Algorithms,* Van Nostrand Rheinhold, New York.

Earl, D. J., and M. W. Deem (2004). Optimal allocation of replicas to processors in parallel tempering simulations, Department of Bioengineering and Department of Physics and Astronomy, Rice University.

Falcioni, M., and M. W. Deem (1999). A biased Monte Carlo scheme for zeolite structure solution, *Journal of Chemical Physics,* **110,** 1754–1766.

Geman, S., and D. Geman (1984). Stochastic relaxation, Gibbs distributions, and the Bayesian restoration of images, *IEEE Transactions on Pattern Analysis and Machine Intelligence,* **PAMI-6,** 721–741.

Geyer, C. J., and E. A. Thompson (1995). Annealing Markov chain Monte Carlo with applications to ancestral inference, *Journal of the American Statistical Association,* **90,** 909–920.

Goldberg, D. (1989). *Genetic Algorithms in Search, Optimization, and Machine Learning,* Addison-Wesley, Reading, MA.

Hajek, B. (1988). Cooling schedules for optimal annealing, *Math. of Oper. Res.,* **13,** 311–329.

Hansmann, U. H. E. (1997). Parallel tempering algorithm for conformational studies of biological molecules, *Chemical Physical Letters,* **281,** 140–150.

Holland, J. H. (1975). *Adaptation in Natural and Artificial Systems,* The University of Michigan, Ann Arbor.

Hwang, F. K., D. S. Richards, and P. Winter (1992). *The Steiner Tree Problem,* North-Holland, Amsterdam.

Karplus, M., and G. A. Petsko (1990). Molecular dynamics simulation in biology, *Nature,* **346,** 631–639.

Kern, W. (1993). On the depth of combinatorial optimization problems, *Discrete and Applied Mathematics,* **43,** 115–129.

Kirkpatrick, S., C. D. Gelatt, Jr., and M. C. Vecchi (1983). Optimization by simulated annealing, *Science,* **220,** 671–680.

Kolinski, A., and J. Skolnick (1994a). Monte Carlo simulations of protein folding. I. Lattice model and interaction scheme, *Proteins: Structure, Function, and Genetics,* **18,** 338–352.

Kolinski, A., and J. Skolnick (1994b). Monte Carlo simulations of protein folding. II. Applications to protein A, ROP and crambin, *Proteins: Structure, Function, and Genetics,* **18,** 353–366.

Kolinski, A., and J. Skolnick (1998). Assembly of protein structure from sparse experimental data: An efficient Monte Carlo model, *Proteins: Structure, Function, and Genetics,* **32,** 475–494.

Kolinski, A., L. Jaroszewski, P. Rotkiewicz, and J. Skolnick (1998). An efficient Monte Carlo model for protein chains: Modeling the short-range correlations between side group centers of mass, *J. Phys. Chem. B,* **102,** 4628–4637.

Liang, F. (2005). Personal communication.

Liang, F., and W. H. Wong (2000). Evolutionary Monte Carlo: Applications to C_p modeling and change point problem, *Statistica Sinica,* **10,** 317–342.

Liang, F., and W. H. Wong (2001a). Evolutionary Monte Carlo for protein folding simulations, *Journal of Chemical Physics,* **115,** 3374–3380.

Liang, F., and W. H. Wong (2001b). Real-parameter evolutionary Monte Carlo with applications to Bayesian mixture models, *Journal of the American Statistical Association,* **96,** 653–666.

Lyubartsov, A. P., A. A. Martsinovski, S. V. Shevkunov, and P. N. Vorontsov-Velyaminov (1992). New approach to Monte Carlo calculation of the free energy method of expanded ensembles, *Journal of Chemical Physics,* **96,** 1776–1783.

Madras, N., and Z. Zheng (2003). On the swapping algorithm, *Random Structures Algorithms,* **22,** 66–97.

Mallows, C. L. (1973). Some comments on C_p, *Technometrics,* **15,** 661–675.

Marinari, E., and G. Parisi (1992). Simulated tempering: A new Monte Carlo scheme, *Europhysics Letters,* **19,** 451–468.

Metropolis, N., A. W. Rosenbluth, A. H. Teller, and E. Teller (1953). Equations of state calculations by fast computing machines, *Journal of Chemical Physics,* **21,** 1087–1092.

Rawn, J. D. (1989). *Biochemistry,* Neil Patterson Publishers, Burlington, N.C.

Schug, A., T. Herges, and W. Wenzel (2003). Reproducuble protein folding with the stochastic tunneling method, *Physical Review Letters,* **91,** 158102-1–158102-4.

Shimada, J., E. L. Kussell, and E. I. Shakhnovich (2001). The folding thermodynamics and kinetics of crambin using all-atom Monte Carlo simulation, *Journal of Molecular Biology,* **308,** 79–95.

Skolnick, J. (2001). Personal communication.

Spall, J. C. (2003). *Introduction to Stochastic Search and Optimization,* Wiley Interscience, New York.

Swope, W. C. (2001). Deep computing for the life sciences, *IBM Systems Journal,* **40,** 248–264.

Weisberg, S. (1985). *Applied Linear Regression,* Wiley, New York.

Wenzel, W., and K. Hamacher (1999). Stochastic tunneling approach to global minimization of complex potential energy landscapes, *Physical Review Letters,* **82,** 3003–3006.

Zheng, Z. (2003). On swapping and simulated tempering algorithms, *Stoch. Proc. Appl.,* **104,** 131–154.

CHAPTER 9

Advanced Concepts in MCMC

Whereas Chapter 6 describes the basics of MCMC, the present chapter introduces several more advanced concepts that broaden its range of application. The account is intended for those readers whose curiosity about MCMC has been aroused in earlier chapters as to what else MCMC has to offer. Topics include

- More on bounds on the convergence rate to equilibrium from an initial state (Section 9.1).
- More on bounds on $\text{var}_\pi \widehat{\lambda}_t$ (Section 9.1).
- A formulation, called a Markov random field, that facilitates sample generation by reducing the dependence on the past to only the values of coordinates in a neighborhood local to the coordinate to be updated (Section 9.3).
- How problem size (dimension) affects convergence (Section 9.5).
- A specialized form of independence nominating kernel, called slice sampling, for which uniform ergodicity holds more generally than for other sample-generating methods (Section 9.6).
- Coupling from the past, a theoretically exact methodology for starting an MCMC experiment, monitoring its sample path, and determining the first (random) iteration on which the path is in equilibrium (Section 9.7).

The descriptions of these and other topics focus more on analytical representations and less on empirical examples than in previous chapters. Although the account demands more conceptualizing on the part of the reader, the reward is immediate for the extra effort. It familiarizes the reader with a broader range of settings and more sophisticated techniques than those described earlier.

9.1 EXPLOITING REVERSIBILITY

Recall that a Markov chain satisfying

$$\pi(\mathbf{x})K(\mathbf{x},\mathbf{y}) = \pi(\mathbf{y})K(\mathbf{y},\mathbf{x}) \qquad \forall(\mathbf{x},\mathbf{y}) \in \mathcal{X}^2$$

is said to be reversible. In addition to ensuring global balance, reversibility facilitates analysis of the dependence in a chain. In particular, it can be shown (e.g., Fishman 1996a, Chapter 5) that for a finite-state space, there exist quantities $1 > \beta_1 \geq \beta_2 \geq \cdots \geq \beta_{|\mathcal{X}|-1} > -1$ and sequences $\{e_l(\mathbf{x},\mathbf{y}),\ (\mathbf{x},\mathbf{y}) \in \mathcal{X}^2\}$ for each $l = 1, \ldots, |\mathcal{X}|-1$ such that the j-step transition kernel has the representation

$$K^{(j)}(\mathbf{x},\mathbf{y}) = \pi(\mathbf{y}) + \sqrt{\frac{\pi(\mathbf{y})}{\pi(\mathbf{x})}} \sum_{l=1}^{|\mathcal{X}|-1} e_l(\mathbf{x},\mathbf{y})\beta_l^j, \qquad (\mathbf{x},\mathbf{y}) \in \mathcal{X}^2, \qquad j \geq 1, \tag{9.1}$$

where

$$\sum_{l=1}^{|\mathcal{X}|-1} e_l(\mathbf{x},\mathbf{y}) = \begin{cases} 1 & \text{if } \mathbf{x} = \mathbf{y} \\ 0 & \text{otherwise.} \end{cases}$$

As a consequence, it can be shown (Diaconis and Stroock 1991) that for all $(\mathbf{x},\mathbf{y}) \in \mathcal{X}^2$

$$\left|K^{(j)}(\mathbf{x},\mathbf{y}) - \pi(\mathbf{y})\right| \leq \frac{1}{2}\sqrt{\frac{1}{\pi(\mathbf{x})} - 1}\beta_{\max}^j \qquad j \geq 1, \tag{9.2}$$

where

$$\beta_{\max} := \max(\beta_1, |\beta_{|\mathcal{X}|-1}|).$$

Given $\mathbf{X}_{j-1}$, (9.2) provides an opportunity to assess how close $\mathbf{X}_j$ is to equilibrium. For some $\epsilon > 0$,

$$\left|K^{(j)}(\mathbf{x},\mathbf{y}) - \pi(\mathbf{y})\right| \leq \epsilon \tag{9.3}$$

holds for all

$$j \geq j^* := \frac{\frac{1}{2}\ln\left[\frac{1}{\pi(\mathbf{x})} - 1\right] - \ln\epsilon}{-\ln\beta_{\max}}. \tag{9.4}$$

This result has an immediate implication for practice. In particular, it reveals the benefit of starting in a modal state $\mathbf{x}$ for which $\pi(\mathbf{x}) = \pi_{\max}$, where

$$\pi_{\max} := \max_{\mathbf{x}\in\mathcal{X}} \pi(\mathbf{x}).$$

Doing so requires the smallest number of iterations in (9.4) to satisfy the error bound (9.3). Hereafter, we call j^* in (9.4) a *mixing time.*

Representation (9.1) also has implications for variance. For large t,

$$t\,\widehat{\text{var}\lambda_t} \approx \sigma_\infty^2 \leq R_0\frac{1+\beta_1}{1-\beta_1}.$$

Note that $\beta_{\max}$ controls the convergence rate to equilibrium, whereas β_1 controls the variance of the sample average.

As illustration of how the properties of reversible chains can be revealing, it is known that for Gibbs sampling (see Section 6.6) with random-coordinate updating on an m-dimensional countably finite state space $\mathcal{X}$, Algorithm RGIBBS induces $\beta_l \geq 0$ for all $l = 1, \ldots, |\mathcal{X}| - 1$ in (9.1) with $\beta_1 \geq 1 - \frac{1}{m}$ (Fishman 1996b). Because $\ln z \approx z$ for small z, this lower bound implies that j^* in (9.4) must grow at least proportionately to m as the dimension of the problem m grows. This explains the frequently observed slow convergence of random-coordinate Gibbs sampling for large m. Also, it can be shown that σ_∞^2 must increase at least linearly with m.

9.2 RAPID MIXING

Although the bound (9.2) characterizes convergence in terms of $\beta_{\max}$ and $\pi(\mathbf{x})$, it tells us nothing about the broader issue of convergence as a function of problem size. For present purposes, we take the dimension m as a measure of problem size. It is easily seen that unit generating time must be at least linear in m. However, size also affects convergence to equilibrium. Suppose there exist nondecreasing polynomial functions of m, $\{p_1(m) \geq 0\}$ and $\{p_2(\mathbf{x}, m)\}$ for each $\mathbf{x} \in \mathcal{X}$ such that

$$\beta_{\max} \leq 1 - \frac{1}{p_1(m)}$$

and

$$\pi(\mathbf{x}) \geq e^{-p_2(\mathbf{x}, m)}.$$

Then it is easily seen that for initial state $\mathbf{X}_0 = \mathbf{x}_0$, the mixing time (9.4) has the upper bounds

$$\begin{aligned} j^* &\leq \frac{\frac{1}{2}\ln\left(e^{p_2(\mathbf{x}_0, m)} - 1\right) - \ln(2\epsilon)}{-\ln\left[1 - \frac{1}{p_1(m)}\right]} \\ &\leq \tfrac{1}{2} p_1(m)\left[p_2(\mathbf{x}_0, m) - \ln(2\epsilon)\right], \end{aligned} \tag{9.5}$$

where the second inequality follows from $-z > \log(1 - z)$ for $0 < z < 1$.

Whenever the mixing time is bounded above by a nondecreasing polynomial in m, we say that the corresponding MCMC procedure induces a *rapidly mixing* stochastic process. Those lacking the property are called *slow mixing*. They have mixing times that increase, at least, exponentially in some function of m, severely limiting their appeal to problems with relatively few dimensions. In practice, knowing that a procedure is slow mixing encourages a potential user to search for an alternative technique that is rapid mixing.

Although rapid mixing rules out exponential growth, it does not guarantee fast convergence. In particular, the presence of higher-order terms in $\{p_1(m)\}$ and $\{p_2(\mathbf{x}, m)\}$ can substantially retard convergence as m increases. Nevertheless, the rapid-mixing characterization is invaluable for avoiding procedures that are unable to attain sufficient accuracy in any reasonable amount of computing time. Section 9.6 returns to this issue.

9.3 MARKOV RANDOM FIELDS

Updating the value of all m coordinates takes time at least proportional to m. In the absence of special structure for π, using the full conditional distributions (9.22) to update the value of a single coordinate may also require computing time proportional to m, thereby leading to a total updating time for the m coordinates at least proportional to m^2. When it applies, the neighborhood condition provides a structure that can significantly reduce this cost.

Neighborhood Condition. Let $\mathcal{M}$ denote the index set of the m coordinates, and let $\mathcal{N}_1, \ldots, \mathcal{N}_m$ denote subsets of $\mathcal{M}$ with the property that

$$i \in \mathcal{N}_l \Longleftrightarrow l \in \mathcal{N}_i \qquad \forall i, l \in \mathcal{M}.$$

If

$$\pi_i(x_i \mid x_l,\ \forall l \neq i) = \pi_i(x_i \mid x_l,\ l \in \mathcal{N}_i) \qquad \forall i \in \mathcal{M}, \tag{9.6}$$

we say that the distribution π satisfies the neighborhood condition.

Sometimes $\mathcal{N}_i$ is called the *neighborhood* of i, and $\mathcal{N} := \{\mathcal{N}_i,\ i \in \mathcal{M}\}$ the *neighborhood system* for $\mathcal{M}$. Unless otherwise stated, we take $\mathcal{M} := \{1, \ldots, m\}$. The neighborhood condition implies that the full conditional distribution (9.6) for coordinate i depends only on the values assumed by its neighbors in $\mathcal{N}_i$. As a consequence, computing time is roughly proportional to $|\mathcal{N}_i|$ for coordinate i, an especially valuable property when $\max_{i \in \mathcal{M}} |\mathcal{N}_i| \ll m$.

A distribution that satisfies the positivity (Section 6.5) and neighborhood conditions is called a *Markov random field*. The formulation provides a way of extending the notion of one-step (Markov) dependence to many dimensions while restricting conditional dependence to a local neighborhood as in (9.6). As illustration, the assumptions in the capture-recapture problem in Example 6.4.1 with state vector $(p_1, \ldots, p_r, \theta)$ indexed from $1, \ldots, r+1$ endow θ with neighborhood $\mathcal{N}_{r+1} = \{1, \ldots, r\}$ and p_i with neighborhood $\mathcal{N}_i = \{\theta\}$ for $i = 1, \ldots, r$. This implies that the generation of each p_i depends only on θ, the one coordinate in its neighborhood, whereas the generation of θ depends on all $m-1 = r$ coordinates $p_1, \ldots, p_r$. However, the neighborhood structure leads to a total generating time proportional to m for all m coordinates. Because (6.30) is everywhere positive on $[0, 1]^r \times \{S_{r+1}, S_{r+1}+1, \ldots\}$, the equilibrium distribution π also satisfies the positivity condition and, therefore, is a Markov random field.

As a second illustration, let $\mathbb{Z}$ denote the set of all integers, let $\mathcal{S} := \{\mathbf{s}_1, \mathbf{s}_2, \ldots\}$, where $\mathbf{s}_i \in \mathbb{Z}^k$ denote the locations of the vertices or *sites* of a k-dimensional infinite *cubic lattice*. Figure 9.1a shows a section for $k = 2$. For $d_{\min} := \min_{i \neq j} |\mathbf{s}_i - \mathbf{s}_j|$, we can think of this structure as a graph $\mathcal{G} = (\mathcal{S}, \mathcal{E})$ imbedded in $\Re^k$, where each edge connects two sites separated by distance $d_{\min}$. That is,

$$\mathcal{E} := \{\{\mathbf{s}, \mathbf{t}\} \in \mathcal{S} \times \mathcal{S}:\ |\mathbf{s} - \mathbf{t}| = d_{\min}\}. \tag{9.7}$$

Then

$$\mathcal{N}_{\mathbf{s}} := \{\mathbf{t} \in \mathcal{S}:\ 0 < |\mathbf{s} - \mathbf{t}| \leq d_{\min}\} \qquad \forall \mathbf{s} \in \mathcal{S}$$

exemplifies one definition of a *nearest-neighbor* neighborhood system as shown in Figure 9.1b for $k = 2$. Alternatively, for $k = 2$ and $d_{\min} = 1$,

$$\mathcal{E} := \{\{\mathbf{s}, \mathbf{t}\} \in \mathcal{S} \times \mathcal{S}: 0 < |\mathbf{s} - \mathbf{t}| \leq 1\} \cup \{\{\mathbf{s}, \mathbf{s} + (1, 1)\},\ \{\mathbf{s}, \mathbf{s} - (1, 1)\},\ \forall \mathbf{s} \in \mathcal{S}\}$$

specifies a *tetrahedral lattice*. Whereas a cubic lattice has square faces, a tetrahedral lattice has trianglular faces. The corresponding nearest-neighbor system is

$$\mathcal{N}_{\mathbf{s}} := \{\mathbf{t} \in \mathcal{S} : 0 < |\mathbf{s} - \mathbf{t}| \leq 1\} \cup \{\mathbf{s}, \mathbf{s} + (1, 1)\} \cup \{\mathbf{s}, \mathbf{s} - (1, 1)\} \qquad \forall \mathbf{s} \in \mathcal{S}, \tag{9.8}$$

as in Figure 9.1c.

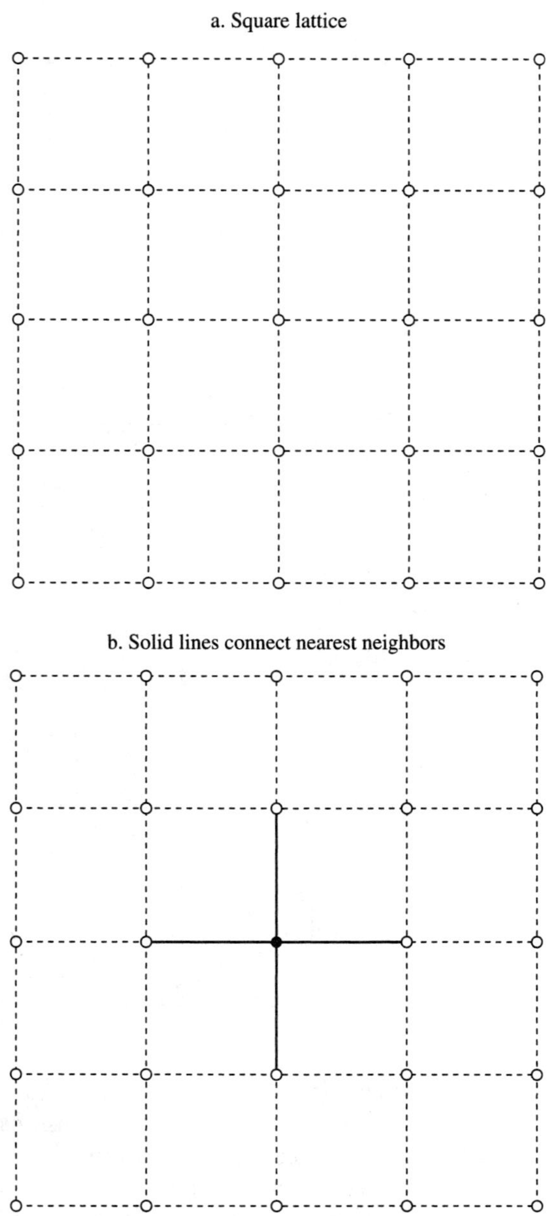

FIGURE 9.1 | *Lattices*

c. Tetrahedral lattice: Solid lines connect nearest neighbors

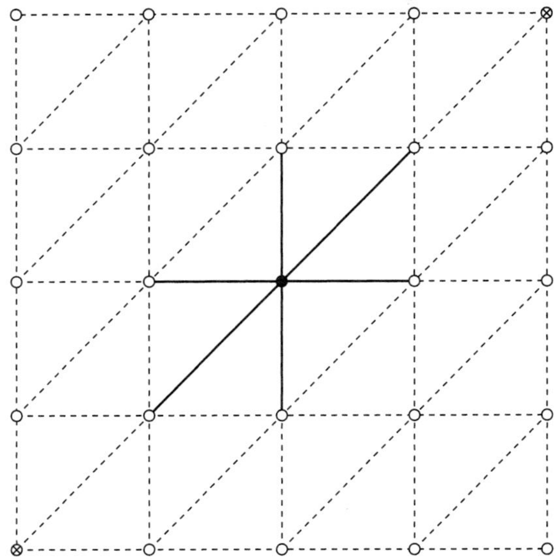

d. Solid lines connect boundary nodes

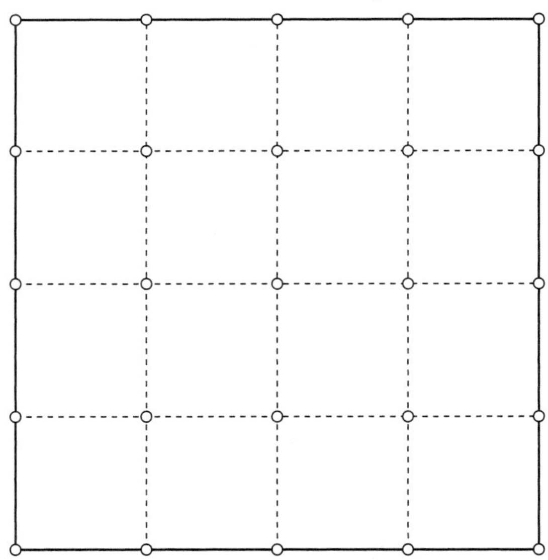

FIGURE 9.1 | *Continued*

Studies in biology, chemistry, and physics frequently employ cubic and tetrahedral lattices in two and three dimensions to model interactions among particles (e.g., Example 9.5.1). Of computational necessity, finite lattices of m sites are used to study these interactions, where m is taken sufficiently large so that insights gained from the study of this large but finite lattice are often expected to apply with negligible error to an infinite lattice. In practice, this assertion may not be true unless specified conditions are satisfied. Section 9.5.1 returns to the issue. This is best seen within the context of Gibbs distributions.

9.3.1 Gibbs Distribution

To facilitate modeling, Markov random fields rely on the concept of a *clique.*

Definition. A clique $\mathcal{C}$ denotes a subset of the index set $\mathcal{M} := \{1, \ldots, m\}$ for which every pair of distinct elements are neighbors in the sense that for all $i \neq l$ and $i, l \in \mathcal{C}$, i belongs to neighborhood $\mathcal{N}_l$, and l to neighborhood $\mathcal{N}_i$. In addition, every singleton $i \in \mathcal{M}$ is a clique.

We now use this concept to describe the representation that the positivity and neighborhood conditions induce.

Theorem 9.3.1

If π obeys the positivity and neighborhood conditions, is bounded from above, and $\varphi < \infty$ as in (6.2), then there exists a nonnegative function $\{H(\mathbf{x}),\ \mathbf{x} \in \mathcal{X}\}$ such that

$$\pi(\mathbf{x}) = \varphi^{-1} \mathrm{e}^{-H(\mathbf{x})} \qquad \mathbf{x} \in \mathcal{X}, \tag{9.9}$$

where

$$\varphi := \sum_{\mathbf{y} \in \mathcal{X}} \mathrm{e}^{-H(\mathbf{y})}, \tag{9.10}$$

$$H(\mathbf{x}) := \sum_{\mathcal{C} \in \mathcal{D}} J_{\mathcal{C}}(\mathbf{x}),$$

$\mathcal{D}$ denotes the set of all cliques of $\mathcal{M}$, and $J_{\mathcal{C}}(\mathbf{x})$ is a function only of those x_s in $\mathbf{x} := (x_1, \ldots, x_m)$ for which $s \in \mathcal{C}$. ■

Theorem 9.3.1 is called the Hammersley-Clifford theorem. Besag (1974) and Grimmett (1973) provide proofs for finite $\mathcal{X}$. The boundedness of φ suffices for the theorem to hold for continuous- and countably infinite-state spaces.

Expression (9.9) is called the *Gibbs distribution.* It originated in statistical physics where:

- $\{H(\mathbf{x})\}$ denotes the *energy function.*
- φ denotes the *partition function.*
- For all $\mathcal{C} \in \mathcal{D}$, $\{J_{\mathcal{C}}(\mathbf{x}) := J_{\mathcal{C}}(x_s : s \in \mathcal{C})\}$ denotes a family of functions indexed on $\mathcal{C}$ and is called the *potential* or *potential function.*
- $\mathcal{X}$ denotes the set of all possible equilibrium states or configurations for interacting bodies in a physical system. These include a ferromagnet, ideal gas, and binary alloy.

Occasionally, $aH(\mathbf{x})/\tau$ replaces $H(\mathbf{x})$, where a denotes the Boltzmann constant and τ denotes absolute temperature. In particular, Chapter 8 makes τ explicit in (8.3) when describing simulated annealing.

The Hammersley-Clifford theorem also provides an important insight into the relationship between the joint distribution π and its full conditional distributions (6.30). Because

$$\pi(\mathbf{x}) = \pi_i(x_i \mid x_l,\ l \in \mathcal{N}_i)\pi_i(\mathbf{x}_{-i})$$

and because for each $i \in \mathcal{M}$,

$$\pi(\mathbf{x}) = \varphi^{-1} \times \mathrm{e}^{-\sum_{\mathcal{C} \ni i} J_{\mathcal{C}}(x_s:\ s\in\mathcal{C})} \times \mathrm{e}^{-\sum_{\mathcal{C} \not\ni i} J_{\mathcal{C}}(x_s:\ s\in\mathcal{C})},$$

where $\mathcal{C} \ni i$ denotes all sets $\mathcal{C}$ that include element i and where the rightmost term is independent of x_s for all $s \in \mathcal{N}_i \cup \{i\}$, we have

$$\pi_i(x_i \mid x_l \in \mathcal{N}_i) = \varphi_i^{-1} \mathrm{e}^{-\sum_{\mathcal{C} \ni i} J_{\mathcal{C}}(x_s:\ s\in\mathcal{C})}, \tag{9.11}$$

where

$$\varphi_i := \sum_{x_i \in \mathcal{X}_i} \mathrm{e}^{-\sum_{\mathcal{C} \ni i} J_{\mathcal{C}}(x_s:\ s\in\mathcal{C})}.$$

Therefore, the alternative but equivalent *product form*

$$\pi(\mathbf{x}) = \varphi^{-1} \prod_{\mathcal{C}} \mathrm{e}^{-J_{\mathcal{C}}(\mathbf{x})} \tag{9.12}$$

reveals that the full conditional distributions contain all the information in the joint distribution up to multiplicative constants. Expression (9.12) is a particularly convenient form for analysis, especially in the study of stochastic networks and reversibility (e.g., Kelly 1979).

Many problems amenable to MCMC sampling admit Gibbs representations for their equilibrium distributions. As illustration, the capture-recapture problem in Example 6.4.1, with continuous $\mathcal{X}$, has cliques

$$\mathcal{C}_l = \begin{cases} \{l\} & \text{if } l = 1, \ldots, r+1 \quad \text{(singletons)} \\ \{l-r-1, r+1\} & \text{if } l = r+2, \ldots, 2r+1 \quad \text{(pairs)} \end{cases} \tag{9.13}$$

and potential

$$J_{\mathcal{C}_l}(\mathbf{x}) = \begin{cases} (N_l + a - 1)\ln x_l - (N_l - b + 1)\ln(1 - x_l) & \text{if } l = 1, \ldots, r \\ \sum_{k=x_l - S_{r+1}+1}^{x_l} \ln k & \text{if } l = r+1 \\ x_{r+1} \ln(1 - x_{l-r-1}) & \text{if } l = r+2, \ldots, 2r+1, \end{cases}$$

where $a, b, N_1, \ldots, N_r, S_{r+1}$ are defined in the example. In addition to the positivity condition in Section 6.5 and the neighborhood condition in Section 9.3, the concept of *translation invariance* or *stationarity* characterizes many Gibbs distributions. When applicable, it reduces all full conditional distributions to a single functional form.

Stationarity Condition. If all coordinates are designated in a common unit, then for all $i \in \mathcal{M}$, $\{\pi_i(x_i \mid x_l,\ l \in \mathcal{N}_i)\}$ is a function of the values $\{x_l,\ l \in \mathcal{N}_i \cup \{i\}\}$ and independent of the particular site or coordinate i, and the distribution π is said to satisfy the *stationarity condition*. That is, the full conditional distributions are identical for all sites. The stationarity property becomes especially important when the space of interest is a regular lattice (e.g., see Section 9.5.1).

9.4 POTTS MODEL

The Potts model is a Gibbs distribution often useful for analyzing problems, especially those concerned with *clustering*. Consider a graph $\mathcal{G} = (\mathcal{M}, \mathcal{E})$ with node set $\mathcal{M} = \{1, \ldots, m\}$, edge set $\mathcal{E} \subseteq \{\{i, j\} \in \mathcal{M} \times \mathcal{M}\}$, and neighborhoods

$$\begin{aligned}\mathcal{N}_i &= \{l \in \mathcal{M} : \{i, l\} \in \mathcal{E}\} \\ &= \text{all nodes that share an edge with node } i \qquad i \in \mathcal{M}.\end{aligned}$$

The *standard Potts model* assumes the Gibbs representation (9.9) with potential

$$J_{\mathcal{C}}(\mathbf{x}) := \beta\delta(\mathbf{x}, \mathcal{C}), \qquad \mathcal{C} \in \mathcal{D}, \tag{9.14}$$

where $J_{\mathcal{C}}$ is a function only of those x_s in $\mathbf{x}$ for which s belongs to the clique $\mathcal{C}$ defined by problem context, $\mathcal{D}$ denotes the collection of all cliques, and

$$\delta(\mathbf{x}, \mathcal{C}) := \begin{cases} 1 & \text{if all } x_s \text{ are equal for } s \in \mathcal{C} \\ 0 & \text{otherwise} \end{cases} \qquad \mathcal{C} \in \mathcal{D},$$

where $-\infty < \beta < \infty$ and x_i takes values in a finite set $\mathcal{X}_0$ for each node $i \in \mathcal{M}$. For example, if $\mathcal{D} = \mathcal{E}$, then each clique $\mathcal{C}$ is an edge. The corresponding partition function (9.10) takes the form

$$\begin{aligned}\varphi &= \sum_{\mathbf{x}\in\mathcal{X}} \exp\left[-\beta \sum_{\mathcal{C}\in\mathcal{D}} \delta(\mathbf{x}, \mathcal{C})\right] \\ &= \sum_{\mathbf{x}\in\mathcal{X}} \prod_{\mathcal{C}\in\mathcal{D}} [1 + w\delta(\mathbf{x}, \mathcal{C})],\end{aligned} \tag{9.15}$$

where

$$\delta(\mathbf{x}, \mathcal{C}) := \begin{cases} 1 & \text{if } \mathbf{x} \in \mathcal{C} \\ 0 & \text{otherwise,} \end{cases}$$

$\mathcal{X} = \mathcal{X}_0^m$ and $w := \mathrm{e}^{-\beta} - 1$ (see Exercise 1).

Suppose there are d cliques $\mathcal{C}_1, \ldots, \mathcal{C}_d$. Let $\mathcal{L} = \{1, \ldots, d\}$, and for $i = 1, \ldots, d$, let $\mathcal{L}_i$ denote the set of all permutations of i elements of $\mathcal{L}$. Then direct expansion of (9.15) yields the *dichromatic polynomial* (see Exercise 2)

$$\varphi = |\mathcal{X}| + \sum_{s=1}^{d} a_s w^s,$$

where

$$a_s := \sum_{(l_1,\ldots,l_s)\in\mathcal{L}_s} \sum_{\mathbf{x}\in\mathcal{X}} \prod_{v=1}^{s} \delta(\mathbf{x}, \mathcal{C}_{l_v}).$$

Partition functions of this form are a topic of considerable interest in graph theory. For example, suppose that $\mathcal{D} = \mathcal{E}$ and that the elements of $\mathcal{X}_0$ denote distinct colors. If $\beta = \infty$, then $w = -1$ so that φ denotes the number of ways of coloring the vertices of the planar graph $\mathcal{G}$ with $|\mathcal{X}_0|$ colors such that no two vertices connected by an edge have the same color. The well-known *four-color problem* of graph theory (i.e., $|\mathcal{X}_0| = 4$) fits within this setting.[1]

9.4.1 Random Cluster Model

One particularly interesting extension of the Potts model in Gray (1994) has $\mathcal{D} = \mathcal{M} \cup \mathcal{E}$, so that both nodes and edges are cliques, and defines the partition function as

$$J_{\mathcal{C}}(\mathbf{x}) := \begin{cases} \alpha_i(x_i) & \text{if } \mathcal{C} = \{i\} \in \mathcal{M} \\ \beta_{\{i,j\}}[1 - \delta(x_i - x_j)] & \text{if } \mathcal{C} = \{i, j\} \in \mathcal{E}. \end{cases} \tag{9.16}$$

Then $\pi(\mathbf{x})$ assumes the form

$$\pi(\mathbf{x}) = \frac{1}{\varphi} \prod_{i=1}^{m} e^{-\alpha_i(x_i)} \prod_{\{i,j\}\in\mathcal{E}} [e^{-\beta_{\{i,j\}}} + (1 - e^{-\beta_{\{i,j\}}})\delta(x_i - x_j)], \tag{9.17}$$

where

$$\varphi = \sum_{\mathbf{x}\in\mathcal{X}} \prod_{i=1}^{m} e^{-\alpha_i(x_i)} \prod_{\{i,j\}\in\mathcal{E}} [e^{-\beta_{\{i,j\}}} + (1 - e^{-\beta_{\{i,j\}}})\delta(x_i - x_j)]. \tag{9.18}$$

Once again, regarding $\mathcal{X}_0$ as a collection of distinct colors provides a convenient way to characterize the problem. In particular, each $\mathbf{x}$ denotes a coloring configuration in which node i has color x_i.

Suppose the objective is to evaluate λ in (2.1) for π in (9.17) for some bounded function $\{g(\mathbf{x}) = \pi(\mathbf{x})h(\mathbf{x}),\ \mathbf{x} \in \mathcal{X}_0^m\}$. For example, for colors y_1 and y_2 in $\mathcal{X}_0$,

$$h(\mathbf{x}) := \sum_{i\in\mathcal{M}} [\delta(x_i - y_1) + \delta(x_i - y_2)]$$

[1]For more details regarding Potts models in their classical graph-theoretic setting, see Baxter (1982) and Wu (1982). The Potts model serves as the basis for function minimization as in Chapter 8.

denotes the number of nodes with color y_1 or y_2, and

$$\lambda := \sum_{\mathbf{x}\in\mathcal{X}_0^m} \pi(\mathbf{x})h(\mathbf{x}) \tag{9.19}$$

denotes the mean number of nodes colored either y_1 or y_2. Because total enumeration would take $|\mathcal{X}_0|^m$ steps, this evaluation is computationally intractable for as few as two colors and $m = 100$ nodes or sites. To estimate λ via MCMC requires the ability to sample colorings randomly from π. The next section addresses this topic with regard to single-coordinate updating and describes its limitations. Section 9.6.1 describes an alternative sampling method that overcomes some of these limitations.

9.5 PROBLEM SIZE

In a problem modeled as a Markov random field, the index set $\mathcal{M}$ is frequently large, leading to methods of analysis that exploit its neighborhood properties to reduce computing time. Nowhere is this more true than in MCMC sampling. In particular, the availability of the full conditional distributions, with evaluations required only within neighborhoods as in (9.11), has made single-coordinate updating via Gibbs sampling a convenient way to generate sample paths. Because the ratio

$$\frac{\pi(\mathbf{x}_i(y))}{\pi(\mathbf{x})} = e^{-\sum_{\mathcal{C}\ni i}[J_{\mathcal{C}}(\mathbf{x}_i(y))-J_{\mathcal{C}}(\mathbf{x})]} \qquad i = 1, \ldots, m$$

reduces evaluation of the acceptance probability in (6.15) to only those cliques that contain site i, the neighborhood formulation also makes single-coordinate updating via Hastings-Metropolis sampling an appealing way to generate sample paths. As illustration, the Potts model has

$$\frac{\pi(\mathbf{x}_i(y_i))}{\pi(\mathbf{x})} = e^{-[\alpha_i(y_i)-\alpha_i(x_i)]} \prod_{l\in\mathcal{N}_i} e^{-\beta_{\{i,l\}}[\delta(x_i-x_l)-\delta(y_i-x_l)]}.$$

Although single-coordinate updating leads to computational efficiency, it also affects dependence among elements on a sample path. In particular, dependence tends to grow with m and $s = |\mathcal{X}_0|$. This increase slows convergence to equilibrium from arbitrarily chosen starting states and increases the sample-path length t required to meet a specified statistical accuracy for sample averages computed from sample-path data. For example, relying on single-coordinate updating on each successive step leaves $m - 1$ coordinate values unchanged. For large m, this severely limits movement through the sample space, even when the m coordinates are independent. When the m coordinates are dependent, as in the present context, the retardation of movement becomes potentially more acute.

In addition to m and s, the forms of the potential $\{J_{\mathcal{C}},\ \mathcal{C} \in \mathcal{D}\}$ and, in particular, its neighborhood system affect dependence. Therefore, it is of interest to familiarize ourselves, at least roughly, with the way in which all these concepts influence our ability to obtain meaningful estimates via single-coordinate updating MCMC sampling. We begin with an example.

EXAMPLE 9.5.1

ISING MODEL

Let the graph $\mathcal{G} = (\mathcal{S}, \mathcal{E})$ denote a finite two-dimensional $r \times r$ cubic lattice with vertex set $\mathcal{S} = \{\mathbf{s}_1, \ldots, \mathbf{s}_m\}$ in $\Re^k$ of size $m = r^2$ and edge set $\mathcal{E}$ of size $|\mathcal{E}| = 2r(r-1)$ as in (9.7). As before, we define $\mathcal{M} := \{1, \ldots, m\}$ as one-to-one with $\mathcal{S}$ so that $\mathcal{E} \subseteq \mathcal{M} \times \mathcal{M}$. Figure 9.1a illustrates the concept with $r = 5$. Define the potential as

$$J_C(\mathbf{x}) := \begin{cases} \alpha x_i & \text{if } C = \{i\} \in \mathcal{M} \\ \beta x_i x_j & \text{if } C = \{i, j\} \in \mathcal{E} \\ 0 & \text{elsewhere,} \end{cases}$$

where $x_i \in \mathcal{X}_0 = \{-1, 1\}$ for all $i \in \mathcal{M}$, $\alpha, \beta > 0$, and define the neighborhood for vertex or site i as $\mathcal{N}_i := \{j : \{i, j\} \in \mathcal{E}\}$, one form of the nearest-neighbor concept of neighborhood (e.g., Figure 9.1b). Then $\mathcal{X} = \mathcal{X}_0^m$ is the configuration space, and

$$\pi(\mathbf{x}) = \varphi^{-1} e^{-\alpha \sum_{i \in \mathcal{M}} x_i - \beta \sum_{\{i,j\} \in \mathcal{E}} x_i x_j} \qquad \mathbf{x} \in \mathcal{X} \tag{9.20}$$

is a special case of the *Ising model*. This model has long been used in the sciences as an elementary representation of how interacting particles arrayed on the vertices of a cubic lattice with edges connecting nearest neighbors can affect magnetization (e.g., Cipra 1987). It is also used as an elementary model of gas and water molecules.

With regard to computation, it leads to

$$\frac{\pi(\mathbf{x}_i(y))}{\pi(\mathbf{x})} = e^{-\alpha(x_i - y) - \beta(y - x_i) \sum_{j \in \mathcal{N}_i} x_j} \qquad y \in \mathcal{X}_0, \ \mathbf{x} \in \mathcal{X}$$

and

$$\pi(\mathbf{x}_i(y)) \mid x_j, \ j \in \mathcal{N}_i) = \frac{e^{-y(\alpha + \beta \sum_{j \in \mathcal{N}_i} x_j)}}{e^{-\alpha + \beta \sum_{j \in \mathcal{N}_i} x_j} + e^{\alpha + \beta \sum_{j \in \mathcal{N}_i} x_j}} \qquad y \in \mathcal{X}_0,$$

both particularly appealing expressions for single-coordinate updating.

Let $\mathcal{M}_I$ denote the collection of interior sites of the lattice, and let $\mathcal{M}_B$ denote the collection of sites on the boundary (e.g., Figure 9.1d), so that $\mathcal{M} = \mathcal{M}_I \cup \mathcal{M}_B$, where $|\mathcal{M}_B| = 4(r-1)$ and $|\mathcal{M}_I| = (r-2)^2$. Let $\mathbf{Y} := (Y_1, \ldots, Y_m)$ denote a random vector with distribution π as in (9.20) and recall that Y_i given $\{Y_j, \ j \in \mathcal{N}_i\}$ is independent of all $\{Y_l, \ l \in \mathcal{M} \setminus \mathcal{N}_i\}$. However, the (unconditional) correlations

$$\eta_{il} := \text{corr}(Y_i, Y_l) \qquad \forall i \neq l \in \mathcal{M}$$

are not necessarily 0. In particular, the stationarity condition holds for all interior sites so that $\eta_{il} = \eta$ for some $\eta \in [-1, 1]$ for all $l \in \mathcal{N}_i$ and $i \in \mathcal{M}_I$. It is known that for fixed α, η is an increasing function of positive β.

How does β affect the correlation between sites that do not form cliques? For example, consider the correlation between the two extremal interior sites in Figure 9.1d. Call them $i^*(m)$ and $l^*(m)$. As the size (m) of the lattice grows, the euclidean distance between these sites increases. How does this affect $\eta_{i^*(m)l^*(m)}$?

Let $\beta_c := \frac{1}{2}\ln\left(1+\sqrt{2}\right) \approx 0.4407$. For $\alpha = 0$, it can be shown that for i^* and l^*, the nearest neighbor correlations $\eta \to \frac{1}{\sqrt{2}} = 0.7071\ldots$ as $m \to \infty$, whereas

$$\eta_{i^*(m)l^*(m)} \to \begin{cases} 0 & \text{if } \beta < \beta_c \\ \eta^* & \text{if } \beta > \beta_c \end{cases} \qquad \text{as } m \to \infty, \tag{9.21}$$

where $\eta^* \to 1$ as $\beta \to \infty$ (Pickard 1987). More generally, the limits (9.21) hold for η_{il} for every pair of sites i and l for which the distance between them increases as $m \to \infty$. Therefore, the choice of β determines whether values at sites that are far away from each other are uncorrelated or correlated. For $\beta < \beta_c$, it can be shown that $\eta_{il} \to 0$ at an exponential rate for any two sites i and l as the distance between them increases.

It is known that for an infinite two-dimensional cubic lattice, a single-coordinate updating transition kernel that is everywhere positive on $\mathcal{X} \times \mathcal{X}$ and reversible induces a unique solution π to the global balance equations (6.3) for $\alpha = 0$ and $\beta < \beta_c$. By contrast, $\alpha = 0$ and $\beta > \beta_c$ admit two solutions. The existence of more than one solution is called *phase transition* and, therefore, is sometimes used to mean nonergodic. For $\alpha \neq 0$, phase transition does not occur, so the chain is ergodic. The critical values of β_c are unknown for $k > 2$. However, it is known that phase transition leads to an infinite number of solutions to the global balance equations. (See Liggett 1985, p. 201.)

For single-coordinate updating via Gibbs sampling with randomized coordinate selection on each step (RCS rule in Section 6.1) and $\beta < \beta_c$, the Markov chain is uniformly ergodic, where the bound in (6.57) has $M^* = c_1 m$ and $\rho = e^{-c_2/m}$ for positive constants c_1 and c_2 independent of m (Frigessi, et al. 1997). This result also holds for single-coordinate updating via Hastings-Metropolis sampling if on each pair of successive iterations $2j-1$ and $2j$, the same randomly selected coordinate is updated (Frigessi, et al. 1993). This is not to suggest that this sampling procedure be employed in preference to independent randomized coordinate selection on each step.

This convergence rate enables us to identify a mixing time j^* such that on any step $j \geq j^*$, the influence of intial conditions is within a specified error tolerance. In particular, for $\epsilon > 0$ and for all $(\mathbf{x}, \mathbf{y}) \in \mathcal{X}^2$,

$$|K^{(j)}(\mathbf{x}, \mathbf{y}) - \pi(\mathbf{y})| \leq \epsilon \qquad \forall j \geq j^*,$$

where $j^* := c_2^{-1}[m \ln m + m \ln(c_1/\epsilon)]$. As m increases for fixed $\beta < \beta_c$, j^* grows as $m \ln m$, the same rate at which the mean number of steps to visit all m sites increases for this sampling method. ■

9.5.1 More General Models

For the more general problems, dependence among sites remains the central issue but is more difficult to characterize. Therefore, our discussion concentrates on what can be said about convergence without tying the discussion explicitly to dependence among sites.

Let $\mathbb{Z}$ denote the set of all integers, $\mathcal{S}$ a subset of $\mathbb{Z}^k$ with $|\mathcal{S}| = m$, $\mathcal{S}_B$ the collection of sites on the boundary of $\mathcal{S}$, $\mathcal{S}_I = \mathcal{S} \backslash \mathcal{S}_B$ the collection of interior sites, and $\mathcal{X} = \mathcal{X}_0^m$ a product space from which each site in $\mathcal{S}$ takes its value. Also, assume the properties:

i. The energy function H has a finite upper bound, implying the positivity condition for π everywhere on $\mathcal{S}_I$.

ii. Each neighborhood $\mathcal{N}_i$ has a constant range independent of m.

iii. The stationarity condition holds for all interior sites in $\mathcal{S}_I$.

iv. Single-coordinate updating is employed on each step.

v. The one-step transition kernel K is reversible.

vi. Coordinate values are fixed at sites on the boundary.

Reversibility (property v) holds for Hastings-Metropolis sampling. Using the RCS rule ensures it for Gibbs sampling. Note that each element in $\mathcal{M} = \{1, \ldots, m\}$ in our earlier notation corresponds one-to-one to a k-dimensional element in $\mathcal{S}$.

In the biological, chemical, and physical sciences, stationarity is an essential property for meaningful inquiry into many problems. For example, the interactions between a particle at one particular site and particles at other sites in its neighborhood must obey the laws of physics independently of location. That is, they need to be translation invariant. Although an infinite lattice provides a convenient concept for accommodating this stationarity condition, of necessity MCMC sampling relies on finite lattices. As a consequence, the same laws do not hold at all sites. As illustration, recall the finite lattice in Figure 9.1d, where for $d_{\min} = 1$ each site in the interior set $\mathcal{M}_I$ with neighborhood system (9.8) has four neighbors, whereas each site in the boundary set $\mathcal{M}_B$ has either two or three neighbors. Hence, the sites in the boundary set differ in character from those in the interior set.

A common method of analysis resolves this issue by fixing the values of all sites in $\mathcal{M}_B$ (property vi) and applies MCMC sampling to all sites in $\mathcal{M}_I$. For small lattices, the procedure leads to results for sites in $\mathcal{M}_I$ that depend on the values assigned to the sites in $\mathcal{M}_B$, sometimes heavily so. One way to reduce the severity of this problem is to work with a finite lattice of sufficient size so that the influence of the fixed boundary values is negligible on those at sites in $\mathcal{M}_I$. Recall that an $r \times r$ cubic lattice has $(r-2)^2$ interior sites and $4(r-1)$ boundary sites so that the relative proportion of boundary sites is of order $\frac{1}{r} = \frac{1}{\sqrt{m}}$.

Because it is not known a priori how big the lattice must be before the influence of boundary values becomes relatively negligible, one approach

1. Performs a sequence of MCMC sampling experiments on successively larger lattices (increasing m).

2. Compares interior-site results for the successive lattices.

3. Ceases iteration on m once the successive results show negligible differences.

4. Uses the results for the largest lattice to characterize the infinite lattice.

The efficacy of this approach relies critically on ergodicity. In particular, it assumes that ergodicity holds for each finite m so that one and only one equilibrium distribution π_m exists. Because single-coordinate updating with randomized coordinate selection induces an everywhere positive one-step transition kernel K on the finite set $\mathcal{X} = \mathcal{X}_0^m$, the ergodic property holds for finite m. However, the validity of the approach also relies on the assumption that the sequence of these equilibrium distributions $\{\pi_m\}$ converges to a unique limiting distribution π as $m \rightarrow \infty$. Because $\mathcal{X}$ is countably infinite in the limit, the ergodic property does not necessarily follow from the property for finite m. The Ising model with $\beta > \beta_c$ exemplifies this.

Although necessary and sufficient conditions are known for ergodicity as $m \rightarrow \infty$, their forms limit their use in practice (e.g., Liggett 1985, Stroock and Zegarliński 1995). Nevertheless, they do provide a basis for studying the influence of size (m) on convergence rates. As a practical matter, it is always advisable to vary the initial conditions on successive experiments that increase m. If ergodicity holds in the limit, then convergence w.p.1 to the same π occurs, regardless of initial conditions.

Sometimes other motivations exist for growing the lattice on successive experiments. For example, although image analysis via MCMC sampling does not require stationarity, the lattice must be sufficiently dense in points to make image resolution acceptable for interpretation. Growing the lattice on successive experiments provides a basis for determining when adequate resolution has been achieved.

As size increases, convergence rates decrease, thereby requiring longer warm-up intervals to dissipate the influence of initial conditions. This becomes especially important when deciding whether it is computationally reasonable to employ the MCMC approach. For a stationary Markov random field with $\mathcal{S} \in \mathbb{Z}^k$, countable-state space $\mathcal{X}_0$, and single-coordinate updating, sufficient conditions are known for π to ensure uniform ergodicity (e.g., Stroock and Zegarliński 1995). These conditions set the stage for studying convergence.

For an arbitrarily selected initial state $\mathbf{x} \in \mathcal{X}$ and Gibbs sampling using the RCS rule, there exist sufficient conditions that ensure that uniform ergodicity (6.57) holds with $M^* = c_1 m$ and $\rho = \mathrm{e}^{-c_2/m}$, where c_1 and c_2 are positive and independent of m (Frigessi, et al. 1997). Therefore, for some $\epsilon > 0$,

$$|K^{(j)}(\mathbf{x}, \mathbf{y}) - \pi(\mathbf{y})| \leq \epsilon \qquad \forall j \geq j^* \tag{9.22}$$

for all $(\mathbf{x}, \mathbf{y}) \in \mathcal{X}^2$, where the mixing time $j^* := c_2^{-1}[m \ln m + m \ln(c_1/\epsilon)]$ and c_1 and c_2 are positive and independent of m (see Exercise 3). Thus, the MCMC procedures are rapidly mixing in m (see Section 9.2) when applied to the Potts model that satisfies the assumption of this section.

For an initial state $\mathcal{X}_0 = \mathbf{x}$ randomly drawn from the distribution $\{\pi_0(\mathbf{z}) := \frac{1}{m}, \mathbf{z} \in \mathcal{X}\}$, results in Kira and Ji (1997) for Gibbs sampling using the RCS rule indicate that if an analogous sufficient condition holds, then j^* increases as $ms \ln(ms^2)$

as $m \to \infty$ and $s \to \infty$, where $s := |\mathcal{X}_0|$ so that rapid mixing holds in s as well as in m. This last result provides a theoretical justification for the slow convergence rates observed in practice for many problems with single-coordinate updating and large m and s.

If $s > 6m$ in a Potts model on the k-dimensional cubic lattice with potential as in (9.17) with $\alpha_i(\mathbf{x}) = 0$ for all $i \in \mathcal{M}$ and nearest-neighbor neighborhoods, then the sufficient condition of Kira and Ji is satisfied and uniform ergodicity holds. However, the conditions in both references are difficult to verify in more general models, limiting their practical use.

9.6 SLICE SAMPLING

Multiple-coordinate updating via Hastings-Metropolis sampling using an independence nominating kernel as in the appendix to Chapter 6 (Section 6.18) provides one attempt to overcome the slow movement through the sample space that single-coordinate updating techniques induce. The efficacy of that approach relies on the user's ability to choose a kernel that has a low unit generating cost and that ensures a relatively high acceptance rate.

Slice sampling (SS) offers an alternative multiple-coordinate updating methodology for which uniform ergodicity holds provided that a specified boundedness condition holds. Swendsen and Wang (1987) describe the approach for generating sample paths for the Ising model. Its most notable feature is that phase transition ($\beta > \beta_c$) does not affect the statistical properties of sample paths in the manner described in Section 9.5.1, thereby overcoming a major limitation of single-coordinate updating techniques. Edwards and Sokal (1988) extend the formulation to Potts models. Besag and Green (1993) describe the concept in more general terms, and Mira and Tierney (1999) describe many of its statistical properties for general- as well as finite-state space models. Sometimes the approach is called the *method of auxiliary variables*.[2] For expository convenience, we focus on the finite-state case.

Suppose that π admits the factorization

$$\pi(\mathbf{x}) = c\pi'(\mathbf{x})\psi(\mathbf{x}), \tag{9.23}$$

where ψ is a positive bounded function,

$$\begin{aligned} \pi'(\mathbf{x}) &> 0 \quad \forall \mathbf{x} \in \mathcal{X}, \\ \textstyle\sum_{\mathbf{z}\in\mathcal{X}} \pi'(\mathbf{z}) &= 1, \end{aligned}$$

and

$$c := \frac{1}{\sum_{\mathbf{z}\in\mathcal{X}} \pi'(\mathbf{z})\psi(\mathbf{z})}.$$

[2]Reciprocal temperature in Section 8.5 is an auxiliary variable.

Let U, given $\mathbf{X} = \mathbf{x}$, be a random variable uniformly distributed on $[0, \psi(\mathbf{x})]$. That is, U has the conditional distribution

$$\pi(u \mid \mathbf{x}) = \begin{cases} \frac{1}{\psi(\mathbf{x})} & \text{if } u \leq \psi(\mathbf{x}) \\ 0 & \text{otherwise.} \end{cases} \tag{9.24}$$

Then $\mathbf{X}$ has the conditional distribution

$$\pi(\mathbf{x} \mid u) = \begin{cases} \frac{\pi'(\mathbf{x})}{\sum_{\mathbf{z} \in \mathcal{X}(u)} \pi'(\mathbf{z})} & \text{if } \psi(\mathbf{x}) \geq u \\ 0 & \text{otherwise,} \end{cases} \tag{9.25}$$

where

$$\mathcal{X}(u) := \{\mathbf{x} \in \mathcal{X} : \ \psi(\mathbf{x}) \geq u\}.$$

Slice sampling uses the conditional distributions (9.24) and (9.25) to sample $(\mathbf{X}_1, U_1), \ldots, (\mathbf{X}_t, U_t)$ from an expanded sample space and then takes $\mathbf{X}_1, \ldots, \mathbf{X}_t$ as the sample path for analysis.

Consider the one-step transition kernel

$$K((\mathbf{x}_1, u_1), (\mathbf{x}_2, u_2)) := \pi(u_2 \mid \mathbf{x}_1)\pi(\mathbf{x}_2 \mid u_2) \qquad \mathbf{x}_2 \in \mathcal{X}(u_2), \ 0 \leq u_2 \leq \psi(\mathbf{x}_1).$$

Given the initial state $\mathbf{X}_0$, on each step $j \geq 1$ we randomly generate U_j given $\mathbf{X}_{j-1}$ and then $\mathbf{X}_j$ given U_j. Note that U_j depends on U_{j-1} only through $\mathbf{X}_{j-1}$. To show that in equilibrium each $\mathbf{X}_j$ has the desired distribution π, first observe that the one-step transition kernel for movement in the $\mathcal{X}$ space is

$$\begin{aligned} K_{SS}(\mathbf{x}, \mathbf{y}) &= \int_0^{\psi(\mathbf{x})} K((\mathbf{x}, u_1), (\mathbf{y}, u_2)) \mathrm{d}u_2 \\ &= \frac{\pi'(\mathbf{y})}{\psi(\mathbf{x})} \int_0^{\min[\psi(\mathbf{x}), \psi(\mathbf{y})]} \left[\frac{1}{\sum_{\mathbf{z} \in \mathcal{X}(u_2)} \pi'(\mathbf{z})} \right] \mathrm{d}u_2, \end{aligned} \tag{9.26}$$

for which detailed balance (6.11) holds (see Exercise 4). Because the positivity of π implies the positivity of π', K_{SS} in (9.26) is everywhere positive on $\mathcal{X}^2$ and, therefore, by Theorem 6.7.1, π is the unique solution of the global balance equation (6.3). Because the positivity of K_{SS} also implies aperiodicity, the Markov chain on $\mathcal{X}$ is ergodic.

Because ψ is a bounded function, uniform ergodicity holds (Mira and Tierney 1999, Theorem 9.3) with the convergence factor in (6.57),

$$\rho = 1 - \frac{\sum_{\mathbf{x} \in \mathcal{X}} \pi'(\mathbf{x})\psi(\mathbf{x})}{\max_{\mathbf{x} \in \mathcal{X}} \psi(\mathbf{x})}.$$

This result implies that a factorization for which $\frac{\max_{\mathbf{x} \in \mathcal{X}} \psi(\mathbf{x})}{\min_{\mathbf{x} \in \mathcal{X}} \psi(\mathbf{x})}$ is close to unity is desirable. For general-state spaces, if ψ is positive then the chain is Harris recurrent (see Section 6.12), and if ψ is bounded from above, it is uniformly ergodic with convergence factor (Mira and Tierney 1999)

$$\rho = 1 - \frac{\sum_{\mathbf{x} \in \mathcal{X}} \pi'(\mathbf{x})\psi(\mathbf{x})}{\max_{\mathbf{x} \in \mathcal{X}} \psi(\mathbf{x})}.$$

Of the many factorizations with form as in (9.23), some may induce a distribution $\{\pi'(\mathbf{x}),\ \mathbf{x} \in \mathcal{X}(u)\}$ from which $\mathbf{X}_1, \mathbf{X}_2, \ldots$ can be generated for each $u \in [0, 1]$ at considerably less cost than from the original π. A key to success in implementing SS is to find a factorization that keeps the convergence factor small while allowing this low-cost unit generation. In this regard, a comparison with Hastings-Metropolis sampling using an independence nominating kernel as in Section 6.10 with nominating kernel $q(\mathbf{x}) = \pi'(\mathbf{x})$ is revealing. Let K_{IHMS} denote the one-step transition kernel there. Then as a consequence of Theorem 6.11.1, SS has faster convergence to equilibrium than IHMS and

$$\lim_{t\to\infty} \frac{\text{var}\lambda_t(\pi, h, K_{\text{SS}})}{\text{var}\lambda_t(\pi, h, K_{\text{IHMS}})} \le 1,$$

thus favoring SS (Mira and Tierney 1999).

When applied to the standard Potts model (see Section 9.4), where

- The collection of sites is the node set $\mathcal{V}$ of a graph $\mathcal{G} = (\mathcal{V}, \mathcal{E})$,
- Coordinates take values in a finite product space $\mathcal{X} = \mathcal{X}_1^m$ where $m = |\mathcal{V}|$,
- $\beta > 0$ in (9.14),
- $|\mathcal{N}_i| :=$ neighborhood size = degree of node i (is independent of m for each $i \in \mathcal{V}$),

SS induces rapid mixing (see Section 9.2) for sufficiently small β (Cooper and Frieze 1999, Theorem 1). Table 9.1 gives the upper bounds $\beta_0 > \beta$ for selected neighborhood sizes.

Specialized results for m-vertex graphs include:

- For trees, the mixing time increases as m for all β.
- For cyclic graphs, the mixing time is as $m \log m$ for all β.
- For complete graphs with $|\mathcal{X}_1| \ge 3$, there are critical values of β for which SS does not induce a rapidly mixing process. For example, this occurs for $\beta = \frac{Q}{m}$, where

$$Q := 2\frac{|\mathcal{X}_1| - 1}{|\mathcal{X}_1| - 2} \ln(|\mathcal{X}_1| - 1) \le .925|\mathcal{X}_1|.$$

TABLE 9.1 | *Rapid-mixing bounds on β for standard Potts model**

	$\max_{i\in\mathcal{M}} \lvert\mathcal{N}_i\rvert$					
	2	3	4	5	6	7
β_0	0.538	0.235	0.146	0.105	0.082	0.067
$1 - e^{-\beta_0}$	0.416	0.209	0.136	0.100	0.079	0.065

*Source: Cooper and Frieze (1999).

- For random graphs $\mathcal{G}_{mp}$, where $p = \Omega\left(m^{-1/3}\right)$, there are values of β for which SS does not induce rapid mixing w.p.1 as $m \to \infty$. For example, $\beta = 2Q/mp$.

Huber (1999) contains related results for β_0 for trees. The notation $p_m = \Omega(f_m)$ denotes the existence of a function f and positive constants c and m_0 such that $p_m \geq cf_m$ for all $m \geq m_0$.

For $\mathcal{V} = \{1, \ldots, m\}$, a cycle graph has $\mathcal{E} = \{(i, i+1);\ i = 1, \ldots, m-1\} \cup \{(m, 1)\}$, so $|\mathcal{N}_i| = 2$. A *tree* is a graph with no cycles. A *complete graph* has an edge between every pair of nodes, so $\mathcal{E} = \{(i,\ j);\ i \neq j = 1, \ldots, m\}$. Therefore, $|\mathcal{N}_i| = m-1$ for all $i \in \mathcal{V}$. A *random graph* $\mathcal{G}_{mp}$ can be thought of as a complete graph where each edge is retained with probability p and deleted with probability $1-p$.

9.6.1 Product Slice Sampling

A more elaborate form of SS is available, and this is actually what Edwards and Sokal (1988) propose. Suppose ψ admits the product form

$$\psi(\mathbf{x}) = \prod_{l=1}^{L} \psi_l(\mathbf{x}) \qquad \mathbf{x} \in \mathcal{X},$$

where each ψ_l is a positive bounded function. Let $\mathbf{U} := (U_1, \ldots, U_L)$ denote a vector of independent random variables, where U_l is from $\mathcal{U}(0, \psi_l(\mathbf{x}))$ so that

$$\pi(\mathbf{u} \mid \mathbf{x}) = \begin{cases} \frac{1}{\prod_{l=1}^{L} \psi_l(\mathbf{x})} & \text{if } u_l \leq \psi_l(\mathbf{x}) \qquad l = 1, \ldots, L \\ 0 & \text{otherwise.} \end{cases}$$

Then $(\mathbf{X}, \mathbf{U})$ has the joint distribution

$$\pi(\mathbf{x}, \mathbf{u}) = \begin{cases} c\pi'(\mathbf{x}) & \text{if } \psi_l(\mathbf{x}) \geq u_l \qquad l = 1, \ldots, L \\ 0 & \text{otherwise.} \end{cases}$$

Also $\mathbf{X}$, given $\mathbf{U}$, has the conditional distribution

$$\pi(\mathbf{x} \mid \mathbf{u}) = \begin{cases} \frac{\pi'(\mathbf{x})}{\sum_{\mathbf{z} \in \mathcal{X}(\mathbf{u})} \pi'(\mathbf{z})} & \text{if } \psi_l(\mathbf{x}) \geq u_l \qquad l = 1, \ldots, L \\ 0 & \text{otherwise,} \end{cases} \tag{9.27}$$

where

$$\mathcal{X}(\mathbf{u}) := \{\mathbf{x} \in \mathcal{X} :\ \psi_l(\mathbf{x}) \geq u_l,\ l = 1, \ldots, L\}.$$

On iteration $j \geq 1$, we first randomly generate $\mathbf{U}_j$ from $\{\pi(\mathbf{u} \mid \mathbf{X}_{j-1})\}$ and then randomly generate $\mathbf{X}_j$ from $\{\pi(\mathbf{x} \mid \mathbf{U}_j)\}$. Sometimes this approach is called *product slice sampling* (PSS). Uniform ergodicity holds (Mira and Tierney 1999, Theorem 9.3) with bound in (6.57)

$$\rho = 1 - \frac{\sum_{\mathbf{x} \in \mathcal{X}} \pi'(\mathbf{x}) \prod_{l=1}^{L} \psi_l(\mathbf{x})}{\prod_{l=1}^{L} \max_{\mathbf{x} \in \mathcal{X}} \psi_l(\mathbf{x})}.$$

Analogous properties hold for general-state spaces.

Because

$$\prod_{l=1}^{L} \max_{\mathbf{x}\in\mathcal{X}} \psi_l(\mathbf{x}) \geq \max_{\mathbf{x}\in\mathcal{X}} \left[\prod_{l=1}^{L} \psi_l(\mathbf{x})\right],$$

PSS has a convergence factor no smaller than for SS. Moreover, its asymptotic variance for a sample average is not necessarily smaller than that for SS (Mira and Tierney 1999). Nevertheless, sample generating considerations may make it preferable to use SS for particular problems. For example, if the constraints $\psi_l(\mathbf{x}) \geq u_l$ for $l = 1, \ldots, L$ are easier to implement than the single global constraint $\psi(\mathbf{x}) \geq u$, the resulting unit generating cost may favor PSS.

EXAMPLE 9.6.1

GENERATING RANDOM CLUSTERS

The special structure of (9.17) for the random cluster model fits well within the framework of PSS. As illustration, we describe sample-path generation for this model. For the factorization (9.23), let

$$\pi'(\mathbf{x}) := \frac{\prod_{i=1}^{m} a_i(x_i)}{\sum_{\mathbf{y}\in\mathcal{X}} \prod_{i=1}^{m} a_i(y_i)}$$

$$\psi_C(x_i, x_j) := \mathrm{e}^{-\beta_C[1-\delta(x_i-x_j)]} \qquad C = \{i, j\} \in \mathcal{E},$$

where

$$a_i(x_i) := \mathrm{e}^{-\alpha_i(x_i)} \qquad i \in \mathcal{M},$$

and, as before, let c be the corresponding normalizing constant. Let $\mathbf{U} := (U_C, C \in \mathcal{E})$, given $\mathbf{X}$, denote a random vector uniformly distributed on the $|\mathcal{E}|$-dimensional hyperrectangle $\prod_{\{i,j\}\in\mathcal{E}}[0, \psi_{\{i,j\}}(x_i, x_j)]$. That is, $\mathbf{U}$, given $\mathbf{X}$, has the conditional distribution

$$\pi(\mathbf{u} \mid \mathbf{x}) = \prod_{C\in\mathcal{E}} \pi_C(u_C \mid \mathbf{x}),$$

where for $C = \{i, j\}$,

$$\pi_C(u_C \mid \mathbf{x}) = \begin{cases} \frac{1}{\psi_C(x_i, x_j)} & \text{if } 0 \leq u_C \leq \psi_C(x_i, x_j) \\ 0 & \text{otherwise.} \end{cases} \tag{9.28}$$

For each edge $\{i, j\} \in \mathcal{E}$, we say that nodes i and j are *bonded* if $U_{\{i,j\}} > \mathrm{e}^{-\beta_{\{i,j\}}}$, and *unbonded* otherwise. By way of interpretation, if $x_i = x_j$, then $U_{\{i,j\}}$ is from $\mathcal{U}(0, 1)$ so that i and j are bonded with probability $\mathrm{e}^{-\beta_{\{i,j\}}}$. If $x_i \neq x_j$, then $U_{\{i,j\}}$ is from $\mathcal{U}\left(0, \mathrm{e}^{-\beta_{\{i,j\}}}\right)$ so that i and j are unbonded with probability 1. Note that unconditionally, $\mathbf{U}$ takes values in the $|\mathcal{E}|$-dimensional unit hypercube $[0, 1]^{|\mathcal{E}|}$.

As a consequence of this formulation, **X**, given **U**, has the conditional distribution (see Exercise 5)

$$\pi(\mathbf{x} \mid \mathbf{u}) = \frac{\prod_{i=1}^{m} a_i(x_i)}{\sum_{\mathbf{y}\in\mathcal{X}(\mathbf{u})} \prod_{i=1}^{m} a_i(y_i)} \qquad \mathbf{x} \in \mathcal{X}(\mathbf{u}), \tag{9.29}$$

where

$$\mathcal{X}(\mathbf{u}) := \mathcal{X}\backslash\left\{\mathbf{x} \in \mathcal{X} : \ x_i \neq x_j \text{ and } u_{\{i,j\}} > \mathrm{e}^{-\beta_{\{i,j\}}}, \ \forall\{i, j\} \in \mathcal{D}\right\}.$$

The introduction of the auxiliary random vector **U** allows us to exploit the special structure of (9.29). Each vector **u** induces a collection of $N(\mathbf{u})$ clusters $\mathcal{H}_1(\mathbf{u}), \ldots, \mathcal{H}_{N(\mathbf{u})}(\mathbf{u})$, where a cluster denotes a collection of sites that, at a minimum, are bonded together pairwise and that are unbonded with all sites in other clusters. Unless clarity demands it, we hereafter suppress the argument **u** in $N(\mathbf{u}), \mathcal{H}_1(\mathbf{u}), \ldots, \mathcal{H}_{N(\mathbf{u})}(\mathbf{u})$.

This definition of clusters implies the empty intersections $\mathcal{H}_i \cap \mathcal{H}_j = \emptyset$ for all $i \neq j$ and the exhaustive union $\cup_{i=1}^{N}\mathcal{H}_i = \mathcal{M}$. Given $\mathbf{U} = \mathbf{u}$, these properties allow us to write

$$\prod_{i=1}^{m} a_i(x_i) = \prod_{l=1}^{N} \prod_{i\in\mathcal{H}_l} a_i(x_i),$$

with $\{x_i, i \in \mathcal{H}_l\}$ all the same color in cluster l. Let W_l denote the color of cluster l, and let $\mathbf{W} := (W_1, \ldots, W_N)$ denote a random vector of cluster colors. Then (9.29) may be rewritten as

$$\pi(\mathbf{x} \mid \mathbf{u}) = \mathrm{pr}(\mathbf{X} = \mathbf{x} \mid \mathbf{U} = \mathbf{u}) = \mathrm{pr}(\mathbf{W} = \mathbf{w} \mid \mathcal{H}_1, \ldots, \mathcal{H}_N) = \prod_{l=1}^{N} \pi_l(w_l \mid \mathbf{u}), \tag{9.30}$$

where

$$\pi_l(w_l \mid \mathbf{u}) := \frac{\prod_{i\in\mathcal{H}_l} a_i(w_l)}{\sum_{w\in\mathcal{X}_0} \prod_{i\in\mathcal{H}_l} a_i(w)}$$

and where $\pi(\mathbf{x} \mid \mathbf{u})$ is subject to the constraints $x_i = w_l$ for all nodes $i \in \mathcal{H}_l$ for each cluster $l = 1, \ldots, N$. Note that the color configuration **X** drawn from $\{\pi(\mathbf{x} \mid \mathbf{u})\}$ depends on **U** through $N, \mathcal{H}_1, \ldots, \mathcal{H}_N$. Expression (9.30) reveals that $W_1, \ldots, W_N$ are independent random variables, a property that considerably simplifies generation.

Algorithm CLUSTER describes a procedure for generating sample paths of color configurations with equilibrium distribution (9.17). ■

ALGORITHM **CLUSTER**

Purpose: To generate sample paths of color configurations with equilibrium distribution (9.17).

Given: Initial state $\mathbf{X}_0$ and sample-path length t.

Method:

$j = 1$.
Until $j > t$:
for each $\mathcal{C} = \{i, k\} \in \mathcal{E}$:
randomly sample $U_{\mathcal{C}}$ from $\mathcal{U}\big(0, \psi_{\mathcal{C}}(X_{i,j-1}, X_{k,j-1})\big)$
determine the number of clusters N and the clusters $\mathcal{H}_1, \ldots, \mathcal{H}_N$
for each cluster $l = 1, \ldots, N$:
randomly sample color W_l from $\big\{\pi_l(w \mid U_{\mathcal{C}},\ \mathcal{C} \in \mathcal{E})\big\}$
for each $i \in \mathcal{H}_l$, $X_{ij} = W_l$
$j = j + 1$.
■

The resulting chain is uniformly ergodic with the convergence factor in (6.57),

$$\rho = 1 - \frac{\sum_{\mathbf{x} \in \mathcal{X}} \prod_{i=1}^{m} \left[a_i(\mathbf{x}_i) e^{-\sum_{\mathcal{C}=\{i,j\}\in\mathcal{E}} \beta_{\mathcal{C}}[1-\delta(x_i - x_j)]}\right]}{\sum_{\mathbf{x} \in \mathcal{X}} \prod_{i=1}^{m} a_i(\mathbf{x}_i)}.$$

Inspection reveals that large (small) $\beta_{\{i,j\}}$ induce fast (slow) convergence.

9.6.2 Partial Decoupling

Empirical studies in Hurn (1997) and Higdon (1998) reveal that PSS can be slow to converge, particularly when the equilibrium distribution π is multimodal. A modification of PSS, called *partial decoupling,* can substantially speed up convergence. For the factorization (9.23), let

$$\begin{aligned} \pi'(\mathbf{x}; \zeta) &:= \frac{\prod_{i=1}^{m} \left[a_i(x_i) e^{-\sum_{\mathcal{C}=\{i,j\}\in\mathcal{E}} (1-\zeta_{\mathcal{C}})\beta_{\mathcal{C}}[1-\delta(x_i - x_j)]}\right]}{\sum_{\mathbf{y} \in \mathcal{X}} \prod_{i=1}^{m} \left[a_i(y_i) e^{-\sum_{\mathcal{C}=\{i,j\}\in\mathcal{E}} (1-\zeta_{\mathcal{C}})\beta_{\mathcal{C}}[1-\delta(y_i - y_j)]}\right]} \\ \psi_{\mathcal{C}}(\mathbf{x}; \zeta) &:= e^{-\sum_{\mathcal{C}=\{i,j\}\in\mathcal{E}} \zeta_{\mathcal{C}}\beta_{\mathcal{C}}[1-\delta(x_i - x_j)]}, \end{aligned} \tag{9.31}$$

where $\zeta := \{\zeta_{\mathcal{C}} \in [0, 1],\ \mathcal{C} \in \mathcal{E}\}$. Then

$$\pi(\mathbf{u} \mid \mathbf{x}; \zeta) = \prod_{\mathcal{C} \in \mathcal{E}} \psi_{\mathcal{C}}^{-1}(\mathbf{x}) I_{[0, \psi_{\mathcal{C}}(\mathbf{x})]}(u_{\mathcal{C}})$$

and

$$\pi(\mathbf{x} \mid \mathbf{u}; \zeta) = \frac{\prod_{l=1}^{N} \prod_{i \in \mathcal{H}_l} \left[a_i(x_i) e^{-\sum_{\mathcal{C}=\{i,j\}\in\mathcal{E}, j \notin \mathcal{H}_l} (1-\zeta_{\mathcal{C}})\beta_{\mathcal{C}}}\right]}{\sum_{\mathbf{y} \in \mathcal{X}(\mathbf{u})} \prod_{l=1}^{N} \prod_{i \in \mathcal{H}_l} \left[a_i(y_i) e^{-\sum_{\mathcal{C}=\{i,j\}\in\mathcal{E}, j \notin \mathcal{H}_l} (1-\zeta_{\mathcal{C}})\beta_{\mathcal{C}}}\right]}.$$

For fixed ζ, PSS induces a uniformly ergodic Markov chain with convergence factor, as in (6.57),

$$\rho = 1 - \frac{\sum_{\mathbf{x} \in \mathcal{X}} \prod_{i=1}^{m} \left[a_i(x_i) e^{-\sum_{\mathcal{C}=\{i,j\}\in\mathcal{E}} \beta_{\mathcal{C}}[1-\delta(x_i - x_j)]}\right]}{\sum_{\mathbf{x} \in \mathcal{X}} \prod_{i=1}^{m} \left[a_i(x_i) e^{-\sum_{\mathcal{C}=\{i,j\}\in\mathcal{E}} (1-\zeta_{\mathcal{C}})\beta_{\mathcal{C}}[1-\delta(x_i - x_j)]}\right]}.$$

Observe that ρ decreases as the $\zeta_{\mathcal{C}}$ decrease in value from 1 to 0, thus increasing the convergence rate. But the ability to sample colors from $\{\pi(\mathbf{x} \mid \mathbf{u}; \zeta)\}$

independently for each cluster, as in Section 9.6.1, is lost once $\zeta_C < 1$ for some $C \in \mathcal{E}$. Indeed, the i.i.d. case for $\zeta_C = 0$, $\forall C \in \mathcal{E}$ implies sampling from the original equilibrium distribution π, the impediment that we set out to overcome.

There are, of course, acceptance-rejection methods, but they inevitably encounter large rejection probabilities for relatively moderate-sized problems.

9.7 COUPLING FROM THE PAST

The graphical techniques in Section 7.3 require little knowledge about a particular problem to come up with an MCMC estimate of its warm-up interval $k - 1$. That is their greatest appeal. However, they have several limitations. They require user intervention during data collection to estimate k. Also, different observers of the same graphs may come up with different estimates of k. Recall that underestimating k leaves a not inconsequential influence caused by initial conditions in the sample data used for parameter estimation, whereas overestimating k is wasteful of data.

Ideally, we would like a procedure that identifies a k such that for all $j \geq k$, $\mathbf{X}_k$ has π for its equilibrium distribution. Although a deterministic k rarely exists that can induce this property, there are procedures, collectively called *coupling from the past* (CFTP), that yield a random sample-path length that comes close to accomplishing this. CFTP makes use of two concepts: *coupling* and starting sampling at a *random point in the past.* Although currently limited in scope of application, CFTP's conceptual importance makes CFTP worthwhile to describe at this point in its development. The present account relies principally on Propp and Wilson (1996) and Corcoran and Tweedie (1998).

We begin by illustrating coupling for two sample paths. Consider an m-dimensional ergodic Markov chain with transition kernel K on a finite-state space $\mathcal{X}$. Let $\mathbf{X}_0^{[1]} \neq \mathbf{X}_0^{[2]}$ be two initial states in $\mathcal{X}$, and let $\mathbf{X}_0^{[1]}, \mathbf{X}_1^{[1]}, \mathbf{X}_2^{[1]}, \ldots$ and $\mathbf{X}_0^{[2]}, \mathbf{X}_1^{[2]}, \mathbf{X}_2^{[2]}, \ldots$ denote the corresponding sample paths. It is assumed that K satisfies global balance. If for all

$$j \geq T := \min\left(l \geq 1 : \mathbf{X}_l^{[1]} = \mathbf{X}_l^{[2]}\right),$$

$\mathbf{X}_j^{[1]} = \mathbf{X}_j^{[2]}$, we say that $\left\{\mathbf{X}_j^{[1]}\right\}$ and $\left\{\mathbf{X}_j^{[2]}\right\}$ have coupled at *coupling time* T. If either $\mathbf{X}_0^{[1]}$ or $\mathbf{X}_0^{[2]}$ is randomly drawn from π, then the coupled chain $\left\{\mathbf{Z}_j = \mathbf{X}_j^{[1]} = \mathbf{X}_j^{[2]}, j \geq T\right\}$ has equilibrium distribution π. This property generalizes to more than two coupled chains.

Although the impracticality of sampling from π limits the value of this property, other properties related to coupling reveal its potential importance for eliminating the influence of initial conditions in an MCMC sampling experiment.

For all subsets $\mathcal{A} \subseteq \mathcal{X}$ and for each $\mathbf{x} \in \mathcal{X}$, an ergodic Markov chain $\{\mathbf{X}_j, j \geq 0\}$ has

$$\text{pr}(\mathbf{X}_j \in \mathcal{A} \mid \mathbf{X}_0 = \mathbf{x}) \to \pi(\mathcal{A}) \quad \text{as } j \to \infty.$$

Extending the indexing to negative as well as positive integers endows the ergodic process $\{\mathbf{X}_j,\ j > -\infty\}$ with the equivalent property. That is, for all $\mathcal{A} \subseteq \mathcal{X}$ and for each $\mathbf{x} \in \mathcal{X}$,

$$\text{pr}(\mathbf{X}_0 \in \mathcal{A} \mid \mathbf{X}_j = \mathbf{x}) \to \pi(\mathcal{A}) \quad \text{as } j \to -\infty. \tag{9.32}$$

This implies that if we begin sampling in the *infinite past*, $\mathbf{X}_0$ on step 0 is independent of the initial state or, equivalently, has the equilibrium distribution π. All *coupling-from-the-past techniques* exploit this property.

Because starting on step $-\infty$ is not possible, Propp and Wilson propose a *reiteration scheme* where each reiteration i starts on a step $j_i < 0$ that is further back in the past than the starting step on the previous reiteration. That is, $j_{i+1} < j_i$. For an equilibrium distribution π on a finite-state space $\mathcal{X} := \{\mathbf{x}_1, \ldots, \mathbf{x}_v\}$, CFTP takes the following form:

1. Each reiteration i begins on step j_i by initializing v sample paths with $\mathbf{X}_{j_i}^{[l]} = \mathbf{x}_l$ on path $l = 1, \ldots, v$.
2. On each step $j_i, j_i + 1, \ldots, 0$, each path is randomly updated.
3. If $\mathbf{X}_0^{[1]} = \cdots = \mathbf{X}_0^{[v]}$, then all paths have coupled prior to or on step 0.
4. The randomizing mechanism of Propp and Wilson ensures that if $\mathbf{X}_0^{[1]} = \cdots = \mathbf{X}_0^{[v]}$, then $\mathbf{X}_0^{[1]}$ has the equilibrium distribution π.
5. If the paths have not coupled before or on step 0, the next reiteration begins on step $j_{i+1} < j_i$.

If $\mathbf{X}_0^{[1]} = \cdots = \mathbf{X}_0^{[v]}$, all paths are said to have *coalesced*. In the remainder of the chapter,

$$T := -\min\left(j \le 0 : \mathbf{X}_j^{[1]} = \cdots \mathbf{X}_j^{[v]}\right)$$

is called the *backward coupling time*.

Algorithm CFTP1 describes the principal steps. Successive reiterations start on steps $j_1 = -1,\ j_2 = -2,\ j_3 = -4,\ j_4 = -8, \ldots$. Section 9.7.3 describes the sense in which this progression is optimal. Observe that on step $j > j_i + 1$ of reiteration i, the same random vector, $\mathbf{U}_j$, is used to update the paths. Section 9.7.2 describes the significance of this reused randomness.

ALGORITHM CFTP1

Purpose: To generate $\mathbf{Z}_0$ randomly from the equilibrium distribution π.

Based on: Propp and Wilson (1996).

Method:

$N = -1$, $T = -\infty$, and $L = 0$.

Until $\mathbf{X}_0^{[1]} = \cdots = \mathbf{X}_0^{[v]}$:

$j = N$

$\mathbf{X}_j^{[l]} = \mathbf{x}_l$ for $l = 1, \ldots, v$

$j = j + 1$
while $j \le 0$:
  if $j \le L$:
    sample $\mathbf{U}_j$ from the uniform distribution on the M-dimensional hypercube $[0, 1]^M$
  on each sample path $l = 1, \ldots, v$:
    $\mathbf{U}_j$ together with $\left\{K\left(\mathbf{X}_{j-1}^{[l]}, \mathbf{y}\right)\right\}$ determine the update $\mathbf{X}_j^{[l]}$
  if $\mathbf{X}_j^{[1]} = \cdots = \mathbf{X}_j^{[v]}$, $T = -j$
  $L = N$
  $N = 2N$
$\mathbf{Z}_0 = \mathbf{X}_0^{[1]} = \cdots = \mathbf{X}_0^{[v]}$ has the equilibrium distribution π. ■

9.7.1 Monotone Markov Chains

Although Algorithm CFTP1 solves the problem of starting in the past, it remains impractical for large v. When it applies, the concept of a *monotone Markov chain* resolves this issue. In particular, we show that the concept leads to a coupling for all v paths without the need to generate them all explicitly. Consider a univariate ergodic Markov chain $\{X_j, j > -\infty\}$ on $\mathcal{X} = \{0, 1, \ldots, v-1\}$ with equilibrium distribution $\pi = \{\pi(x) > 0,\ x \in \mathcal{X}\}$ and transition kernel $K = \{K(x, y), (x, y) \in \mathcal{X}^2\}$. For an old value x in $\mathcal{X}$ and a number $w \in [0, 1]$, let

$$y := \eta(x, w) \in \mathcal{X}$$

denote an assignment rule that determines a unique new value y in $\mathcal{X}$ that satisfies

$$\mathrm{pr}[\eta(x, U) = y] = K(x, y), \tag{9.33}$$

where U is from $\mathcal{U}(0, 1)$. Moreover, assume that η is monotone nondecreasing in its first argument, so

$$x \le y \Longleftrightarrow \eta(x, w) \le \eta(y, w). \tag{9.34}$$

As illustration, suppose that K is chosen to preserve the ordering

$$x \le y \Longleftrightarrow \sum_{i=0}^{z} K(x, i) \ge \sum_{i=0}^{z} K(y, i) \quad z \in \mathcal{X}. \tag{9.35}$$

Then

$$\eta(x, w) := \min\left(z : \sum_{i=0}^{z} K(x, i) \ge w\right) \tag{9.36}$$

satisfies (9.35) and suggests the inverse-transform approach (see Section 3.3) for sampling $X_j^{[l]}$ given $X_{j-1}^{[l]}$. A Markov chain satisfying the ordering (9.35) is said to be *monotone*. Note that in the illustration $\{\eta(x, u),\ 0 \le u < 1\}$ is just the inverse of the distribution function corresponding to the p.m.f. $\{K(x, y),\ y \in \mathcal{X}\}$.

Let j_1 denote a negative integer, $\{U_j,\ j_1 < j \le 0\}$ a sequence of i.i.d. random variables from $\mathcal{U}(0,1)$, and suppose we generate each path $l = 1, \dots, v$ as

$$X_{j_1}^{[l]} = l - 1$$

$$X_j^{[l]} = \eta\left(X_{j-1}^{[l]}, U_j\right) \quad j = j_1 + 1, j_1 + 2, \dots, 0.$$

Then the ordering (9.35) ensures that for all $0 \ge j \ge j_1$,

$$X_j^{[1]} \le X_j^{[2]} \le \cdots \le X_j^{[v]} \quad \text{w.p.1},$$

so the sample paths $\left\{X_j^{[l]},\ j_1 < j \le 0\right\}$ for $l = 2, \dots, v-1$ are *sandwiched* between the lowest path $\left\{X_j^{[1]},\ j_1 < j \le 0\right\}$ and the highest path $\left\{X_j^{[v]},\ j_1 < j \le 0\right\}$. Therefore, if there exists a $-T \in \{j_1 + 1, \dots, 0\}$ such that $X_T^{[1]} = X_T^{[v]}$, then for all $j \ge -T$, $X_j^{[1]} = \cdots = X_j^{[v]}$ w.p.1. This implies that all paths have coalesced and $X_0^{[1]} = \cdots = X_0^{[v]}$ has the equilibrium distribution π. Most importantly, only sample path 1 with $X_{j_1}^{[1]} = 0$ and path v with $X_{j_1}^{[v]} = v - 1$ need be generated to produce this result.

This sandwiching of sample paths extends to multivariable Markov chains, provided that a more general *partial ordering* than (9.35) exists. For expository convenience, we continue to focus on the univariate case. Section 9.7.5 discusses multivariable Markov chains and partial ordering.

The monotone property eliminates the problem of size. Together with reiteration, it provides a basis for an implementable algorithm. However, this time we generalize the scheme in Algorithm CFTP1 to begin on steps $j_1 > j_2 > \cdots$ on successive reiterations, where $j_1 < 0$. Let reiteration 1 denote the first attempt to realize coupling, as described above, and suppose the generated sample path ends with $X_0^{[1]} \ne X_0^{[v]}$. Then on reiteration 2 one chooses a negative integer $j_2 < j_1$, assigns $X_{j_2}^{[1]} = 0$ and $X_{j_2}^{[v]} = v - 1$ as the initial states, and generates $X_{j_2+1}^{[1]}, \dots, X_0^{[1]}$ and $X_{j_2+1}^{[v]}, \dots, X_0^{[v]}$, retaining the uniform deviates $U_{j_1+1}, \dots, U_0$ from the first reiteration on steps $j_1+1, \dots, 0$ on the corresponding steps on a second reiteration. If the second reiteration ends with $X_0^{[1]} = X_0^{[v]}$, then the paths have coalesced and $X_0^{[1]}$ has the p.m.f. π. If not, we choose a $j_3 < j_2$ and repeat the process on a third reiteration, retaining $U_{j_2+1}, \dots, U_{j_1}, U_{j_1+1}, \dots, U_0$ on steps $j_2 + 1, \dots, 0$. This procedure continues until some reiteration b with starting point $j_b < j_{b-1} < \cdots < j_1$ ends with $X_0^{[1]} = X_0^{[v]}$. Algorithm CFTP2 describes the steps for $j_b := 2j_{b-1}$ for $b = 2, 3, \dots$ and $j_1 := -1$ so that the final number $j_b = 2^{b-1} j_1$. Note that this is precisely the starting protocol in Algorithm CFTP1.

ALGORITHM | **CFTP2**

Purpose: To draw a sample X_0 from $\{\pi(x);\ x = 0, 1, \dots, v-1\}$.

Output: X_0.

Based on: Propp and Wilson (1996).

Method:

$b = 1,\ j_0 = 0$, and $j_1 = -1$.
Repeat:
  $j = j_b,\ X_j^{[1]} = 0$, and $X_j^{[v]} = v - 1$
  repeat:
    $j = j + 1$
    if $j \le j_{k-1}$, sample U_j from $\mathcal{U}(0, 1)$ and store it
    $X_j^{[1]} = \eta\left(X_{j-1}^{[1]}, U_j\right)$
    $X_j^{[v]} = \eta\left(X_{j-1}^{[v]}, U_j\right)$
    if $X_j^{[1]} = X_j^{[v]},\ T = -j$
  until $j > 0$.
  $b = b + 1$
  $j_b = 2j_{b-1}$
until $X_0^{[1]} = X_0^{[v]}$.
$X_0 = X_0^{[1]}$.
Return X_0. ■

It is important to note that regarding $X_{-T}^{[1]}$ as being from the equilibrium distribution would be inappropriate. Whereas the entire sample sequence $U_{j_b}, \ldots, U_0$ determines $X_0^{[1]}$, the subsequence $U_{j_b}, \ldots, U_{-T}$, given $U_{-T+1}, \ldots, U_0$, determines $X_{-T+1}^{[1]}$, an important distinction.

9.7.2 Reusing Randomness

In what follows, we retain the simplicity of the univariate Markov chain on the space $\mathcal{X} := \{0, 1, \cdots, v-1\}$, but we emphasize that analogous results hold for multivariable Markov chains. To explain why we need to retain $U_{j_{b-1}+1}, \ldots, U_0$ on reiteration b for use on steps $j > j_b$ on later reiterations, we rely on inequality (9.35). Suppose we generate all v sample paths. Recall that paths l and l' couple on step j if and only if $X_{j-1}^{[l]} \neq X_{j-1}^{(l')}$ and $X_j^{[l]} = X_j^{[l']}$, implying that $X_j^{[l]} = X_j^{[l+1]} = \cdots = X_j^{[l'-1]} = X_j^{[l']}$ for $l' > l$. For integers $a < c < 0$, let

$$N_{bac} := \text{number of couplings that occur on reiteration } b \text{ on steps } a+1 \text{ through } c.$$

Then for each reiteration $b \ge 1$, $N_{b+1, j_{b+1}, 0} \ge N_{b j_b 0}$ w.p.1. As proof, let

$$A_{2j_2j_1} := \text{number of couplings that occur on steps } j_2+1, \ldots, j_1 \text{ on reiteration 2 that were also coupled on steps } j_1+1, \ldots, 0 \text{ on reiteration 1.}$$

Note that

$$N_{2j_20} = N_{2j_2j_1} + N_{2j_10}.$$

Because $U_{j_1+1}, \ldots, U_0$ are used on steps $j_1+1, \ldots, 0$ on both reiterations 1 and 2, it follows that

$$N_{1j_10} = A_{2j_2j_1} + N_{2j_10},$$

so

$$N_{2j_20} \geq A_{2j_2j_1} + N_{2j_10} = N_{1j_10} \quad \text{w.p.1.}$$

These results imply that if $N_{bj_b} = v$ for some reiteration b, then any subsequent reiteration, including one starting in the infinite past at $j = -\infty$, has complete coalescence by step 0 w.p.1. Together with (9.32), this implies that on step 0, all the coalesced paths are in the same steady state w.p.1.

9.7.3 Total Number of Steps

Recall that Algorithm CFTP2 employs $j_1 := -1$ and $j_{b+1} := 2j_b$ for $b = 1, 2, \ldots$ to move the starting point of iteration b back into the past. Let us now consider the more general scheme $j_1 := -1$ and $j_{b+1} := \theta j_b$ for some positive integer θ and $b = 1, 2, \ldots$. If coalescence first occurs on random reiteration B, then the algorithm takes

$$\begin{aligned} C &:= -j_1 - j_2 - \cdots - j_B \\ &= 1 + \theta + \theta^2 + \cdots + \theta^{B-1} \\ &= (\theta^B - 1)/(\theta - 1) \end{aligned} \tag{9.37}$$

steps in total to achieve a steady state on step 0. Because there are two paths, the total number of updates is $2C$.

Because B is the smallest integer such that $\theta^{B-1} > T$, it follows that $T \geq \theta^{B-2}$. Therefore,

$$C = \frac{\theta^B - 1}{\theta - 1} < \frac{\theta^B}{\theta - 1} = \frac{\theta^{B-2}}{\theta - 1}\theta^2 \leq \frac{\theta^2}{\theta - 1}T. \tag{9.38}$$

Because $C > T$, the total number of steps lies in the interval

$$T < C < \frac{\theta^2}{\theta - 1}T,$$

whose length, $\frac{\theta^2}{\theta-1}T - T$, is minimized for $\theta = 2$.

Propp and Wilson also give a rationale for anticipating that EC is minimized by taking $\theta = \mathrm{e} = 2.71828\ldots$. They also show that the assignment $\theta = 2$ gives EC close to the minimal expectation. We describe the probabilistic character of C by first investigating the probabilistic properties of the backward coupling time T.

Theorem 9.7.1

Suppose a reiteration begins at some initial step $J = -\theta^{B-1} < 0$, and let $\mathcal{W}_J$ denote the assignment of the initial states $X_J^{[1]} = 0$ and $X_J^{[v]} = v - 1$. Then

$$\text{pr}(T > k \mid \mathcal{W}_J) \geq \max_{0 \leq i \leq v-1} \left| K^{(k)}(0, i) - K^{(k)}(v - 1, i) \right| \qquad k = 1, 2, \ldots, -J.$$

■

PROOF The inequality pr(A, B) $\leq$ min [pr(A), pr(B)] and the definition

$$K^{(k)}(l-1,i) := \mathrm{pr}\big(X_{J+k}^{[l]} = i \,|\, X_J^{[l]} = l-1\big)$$

imply

$$\begin{aligned}
\mathrm{pr}\,(T \leq k \mid \mathcal{W}_J) &= \sum_{i=0}^{v-1} \mathrm{pr}\big(X_{J+k}^{[1]} = X_{J+k}^{[v]} = i \big| \mathcal{W}_J\big) \\
&= \sum_{i=0}^{v-1} \mathrm{pr}\big(X_{J+k}^{[1]} = i,\, X_{J+k}^{[v]} = i \big| \mathcal{W}_J\big) \\
&\leq \sum_{i=0}^{v-1} \min\Big[\mathrm{pr}\big(X_{J+k}^{[1]} = i \big| \mathcal{W}_J\big), \mathrm{pr}\big(X_{J+k}^{[v]} = i \big| \mathcal{W}_J\big)\Big] \\
&\leq \sum_{i=0}^{v-1} \min\big[K^{(k)}(0,i), K^{(k)}(v-1,i)\big].
\end{aligned}$$

Moreover, the identity $\min(a, c) = (a + c - |a - c|)/2$ implies

$$\begin{aligned}
\sum_{i=0}^{v-1} \min\big[K^{(k)}(0,i), K^{(k)}(v-1,i)\big] &= 1 - \frac{1}{2}\sum_{i=0}^{v-1} \big|K^{(k)}(0,i) - K^{(k)}(v-1,i)\big| \\
&= 1 - \max_{0\leq i\leq v-1} \big|K^{(k)}(0,i) - K^{(k)}(v-1,i)\big|,
\end{aligned}$$

which proves the theorem. ■

Theorem 9.7.2

For a uniformly ergodic stochastic process, there exist constants $M > 0$ and $\rho \in [0, 1)$ such that

$$\mathrm{pr}(T > k) \leq M\rho^k \qquad k = 1, 2, \ldots, -J. \tag{9.39}$$

■

PROOF We have

$$\begin{aligned}
\mathrm{pr}(T > k \mid \mathcal{W}_J) &= \mathrm{pr}\big(X_{J+k}^{[1]} \neq X_{J+k}^{[v]} \big| \mathcal{W}_J\big) \\
&= \mathrm{pr}\big(X_{J+k}^{[v]} - X_{J+k}^{[1]} \geq 1 \big| \mathcal{W}_J\big) \\
&\leq \mathrm{E}\big(X_{J+k}^{[v]} - X_{J+k}^{[1]} \big| \mathcal{W}_J\big) \\
&= \big|\mathrm{E}\big(X_{J+k}^{[v]} \big| X_J^{[1]} = 0\big) - \mathrm{E}\big(X_{J+k}^{[v]} \big| X_J^{[v]} = v-1\big)\big| \\
&\leq v \sum_{i=0}^{v-1} \big|K^{(k)}(0,i) - K^{(k)}(v-1,i)\big|,
\end{aligned}$$

where the first inequality follows from *Markov's inequality*, $\text{pr}(Z \geq z) \leq \text{E}Z/z$, for a nonnegative random variable Z and positive z. By the triangle inequality,

$$\max_{0\leq i\leq v-1} |K^{(k)}(0,i) - K^{(k)}(v-1,i)| \leq \max_{0\leq i\leq v-1} |K^{(k)}(0,i) - \pi(i)| + \max_{0\leq i\leq v-1} |K^{(k)}(v-1,i) - \pi(i)|.$$

By Theorem 6.8.3, there exist constants $M_l > 0$ and $\rho \in [0, 1)$ such that for all $l \in \mathcal{X}$

$$\max_{0\leq i\leq v-1} |K^{(k)}(l,i) - \pi(i)| \leq M_l \rho^k.$$

Then by Theorem 9.7.2, (9.39) obtains with $M = v \max(M_0, M_{v-1})$. ■

Because (9.38) implies

$$\text{pr}\left(\frac{T\theta^2}{\theta - 1} > \frac{k\theta^2}{\theta - 1}\right) \geq \text{pr}\left(C > \frac{k\theta^2}{\theta - 1}\right),$$

it follows from Theorems 9.7.1 and 9.7.2 that for some $c > 0$,

$$\text{pr}(C > c) \leq M\rho^{c(\theta-1)/\theta^2}.$$

That is, the right-tail probability of the d.f. of C, the total number of steps until coalescence, converges geometrically as c increases. For example, the optimal $\theta = 2$ gives

$$\text{pr}(C \geq c) \leq M\rho^{c/4}.$$

9.7.4 Saving Space

Storing all the U_j when executing Algorithms CFTP1 and CFTP2 is inconvenient and unnecessary. As Chapter 4 reveals, a one-step recursion

$$Z_i = f(Z_{i-1}) \quad i \geq 0$$

usually provides the pseudorandom integers $1 \leq Z_i < M$ for $i = 1, 2, \ldots$ that yield the pseudorandom uniform deviates $U_i := Z_i/M$ for $i = 1, 2, \ldots$ on [0,1). Examples include the one-step linear congruential generator (see Section 4.1) and the one-step inversive generator (e.g., see Fishman 1996a, Section 7.17.2). Alternatively, the *Mersenne Twister* in Section 4.7.1 relies on an N-step recursion

$$Z_i = f(Z_{i-1}, \ldots, Z_{i-N}) \qquad i \geq 1,$$

thereby requiring that N sequential seeds be stored. For example, the Mersenne twister MT19927 has $N = 624$. For convenience of exposition, we use the one-step recursion here, but its extension to an N-step recursion is immediate.

Suppose we take Z_0 as the seed on reiteration 1, Z_{-j_1} as the seed on reiteration 2, Z_{-j_2} as the seed on reiteration 3, and so on. This leads to the mapping:

$$\begin{array}{llllll}
\text{Reiteration} & & \overbrace{\qquad\qquad\qquad}^{1} & \overbrace{\qquad\qquad\qquad}^{2} & & \overbrace{\qquad\qquad\qquad}^{b} \\
\text{Seed} & Z_0, & Z_1, \ldots, Z_{-j_1}, & Z_{-j_1+1}, \ldots, Z_{-j_2}, & \ldots, & Z_{-j_{b-1}+1}, \ldots, Z_{-j_b} \\
 & & \downarrow & \downarrow & & \downarrow \\
\text{Deviate} & & U_{j_1+1}, \ldots, U_0, & U_{j_2+1}, \ldots, U_{j_1}, & \ldots, & U_{j_b+1}, \ldots, U_{j_{b-1}}
\end{array}$$

Algorithm CFTP3 incorporates steps into Algorithm CFTP2 that store only the seeds $Z_0, Z_{-j_1}, Z_{-j_2}, \ldots$ and rely on recursion to reproduce the U_j as required. For multiple recursive generators, N seeds would be stored for each iteration.

ALGORITHM

CFTP3

Purpose: To draw a random sample X_0 from $\{\pi(x);\ x = 0, 1, \ldots, v-1\}$.

Input: Pseudorandom number generator $\{Z_0,\ Z_i = f(Z_{i-1});\ i = 1, 2, \ldots\}$.

Output: X_0.

Definition: $W_k := Z_{-j_k}$.

Method:

$b = 1,\ j_0 = 0,\ j_1 = -1$, and $W_0 = Z_0$.

Repeat:

  $j = j_b,\ l = b,\ X_j^{[1]} = 0$, and $X_j^{[v]} = v - 1$

  $W = W_{b-1}$

  Repeat:

    $j < j + 1$

    $W = f(W)$

    $U_j = W/M$

    $X_j^{[1]} = \eta\left(X_{j-1}^{[1]}, U_j\right)$

    $X_j^{[v]} = \eta\left(X_{j-1}^{[v]}, U_j\right)$

    if $X_j^{[1]} = X_j^{[v]},\ T = -j$

    if $j = j_{l-1},\ l = l - 1$, and $W = W_l$

  until $j > 0$.

  $b = b + 1$

  $j_b = 2j_{b-1}$

  $W_b = W$

until $X_0^{[1]} = X_0^{[v]}$.

$X_0 = X_0^{[1]}$.

Return X_0. ■

9.7.5 Independence Hastings-Metropolis Sampling

A CFTP construction in Corcoran and Tweedie (1998) relies on Hastings-Metropolis (HM) sampling and applies to continuous as well as discrete state spaces. On step $j \geq 1$, with $\mathbf{X}_{j-1} = \mathbf{x} \in \mathcal{X}$, $\mathbf{Y}_j$, the HM method samples from a nominating distribution $\{q(\mathbf{x}, \mathbf{y}),\ \mathbf{y} \in \mathcal{X}\}$, samples U_j from $\mathcal{U}(0, 1)$, and sets

$$\mathbf{X}_j = \eta(\mathbf{x}, \mathbf{Y}_j, U_j),$$

where

$$\eta(\mathbf{x}, \mathbf{y}, w) := \begin{cases} \mathbf{y} & \text{if } w \leq \alpha(\mathbf{x}, \mathbf{y}) \\ \mathbf{x} & \text{otherwise} \end{cases} \qquad 0 \leq w \leq 1, \quad (\mathbf{x}, \mathbf{y}) \in \mathcal{X}^2 \tag{9.40}$$

as the assignment rule. Recall that HM sampling has the acceptance ratio

$$\alpha(\mathbf{x}, \mathbf{y}) = \min\left[1, \frac{\pi(\mathbf{y})q(\mathbf{y}, \mathbf{x})}{\pi(\mathbf{x})q(\mathbf{x}, \mathbf{y})}\right],$$

and suppose we choose a nominating distribution $\{q(\mathbf{x}, \mathbf{y}) := q(\mathbf{y}) > 0,\ \mathbf{y} \in \mathcal{X}\}$ for each $\mathbf{x} \in \mathcal{X}$ (see Section 6.10.1), where $\{q(\mathbf{y})\}$ ensures the *partial ordering* ($\preceq$)

$$\mathbf{x} \preceq \mathbf{z} \Rightarrow \alpha(\mathbf{x}, \mathbf{y}) \leq \alpha(\mathbf{z}, \mathbf{y}) \qquad \forall \mathbf{x} \in \mathcal{X}. \tag{9.41}$$

Assume that there exists a finite, minimal element $\mathbf{x}_{\min}$ such that $\mathbf{x}_{\min} \preceq \mathbf{x}$ for all $\mathbf{x} \in \mathcal{X}$. As in Section 9.7.1, consider a reiterative approach where sampling on reiteration b begins on step $j_b + 1$ for $j_1 := -1$ and $j_{i+1} := 2j_i$ for $i = 1, 2, \ldots$. Path 1 begins in state $\mathbf{X}_b^{[1]} = \mathbf{x}_{\min}$, and paths 2 through v each assume one of the remaining $v - 1$ distinct elements $\mathbf{x}$ in $\mathcal{X}$. On step j, $\mathbf{Y}_j$ is randomly drawn from $\{q(\mathbf{y})\}$, and U_j from $\mathcal{U}(0, 1)$. Path l is updated as

$$\mathbf{X}_j^{[l]} = \eta\left(\mathbf{X}_{j-1}^{[l]}, \mathbf{Y}_j, U_j\right)$$

for $l = 1, \ldots, v$. Most notably, if $U_j \leq \alpha\left(\mathbf{X}_j^{[1]}, \mathbf{Y}_j\right)$, then $\mathbf{X}_j^{[1]} = \cdots = \mathbf{X}_j^{[v]} = \mathbf{Y}_j$ so that all v chains are coupled.

If coupling has not occurred by step 0, then reiteration is performed. On reiteration $b + 1$, $(\mathbf{Y}_{j_b+1}, U_{j_b}), \ldots, (\mathbf{Y}_0, U_0)$ from reiteration b are reused on steps $j_b + 1, \ldots, 0$. Together with the ordering (9.41), this reuse of the 2-tuples ensures that if success first occurs for path 1 on step $-T$ of reiteration b, then

P38. $\mathbf{X}_j^{[1]} = \mathbf{x}_{\min} \preceq \mathbf{X}_j^{[l]}$ for $l = 2, \ldots, v$ and $j_b < j < -T$.

P39. $\mathbf{X}_j^{[1]} = \mathbf{X}_j^{[2]} = \cdots = \mathbf{X}_j^{[v]}$ for $-T \leq j \leq 0$.

P40. On reiteration $b' > b$, all paths coalesce on or before step $-T$.

P41. This last property holds for a reiteration starting at $j = -\infty$.

P42. $\mathbf{X}_0^{[1]}$ has the equilibrium distribution π.

Property P38 follows from the observation that $U_j - T > \alpha\left(\mathbf{X}_{j-1}^{[1]}, \mathbf{Y}_j\right)$ and that $\mathbf{X}_j^{[l]} > \mathbf{X}_j^{[1]} = \mathbf{x}_{\min}$ for $l = 2, \ldots, v$ and $j_b < j < -T \leq 0$. Property P37 follows from the observation that $\alpha\left(\mathbf{X}_{j-1}^{[1]}, \mathbf{Y}_j\right) = \cdots = \alpha\left(\mathbf{X}_{j-1}^{[v]}, \mathbf{Y}_j\right)$ for $-T < j \leq 0$. The repeated use of $(\mathbf{Y}_j, U_j)$ on step j on all reiterations ensures properties P40 and P41.

Theorem 9.7.3 (Corcoran and Tweedie 1998)

A necessary and sufficient condition for the ordering (9.41) to hold is

$$\mathbf{x} \preceq \mathbf{y} \Longleftrightarrow \frac{\pi(\mathbf{x})q(\mathbf{y})}{\pi(\mathbf{y})q(\mathbf{x})} \geq 1. \tag{9.42}$$

■

PROOF We distinguish two cases. If $\mathbf{X}_{j-1}^{[1]} = \mathbf{x}_{\min} \leq \mathbf{x} \leq \mathbf{X}_{j-1}^{[l]} = \mathbf{z}$ for $l > 1$, then the ordering (9.42) gives

$$\frac{\pi(\mathbf{x}_{\min})q(\mathbf{x})}{\pi(\mathbf{x})q(\mathbf{x}_{\min})} \geq 1 \text{ and } \frac{\pi(\mathbf{x})q(\mathbf{z})}{\pi(\mathbf{x}_{\min})q(\mathbf{x})} \geq 1,$$

which implies $\alpha(\mathbf{x}_{\min}, \mathbf{x}) \leq \alpha(\mathbf{z}, \mathbf{x}) = 1$. Therefore, acceptance on path 1 implies acceptance w.p.1 on each path l.

If $\mathbf{X}_{j-1}^{[1]} = \mathbf{x}_{\min} \leq \mathbf{X}_{j-1}^{[l]} = \mathbf{z} \leq \mathbf{x}$ for $l \geq 2$, then

$$\frac{\pi(\mathbf{x}_{\min})q(\mathbf{z})}{\pi(\mathbf{z})q(\mathbf{x}_{\min})} \geq 1, \quad \frac{\pi(\mathbf{x}_{\min})q(\mathbf{x})}{\pi(\mathbf{x})q(\mathbf{x}_{\min})} \geq 1, \text{ and } \frac{\pi(\mathbf{z})q(\mathbf{x})}{\pi(\mathbf{x})q(\mathbf{z})} \geq 1,$$

so that $q(\mathbf{x}_{\min})/\pi(\mathbf{x}_{\min}) \leq q(\mathbf{z})/\pi(\mathbf{z})$ and, therefore, $\alpha(\mathbf{x}_{\min}, \mathbf{x}) \leq \alpha(\mathbf{z} \leq \mathbf{x}) \leq 1$, again implying that acceptance occurs w.p.1 on path l if acceptance occurs on path 1. ■

Because $U_j \leq \alpha(\mathbf{X}_{j-1}^{[1]}, \mathbf{Y}_j)$ implies acceptance on all paths, it is unnecessary to simulate more than two paths, namely one with $\mathbf{X}_{j_b}^{[1]} = \mathbf{x}_{\min}$ and the other with $\mathbf{X}_{j_b}^{[2]} = \mathbf{x} \neq \mathbf{x}_{\min}$, where $\mathbf{x}$ is chosen arbitrarily or randomly; for example, in the countably finite case from $\{q(\mathbf{x})/[1 - q(\mathbf{x}_{\min})], \mathbf{x} \in \mathcal{X}\backslash\{\mathbf{x}_{\min}\}\}$. Algorithm CFTP4 describes the steps.

ALGORITHM CFTP4

Purpose: To sample $\mathbf{X}_0$ from π.

Input: Minimal state $\mathbf{x}_{\min}$ and initial state $\mathbf{x}$ for path 2.

Output: $\mathbf{X}_0$.

Method:

$b \leftarrow 1$ and $j_1 \leftarrow -1$.

Repeat:

$j \leftarrow j_b, \mathbf{X}_j^{[1]} \leftarrow \mathbf{x}_{\min}$, and $\mathbf{X}_j^{[2]} \leftarrow \mathbf{x}$

Repeat:
$j \leftarrow j+1$
if $j \leq j_{b-1}$, sample and store $\mathbf{Y}_j$ from $\{q(\mathbf{y})\}$ and U_j from $\mathcal{U}(0,1)$
if $U_j \leq \alpha\left(\mathbf{X}_{j-1}^{[1]}, \mathbf{Y}_j\right)$, $\mathbf{X}_j^{[1]} \leftarrow \mathbf{Y}_j$ and $\mathbf{X}_j^{[2]} \leftarrow \mathbf{Y}_j$
else:
$\mathbf{X}_j^{[1]} \leftarrow \mathbf{X}_{j-1}^{[1]}$
if $U_j \leq \alpha\left(\mathbf{X}_{j-1}^{[2]}, \mathbf{Y}_j\right)$, $\mathbf{X}_j^{[2]} \leftarrow \mathbf{Y}_j$
else:
$\mathbf{X}_j^{[2]} \leftarrow \mathbf{X}_{j-1}^{[2]}$
until $j > 0$.
$b \leftarrow b+1$
$j_b \leftarrow 2j_{b-1}$
until $\mathbf{X}_0^{[1]} = \mathbf{X}_0^{[2]}$.
$\mathbf{X}_0 \leftarrow \mathbf{X}_0^{[1]}$.
Return $\mathbf{X}_0$. ■

Storing the pseudorandom number seeds for step $j = j_b$ in place of $(\mathbf{Y}_j, U_j)$ for $j = j_b+1, \ldots, j_{b-1}$ for each reiteration b with $j_0 := 0$, as in Algorithm CFTP3, reduces the space requirement. Observe that

$$T := -\min[-\infty < j \leq 0 : U_j \leq \alpha(\mathbf{x}_{\min}, \mathbf{Y}_j)]$$

is now the backward coupling time. Because the ordering (9.42) implies

$$\mathrm{pr}(T = -j \mid \mathbf{X}_{j-1}^{[1]} = \mathbf{x}_{\min}) = \int_{\mathcal{X}} \alpha(\mathbf{x}_{\min}, \mathbf{y}) q(\mathbf{y}) d\mathbf{y} = \frac{q(\mathbf{x}_{\min})}{\pi(\mathbf{x}_{\min})},$$

and because the successive tests prior to the first success are independent, it follows that

$$\mathrm{pr}(T = -j) = \frac{q(\mathbf{x}_{\min})}{\pi(\mathbf{x}_{\min})} \left[1 - \frac{q(\mathbf{x}_{\min})}{\pi(\mathbf{x}_{\min})}\right]^{-j-1} \qquad j = -1, -2, \ldots$$

so that T has the geometric distribution $\mathcal{G}e(1 - q(\mathbf{x}_{\min})/\pi(\mathbf{x}_{\min}))$ with $\mathrm{E}T = \pi(\mathbf{x}_{\min})/q(\mathbf{x}_{\min})$. Because (9.41) and (3.10) imply $\kappa(\pi, q) = \frac{\pi(\mathbf{x}_{\min})}{q(\mathbf{x}_{\min})}$, $\mathrm{E}T = \kappa(\pi, q)$, thus relating the backward coupling time to the convergence rate for uniform ergodicity. From Section 6.10.1, recall that a Markov chain generated by independent HM sampling is uniformly ergodic with convergence factor $\rho = 1 - \frac{1}{\kappa(\pi,q)}$, provided that $\kappa(\pi, q)$ in (3.10) is bounded.

An alternative approach to CFTP in Murdoch and Green (1997) focuses on the decomposition of the transition kernel to achieve $\mathbf{X}_0^{[1]}$ from π and on a partition of the state space $\mathcal{X}$ to increase computational efficiency.

9.8 LESSONS LEARNED

- For finite-state Markov chains, reversibility provides a basis for characterizing the ways in which errors of estimation converge. In particular, this characterization suggests that starting a sample path in the modal state minimizes a bound on error caused by initial conditions.
- The neighborhood structure of Markov random fields (MRF) randomly updates a coordinate by conditioning only on the values of coordinates that are in its neighborhood.
- An MRF has an equilibrium distribution whose form is that of a Gibbs distribution. Together with the neighborhood property, this form facilitates sample generation.
- For some sample-generating methods, the values of parameters in an MRF may induce a fixed correlation among all sites. Slice sampling precludes this correlation structure for the Ising model, a particular form of MRF. More generally, slice sampling ensures uniform ergodicity under relatively mild conditions.
- Slice sampling is particularly appealing for generating random clusters.
- Coupling from the past (CFTP) is a theoretically exact technique for initiating an MCMC sampling experiment in a way such that the sample path is in equilibrium after a random number of iterations.
- CFTP is most easily implemented for monotone Markov processes.

9.9 EXERCISES

1. In (9.15), show that

$$\sum_{\mathbf{x}\in\mathcal{X}} \exp\left[-\beta \sum_{\mathcal{C}\in\mathcal{D}} \delta(\mathbf{x},\mathcal{C})\right] = \sum_{\mathbf{x}\in\mathcal{X}} \prod_{\mathcal{C}\in\mathcal{D}} [1 + w\delta(\mathbf{x},\mathcal{C})],$$

where $w := \mathrm{e}^{-\beta} - 1$.

2. For φ given in (9.15), show that

$$\varphi = |\mathcal{X}| + \sum_{s=1}^{d} a_s w^s.$$

3. Show that

$$j^* := c_2^{-1}[m \ln m + m \ln(c_1/\epsilon)]$$

is the mixing time for (9.22).

4. Show that the one-step transition kernel (9.26) satisfies detailed balance.

5. In Example 9.6.1 show that $\mathbf{X}$, given $\mathbf{U}$, has the conditional distribution (9.29).

9.10 REFERENCES

Baxter, R. J. (1982). *Exactly Solved Models in Statistical Mechanics,* Academic Press, New York.

Besag, J. (1974). Spatial interaction and the statistical analysis of lattice systems, *Journal of the Royal Statistical Society B,* **36,** 192–225.

Besag, J., and P. Green (1993). Spatial statistics and Bayesian computation, *Journal of the Royal Statistical Society B,* **55,** 25–37.

Cipra, B. A., (1987). An introduction to the Ising model, *American Mathematical Monthly,* **94,** 937–959.

Cooper, C., and A. M. Frieze (1999). Mixing properties of the Swendsen-Wang process in classes of graphs, *Random Structures and Algorithms,* **15,** 242–261.

Corcoran, J. N., and R. L. Tweedie (1998). Perfect sampling from independent Metropolis-Hastings chains, Statistics Department, Colorado State University, Fort Collins.

Diaconis, P., and D. Stroock (1991). Geometric bounds for eigenvalues of Markov chains, *Annals of Applied Probability,* **1,** 36–66.

Edwards, R. G., and A. D. Sokal (1988). Generalization of the Fortuin-Kasteleyn-Swendsen-Wang representation and Monte Carlo algorithm, *Physical Review,* **38,** 2009–2012.

Fishman, G. S. (1996a) *Monte Carlo: Concepts, Algorithms, and Applications,* Springer-Verlag, New York.

Fishman, G. S. (1996b). Coordinate selection rules for Gibbs sampling, *Annals of Applied Probability,* **6,** 444–465.

Frigessi, A., P. DiStefano, C. R. Hwang, and S. J. Sheu (1993). Convergence rates of the Gibbs sampler, the Metropolis algorithm and other single site updating dynamics, *Journal of the Royal Statistical Society B,* **55,** 205–219.

Frigessi, A., F. Martinelli, and J. Stander (1997). Computational complexity of Markov chain Monte Carlo methods for finite Markov random fields, *Biometrika,* **84,** 1–18.

Gray, A. J. (1994). Simulating posterior Gibbs distributions: A comparison of the Swendsen-Wang and Gibbs sampler methods, *Statistics and Computing,* **4,** 189–201.

Grimmett, G. R. (1973). A theorem about random fields, *Bulletin of the London Mathematical Society,* **5,** 81–84.

Higdon, D. M. (1998). Auxiliary variable methods for Markov chain Monte Carlo with applications, *Journal of the American Statistical Association,* **93,** 585–595.

Huber, M. (1999). Efficient exact sampling from the Ising model using Swendsen-Wang, *Proceedings of the 10th Annual ACM-SIAM Symposium on Discrete Algorithms, Baltimore.*

Hurn, A. (1997). Difficulties with auxiliary variables in Markov chain Monte Carlo applications, *Statistics and Computing,* **7,** 35–44.

Kelly, F. (1979). *Stochastic Networks and Reversibility,* Wiley, New York.

Kira, E., and C. Ji (1997). Rates of convergence for the Gibbs sampler, *Markov Processes and Related Fields,* **3,** 89–102.

Liggett, T. M. (1985). *Interacting Particle Systems,* Springer-Verlag, New York.

Mira, A., and L. Tierney (1999). On the use of auxiliary variables in Markov chain Monte Carlo sampling, School of Statistics, University of Minnesota, Minneapolis.

Murdoch, D. J., and P. J. Green (1997). Exact sampling from a continuous state space, Mathematics Department, University of Bristol, United Kingdom.

Pickard, D. K. (1987). Inference for discrete Markov fields: The simplest nontrivial case, *J. Amer. Statistical Association,* **82,** 90–96.

Propp, J. G., and D. B. Wilson (1996). Exact sampling with coupled Markov chains and applications, *Random Structures and Algorithms,* **9,** 223–252.

Stroock, D., and B. Zegarliński (1995). On the ergodic properties of Glauber dynamics, *J. Statistical Physics,* **81,** 1007–1019.

Swendsen, R. H., and J. S. Wang (1987). Nonuniversal critical dynamics in Monte Carlo simulations, *Physical Review Letters,* **58,** 86–88.

Wu, F. Y. (1982). The Potts model, *Reviews of Modern Physics,* **54,** 235–268.

Glossary

Notation	*Definition*

General

$a \approx b$	a approximately equals b.
$a \doteq b$	a equals b rounded to the nearest integer.
$a \vee b$	$\max(a, b)$
$a \to b$	a converges to b.
$x \gg y$	x is much greater than y.
$x \ll y$	x is much less than y.
$(x)^+$	$\begin{cases} x & \text{if } x \geq 0 \\ 0 & \text{elsewhere} \end{cases}$
$A \Rightarrow B$	A implies B.
$A \Leftrightarrow B$	A implies B, and B implies A.
$\lfloor Z \rfloor$	Largest integer less than or equal to Z.
$\lceil Z \rceil$	Smallest integer greater than or equal to Z.
$\Gamma(\alpha)$	$\int_0^\infty x^{\alpha-1}e^{-x}dx \quad \alpha > 0$ (gamma function).
$O(z)$	Denotes a function $\{v(z)\}$ for which there exist constants $c > 0$ and z_0 such that $v(z) \leq cu(z)$ for all $z \geq z_0$.

Modulo operations

$a = b \pmod M$	a is congruent to b modulo M if M divides $a - b$.
$\oplus$	$0 \oplus 0 = 0,\ 1 \oplus 0 = 1,\ 0 \oplus 1 = 1,\ 1 \oplus 1 = 0$.

Intervals and sets

$[a, b]$	All points x in the closed continuous interval $a \leq x \leq b$.
$[a, b)$	All points x in the half-open continuous interval $a \leq x < b$.
$(a, b]$	All points x in the half-open continuous interval $a < x \leq b$.
(a, b)	All points x in the open continuous interval $a < x < b$.
$\Re$	$(-\infty, \infty)$ (one-dimensional euclidean space).
$\Re^m$	$(-\infty, \infty)^m$ (m-dimensional euclidean space for $m \geq 1$).
Countable set	If $\mathcal{X}$ is a collection of discrete elements, then $\mathcal{X}$ is countable.
Finite set	If $\mathcal{X}$ contains a finite number of elements, then $\mathcal{X}$ is a finite set.
Countably infinite	If $\mathcal{X}$ consists of an infinite number of discrete elements, then $\mathcal{X}$ is countably infinite.

Notation	*Definition*
Intervals and sets (*continued*)	
Uncountable	If $\mathcal{X}$ consists of continuous subsets of $\Re^m$, then $\mathcal{X}$ is uncountable.
Product space	$\mathcal{X} = \mathcal{X}_1 \times \cdots \times \mathcal{X}_m$ denotes a product space such that for each $i = 1, \ldots, m$, x_i in $\mathbf{x} := (x_1, \ldots, x_m) \in \mathcal{X}$ can assume all values in $\mathcal{X}_i$ independent of the values that x_j assumes for all $j \neq i$.
$\emptyset$	Empty set.
$\mathcal{A} \subset \mathcal{B}$	$\mathcal{A}$ is strictly a subset of $\mathcal{B}$.
$\mathcal{A} \subseteq \mathcal{B}$	$\mathcal{A}$ is either a subset of $\mathcal{B}$ or is the set $\mathcal{B}$.
$\mathcal{A} \supset \mathcal{B}$	Set $\mathcal{A}$ contains set $\mathcal{B}$ strictly as a subset.
$\mathcal{A} \supseteq \mathcal{B}$	Set $\mathcal{A}$ contains set $\mathcal{B}$.
$\mathbf{x} \in \mathcal{X}$	$\mathbf{x}$ belongs to $\mathcal{X}$.
$\mathcal{X} \ni \mathbf{x}$	Set $\mathcal{X}$ contains element $\mathbf{x}$.
$\lvert\mathcal{X}\rvert$	Size of $\mathcal{X}$. If $\mathcal{X}$ is finite, then $\lvert\mathcal{X}\rvert$ is the number of elements in $\mathcal{X}$. If continuous, then $\lvert\mathcal{X}\rvert = \int_{\mathcal{X}} d\mathbf{x}$.
$\mathcal{X} \backslash \mathcal{Y}$	Complement of $\mathcal{Y}$. If $\mathcal{Y} \subseteq \mathcal{X}$, then $\mathcal{X} \backslash \mathcal{Y}$ denotes the elements that remain in $\mathcal{X}$ after deleting those in $\mathcal{Y}$.
$\mathcal{A} \cap \mathcal{B}$	Intersection of sets $\mathcal{A}$ and $\mathcal{B}$. All elements that are in both sets $\mathcal{A}$ and $\mathcal{B}$.
$\mathcal{A} \cup \mathcal{B}$	Union of sets $\mathcal{A}$ and $\mathcal{B}$. All elements that are in either set $\mathcal{A}$ or set $\mathcal{B}$.

Notation	*Definition*
Probability	
d.f.	Distribution function.
e.d.f.	Empirical distribution function.
$\text{pr}(X = x)$	Probability that X equals x.
p.d.f.	Probability density function.
p.m.f.	Probability mass function.
$\{\Phi(x)\}$	$\frac{1}{\sqrt{2\pi}} \int_{-\infty}^{x} e^{-z^2/2} dz, \quad -\infty < x < \infty$ (standard normal distribution function).
$\{\Phi^{-1}(u)\}$	Smallest x for which $\Phi(x) \geq u, \quad 0 \leq u \leq 1$ (inverse standard normal distribution function).
w.p.1	With probability 1.
$X_j \xrightarrow{d} F$ as $j \to \infty$	The distribution of X_j converges to distribution F as $j \to \infty$.
$\mathcal{B}er(p)$	Bernoulli p.m.f.
$\mathcal{B}e(\alpha, \beta)$	Beta p.d.f.
$\mathcal{B}(n, p)$	Binomial p.m.f.
$\mathcal{C}(\alpha, \beta)$	Cauchy p.d.f.
$\mathcal{E}(\beta)$	Exponential p.d.f.
$\mathcal{G}(\alpha, \beta)$	Gamma p.d.f.
$\mathcal{G}e(p)$	Geometric p.m.f.

Notation	*Definition*
Probability (*continued*)	
$\mathcal{LN}(\eta, \sigma^2)$	Lognormal p.d.f.
$\mathcal{NB}(r, p)$	Negative binomial p.m.f.
$\mathcal{N}(\mu, \sigma^2)$	Normal p.d.f.
$\mathcal{P}a(\alpha, \beta)$	Pareto p.d.f.
$\mathcal{P}(\mu)$	Poisson p.m.f.
$\mathcal{S}(n)$	Student t p.d.f.
$\mathcal{U}(\alpha, \beta)$	Uniform p.d.f.
$\mathcal{W}(\alpha, \beta)$	Weibull p.d.f.

Partial Solutions to Selected Chapter Exercises

All entries are randomly-generated Monte Carlo sample solutions and are, therefore, approximate.

Chapter 2

Exercise	$\bar{\lambda}_n$	s.e.
H.1	0.2209	1.752×10^{-4}
H.3	1.366	9.515×10^{-4}
H.5	0.2520	4.358×10^{-3}
H.7	0.3558	4.295×10^{-3}
H.9	0.2512	6.659×10^{-4}
H.11b	152,165	155
H.15a	$\bar{\lambda}_n$ (0) = 110,603	3.336
H.17b	0.06447 (Bernoulli sampling)	2.456×10^{-4}
H.19b	0.06180	2.408×10^{-4}

Chapter 5

Exercise	Number of Strata n	$\tilde{\lambda}_n$	s.e.
H.1b	4	0.22104	6.5448×10^{-5}
	16	0.22089	5.5064×10^{-5}
	64	0.22097	5.4324×10^{-5}
	256	0.22093	5.4241×10^{-5}
H.3b	4	1.3656	5.2111×10^{-4}
	16	1.3658	4.8654×10^{-4}
	64	1.3663	4.8426×10^{-4}
	256	1.3658	4.8469×10^{-4}
H.5c	9	0.25207	4.2890×10^{-3}
	36	0.24522	3.9802×10^{-3}
	144	0.24538	3.1258×10^{-3}
	576	0.24024	2.4694×10^{-3}

All entries are randomly-generated Monte Carlo sample solutions and are, therefore, approximate.

(Chapter 5 continued)

Exercise	$\dot{\lambda}_n$	s.e.
H.7a	1.3677	8.1236×10^{-4}
H.9a	0.35565	6.7182×10^{-4}
H.11a	0.25040	5.5781×10^{-4}

Exercise	CV	W	$E(CV)$	$\breve{\lambda}_n$	s.e.
H.15a	$\frac{XE^{XY}}{W}$	$\frac{2}{\pi}$	$e^{\pi/2} - \frac{\pi}{2} - 1$	1.3646	8.2414×10^{-4}
H.17a	$W^{-1}X^{1/2}(Y^{3/2} + 2Y)$	1	$\frac{4}{3} + 2\Gamma(4/3)$	0.35081	2.2391×10^{-3}
H.19a	$\frac{X}{W}$	1	$\frac{1}{4}$	0.25111	3.1684×10^{-4}

Exercise	ζ_1	ζ_2	K	$\breve{\lambda}_n$	s.e.
H.21b	0.03	0.2	35	5.1139	3.776×10^{-4}
		0.3	35	5.2305	7.740×10^{-4}
		0.4	35	5.4628	1.289×10^{-3}
	0.07	0.2	35	5.2331	3.928×10^{-4}
		0.3	35	5.3333	7.686×10^{-4}
		0.4	35	5.5448	1.290×10^{-4}

Exercise	$\breve{\lambda}_n$	s.e.
H.23b, c	0.39186	4.0208×10^{-3}
H.25b, c	0.25063	6.3237×10^{-4}

Chapter 6

Exercise	$\widehat{\lambda}_t$
H.1d	0.2212
H.3d	1.3652
H.5c	1.6191
H.7e	0.2211
H.9e	1.3677
H.11c	1.6202
H.13	$\widehat{\beta}_s = 1.1568$

All entries are randomly-generated Monte Carlo sample solutions and are, therefore, approximate.

Chapter 7

Exercise	$\widehat{\lambda_t}$	s.e.
H.1	0.2205	3.9743×10^{-4}
H.3	1.3663	1.6794×10^{-3}
H.5	1.6164	2.6533×10^{-3}
H.7	0.2208	3.6307×10^{-4}
H.9	1.3658	9.6556×10^{-4}
H.11	1.6203	1.2140×10^{-4}
H.23	0.2217	4.638×10^{-4}
H.25	1.365	1.593×10^{-3}
H.27	1.617	3.106×10^{-3}
H.29	0.2213	4.352×10^{-4}
H.31	1.365	1.124×10^{-3}
H.33	1.618	1.810×10^{-3}
H.43	$\widehat{\beta_s} = 1.1568$	

Chapter 8

Exercise	Initial Restoration Cost	Minimal Restoration Cost
H.1d	319,150	232,050
H.3c	316,750	232,050
H.5	Minimal Distance 11.9533	

Name Index

Subject Index

P

Q

R